Flux Pinning in Superconductors

Teruo Matsushita

Flux Pinning in Superconductors

 Springer

Professor Dr. Teruo Matsushita
Kyushu Institute of Technology
Department of Computer Science and Electronics
Kawazu 680-4
820-8502 Iizuka, Japan
E-mail: matsusita@cse.kyutech.ac.jp

Library of Congress Control Number: 2006933137

ISBN-10 3-540-44514-5 Springer Berlin Heidelberg New York
ISBN-13 978-3-540-44514-2 Springer Berlin Heidelberg New York

This work is subject to copyright. All rights are reserved, whether the whole or part of the material is concerned, specifically the rights of translation, reprinting, reuse of illustrations, recitation, broadcasting, reproduction on microfilm or in any other way, and storage in data banks. Duplication of this publication or parts thereof is permitted only under the provisions of the German Copyright Law of September 9, 1965, in its current version, and permission for use must always be obtained from Springer. Violations are liable for prosecution under the German Copyright Law.

Springer is a part of Springer Science+Business Media
springer.com
© Springer-Verlag Berlin Heidelberg 2007

The use of general descriptive names, registered names, trademarks, etc. in this publication does not imply, even in the absence of a specific statement, that such names are exempt from the relevant protective laws and regulations and therefore free for general use.

Typesetting: by the author and techbooks using a Springer LaTeX macro package
Cover design: *design & production* GmbH, Heidelberg

Printed on acid-free paper   SPIN: 11806752   56/techbooks   5 4 3 2 1 0

Preface

Superconductivity is now a considerable focus of attention as one of the technologies which can prevent environmental destruction by allowing energy to be used with high efficiency. The possibility of practical applications of superconductivity depends on the maximum current density which superconductors can carry, the value of losses which superconductors consume, the maximum magnetic field strength in which superconductors can be used, etc. These factors are directly related to the flux pinning of quantized magnetic flux lines in superconductors. This book extensively describes related subjects, from the fundamental physics of flux pinning to electromagnetic phenomena caused by flux pinning events, which will be useful for anyone who wants to understand applied superconductivity.

The Japanese edition was published for this purpose in 1994. Since then, there has been significant progress in the research and development of high-temperature superconductors. In particular, the new superconductor MgB_2 was discovered in 2001, followed by steady improvements in the superconducting properties necessary for applications. On the other hand, there are no essential differences in the flux pinning phenomena between these new superconductors and metallic superconductors. Hence, the framework of the previous Japanese edition was kept unchanged, while new description was added on these new superconductors in the English edition.

In the following the content of each chapter is briefly introduced.

In Chapter 1 various fundamental superconducting properties which determine the flux pinning and electromagnetic phenomena in type II superconductors are described, based on the Ginzburg-Landau theory. In particular, it is shown that the center of a quantized flux line must be in the normal state so that the Josephson current does not diverge due to the singularity in the gradient of the phase of the superconducting order parameter there. This causes a loss due to the motion of normal electrons in the core that is driven by the electric field, which is induced when flux lines are forced to move by the Lorentz force. At the same time such a structure of the core contributes to the flux pinning event. The role of the kinetic energy in determination of the upper critical field is also shown. This will help the readers to understand the kinetic energy pinning mechanism for the artificial Nb pinning centers introduced into Nb-Ti, which is discussed in Chapter 6.

In Chapter 2 the critical state model, which is needed to understand the irreversible electromagnetic phenomena in superconductors, is described. The mechanism of the irreversibility is introduced on the basis of the ohmic electric resistivity, which is induced when a flux line is driven by the Lorentz force. On the other hand, the losses in superconductors are non-ohmic ones with a hysteretic nature. The reason for this will also be discussed. The critical state model provides the relationship between the current density and the electric field strength, and the electromagnetic phenomena in superconductors are described by the Maxwell equations coupled with this relationship. It is shown that the critical state model can describe irreversible magnetizations and AC losses in superconductors. The effect of superconductor diamagnetism will also be an important topic.

Various electromagnetic phenomena are introduced in Chapter 3. These include geometrical effects and dynamic phenomena which were not treated in Chapter 2. The rectifying effect in the DC current-voltage characteristics in a superposed AC magnetic field, flux jumps, surface irreversibility, and DC susceptibility in a varying temperature are also included. In addition, it is shown that an abnormal reduction in losses occurs, deviating from the prediction of the critical state model when an AC magnetic field is applied to a superconductor smaller than the pinning correlation length called Campbell's AC penetration depth. This is attributed to the reversible motion of flux lines limited within pinning potential wells, being in contrast with the hysteresis loss which results from the flux motion involved in dropping into and jumping out of the pinning potential wells. In high-temperature superconductors the superconducting current sustained by flux pinning appreciably decays with time due to the thermal agitation of flux lines. This phenomenon, which is called flux creep, is also discussed. In extreme cases the critical current density is reduced to zero at some magnetic field called the irreversibility field. The principles used to determine the irreversibility field are described, and the result is applied to high-temperature superconductors in Chapter 8.

In Chapter 4 various phenomena which are observed when the transport current is applied to a long superconducting cylinder or tape in a longitudinal magnetic field are introduced, and the force-free model, which assumes a current flow parallel to the flux lines, is explained. Although this model insists that the force-free state is intrinsically stable, the observed critical current density in a longitudinal magnetic field depends on the flux pinning strength, similarly to the case in a transverse magnetic field, indicating that the force-free state is unstable without the pinning effect. From the energy increase caused by introducing a distortion due to the parallel current to the flux line lattice the restoring torque is derived, and the critical current density is predicted to be determined by the balance between this torque and the moment of pinning forces. The resultant rotational motion of flux lines explains the observed break in Josephson's formula on the induced electric field. A peculiar helical structure of the electric field with a negative region in the resistive state can also be explained by the flux motion induced by the restoring torque.

The critical current density is a key parameter which determines the applicability of superconductors to various fields, and hence the measurement of this parameter is very important. In Chapter 5 various measurement methods are reviewed, including transport and magnetic ones. Among them, it is shown that distributions of magnetic flux and current inside the superconductor can be measured by using Campbell's method, which is also useful for analyzing the reversible motion of flux lines discussed in Chapter 3. However, if AC magnetic methods including Campbell's method are used for superconductors smaller than the pinning correlation length, the critical current densities are seriously overestimated. The reason for the overestimation is discussed, and a method of correction is proposed.

Mechanisms of pinning interactions between various defects and individual flux lines are reviewed, and the elementary pinning force, the maximum strength of each defect, is theoretically estimated in Chapter 6. These include the condensation energy interaction, the elastic interaction, the magnetic interaction and the kinetic energy interaction. In particular, the reason why the flux pinning strength of thin normal α-Ti layers in Nb-Ti is not weak in spite of a remarkable proximity effect, is discussed. The kinetic energy interaction is proposed as the pinning mechanism responsible for the very high critical current density achieved by Nb layers introduced artificially into Nb-Ti. The shape of pinning centers, which contributes to the enhancement of the pinning efficiency, is also discussed.

In Chapter 7 the summation problems which relate the global pinning force density to the elementary pinning forces and number densities of pinning centers are discussed. The summation theories are reviewed historically according to their development, since the fundamental issue of threshold value of the elementary force, which is deeply associated with the nature of hysteresis loss by the pinning interaction, was proposed first by the statistical theory. Then, the consistency of this theory with the dynamic theory is shown. The fundamental threshold problem was resolved by Larkin and Ovchinnikov, who showed that a long-range order does not exist in the flux line lattice. However, quantitative disagreements are sometimes found between their theory and experiments, and the instability of flux motion related to the hysteresis loss is not clearly described in this theory. In the coherent potential approximation theory the statistical method is used, taking into account the lack of long range order, and the compatibility of the threshold problem and the instability of flux motion is obtained. A detailed comparison is made between the theories and experiments. The saturation phenomenon observed for commercial metallic superconductors at high fields is explained, and experimental results are compared with Kramer's model, etc. The theoretical pinning potential energy, which is important for the analysis of flux creep, is also derived.

In Chapter 8 the various properties of high-temperature superconductors are discussed. These superconductors show significant anisotropy due to the two-dimensional crystal structure composed of superconducting CuO_2 layers and insulating block layers. This makes the states of flux lines extremely complicated. Various phase transitions of the flux line system to be pinned and the mechanisms responsible are reviewed. In particular, the transitions in which the pinning plays an important role, i.e., the order-disorder transition associated with the peak effect of critical current density and the glass-liquid transition associated with the irreversibility field, are discussed in detail. The influences on these transitions, not only of the flux pinning strength and the anisotropy of the superconductor, but also of the electric field and the specimen size are discussed. Y-123, Bi-2212, and Bi-2223, which have been developed for applications, are at the focus of the discussion, and their pinning properties and recent progress are introduced.

Superconducting MgB_2 was discovered in 2001. Since this superconductor has a critical temperature considerably higher than those of metallic superconductors and is not seriously influenced by weak links and flux creep as in high-temperature superconductors, applications of this superconductor are expected in the future. In fact, the critical current density was improved significantly within a very short period after the discovery. In Chapter 9 the pinning property introduced by grain boundaries in MgB_2 is reviewed, and the mechanism which determines the present critical current density is discussed. Then, the matters to overcome are summarized as topics for further improvements. Finally the potential for realization of the improvement is discussed by comparing the condensation energy of this material with the values of Nb-Ti and Nb_3Sn.

Thus, this book deals with the flux pinning mechanisms, the fundamental physics needed for understanding the flux pinning, and various electromagnetic phenomena caused by the flux pinning.

On the other hand, it is effective to focus on one matter which is described in many chapters to give a comprehensive understanding. The size of superconductors is a focus of attention, for example. When the superconductor is smaller than the pinning correlation length, pinning at a lower dimension with a higher efficiency occurs, resulting in the disappearance of the peak effect of the critical current density. In this case the irreversibility field is smaller than the bulk value due to a smaller pinning potential. At the same time the flux motion becomes reversible in the electromagnetic phenomena, resulting in a significant reduction in AC losses. This also causes a serious overestimation of the critical current density from AC magnetic measurements. It is also worth noting that the concept of irreversible thermodynamics on minimization of energy dissipation appears in various pinning phenomena discussed in this book. Another example may be found in a contrast between the flux motion driven by the Lorentz force and the rotational flux motion driven by the force-free torque treated in Chapter 4. The former shows an analogy with mechanical systems, but the latter does not, and the flux motion is perpendicular to the energy flow. This contrast is associated with the fact that the force-free torque is not the moment of forces.

In this book appendices are included to assist the understanding of readers. Many exercises and detailed answers will also be useful for better understanding.

Finally the author would like to thank Ms. T. Beppu for drawing all electronic figures and for assistance in making electronic files. The author acknowledges also Dr. T.M. Silver at Wollongong University, Prof. E.W. Collings of Ohio State University, and Dr. L. Cooley of Brookhaven National Laboratory for correction of the English in the book.

Contents

1 **Introduction** 1
   1.1 Superconducting Phenomena 1
   1.2 Kinds of Superconductors 3
   1.3 London Theory 6
   1.4 Ginzburg-Landau Theory 9
   1.5 Magnetic Properties 16
      1.5.1 Quantization of Magnetic Flux 16
      1.5.2 Vicinity of Lower Critical Field 17
      1.5.3 Vicinity of Upper Critical Field 24
   1.6 Surface Superconductivity 30
   1.7 Josephson Effect 32
   1.8 Critical Current Density 34
   1.9 Flux Pinning Effect 38
   References 40

2 **Fundamental Electromagnetic Phenomena in Superconductors** 41
   2.1 Equations of Electromagnetism 41
   2.2 Flux Flow 45
   2.3 Mechanism of Hysteresis Loss 51
   2.4 Characteristic of the Critical State Model and its Applicable Range 54
   2.5 Irreversible Phenomena 55
   2.6 Effect of Diamagnetism 66
   2.7 AC Losses 76
   References 82

3 **Various Electromagnetic Phenomena** 85
   3.1 Geometrical Effect 85
      3.1.1 Loss in Superconducting Wire due to AC Current 85

XII   Contents

        3.1.2   Loss in Superconducting Wire of Ellipsoidal Cross Section and Thin Strip due to AC Current 89
        3.1.3   Transverse Magnetic Field 90
        3.1.4   Rotating Magnetic Field 92
  3.2   Dynamic Phenomena 95
  3.3   Superposition of AC Magnetic Field 97
        3.3.1   Rectifying Effect 97
        3.3.2   Reversible Magnetization 100
        3.3.3   Abnormal Transverse Magnetic Field Effect 102
  3.4   Flux Jump 103
  3.5   Surface Irreversibility 108
  3.6   DC Susceptibility 119
  3.7   Reversible Flux Motion 125
  3.8   Flux Creep 138
  References 152

4 Longitudinal Magnetic Field Effect 155
  4.1   Outline of Longitudinal Magnetic Field Effect 155
  4.2   Flux-Cutting Model 161
  4.3   Stability of the Force-Free State 168
  4.4   Motion of Flux Lines 175
  4.5   Critical Current Density 186
  4.6   Generalized Critical State Model 191
  4.7   Resistive State 194
  References 207

5 Measurement Methods for Critical Current Density 209
  5.1   Four Terminal Method 209
  5.2   DC Magnetization Method 212
  5.3   Campbell's Method 213
  5.4   Other AC Inductive Methods 221
        5.4.1   Third Harmonic Analysis 221
        5.4.2   AC Susceptibility Measurement 225
  References 231

6 Flux Pinning Mechanisms 233
  6.1   Elementary Pinning and the Summation Problem 233
  6.2   Elementary Pinning Force 234
  6.3   Condensation Energy Interaction 237
        6.3.1   Normal Precipitates 237
        6.3.2   Grain Boundary 245
  6.4   Elastic Interaction 253
  6.5   Magnetic Interaction 258
  6.6   Kinetic Energy Interaction 259
  6.7   Improvement of Pinning Characteristics 261
  References 264

7 Flux Pinning Characteristics ... 267
7.1 Flux Pinning Characteristics ... 267
7.2 Elastic Moduli of Flux Line Lattice ... 271
7.3 Summation Problem ... 275
7.3.1 Statistical Theory ... 275
7.3.2 Dynamic Theory ... 283
7.3.3 Larkin-Ovchinnikov Theory ... 286
7.3.4 Coherent Potential Approximation Theory ... 292
7.4 Comparison with Experiments ... 297
7.4.1 Qualitative Comparison ... 298
7.4.2 Quantitative Comparison ... 306
7.4.3 Problems in Summation Theories ... 307
7.5 Saturation Phenomenon ... 310
7.5.1 Saturation and Nonsaturation ... 310
7.5.2 The Kramer Model ... 313
7.5.3 Model of Evetts et al. ... 316
7.5.4 Comparison Between Models and Experiments ... 316
7.5.5 Avalanching Flow Model ... 320
7.6 Peak Effect and Related Phenomena ... 322
7.7 Pinning Potential Energy ... 331
References ... 337

8 High-Temperature Superconductors ... 341
8.1 Anisotropy of Superconductors ... 341
8.2 Phase Diagram of Flux Lines ... 345
8.2.1 Melting Transition ... 345
8.2.2 Vortex Glass-Liquid Transition ... 346
8.2.3 Order-Disorder Transition ... 352
8.2.4 Phase Diagram of Flux Lines in Each Superconductor ... 357
8.2.5 Size Effect ... 359
8.2.6 Other Theoretical Predictions ... 360
8.3 Weak Links of Grain Boundaries ... 361
8.4 Electromagnetic Properties ... 365
8.4.1 Anisotropy ... 365
8.4.2 Differences in the Size Effect due to the Dimensionality ... 367
8.4.3 Flux Creep ... 372
8.4.4 E-J Curve ... 374
8.4.5 Josephson Plasma ... 378
8.5 Irreversibility Field ... 380
8.5.1 Analytic Solution of Irreversibility Field ... 380
8.5.2 Effect of Distribution of Pinning Strength ... 381
8.5.3 Comparison with Flux Creep-Flow Model ... 383
8.5.4 Relation with G-L Transition ... 391
8.6 Flux Pinning Properties ... 394
8.6.1 Y-123 ... 394

            8.6.2   Bi-2223 402
            8.6.3   Bi-2212 406
    References 409

9 MgB$_2$ 413
    9.1 Superconducting Properties 413
    9.2 Flux Pinning Properties 415
            9.2.1   Wires and Bulk Materials 415
            9.2.2   Thin Films 428
    9.3 Possibility of Improvements in the Future 430
    References 433

A Appendix 435
    A.1 Description of Equilibrium State 435
    A.2 Magnetic Properties of a Small Superconductor 437
    A.3 Minimization of Energy Dissipation 439
    A.4 Partition of Pinning Energy 440
    A.5 Comments on the Nonlocal Theory of the Elasticity
         of the Flux Line Lattice 441
    A.6 Avalanching Flux Flow Model 448
    A.7 Josephson Penetration Depth 452
    A.8 On the Transverse Flux Bundle Size 452
    References 461

Answers to Exercises 463

Index 499

1
Introduction

1.1 Superconducting Phenomena

The superconductivity is a phenomenon that was discovered first for mercury by Kamerlingh-Onnes in 1911 and has been found for various elements, alloys and compounds. One of the features of superconductivity is that the electrical resistance of a material suddenly drops to zero as the temperature decreases through a transition point; such a material is called superconductor. Advantage is taken of this property in the application of the superconducting phenomenon to technology. Later it was found that the origin of the zero electrical resistivity is not the perfect conductivity as such but the perfect diamagnetism, i.e., the ability of the superconductor to completely exclude a weak applied magnetic field. A related phenomenon is the complete expulsion of a weak applied magnetic field as the temperature decreases through the transition point. These diamagnetic phenomena are called the Meissner effect. As will be mentioned later, the perfect diamagnetism is broken at sufficiently high magnetic fields. There are two alternative ways in which this break down can take place depending on whether the superconductor is "type-1" or "type-2." In type-2 materials the superconductivity can be maintained up to very high fields even after the break down. Such superconductors are therefore suitable for use in high-field devices such as magnets, motors, and generators.

Another characteristic of superconductivity is the existence of a gap, just below the Fermi energy, of the energy of the conduction electrons. It turns out that the electron energy in the superconducting state is lower than that in the ground state of the normal state; the difference in the energy per electron between the two states is the energy gap. The size of the energy gap in the superconductor can be measured using the absorption of microwave radiation in the far infrared range, or the tunneling effect of a junction composed of a superconductor and a normal metal separated by a thin insulating layer. In the case of a sufficiently small excitation, the energy gap provides a barrier against the transition of electrons from the superconducting state to the normal state. That is, even when the electrons are scattered by lattice defects, impurities, or thermally oscillating ions, the energy may not be dissipated and hence, electrical resistance may not appear. It was theoretically proved by Bardeen, Cooper and Schrieffer in 1957 that the electrons in the vicinity of the Fermi level exist in so-called Cooper pairs whose condensation yields the superconducting state. This is essentially the BCS theory of superconductivity.

Another essential property of type-2 superconductors is embodied in the so-called Josephson effect. In a junction composed of two superconductors separated by a thin insulating layer, the local property of type-2 superconductor can be directly observed without being averaged. The DC Josephson effect that predicts the superconducting tunneling current is not the tunneling of normal electrons but the tunneling of the Cooper pairs described by a macroscopic wave function. The effect demonstrates that the superconducting state is a coherent state in which the phase of a macroscopic wave function, which is introduced later as the order parameter of the Ginzburg-Landau (G-L) theory, is uniform in the superconductor. In this state the quantum mechanical property is maintained up to a macroscopic scale and the gauge-invariant relation is kept between the macroscopic wave function and the vector potential. This leads to the macroscopic quantization of magnetic flux through the quantization of the total angular momentum, and this phenomenon can be directly seen from the interference of the superconducting tunneling current due to the magnetic field. Another important result, the AC Josephson effect, describes the relation between the time variation rate of the phase of macroscopic wave function and the voltage which in this case appears across the junction. This voltage comes from the motion of the quantized magnetic flux and is identical to the voltage observed in a type-2 superconductor in the flux flow state as will be shown later.

The superconducting state transforms into the normal state at temperatures above the critical temperature and magnetic fields above the critical field. Transitions from the superconducting state to the normal state and vice versa are phase transitions comparable, for example, to the transition between ferromagnetism and paramagnetism. From a microscopic viewpoint, the Cooper-pair condensation of electrons (which can be compared to a Bose-Einstein condensation of Bose particles) results in the superconducting state and the electron energy gap that exists between the superconducting and normal states. From a macroscopic viewpoint, on the other hand, the superconducting state is a thermodynamic phase and thermodynamics is useful in the description of the phenomenon. Finally, since the electron state is coherent in the superconducting state, the G-L theory, in which the order parameter defined as a superposed wave function of coherent superconducting electrons is used. Among its many applications it is suitable for describing the magnetic properties of type-2 superconductors.

In 1986 a La-based copper-oxide superconductor with a higher critical temperature than metallic superconductors was discovered by Bednortz and Müller. Taking advantage of this break through, numerous so-called "high-temperature" superconductors with even higher critical temperatures but containing Y, Bi, Tl and Hg instead of La were discovered. The exact mechanism of superconductivity in these materials is not yet understood. We have yet to wait for a suitable microscopic explanation. However, the macroscopic electromagnetic properties of high-temperature superconductors have been found to be describable phenomenologically in a manner comparable to those previously applied to metallic superconductors. In this description, the characteristic features of high-temperature superconductors are a large two-dimensional anisotropy originating in the crystal structure and a strong fluctuation effect. The latter feature results from a short coherence length in associating with the high critical temperature, the quasi-two-dimensionality itself, and the condition of high temperature. It was shown theoretically that, as a result of the fluctuation effect, the phase boundary between the superconducting and normal states derived using G-L theory within a mean field approximation is not clear. It follows that a G-L description would be correct only in the region far from the phase boundary. However, because these materials have such high upper critical fields G-L theory is still valid over a wide practical range of temperature and magnetic field.

This book is based on the G-L theory that describes the superconductivity phenomenologically and the Maxwell theory that is the foundation of the electromagnetism. The SI units and the ***E-B*** analogy are used.

1.2 Kinds of Superconductors

There are two kinds of superconductor – type-1 and type-2. These are classified with respect to their magnetic properties.

The magnetization of a type-1 superconductor is shown in Fig. 1.1(a). When the external magnetic field H_e is lower than some critical field H_c, the magnetization is given by

$$M = -H_e \tag{1.1}$$

and the superconductor shows a perfect diamagnetism ($B = 0$). It is in the Meissner state. The transition from the superconducting state to the normal state occurs at $H_e = H_c$ with a discontinuous variation in the magnetization to $M = 0$ (i.e. $B = \mu_0 H_e$ with μ_0 denoting the permeability of vacuum). For a type-2 superconductor, on the other hand, the perfect diamagnetism given by Eq. (1.1) is maintained only up to the lower critical field, H_{c1}, and then the magnetization varies continuously with the penetration of magnetic flux as shown in Fig. 1.1(b) until the diamagnetism disappears at the upper critical field, H_{c2}, where the normal state starts. The partially diamagnetic state between H_{c1} and H_{c2} is called the mixed state. Since the magnetic flux in the superconductor is quantized in the form of "vortices" in this state, it is also called the vortex state.

It is empirically known that the critical field of type-1 superconductors varies with temperature according to

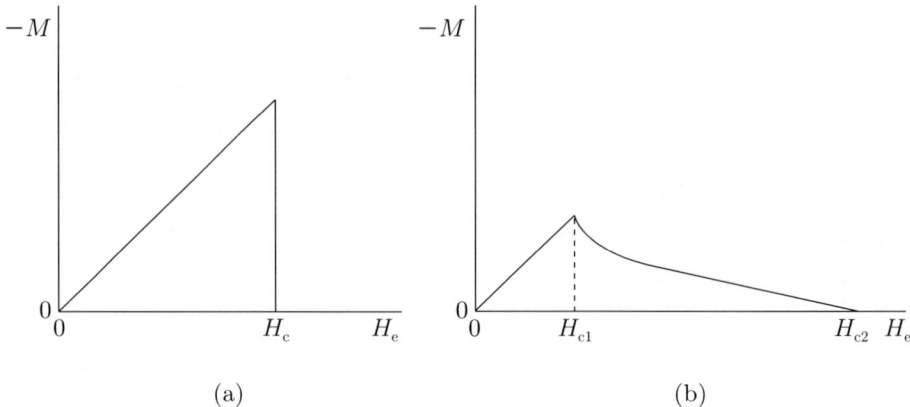

Fig. 1.1. Magnetic field dependence of magnetization for **(a)** type-1 superconductor and **(b)** type-2 superconductor

$$H_c(T) = H_c(0)\left[1 - \left(\frac{T}{T_c}\right)^2\right]. \tag{1.2}$$

The lower and upper critical fields of type-2 superconductors show similar temperature dependences. Obviously they reduce to zero at the critical temperature, T_c. Strictly speaking for type-2 superconductors, whereas the thermodynamic critical field, H_c, shows the temperature dependence of Eq. (1.2), that of H_{c2} deviates from this relationship for some superconductors. Especially in high-temperature superconductors and MgB_2 these critical fields have almost linear temperature dependences even in the low temperature region. The phase diagrams of type-1 and type-2 superconductors on the temperature-magnetic field plane are shown in Fig. 1.2(a) and 1.2(b). The superconducting parameters of various superconductors are listed in Table 1.1. Here H_c in type-2 superconductors is the thermodynamic critical field. Since H_{c1} and H_{c2} in high-temperature superconductors and MgB_2 are significantly different depending on the direction of magnetic field with respect to the crystal axes, the doping state of carriers and the electron mean free path, only the value of H_c in the optimally doped state is given in the table. The details of the anisotropy and the dependence on such factors for the critical fields in these superconductors will be described in Sects. 8.1 and 9.1.

The practical superconducting materials, Nb-Ti and Nb_3Sn, belong to the type-2 class. Their upper critical fields are very high and hence, their superconducting state can be maintained up to high fields. In high-temperature superconductors the upper critical fields are extremely high and it is considered that a clear phase transition to the normal state does not occur due to the effect of significant fluctuation at the phase boundary, $H_{c2}(T)$, derived from the G-L theory.

1.2 Kinds of Superconductors

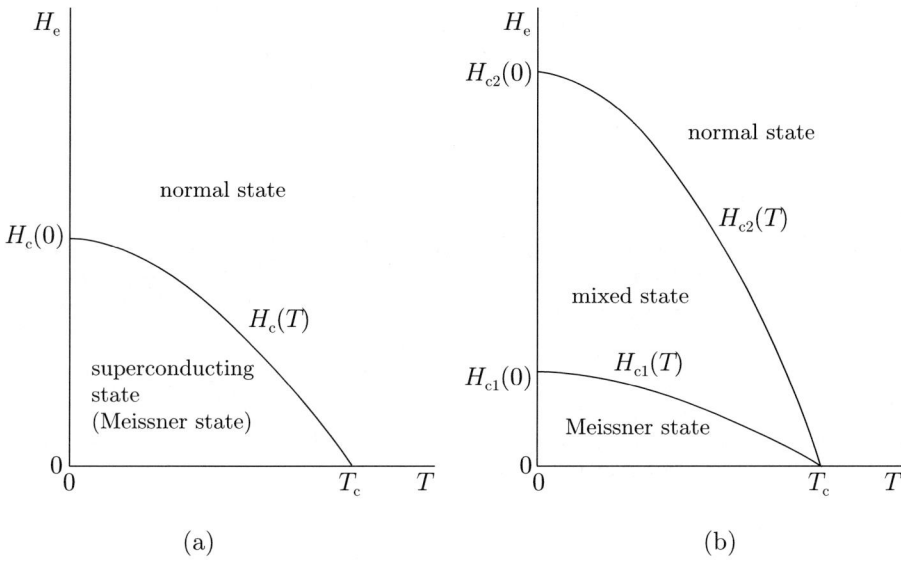

Fig. 1.2. Phase diagram on the magnetic field vs temperature plane for (a) type-1 superconductor and (b) type-2 superconductor

Table 1.1. Critical parameters of various superconductors

	Superconductor	T_c (K)	$\mu_0 H_c(0)$ (mT)	$\mu_0 H_{c1}(0)$ (mT)	$\mu_0 H_{c2}(0)$ (T)
type-1	Hg(α)	4.15	41	–	–
	In	3.41	28	–	–
	Pb	7.20	80	–	–
	Ta	4.47	83	–	–
type-2	Nb	9.25	199	174	0.404
	Nb$_{37}$Ti$_{63}$	9.08	253		15
	Nb$_3$Sn	18.3	530		29
	Nb$_3$Al	18.6			33
	Nb$_3$Ge	23.2			38
	V$_3$Ga	16.5	630		27
	V$_3$Si	16.9	610		25
	PbMo$_6$S$_8$	15.3			60
	MgB$_2$	39	660		
	YBa$_2$Cu$_3$O$_7$	93	1270		
	(Bi,Pb)$_2$Sr$_2$Ca$_2$Cu$_3$O$_x$	110			
	Tl$_2$Ba$_2$Ca$_2$Cu$_3$O$_x$	127			
	HgBa$_2$CaCu$_2$O$_x$	128	700		
	HgBa$_2$Ca$_2$Cu$_3$O$_x$	138	820		

1.3 London Theory

The fundamental electromagnetic properties of superconductors, such as the Meissner effect, can be described by a phenomenological theory first propounded by the London brothers in 1935, even before the discovery of type-2 superconductors. Fortunately this theory turned out to be a good approximation for type-2 superconductors with high upper critical fields, or with large values of G-L parameter; several important characteristics of such superconductors can be derived from this theory. So, we shall here briefly introduce the classic London theory.

A steady persistent current can flow through superconductors. Hence, the classical equation of motion of superconducting electrons should be one that can describe this state. In other words, the deviation from the steady motion, i.e., the acceleration of superconducting electrons is done only by the force due to the electric field. Hence, the equation of motion is given by

$$m^* \frac{d\boldsymbol{v}_\mathrm{s}}{dt} = -e^* \boldsymbol{e} , \tag{1.3}$$

where m^*, $\boldsymbol{v}_\mathrm{s}$ and $-e^*$ are the mass, the velocity and the electric charge ($e^* > 0$) of the superconducting electron, and $\boldsymbol{e}$ is the electric field. If the number density of superconducting electrons is represented by n_s, the superconducting current density is written as

$$\boldsymbol{j} = -n_\mathrm{s} e^* \boldsymbol{v}_\mathrm{s} . \tag{1.4}$$

Substitution of this into Eq. (1.3) leads to

$$\boldsymbol{e} = \frac{m^*}{n_\mathrm{s} e^{*2}} \cdot \frac{d\boldsymbol{j}}{dt} . \tag{1.5}$$

If the magnetic field and the magnetic flux density are denoted by $\boldsymbol{h}$ and $\boldsymbol{b}$, respectively, the Maxwell equations are

$$\nabla \times \boldsymbol{e} = -\frac{\partial \boldsymbol{b}}{\partial t} \tag{1.6}$$

and

$$\nabla \times \boldsymbol{h} = \boldsymbol{j} , \tag{1.7}$$

where the displacement current is neglected in Eq. (1.7). From these equations and with

$$\boldsymbol{b} = \mu_0 \boldsymbol{h} , \tag{1.8}$$

the rotation (curl) of Eq. (1.5) is written as

$$\frac{\partial}{\partial t} \left(\boldsymbol{b} + \frac{m^*}{\mu_0 n_\mathrm{s} e^{*2}} \nabla \times \nabla \times \boldsymbol{b} \right) = 0 . \tag{1.9}$$

1.3 London Theory

Thus, the quantity in the parenthesis on the left hand side of Eq. (1.9) is a constant. The London brothers showed that, when this constant is zero, the Meissner effect can be explained. That is,

$$\boldsymbol{b} + \frac{m^*}{\mu_0 n_s e^{*2}} \nabla \times \nabla \times \boldsymbol{b} = 0 \, . \tag{1.10}$$

Equations (1.5) and (1.10) are called the London equations. Replacing $\nabla \times \nabla \times \boldsymbol{b}$ by $-\nabla^2 \boldsymbol{b}$ (since $\nabla \cdot \boldsymbol{b} = 0$), Eq. (1.10) may be written

$$\nabla^2 \boldsymbol{b} - \frac{1}{\lambda^2} \boldsymbol{b} = 0 \, , \tag{1.11}$$

where λ is a quantity with the dimension of length defined by

$$\lambda = \left(\frac{m^*}{\mu_0 n_s e^{*2}} \right)^{1/2} . \tag{1.12}$$

Let us assume a semi-infinite superconductor of thickness $x \geq 0$. When an external magnetic field H_e is applied along the z-axis parallel to the surface ($x = 0$), it is reasonable to assume that the magnetic flux density has only a z-component which varies only along the x-axis. Then, Eq. (1.11) reduces to

$$\frac{d^2 b}{dx^2} - \frac{b}{\lambda^2} = 0 \, . \tag{1.13}$$

This can be easily solved; and under the conditions that $b = \mu_0 H_e$ at $x = 0$ and is finite at infinity, we have

$$b(x) = \mu_0 H_e \exp\left(-\frac{x}{\lambda} \right) . \tag{1.14}$$

This result shows that the magnetic flux penetrates the superconductor only a distance of the order of λ from the surface (see Fig. 1.3). The characteristic distance λ is called the penetration depth. Since the "superconducting electron" is by now well known to be an electron pair, we assign a double electronic charge to e^*, i.e., $e^* = 2e = 3.2 \times 10^{-19}$ C. As for the mass of the superconducting electron, m^*, we assume also a double electron mass in spite of an ambiguity in the mass. Thus $m^* = 2m = 1.8 \times 10^{-30}$ kg. If we substitute a typical free-electron number density for n_s viz. 10^{28} m^{-3}, then $\lambda \simeq 37$ nm from Eq. (1.12) and the above quantities. Observed values of λ are of the same order of magnitude as this estimation. Thus, the magnetic flux dose not penetrate much below the surface of the superconductor, thereby explaining the Meissner effect. From Eqs. (1.7), (1.8) and (1.14) it is found that the current is also localized and flows along the y-axis according to

$$j(x) = \frac{H_e}{\lambda} \exp\left(-\frac{x}{\lambda} \right) . \tag{1.15}$$

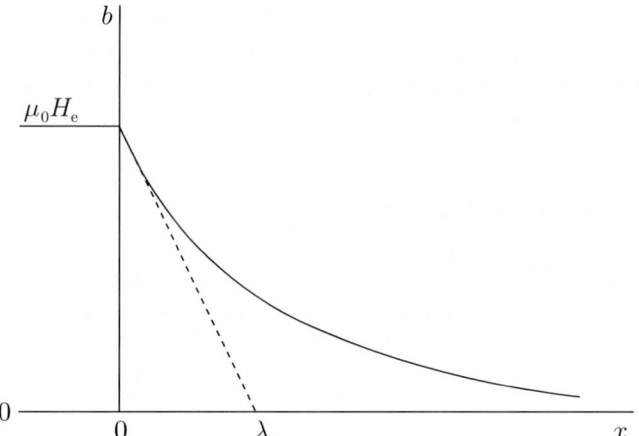

Fig. 1.3. Magnetic flux distribution near the surface of superconductor in the Meissner state

This so-called Meissner current shields the external magnetic field thereby supporting the Meissner effect.

We note that the London equations, (1.5) and (1.10), may be derived just from

$$\boldsymbol{j} = -\frac{n_s e^{*2}}{m^*}\boldsymbol{A}, \tag{1.16}$$

where $\boldsymbol{A}$ is the magnetic vector potential. That is, Eqs. (1.5) and (1.10) respectively can be obtained by differentiation with respect to time and taking the rotation of Eq. (1.16). Equation (1.16) means that the current density of an arbitrary point is determined by the local vector potential at that point. On the other hand, superconductivity is a nonlocal phenomenon and the wave function of electrons is spatially extended. The electrons that contribute to the superconductivity are those within an energy range of the order of $k_B T$ at the Fermi level with k_B denoting the Boltzmann constant. Hence, the uncertainty of the momentum of electrons is $\Delta p \sim k_B T / v_F$ where v_F is the Fermi velocity. Hence, the spatial extent of the wave function of electrons is estimated from the uncertainty principle as

$$\xi_0 \sim \frac{\hbar}{\Delta p} \sim \frac{\hbar v_F}{k_B T_c}, \tag{1.17}$$

where $\hbar = h_P/2\pi$ with h_P denoting Planck's constant. The characteristic length ξ_0 is called the coherence length.

The London theory predicts that physical quantities such as the magnetic flux density and the current density vary within a characteristic distance λ. Hence, λ is required to be sufficiently long with respect to ξ_0 that the local approximation remains valid. That is, the London theory is a good approximation for superconductors in which $\lambda \gg \xi_0$. Such superconductors are typical type-2 superconductors. In this book the London theory will be used to discuss the structure of quantized magnetic flux in type-2 superconductors (Sect. 1.5) and to derive the induced electric field due to the motion of quantized magnetic flux (Sect. 2.2).

1.4 Ginzburg-Landau Theory

Although the London theory explains the Meissner effect, it is unable to deal with the coexistence of magnetic field and superconductivity such as in the intermediate state of type-1 superconductors or the mixed state of type-2. The theory of Ginzburg and Landau (G-L theory) [1] was proposed for the purpose of treating the intermediate state. This theory is based on the deep insight of Ginzburg and Landau on the essence of superconductivity, namely that the superconducting state is such that the phase of the electrons is coherent on a macroscopic scale. The order parameter defined in the theory is, originally a thermodynamic quantity, which now has the property of a mean wave function describing the coherent motion of the center of a group of electrons. This wave function is comparable to the electron wave function of quantum mechanics.

We define the order parameter, Ψ, as a complex number and assume that the square of its magnitude $|\Psi|^2$ gives the number density of superconducting electrons. The free energy of a superconductor depends on this density of superconducting electrons, and hence, is a function of $|\Psi|^2$. In the vicinity of the transition point $|\Psi|^2$ is expected to be sufficiently small and it is expected that the free energy can be expanded as a power series of $|\Psi|^2$:

$$\text{const.} + \alpha |\Psi|^2 + \frac{\beta}{2} |\Psi|^4 + \dots . \tag{1.18}$$

For the purpose of describing the phase transition between the superconducting and normal states, the expansion up to the $|\Psi|^4$ term is sufficient, as will be shown later.

It is speculated that the order parameter varies spatially due to existence of the magnetic field. By analogy with quantum mechanics, this should lead to a kinetic energy. The expected value of the kinetic energy density is written in terms of the momentum operator known in the quantum mechanics as

$$\frac{1}{2m^*} \Psi^* (-i\hbar \nabla + 2e\boldsymbol{A})^2 \Psi , \tag{1.19}$$

where Ψ^* is the complex conjugate of Ψ and m^* is the mass of the superconducting electron, the Cooper pair, and we used the fact that the electric charge of the Cooper pair is $-2e$. The operator of the momentum takes the well known form containing the vector potential, $\boldsymbol{A}$, so that the Lorentz force on a moving charge is automatically derived. From the Hermitian property of the operator the kinetic energy density in Eq. (1.19) is rewritten as

$$\frac{1}{2m^*}|(-i\hbar\nabla + 2e\boldsymbol{A})\Psi|^2 \ . \tag{1.20}$$

Thus, the free energy density in the superconducting state including the energy of magnetic field is given by

$$F_\mathrm{s} = F_\mathrm{n}(0) + \alpha|\Psi|^2 + \frac{\beta}{2}|\Psi|^4 + \frac{1}{2\mu_0}(\nabla \times \boldsymbol{A})^2$$

$$+ \frac{1}{2m^*}|(-i\hbar\nabla + 2e\boldsymbol{A})\Psi|^2 \ , \tag{1.21}$$

where $F_\mathrm{n}(0)$ is the free energy density in the normal state in the absence of the magnetic field.

For simplicity we will first treat the case where the magnetic field is not applied. We may put $\boldsymbol{A} = 0$ without losing generality. Then, since the order parameter does not vary spatially, Eq. (1.21) reduces to

$$F_\mathrm{s}(0) = F_\mathrm{n}(0) + \alpha|\Psi|^2 + \frac{\beta}{2}|\Psi|^4 \ . \tag{1.22}$$

It is necessary for a nonzero equilibrium value of $|\Psi|^2$ to be obtained when the temperature T is lower than the critical value, T_c. This leads to $\alpha < 0$ and $\beta > 0$. From the condition that the derivative of $F_\mathrm{s}(0)$ with respect to $|\Psi|^2$ is zero, we find as the equilibrium value of $|\Psi|^2$

$$|\Psi|^2 = -\frac{\alpha}{\beta} \equiv |\Psi_\infty|^2 \ . \tag{1.23}$$

Substitution of this into Eq. (1.22) leads to the free energy density in the equilibrium state:

$$F_\mathrm{s}(0) = F_\mathrm{n}(0) - \frac{\alpha^2}{2\beta} \ . \tag{1.24}$$

At $T = T_\mathrm{c}$ the transition from the superconducting state to the normal one takes place and $|\Psi_\infty|^2$ becomes zero. Thus α is zero at that temperature. The variation of α with temperature in the vicinity of T_c is assumed to be proportional to $(T - T_\mathrm{c})$. α takes a positive value at $T > T_\mathrm{c}$ and the free energy density given by Eq. (1.22) is minimum at $|\Psi|^2 = 0$. Such variations in the free energy density and the equilibrium value of $|\Psi|^2$ near T_c are shown in Figs. 1.4 and 1.5, respectively. As shown in the above the phase transition can be explained by the expansion of the free energy density up to the term of the order of $|\Psi|^4$.

Now the phase transition in a magnetic field is treated. We assume that the superconductor is type-1 of a sufficient size. Hence, the superconductor shows the Meissner effect, and a magnetic field does not exist inside it except in a region of about λ from the surface when it is in the superconducting state. Such a surface region can be neglected in a large superconductor, and the spatial variation in the order parameter can be disregarded. The equilibrium

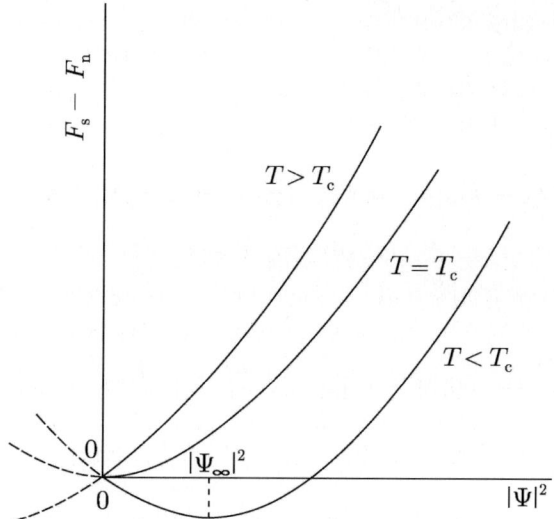

Fig. 1.4. Variation in the free energy density vs. $|\Psi|^2$ at various temperatures

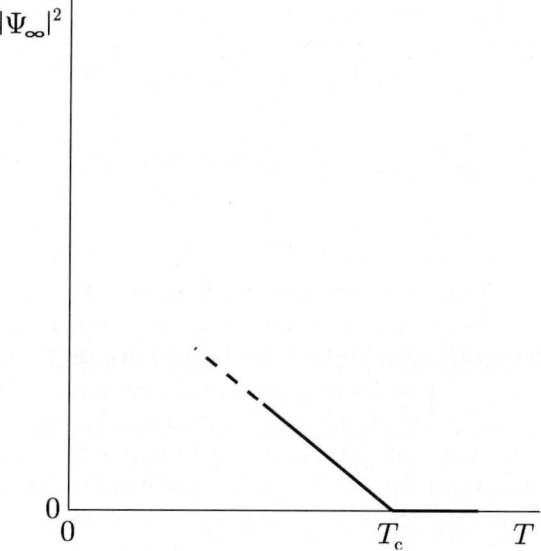

Fig. 1.5. Variation in the equilibrium value of the order parameter, $|\Psi_\infty|^2$, with temperature

state of the superconductor in the magnetic field is determined by minimizing the Gibbs free energy density. If the external magnetic field and the magnetic flux density inside the superconductor are denoted by H_e and B, respectively, the Gibbs free energy density is given by $G_\mathrm{s}(H_\mathrm{e}) = F_\mathrm{s} - BH_\mathrm{e}$. If we note that $B = 0$ and $|\Psi|^2$ is given by Eq. (1.23) in the superconducting state, we have

$$G_\mathrm{s}(H_\mathrm{e}) = F_\mathrm{n}(0) - \frac{\alpha^2}{2\beta} . \tag{1.25}$$

On the other hand, in the normal state, $|\Psi|^2 = 0$ and $B = \mu_0 H_\mathrm{e}$ lead to

$$G_\mathrm{n}(H_\mathrm{e}) = F_\mathrm{n}(0) + \frac{B^2}{2\mu_0} - BH_\mathrm{e} = F_\mathrm{n}(0) - \frac{1}{2}\mu_0 H_\mathrm{e}^2 . \tag{1.26}$$

Since G_s and G_n are the same at the transition point, $H_\mathrm{e} = H_\mathrm{c}$, we have

$$\frac{\alpha^2}{\beta} = \mu_0 H_\mathrm{c}^2 . \tag{1.27}$$

In the vicinity of T_c, β does not appreciably change with temperature and α is approximately proportional to H_c. That is, we have $\alpha \simeq 2(\mu_0\beta)^{1/2} H_\mathrm{c}(0)(T - T_\mathrm{c})/T_\mathrm{c}$. Thus, it is found that the above assumption on the temperature dependence of α is satisfied. From Eqs. (1.25)–(1.27) we obtain

$$G_\mathrm{s}(H_\mathrm{e}) = G_\mathrm{n}(H_\mathrm{e}) - \frac{1}{2}\mu_0(H_\mathrm{c}^2 - H_\mathrm{e}^2) . \tag{1.28}$$

This result shows that $G_\mathrm{s}(H_\mathrm{e}) < G_\mathrm{n}(H_\mathrm{e})$ and the superconducting state occurs for $H_\mathrm{e} < H_\mathrm{c}$ and the normal state occurs for $H_\mathrm{e} > H_\mathrm{c}$. That is, the transition in the magnetic field is explained by this equation. Especially when $H_\mathrm{e} = 0$, the above equation leads to

$$G_\mathrm{s}(0) = G_\mathrm{n}(0) - \frac{1}{2}\mu_0 H_\mathrm{c}^2 . \tag{1.29}$$

The maximum difference of the free energy density between the superconducting and normal states, $(1/2)\mu_0 H_\mathrm{c}^2$, is called the condensation energy density.

When the superconductor coexists with the magnetic field, $\Psi(\boldsymbol{r})$ and $\boldsymbol{A}(\boldsymbol{r})$ are determined so that the free energy, $\int F_\mathrm{s} dV$, is minimized. Hence, the variations of $\int F_\mathrm{s} dV$ with respect to $\Psi^*(\boldsymbol{r})$ and $\boldsymbol{A}(\boldsymbol{r})$ are required to be zero and the following two equations are derived:

$$\frac{1}{2m^*}(-i\hbar\nabla + 2e\boldsymbol{A})^2\Psi + \alpha\Psi + \beta|\Psi|^2\Psi = 0 , \tag{1.30}$$

$$\boldsymbol{j} = \frac{i\hbar e}{m^*}(\Psi^*\nabla\Psi - \Psi\nabla\Psi^*) - \frac{4e^2}{m^*}|\Psi|^2\boldsymbol{A} , \tag{1.31}$$

with

1.4 Ginzburg-Landau Theory

$$\boldsymbol{j} = \frac{1}{\mu_0} \nabla \times \nabla \times \boldsymbol{A} \ . \tag{1.32}$$

The above Eqs. (1.30) and (1.31) are called the Ginzburg-Landau equations, or the G-L equations. In the derivation, the Coulomb gauge, $\nabla \cdot \boldsymbol{A} = 0$, and the condition,

$$\boldsymbol{n} \cdot (-i\hbar\nabla + 2e\boldsymbol{A})\Psi = 0 \ , \tag{1.33}$$

on the surface were used. In the above, $\boldsymbol{n}$ is a unit vector normal to the surface and the condition of Eq. (1.33) implies that current does not flow across the surface. This is fulfilled for the case where the superconductor is facing a vacuum or an insulating material. On the other hand, if the superconductor is facing a metal, the right hand side in Eq. (1.33) is replaced by $ia\Psi$ with a being a real number [2].

The electromagnetic properties in the superconductor are determined by two characteristic lengths, i.e. λ, the penetration depth of magnetic field and ξ, the coherence length. These are related to the spatial variations in the magnetic flux density B and the order parameter Ψ. Here we shall derive these quantities from the G-L theory.

We assume that a weak magnetic field is applied to the superconductor. In this case the variation in the order parameter is expected to be small, and hence, the approximation, $\Psi = \Psi_\infty$, may be allowed. Then, Eq. (1.31) reduces to

$$\boldsymbol{j} = -\frac{4e^2}{m^*}|\Psi_\infty|^2 \boldsymbol{A} \ . \tag{1.34}$$

This is similar to Eq. (1.16) of the London theory. If we recognize that $e^* = 2e$, it follows that $|\Psi_\infty|^2$ corresponds to n_s. Thus, the G-L theory is a more general theory that reduces to the London theory when the order parameter does not vary in space. Hence, the Meissner effect can be derived in the same manner as in Sect. 1.3 and the penetration depth is given by

$$\lambda = \left(\frac{m^*}{4\mu_0 e^2 |\Psi_\infty|^2}\right)^{1/2} \ . \tag{1.35}$$

Near T_c, $|\Psi_\infty|^2$ is proportional to $(T_\mathrm{c} - T)$ and λ varies proportionally to $(T_\mathrm{c} - T)^{-1/2}$ and diverges at $T = T_\mathrm{c}$. In terms of λ, the coefficients of α and β are expressed as

$$\alpha = -\frac{(2e\mu_0 H_\mathrm{c} \lambda)^2}{m^*} \ , \tag{1.36}$$

$$\beta = \frac{16e^4 \mu_0^3 H_\mathrm{c}^2 \lambda^4}{m^{*2}} \ . \tag{1.37}$$

Next we shall discuss the spatial variation in the order parameter, Ψ. We treat the case where the magnetic field is not applied and hence that $\boldsymbol{A} = 0$. For simplicity we assume that Ψ varies only along the x-axis. If we normalize the order parameter according to

$$\psi = \frac{\Psi}{|\Psi_\infty|}, \qquad (1.38)$$

Eq. (1.30) reduces to

$$\xi^2 \frac{d^2\psi}{dx^2} + \psi - |\psi|^2 \psi = 0, \qquad (1.39)$$

where ξ is a characteristic length called the coherence length and is given by

$$\xi = \frac{\hbar}{(2m^*|\alpha|)^{1/2}}. \qquad (1.40)$$

We can choose a real function for ψ in Eq. (1.39). Suppose that the order parameter varies slightly from its equilibrium value such that $\psi = 1 - f$, where $f \ll 1$. Within this range, Eq. (1.39) reduces to

$$\xi^2 \frac{d^2 f}{dx^2} - 2f = 0 \qquad (1.41)$$

and hence

$$f \sim \exp\left(-\frac{\sqrt{2}|x|}{\xi}\right). \qquad (1.42)$$

This shows that the order parameter varies in space within a distance comparable to ξ. From Eqs. (1.36) and (1.40) the coherence length can also be expressed as

$$\xi = \frac{\hbar}{2\sqrt{2} e \mu_0 H_c \lambda}. \qquad (1.43)$$

It turns out from Eq. (1.40) or (1.43) that ξ also increases in proportion to $(T_c - T)^{-1/2}$ in the vicinity of T_c. On the other hand, the coherence length in the BCS theory [3] is given by

$$\xi_0 = \frac{\hbar v_F}{\pi \Delta(0)} = 0.18 \frac{\hbar v_F}{k_B T_c} \qquad (1.44)$$

and does not depend on temperature. $\Delta(0)$ is the energy gap at $T = 0$. In spite of such a difference, the two coherence lengths are related to each other. Since the superconductivity is nonlocal, this relation changes with the electron mean free path, l. In the vicinity of T_c, the coherence length in the G-L theory becomes [4]

$$\xi(T) = 0.74 \frac{\xi_0}{(1-t)^{1/2}}; \qquad l \gg \xi_0, \qquad (1.45a)$$

$$= 0.85 \frac{(\xi_0 l)^{1/2}}{(1-t)^{1/2}}; \qquad l \ll \xi_0, \qquad (1.45b)$$

where $t = T/T_c$. Equations (1.45a) and (1.45b) correspond to the cases of "clean" and "dirty" superconductors, respectively. It is seen that $\xi(T)$ is comparable to ξ_0 in a clean superconductor and is much smaller than ξ_0 in a dirty one.

The penetration depth is also influenced by the electron mean free path l owing to the nonlocal nature of the superconductivity. We call the penetration depth given by Eq. (1.35) the London penetration depth and denote it by λ_L. If λ_L is sufficiently longer than ξ_0 and l, we have $\lambda = \lambda_L$ in a clean superconductor ($l \gg \xi_0$) and $\lambda \simeq \lambda_L(\xi_0/l)^{1/2}$ in a dirty superconductor ($l \ll \xi_0$). In a superconductor where $\xi_0 \gg \lambda_L$, i.e. in a "Pippard superconductor," we have $\lambda \simeq 0.85(\lambda_L^2 \xi_0)^{1/3}$.

The ratio of the two characteristic lengths in G-L theory defined by

$$\kappa = \frac{\lambda}{\xi} \tag{1.46}$$

is called the G-L parameter. According to the G-L theory, λ and ξ have the same temperature dependences, and hence κ is independent of temperature. As a matter of fact, κ decreases slightly with increasing temperature. The G-L parameter is important in describing the magnetic properties of superconductors. In particular the classification into type-1 and type-2 superconductors is determined by the value of this parameter. Also the upper critical field of the type-2 superconductor depends on this parameter.

We next go on to discuss the occurrence of superconductivity in a bulk superconductor in a magnetic field sufficiently high that the higher order term, $\beta|\Psi|^2\Psi$, in Eq. (1.30) can be neglected. We assume that the external magnetic field H_e is applied along the z-axis. The magnetic flux density in the superconductor is taken to be uniform in space, $b \simeq \mu_0 H_e$, and hence the vector potential is given by

$$\boldsymbol{A} = \mu_0 H_e x \boldsymbol{i}_y , \tag{1.47}$$

where $\boldsymbol{i}_y$ is a unit vector directed along the y-axis. In the above the choice of x-axis direction is not important in a bulk superconductor and hence generality is still maintained even under Eq. (1.47). Since $\boldsymbol{A}$ depends only on x, it is reasonable to assume that Ψ also depends only on x. Hence, Eq. (1.30) reduces to

$$-\frac{\hbar^2}{2m^*} \cdot \frac{d^2\Psi}{dx^2} + \frac{2e^2\mu_0^2}{m^*}(H_e^2 x^2 - 2H_c^2 \lambda^2)\Psi = 0 . \tag{1.48}$$

This equation has the same form as the well-known Schrödinger equation for a one-dimensional harmonic oscillator. It has solutions only when the condition

$$\left(n + \frac{1}{2}\right)\hbar H_e = 2e\mu_0 H_c^2 \lambda^2 \tag{1.49}$$

is satisfied with n being nonnegative integers. The maximum value of H_e is obtained for $n = 0$, corresponding to the maximum field within which the superconductivity can exist, i.e., the upper critical field, H_{c2}. Thus, we have

$$H_{c2} = \frac{4e\mu_0 H_c^2 \lambda^2}{\hbar} . \tag{1.50}$$

Using Eqs. (1.43) and (1.46), the upper critical field may also be written as

$$H_{c2} = \sqrt{2}\kappa H_c .\tag{1.51}$$

Hence, the superconducting state may exist in a magnetic field higher than the critical field H_c for a superconductor with κ larger than $1/\sqrt{2}$. Such is the type-2 superconductor. In this case, the superconductor is in the mixed state and no special characteristic phenomena take place at $H_e = H_c$. That is, H_c cannot be directly measured experimentally. Since it is related to the condensation energy density, H_c is called the thermodynamic critical field in type-2 superconductors. If we now introduce the flux quantum, $\phi_0 = h_{\rm P}/2e$, to be considered in the next section, the upper critical field may be rewritten in the form

$$H_{c2} = \frac{\phi_0}{2\pi\mu_0\xi^2} .\tag{1.52}$$

This relationship is used in estimating the coherence length from an observed value of H_{c2}.

1.5 Magnetic Properties

A characteristic feature of type-2 superconductors in a magnetic field is that the magnetic flux is quantized on a macroscopic scale. In this book we refer to the quantized magnetic flux as a flux line. The flux lines are isolated from each other at sufficiently low magnetic fields. On the other hand, at high magnetic fields these overlap and interact to form a flux line lattice. In this section such quantization of magnetic flux is discussed in terms of the G-L theory. The superconductor's magnetic properties are discussed in terms of the internal structure of the flux line at low fields and the structure of the flux line lattice at high fields.

1.5.1 Quantization of Magnetic Flux

We suppose a superconductor in a sufficiently weak magnetic field. For simplicity we assume the magnetic flux to be localized at an certain region inside the superconductor. This assumption pre-supposes the quantization of the magnetic flux. It will be shown later that the assumption is actually fulfilled and hence the treatment is self-consistent. Consider a closed loop, C, enclosing the region in which the magnetic flux is localized. The distance between the localized magnetic flux and C is assumed to be sufficiently long to enable the magnetic flux density and the current density to be zero on C. If we write

$$\Psi = |\Psi|\exp(i\phi) \tag{1.53}$$

with ϕ denoting the phase of the order parameter, Eq. (1.31) reduces to

1.5 Magnetic Properties

$$j = -\frac{2\hbar e}{m^*}|\Psi|^2 \nabla\phi - \frac{4e^2}{m^*}|\Psi|^2 A \ . \tag{1.54}$$

In the above the first term represents the current caused by the gradient of the phase of the order parameter, i.e., the Josephson current. On the loop C, $j = 0$ and hence

$$A = -\frac{\hbar}{2e}\nabla\phi \ . \tag{1.55}$$

Integration of this over C leads to

$$\oint_C A \cdot \mathrm{d}s = \int b \cdot \mathrm{d}S = \Phi \ , \tag{1.56}$$

where Φ is the magnetic flux that interlinks with C. If we substitute the right hand side of Eq. (1.55) for A in Eq. (1.56) the latter becomes

$$\Phi = -\frac{\hbar}{2e}\oint_C \nabla\phi \cdot \mathrm{d}s = -\frac{\hbar}{2e}\Delta\phi \ , \tag{1.57}$$

where $\Delta\phi$ is a variation in the phase after one circulation on C. From the mathematical requirement that the order parameter should be a single-valued function, $\Delta\phi$ must be integral multiple of 2π. That is,

$$\Phi = n\phi_0 \ , \tag{1.58}$$

where n is an integer and

$$\phi_0 = \frac{h_\mathrm{P}}{2e} = 2.0678 \times 10^{-15} \text{ Wb} \ , \tag{1.59}$$

where ϕ_0 is the unit of the magnetic flux and called the flux quantum. Thus we have shown that the magnetic flux in superconductors is quantized. In the above the curvilinear integral of $\nabla\phi$ on the closed loop C is not zero, since $\nabla\phi$ has a singular point at the center of the flux line. This will be discussed in Subsect. 1.5.2.

In the beginning of the above proof we assumed that the magnetic flux was localized in a certain region of a superconductor. This condition is fulfilled at low fields wherein the magnetic flux density decreases as $\exp(-r/\lambda)$ with increasing distance r from the center of the isolated flux line, as will be shown later in Eq. (1.62b). At high fields, on the other hand, the flux lines are not localized and there exists a pronounced overlap of the magnetic flux. Under this condition a flux line lattice is formed. But even in this case the magnetic flux is quantized in each unit cell. The proof of this quantization is Exercise 1.3.

1.5.2 Vicinity of Lower Critical Field

Near the lower critical field, the density of magnetic flux penetrating the superconductor is low and the spacing between the flux lines is large. In this subsection we shall discuss the structure of isolated flux line for the case of typical type-2 superconductor with the large G-L parameter, κ. In this case the London theory can be used. It should be noted that Eq. (1.10) holds correct only in the region greater than a distance ξ from the center of the flux line in which $|\Psi|$ is approximately constant. As will be shown later, $|\Psi|$ is zero at the center and varies spatially within a region of radius ξ, known as the core. Equation (1.10) cannot be used in the core region. In fact, if we assume that this equation is valid within the entire region, an incorrect result is obtained. This can be seen by integrating Eq. (1.10) within a sufficiently wide area including the isolated flux line. From Stokes' theorem the surface integral of the second term in Eq. (1.10) is transformed into the integral of the current on the closed loop that surrounds the area. This integral is zero, since the current density is zero at the position sufficiently far from the flux line. This implies that the total magnetic flux in this area is zero. Hence, some modifications are necessary to enable the contribution from the core to the magnetic flux to be equal to ϕ_0. In the case of superconductor with $\kappa \gg 1$, the area of the core is very narrow in comparison with the total area of the flux line. Hence, we assume most simply that the magnetic structure is described by

$$\boldsymbol{b} + \lambda^2 \nabla \times \nabla \times \boldsymbol{b} = \boldsymbol{i}_z \phi_0 \delta(\boldsymbol{r}) \tag{1.60}$$

in the region outside the core that occupies most of the area. In the above it is assumed that the magnetic field is applied along the z-axis and $\boldsymbol{i}_z$ is a unit vector in that direction. $\boldsymbol{r}$ is a vector in the x-y plane and the center of the flux line exists at $\boldsymbol{r} = 0$. $\delta(\boldsymbol{r})$ is a two-dimensional delta function. The coefficient, ϕ_0, on the right hand side comes from the requirement that the total amount of the magnetic flux of one flux line is ϕ_0. Equation (1.60) is called the modified London equation.

The solution of this equation is given by

$$b(r) = \frac{\phi_0}{2\pi\lambda^2} K_0\left(\frac{r}{\lambda}\right), \tag{1.61}$$

where K_0 is the modified Bessel function of the zeroth order. This function diverges at $r \to 0$. Since the magnetic flux density should have a finite value, the modified London equation still does not hold correct in the region of $r < \xi$. Outside the core, Eq. (1.61) is approximated by

$$b(r) \simeq \frac{\phi_0}{2\pi\lambda^2} \left(\log\frac{\lambda}{r} + 0.116\right); \qquad \xi \ll r \ll \lambda \tag{1.62a}$$

$$\simeq \frac{\phi_0}{2\pi\lambda^2} \left(\frac{\pi\lambda}{2r}\right)^{1/2} \exp\left(-\frac{r}{\lambda}\right); \qquad r \gg \lambda, \tag{1.62b}$$

in terms of elementary functions. The current density flowing around the flux line has only the azimuthal component:

$$j(r) = -\frac{1}{\mu_0} \cdot \frac{\partial b}{\partial r} = \frac{\phi_0}{2\pi\mu_0\lambda^3} K_1\left(\frac{r}{\lambda}\right), \tag{1.63}$$

where K_1 is the modified Bessel function of the first order. In particular, the above equation reduces to

$$j(r) = \frac{\phi_0}{2\pi\mu_0\lambda^2 r} \tag{1.64}$$

in the region of $\xi \ll r \ll \lambda$.

In the region of $r < \xi$, the order parameter varies in space. We shall discuss the structures of the order parameter and the magnetic flux density in this region by solving the G-L equations. From symmetry it is reasonable to assume that $|\Psi|$ is a function only of r, the distance from the center of the flux line. Hence we write $\Psi/|\Psi_\infty| = f(r)\exp(i\phi)$ such that when r becomes sufficiently large, $f(r)$ approaches 1. It can be shown according to the argument of Subsect. 1.5.1 that the variation of the phase when circulating once around a circle with radius of r should be 2π (recognizing that the number of flux lines inside the circle is 1). Hence, ϕ is a function of the azimuthal angle, θ; the simplest function satisfying this condition is

$$\phi = -\theta . \tag{1.65}$$

In this case, we easily obtain

$$\nabla\phi = -\frac{1}{r}\boldsymbol{i}_\theta . \tag{1.66}$$

This shows that the center of the flux line is a singular point at which this function is not differentiable. The relation of $\nabla \times \nabla\phi = 0$ is satisfied except at the singular point, and it can be expressed as

$$\nabla \times \nabla\phi = -2\pi\boldsymbol{i}_z\delta(\boldsymbol{r}) \tag{1.67}$$

over all space.

From Eq. (1.65) we have

$$\frac{\Psi}{|\Psi_\infty|} = f(r)\exp(-i\theta) . \tag{1.68}$$

It is assumed that the vector potential $\boldsymbol{A}$ is also a function only of r. Then, it turns out that $\boldsymbol{A}$ has only the θ-component, A_θ. That is, the relation of $b(r) = (1/r)(\partial/\partial r)(rA_\theta)$ results in

$$A_\theta = \frac{1}{r}\int_0^r r'b(r')\mathrm{d}r' . \tag{1.69}$$

In the case of high-κ superconductors, since b cannot vary in space in the region of $r < \xi$, we have

$$A_\theta \simeq \frac{b(0)}{2}r . \tag{1.70}$$

Substitution of Eqs. (1.68) and (1.70) into Eq. (1.30) leads to

$$f - f^3 - \xi^2 \left[\left(\frac{1}{r} - \frac{\pi b(0) r}{\phi_0} \right)^2 f - \frac{1}{r} \cdot \frac{d}{dr}\left(r \frac{df}{dr}\right) \right] = 0 \ . \tag{1.71}$$

In the normal core f is considered to be sufficiently small. In fact, it is seen that a nearly constant solution for f does not exist. Hence, we shall assume that $f = cr^n$ with $n > 0$. The dominant terms of the lowest order are those of the order of r^{n-2}. If we take notice of these terms, Eq. (1.71) leads to

$$r^{n-2}(1 - n^2) = 0 \ . \tag{1.72}$$

We obtain $n = 1$ from this equation. In the next place we assume as $f = cr(1 + dr^m)$. If the next dominant terms are picked up, we have

$$1 + \frac{b(0)}{\mu_0 H_{c2}} - d\xi^2[1 - (m+1)^2]r^{m-2} = 0 \ , \tag{1.73}$$

where Eq. (1.52) is used. From this equation we obtain $m = 2$ and the value of d. Finally we obtain [5]

$$f \simeq cr\left[1 - \frac{r^2}{8\xi^2}\left(1 + \frac{b(0)}{\mu_0 H_{c2}}\right)\right] \ . \tag{1.74}$$

It is seen that the order parameter is zero at the center of the core. This is the important feature which proves the current density at the flux line center dose not diverge (see Eqs. (1.54) and (1.66)). Hence, the region of $r \lesssim \xi$ is sometimes called the normal core. At low fields $b(0)/\mu_0 H_{c2}$ is small and may be disregarded, in which case f takes on a maximum value at $r = (8/3)^{1/2}\xi = a_0$. This maximum value should be comparable to 1 at a position sufficiently far from the center, hence $c \sim 1/\xi$. If we approximate as

$$f \simeq \tanh\left(\frac{r}{r_n}\right) \tag{1.75}$$

with $c \simeq 1/r_n$, numerical calculation [6] allows the length to be derived:

$$r_n = \frac{4.16\xi}{\kappa^{-1} + 2.25} \ , \tag{1.76}$$

which reduces to 1.8ξ in high-κ superconductors. Therefore, in a strict sense the solutions of the London equation, Eqs. (1.61) and (1.63), are correct only for $r \gtrsim 4\xi$. The structures of the magnetic flux density and the order parameter in the flux line are schematically shown in Fig. 1.6. Since the magnetic flux density in the central part of the core cannot vary steeply in space, its value is approximately given by $(\phi_0/2\pi\lambda^2) \log \kappa$. It will be shown later that this is close to $2\mu_0 H_{c1}$.

We go on to calculate the energy per unit length of the isolated flux line in a bulk high-κ superconductor. From Eq. (1.74) we write approximately

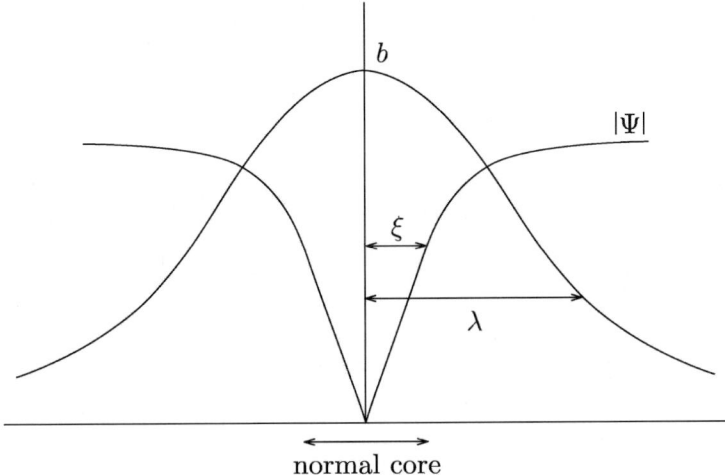

Fig. 1.6. Spatial variations of the order parameter and the magnetic flux density in an isolated flux line

$f \simeq (3r/2a_0) - (r^3/2a_0^3)$ at low fields. This implies that the core occupies the region $r < a_0$. Outside the core, the important terms in the G-L free energy in Eq. (1.21) are the magnetic field energy and the kinetic energy. The kinetic energy density is found to be written as the energy density of the current, $(\mu_0/2)\lambda^2 \boldsymbol{j}^2$, with the replacement of Ψ by Ψ_∞ and the use of Eq. (1.34), the London theory. Hence, the contribution from the outside of the core to the energy of a unit length of the flux line is given by

$$\epsilon' = \int \left(\frac{\boldsymbol{b}^2}{2\mu_0} + \frac{\mu_0}{2}\lambda^2 \boldsymbol{j}^2 \right) dV' = \frac{1}{2\mu_0} \int [\boldsymbol{b}^2 + \lambda^2 (\nabla \times \boldsymbol{b})^2] dV' . \quad (1.77)$$

In the above $\int dV'$ is a volume integral per unit length of the flux line except the area $|\boldsymbol{r}| \leq a_0$. From the condition that the variation of the kernel of the integral of Eq. (1.77) with respect to $\boldsymbol{b}$ is zero, the London equation is derived. Integrating the second term partially, Eq. (1.77) becomes

$$\epsilon' = \frac{1}{2\mu_0} \int (\boldsymbol{b} + \lambda^2 \nabla \times \nabla \times \boldsymbol{b}) \cdot \boldsymbol{b}\, dV' + \frac{\lambda^2}{2\mu_0} \int [\boldsymbol{b} \times (\nabla \times \boldsymbol{b})] \cdot d\boldsymbol{S} . \quad (1.78)$$

It is found from Eq. (1.60) that the first integral is zero. The second integral is carried out on the surfaces of $|\boldsymbol{r}| = a_0$ and $|\boldsymbol{r}| = R(R \to \infty)$. It is easily shown that the latter surface integral at infinity is zero. The former integral on the core surface can be approximately calculated using Eqs. (1.62a) and (1.64). As a result we have

$$\epsilon' \simeq \frac{\lambda^2}{2\mu_0} \cdot \frac{\phi_0}{2\pi\lambda^2} \left(\log \frac{\lambda}{a_0} + 0.116 \right) \frac{\phi_0}{2\pi\lambda^2 a_0} \cdot 2\pi a_0$$
$$= \frac{\phi_0^2}{4\pi\mu_0\lambda^2}(\log\kappa - 0.374) = 2\pi\mu_0\xi^2 H_c^2(\log\kappa - 0.374) \ . \quad (1.79)$$

The contributions from inside the core to the energy are: $0.995\pi\mu_0 H_c^2\xi^2$ from the variation in the order parameter and $(8/3)\pi\mu_0 H_c^2\xi^2(\log\kappa/\kappa)^2$ from the magnetic field. These are about $(2\log\kappa)^{-1}$ and $4\log\kappa/3\kappa^2$ times as large as the energy given by Eq. (1.79). Hence, the second term is found to be very small especially in high-κ superconductors. If this term is disregarded, the energy of a unit length of the flux line becomes

$$\epsilon = 2\pi\mu_0 H_c^2\xi^2(\log\kappa + 0.124) \ . \quad (1.80)$$

According to the rigorous calculation of Abrikosov [7] the number in the second term in the above equation is 0.081.

We shall estimate the lower critical field, H_{c1}, from the above result. The Gibbs free energy is continuous during the transition at $H_e = H_{c1}$. The volume of the superconductor is denoted by V. The Gibbs free energy before and after the penetration of a flux line is given by

$$VG_s = VF_s \quad (1.81)$$

and

$$VG_s = VF_s + \epsilon L - H_{c1}\int b\mathrm{d}V = VF_s + \epsilon L - H_{c1}\phi_0 L \ , \quad (1.82)$$

respectively. In the above F_s is the Helmholtz free energy density before the penetration of the flux line and L is the length of the flux line in the superconductor. The second term in Eq. (1.82) is a variation in the energy due to the formation of the flux line and the third term is for the Legendre transformation. Comparing Eqs. (1.81) and (1.82), we have

$$H_{c1} = \frac{\epsilon}{\phi_0} = \frac{H_c}{\sqrt{2}\kappa}(\log\kappa + 0.081) \ , \quad (1.83)$$

where the correct expression by Abrikosov was used for ϵ. This equation can be used for superconductors with high κ values to which the London theory is applicable.

Here we shall calculate the magnetization in the vicinity of H_{c1}. In this case the spacing between the flux lines is so large that the magnetic flux density $b(\mathbf{r})$ is approximately given by a superposition of the magnetic flux density of the isolated flux lines, $b_i(\mathbf{r})$:

$$b(\mathbf{r}) = \sum_n b_i(\mathbf{r} - \mathbf{r}_n) \ , \quad (1.84)$$

where $\boldsymbol{r}_n$ denotes the position of n-th flux line. The free energy density in this state is again given by Eq. (1.78) and we have

$$F = \frac{\phi_0}{2\mu_0} \sum_{m \neq n} \sum b_i(\boldsymbol{r}_m - \boldsymbol{r}_n) + \frac{B}{\phi_0}\epsilon$$

$$= \frac{B}{2\mu_0} \sum_{n \neq 0} b_i(\boldsymbol{r}_0 - \boldsymbol{r}_n) + \frac{B}{\phi_0}\epsilon , \quad (1.85)$$

where the summation with respect to m is taken within a unit area and the summation with respect to n is taken in the entire region of the superconductor. The first term in Eq. (1.85) is the interaction energy among the flux lines and the second term is the self energy of the flux lines, and B is the mean magnetic flux density. In the surface integral around the n-th core in the derivation of the self energy, the contributions from other flux lines are neglected, since these are sufficiently small in the vicinity of H_{c1}. Substituting for b_i using Eq. (1.61) results in

$$F = \frac{\phi_0 B}{4\pi\mu_0\lambda^2} \sum_{n \neq 0} K_0\left(\frac{|\boldsymbol{r}_0 - \boldsymbol{r}_n|}{\lambda}\right) + BH_{c1} . \quad (1.86)$$

We treat the case of triangular flux line lattice and assume the spacing of flux lines given by

$$a_f = \left(\frac{2\phi_0}{\sqrt{3}B}\right)^{1/2} \quad (1.87)$$

to be sufficiently large. If we take account only the interactions from the six nearest neighbors, the Gibbs free energy density is given by

$$G = F - BH_e = \frac{3\phi_0 B}{2\pi\mu_0\lambda^2} \left(\frac{\pi\lambda}{2a_f}\right)^{1/2} \exp\left(-\frac{a_f}{\lambda}\right) - B(H_e - H_{c1}) , \quad (1.88)$$

where H_e is the external magnetic field. The magnetic flux density B at which G is minimum is obtained from the relation:

$$B^{-1/4}\left[1 + \frac{5}{2}\left(\frac{\sqrt{3}\lambda^2}{2\phi_0}\right)^{1/2} B^{1/2}\right] \exp\left[-\left(\frac{2\phi_0}{\sqrt{3}\lambda^2 B}\right)^{1/2}\right]$$

$$= 3.2\mu_0(H_e - H_{c1})\left(\frac{\lambda^2}{\phi_0}\right)^{5/4} . \quad (1.89)$$

The exact solution of this equation can be obtained only by numerical calculation. However, if we notice that the variation in B is mostly within the exponential function, the B in the prefactor can be approximately replaced by ϕ_0/λ^2, and we have

$$B \sim \frac{2\phi_0}{\sqrt{3}\lambda^2}\left\{\log\left[\frac{\phi_0}{\mu_0(H_e - H_{c1})\lambda^2}\right]\right\}^{-2} . \quad (1.90)$$

It is seen from this equation that B increases rapidly from zero at $H_e = H_{c1}$.

1.5.3 Vicinity of Upper Critical Field

An overlap of the magnetic flux is pronounced and the spacing between the cores is small near the upper critical field. Hence, the London theory cannot be used and an analysis using the G-L theory is necessary. In such a high field the order parameter Ψ is sufficiently small and the higher order term $\beta|\Psi|^2\Psi$ in Eq. (1.30) can be neglected. Due to the pronounced flux overlap the magnetic flux density can be regarded as approximately uniform in the superconductor. We assume that the magnetic field is directed along the z-axis. Then to a first approximation the vector potential can be expressed as in Eq. (1.47). If we write

$$\Psi(x,y) = e^{-iky}\Psi'(x)\,, \tag{1.91}$$

Ψ' obeys Eq. (1.48) with x replaced by $x - x_0$ where

$$x_0 = \frac{\hbar k}{2\mu_0 e H_e}\,. \tag{1.92}$$

The maximum field at which this equation has a solution is H_{c2}. In this case Eq. (1.92) reduces to $x_0 = k\xi^2$. We are interested in the phenomenon at the external magnetic field slightly smaller than H_{c2}, and then, we approximate as $h = H_{c2}$ in the beginning. The equation for Ψ' reduces to

$$-\xi^2 \frac{d^2\Psi'}{dx^2} + \left[\left(\frac{x}{\xi} - k\xi\right)^2 - 1\right]\Psi' = 0\,. \tag{1.93}$$

It can be shown easily that Ψ' has a solution of the form:

$$\Psi' \sim \exp\left[-\frac{1}{2}\left(\frac{x}{\xi} - k\xi\right)^2\right]\,. \tag{1.94}$$

Since the number k is arbitrary, Ψ becomes

$$\Psi = \sum_n C_n e^{-inky} \exp\left[-\frac{1}{2}\left(\frac{x}{\xi} - nk\xi\right)^2\right]\,. \tag{1.95}$$

This corresponds to the assumption of a periodic order parameter, i.e., a periodic arrangement of flux lines. This is because such a periodic structure is expected to be favorable with respect to the energy. One of the lattices with the high periodicity is the triangular lattice. This lattice is obtained by putting $C_{2m} = C_0$ and $C_{2m+1} = iC_0$. It is rather difficult to see that Eq. (1.95) represents a triangular lattice. Let us make the transformation

$$x = \frac{\sqrt{3}}{2}X\,, \qquad y = \frac{X}{2} + Y \tag{1.96}$$

and expand $|\Psi|^2$ into a double Fourier series. After a calculation we obtain

$$|\Psi|^2 = |C_0|^2 3^{-1/4} \sum_{m,n} (-1)^{mn} \exp\left[-\frac{\pi}{\sqrt{3}}(m^2 - mn + n^2)\right]$$
$$\times \exp\left[\frac{2\pi i}{a_f}(mX + nY)\right]. \quad (1.97)$$

In the above we used $k = 2\pi/a_f$ and $a_f^2 = 4\pi\xi^2/\sqrt{3}$, where the latter relation is correct at $h = H_{c2}$. The derivation of Eq. (1.97) is Exercise 1.5. The structure of $|\Psi|^2$ for the triangular lattice was derived by Kleiner et al. [8] Their result is shown in Fig. 1.7. If we pick up only the main terms satisfying $m^2 - mn + n^2 \leq 1$ and rewrite in terms of the original coordinates, Eq. (1.97) reduces to

$$|\Psi|^2 = |C_0|^2 3^{-1/4} \left\{ 1 + 2\exp\left(-\frac{\pi}{\sqrt{3}}\right) \left[\cos\frac{2\pi}{a_f}\left(\frac{2}{\sqrt{3}}x\right) \right.\right.$$
$$\left.\left. + \cos\frac{2\pi}{a_f}\left(\frac{x}{\sqrt{3}} - y\right) - \cos\frac{2\pi}{a_f}\left(\frac{x}{\sqrt{3}} + y\right) \right]\right\}. \quad (1.98)$$

If we replace the factor, $2\exp(-\pi/\sqrt{3}) \simeq 0.326$, in the above equation by $1/3$, it is found that $|\Psi|^2$ is zero at $(x,y) = (\sqrt{3}(p\pm 1/4)a_f, (q\mp 1/4)a_f)$ with p and q denoting integers.

The set of the order parameter given by Eq. (1.95), which is denoted by Ψ_0, and $\bm{A} = \mu_0 H_{c2} x \bm{i}_y = \bm{A}_0$ satisfy the linearized G-L equation at $H_e = H_{c2}$. The corrections to these quantities are written as

$$\Psi_1 = \Psi - \Psi_0, \qquad \bm{A}_1 = \bm{A} - \bm{A}_0. \quad (1.99)$$

Here we shall estimate the deviation of the magnetic flux density from the uniform distribution, $b = \mu_0 H_{c2}$, that was assumed at the beginning. Substituting Eq. (1.95) into Eq. (1.31), we find

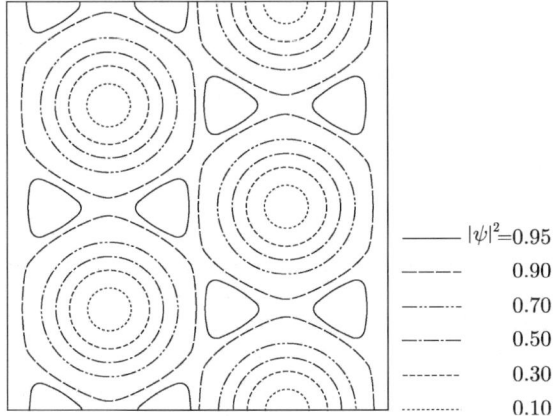

Fig. 1.7. Contour diagram of normalized $|\Psi|^2$ of the triangular flux line lattice (from Kleiner et al. [8])

$$\frac{\partial^2 A_y}{\partial x \partial y} = -\frac{\mu_0 \hbar e}{m^*} \cdot \frac{\partial}{\partial y}|\Psi_0|^2 \tag{1.100}$$

for the x-component of the magnetic flux density. Hence, we have

$$b = \mu_0 H_0 - \frac{\mu_0 H_{c2}|\Psi_0|^2}{2\kappa^2|\Psi_\infty|^2} \tag{1.101}$$

and

$$\boldsymbol{A} = \left(\mu_0 H_0 x - \frac{\mu_0 H_{c2}}{2\kappa^2|\Psi_\infty|^2}\int|\Psi_0|^2 \mathrm{d}x\right)\boldsymbol{i}_y, \tag{1.102}$$

where H_0 is an integral constant. It will be shown later that H_0 is equal to the external magnetic field, H_e. Equation (1.101) shows that the local magnetic flux density also varies periodically in the superconductor and becomes maximum where Ψ is zero. Such spatial structures of the magnetic flux density and the density of superconducting electrons, $|\Psi|^2$, are represented in Fig. 1.8. Figure 1.9 is a photograph of the flux line lattice in a superconducting Pb-Tl specimen obtained by the decoration technique.

Since $\boldsymbol{A}_1$ is already obtained from Eq. (1.102) and $\boldsymbol{A}_0 = \mu_0 H_{c2} x \boldsymbol{i}_y$, we shall derive the equation for Ψ_1, a small quantity. The term, $|\Psi|^2 \Psi$, is also a small quantity. Substituting Eq. (1.99) into Eq. (1.30), we have

$$\frac{1}{2m^*}(-i\hbar\nabla + 2e\boldsymbol{A}_0)^2\Psi_1 + \alpha\Psi_1$$
$$= \frac{i\hbar e}{m^*}[\nabla\cdot(\boldsymbol{A}_1\Psi_0) + \boldsymbol{A}_1\cdot\nabla\Psi_0] - \frac{4e^2}{m^*}\boldsymbol{A}_0\cdot\boldsymbol{A}_1\Psi_0 - \beta|\Psi_0|^2\Psi_0. \tag{1.103}$$

This inhomogeneous equation for Ψ_1 has a solution only if the inhomogeneous term on the right hand side is orthogonal to the solution of the corresponding

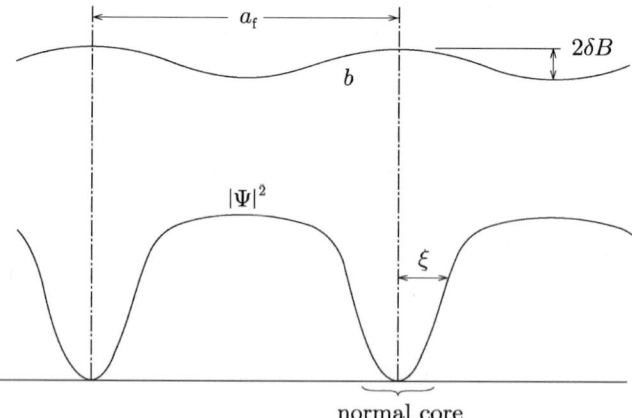

Fig. 1.8. $|\Psi|^2$ and magnetic flux density in the flux line lattice. a_f represents the flux line lattice spacing

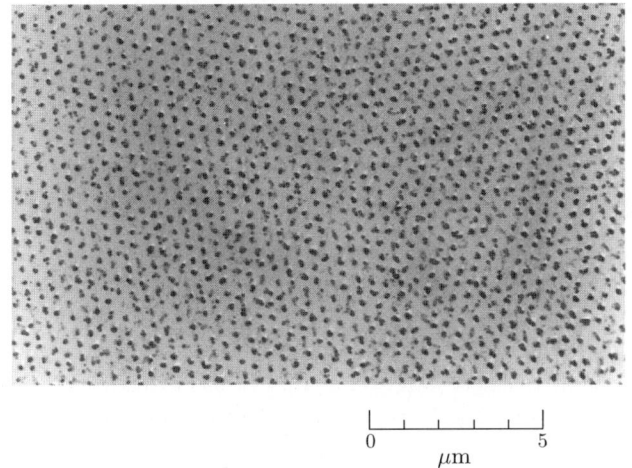

Fig. 1.9. Flux line lattice in Pb-1.6wt%Tl observed by decoration technique after field cooled to 1.2 K at 35 mT and then field is removed. A grain boundary of flux line lattice can be seen. (Courtesy of Dr. B. Obst in Research Center in Karlsruhe)

homogeneous equation, i.e., Ψ_0. It means that the integral of the product of the right hand side and Ψ_0^* is zero. This leads to

$$\langle \boldsymbol{A}_1 \cdot \boldsymbol{j} \rangle - \beta \langle |\Psi_0|^4 \rangle = 0 , \qquad (1.104)$$

where $\langle \ \rangle$ denotes a spatial average. In the derivation of the above equation a partial integral was carried out and the surface integral of less importance was neglected. $\boldsymbol{j}$ in Eq. (1.104) is the current density that we obtain when Ψ_0 and $\boldsymbol{A}_0$ are substituted into Eq. (1.31). From Eq. (1.101) it is given by

$$\boldsymbol{j} = -\frac{H_{c2}}{2\kappa^2 |\Psi_\infty|^2} \nabla \times (|\Psi_0|^2 \boldsymbol{i}_z) . \qquad (1.105)$$

A partial integration of Eq. (1.104) leads to

$$\frac{H_{c2}}{2\kappa^2 |\Psi_\infty|^2} \langle |\Psi_0|^2 (\nabla \times \boldsymbol{A}_1)_z \rangle + \beta \langle |\Psi_0|^4 \rangle = 0 . \qquad (1.106)$$

From Eqs. (1.99) and (1.102) we have

$$(\nabla \times \boldsymbol{A}_1)_z = -\mu_0 (H_{c2} - H_0) - \frac{\mu_0 H_{c2} |\Psi_0|^2}{2\kappa^2 |\Psi_\infty|^2} . \qquad (1.107)$$

Hence, Eq. (1.106) reduces to

$$\left(1 - \frac{H_0}{H_{c2}}\right) |\Psi_\infty|^2 \langle |\Psi_0|^2 \rangle - \left(1 - \frac{1}{2\kappa^2}\right) \langle |\Psi_0|^4 \rangle = 0 . \qquad (1.108)$$

Using this relation, the mean magnetic flux density is obtained from Eq. (1.101) in the form

$$B = \langle b \rangle = \mu_0 H_0 - \frac{\mu_0(H_{c2} - H_0)}{(2\kappa^2 - 1)\beta_A},\qquad(1.109)$$

where

$$\beta_A = \frac{\langle |\Psi_0|^4 \rangle}{\langle |\Psi_0|^2 \rangle^2}\qquad(1.110)$$

is a quantity independent of H_0.

Now we shall calculate the free energy density. If we take zero for $F_n(0)$, the mean value of the free energy density given by Eq. (1.21) is calculated as

$$\langle F_s \rangle = \left\langle \frac{b^2}{2\mu_0} - \frac{\mu_0 H_c^2 |\Psi|^4}{2|\Psi_\infty|^4} \right\rangle,\qquad(1.111)$$

where Eq. (1.30) was used. If we approximately substitute Ψ_0 into Ψ and eliminate H_0 by the use of Eqs. (1.101), (1.108) and (1.109), Eq. (1.111) becomes

$$\langle F_s \rangle = \frac{B^2}{2\mu_0} - \frac{(\mu_0 H_{c2} - B)^2}{2\mu_0[(2\kappa^2 - 1)\beta_A + 1]}.\qquad(1.112)$$

It is found from this equation that β_A should take on a minimum value in order to minimize the free energy. Initially Abrikosov [7] thought that the square lattice was most stable and obtained $\beta_A = 1.18$ for this. Later Kleiner et al. [8] showed that the triangular lattice was most stable with $\beta_A = 1.16$. However, the difference between the two lattices is small.

When Eq. (1.112) is differentiated with respect to B, we have

$$\frac{\partial \langle F_s \rangle}{\partial B} = \frac{(2\kappa^2 - 1)\beta_A B + \mu_0 H_{c2}}{\mu_0[(2\kappa^2 - 1)\beta_A + 1]} = H_0,\qquad(1.113)$$

where Eq. (1.109) is used. Since the derivative of the free energy with respect to the internal variable B gives the corresponding external variable, i.e., the external magnetic field H_e, it follows that H_0 is the external magnetic field as earlier stated. The magnetization then becomes

$$M = \frac{B}{\mu_0} - H_e = -\frac{H_{c2} - H_e}{(2\kappa^2 - 1)\beta_A}.\qquad(1.114)$$

This result suggests that the diamagnetism decreases linearly with increasing magnetic field and reduces to zero at $H_e = H_{c2}$ with the transition to the normal state. The magnetic susceptibility, dM/dH_e, is of the order of $1/2\kappa^2 \beta_A$ and takes a very small value for a type-2 superconductor with a high κ value. According to Eq. (1.101), the deviation of the local magnetic flux density from its mean value is given by

$$\delta B = \frac{\mu_0 H_{c2} \langle |\Psi_0|^2 \rangle}{2\kappa^2 |\Psi_\infty|^2} = -\mu_0 M\qquad(1.115)$$

(see Fig. 1.8: note that $b = \mu_0 H_e$ at the point where $|\Psi_0|^2 = 0$ and that b is minimum at the point where $|\Psi_0|^2$ in Eq. (1.98) takes on a maximum value, $2\langle|\Psi_0|^2\rangle$). Hence, the magnetic flux density is almost uniform and the spatial variation is very small in a high-κ superconductor. For example, the relative fluctuation of the magnetic flux density at $H_e = H_{c2}/2$ is $\delta B/B \sim 1/2\kappa^2 \beta_A$ and takes a value as small as 10^{-4} in Nb-Ti with $\kappa \simeq 70$.

Here we shall argue the transition at H_{c2} from another viewpoint. Since the transition in a magnetic field is treated, the Gibbs free energy density, $G_s = F_s - H_e B$ is suitable. The local magnetic flux density b is given by Eq. (1.101) and a part of the energy reduces to

$$\frac{1}{2\mu_0}\langle b^2 \rangle - H_e B = -\frac{1}{2}\mu_0 H_e^2 , \tag{1.116}$$

where the equation, $H_e = H_0$, was used and the small term proportional to $(b - \mu_0 H_e)^2$ was neglected. Hence, using the expression on the kinetic energy density shown in Exercise 1.1, the Gibbs free energy density is rewritten as

$$G_s = \alpha|\Psi|^2 + \frac{\hbar^2}{2m^*}(\nabla|\Psi|)^2 + \frac{\mu_0}{2}\lambda^2 \left(\frac{|\Psi_\infty|}{|\Psi|}\right)^2 j^2 - \frac{1}{2}\mu_0 H_e^2 \tag{1.117}$$

in the vicinity of the transition point. The first term is the condensation energy density and has a constant negative value. Thus, it can be understood that the transition to the normal state at H_{c2} occurs, since the kinetic energy given by the second and third terms consumes the gain of condensation energy. We shall ascertain that this speculation is correct. For this purpose the approximate solution of $|\Psi|^2$ of Eq. (1.98) in the vicinity of H_{c2} is used: the quantity in $\{\cdots\}$ is represented by g, for simplicity. Hence, we have $|\Psi|^2/|\Psi_\infty|^2 = g\langle|\psi|^2\rangle$ with $\Psi/|\Psi_\infty| = \psi$. Since the error around the zero points of Ψ in this expression is large, the factor of $2\exp(-\pi/\sqrt{3})$ in front of $[\cdots]$ is replaced by $1/3$ so that the zero points are reproduced. Rewriting as $(\nabla|\Psi|)^2 = (\nabla|\Psi|^2)^2/4|\Psi|^2$, the second term of Eq. (1.117) leads to

$$\frac{\hbar^2}{2m^*}(\nabla|\Psi|)^2 = \frac{1}{4}\mu_0 H_c^2 \xi^2 \langle|\psi|^2\rangle \frac{(\nabla g)^2}{g} . \tag{1.118}$$

After a calculation using Eq. (1.101), the third term of Eq. (1.117) leads to

$$\frac{\mu_0}{2}\lambda^2 \left(\frac{|\Psi_\infty|}{|\Psi|}\right)^2 j^2 = \frac{1}{4}\mu_0 H_c^2 \xi^2 \langle|\psi|^2\rangle \frac{(\nabla g)^2}{g} . \tag{1.119}$$

Thus, it is found that the second and third terms are the same. Hence, Eq. (1.117) can be written as

$$G_s = \mu_0 H_c^2 [-|\psi|^2 + 2\xi^2(\nabla|\psi|)^2] - \frac{1}{2}\mu_0 H_e^2 . \tag{1.120}$$

Since $B = \mu_0 H_e$ in the normal state, the third term of Eq. (1.120) is the same as the Gibbs free energy density in the normal state, G_n. Hence, the transition point, H_{c2}, is given by the magnetic field at which the sum of the first and second terms reduces to zero. This condition is given by

$$\langle -|\psi|^2 + 2\xi^2(\nabla|\psi|)^2 \rangle = \langle |\psi|^2 \rangle \left[-1 + \frac{\xi^2}{2} \left\langle \frac{(\nabla g)^2}{g} \right\rangle \right] = 0 \ . \tag{1.121}$$

A numerical calculation leads to $\langle (\nabla g)^2/g \rangle = 14.84/a_{\rm f}^2$, and the flux line lattice spacing at H_{c2} is obtained: $a_{\rm f}^2 = 7.42\xi^2$. Thus, from Eq. (1.52) and the relationship of $a_{\rm f} = (2\phi_0/\sqrt{3}B)^{1/2}$, we have [9]

$$H_{\rm e} = \frac{B}{\mu_0} = 0.98 H_{c2} \ . \tag{1.122}$$

Thus, it is found that the transition point can also be obtained fairly correctly even by such a simple approximation.

In the above the magnetic properties of type-2 superconductors are described using the G-L theory. Especially the fundamental properties are determined by the two physical quantities, $H_{\rm c}$ and κ. That is, the critical fields, H_{c1}(Eq. (1.83)) and H_{c2}(Eq. (1.51)), and the magnetization in their vicinities given by Eqs. (1.90) and (1.114) are described only by the two quantities (note that $\phi_0/\lambda^2 = 2\sqrt{2}\pi\mu_0 H_{\rm c}/\kappa$ in Eq. (1.90)). In addition, from the argument on thermodynamics we have the general relation

$$-\int_0^{H_{c2}} \mu_0 M(H_{\rm e}) {\rm d}H_{\rm e} = \frac{1}{2}\mu_0 H_{\rm c}^2 \ . \tag{1.123}$$

In the above we assumed that κ is a general parameter decreasing slightly with increasing temperature. Strictly speaking, the κ values defined by Eqs. (1.51)(κ_1), (1.114)(κ_2) and (1.83)(κ_3) are slightly different.

1.6 Surface Superconductivity

In the previous section the magnetic properties and the related superconducting order parameter in a bulk superconductor were investigated. In practice, the superconductor has a finite size and the surface. A special surface property different from that of the bulk is expected. Here we assume a semi-infinite type-2 superconductor occupying $x \geq 0$ with the magnetic field applied parallel to the surface along the z-axis for simplicity. On the surface where the superconductor is facing vacuum or an insulating material, the boundary condition on the order parameter is given by Eq. (1.33). Under this condition the vector potential $\boldsymbol{A}$ can be chosen so that it contains only the y-component. Hence, the above boundary condition may be written

$$\left. \frac{\partial \Psi}{\partial x} \right|_{x=0} = 0 \ . \tag{1.124}$$

1.6 Surface Superconductivity

We shall solve again the linearized G-L equation (by ignoring the small term to the third power of Ψ). We assume the order parameter of the form [10]

$$\Psi = e^{-iky}e^{-ax^2} \tag{1.125}$$

referring to Eqs. (1.91) and (1.94). This order parameter satisfies the condition (1.124). In the following we shall obtain approximate values of the parameters, k and a, by the variation method. Under the present condition in which the external variable is given, the quantity to be minimized is the Gibbs free energy density; this is given by the free energy density in Eq. (1.21) minus BH_e. If the small term proportional to the fourth power of Ψ is neglected, the Gibbs free energy per unit length in the directions of the y- and z-axes measured from the value in the normal state is given by

$$G = \frac{1}{2m^*} \int_0^\infty \left[|(-i\hbar\nabla + 2e\boldsymbol{A})\Psi|^2 - \frac{\hbar^2}{\xi^2}|\Psi|^2 \right] dx \tag{1.126}$$

under the approximation $A_y = \mu_0 H_e x$. After substitution of Eq. (1.125) into this equation and a simple calculation we have

$$G = \frac{\hbar^2}{4m^*} \left[\left(\frac{\pi}{2a}\right)^{1/2} \left(k^2 - \frac{1}{\xi^2}\right) - \frac{2e\mu_0 H_e k}{\hbar a} \right.$$
$$\left. + \left(\frac{\pi}{2a^3}\right)^{1/2} \left(a^2 + \frac{e^2\mu_0^2 H_e^2}{\hbar^2}\right) \right]. \tag{1.127}$$

When minimizing this with respect to k, we obtain

$$k = \left(\frac{2}{\pi a}\right)^{1/2} \frac{e\mu_0 H_e}{\hbar}, \tag{1.128}$$

after which, G becomes

$$G_e = \frac{\hbar^2}{4m^*} \left(\frac{\pi}{2}\right)^{1/2} \left[a^{1/2} - \frac{1}{\xi^2} a^{-1/2} + \frac{e^2\mu_0^2 H_e^2}{\hbar^2}\left(1 - \frac{2}{\pi}\right) a^{-3/2} \right]. \tag{1.129}$$

From the requirements that G_e is minimum with respect to a and that $G_e = 0$ at the transition point, we obtain a and the critical value of H_e denoted by H_{c3} as [10]

$$a = \frac{1}{2\xi^2}, \tag{1.130}$$

$$H_{c3} = \frac{\hbar}{2\xi^2 e\mu_0}\left(1 - \frac{2}{\pi}\right)^{-1/2} \simeq 1.66 H_{c2}. \tag{1.131}$$

The exact calculation was carried out by Saint-James and de Gennes [11] who obtained

$$H_{c3} = 1.695 H_{c2}. \tag{1.132}$$

The surface critical field H_{c3} depends on the angle between the surface and the magnetic field. H_{c3} decreases from the value given by Eq. (1.132) with increasing angle and reduces to the bulk upper critical field H_{c2} at the angle normal to the surface.

1.7 Josephson Effect

It was predicted by Josephson [12] that a DC superconducting tunneling current can flow between superconductors separated by a thin insulating layer. This is the DC Josephson effect. The intuitive picture of this effect was given by Eq. (1.54), based on phenomenological theory. That is, it was expected from the first term in this equation that, if a phase difference occurs between the order parameters of superconductors separated by an insulating layer, a superconducting tunneling current proportional to that phase difference flows across the insulating barrier. Here we suppose a Josephson junction as schematically shown in Fig. 1.10 and assume that the physical quantities vary only along the x-axis along which the current flows. If we assume that the order parameter is constant and that the gradient of the phase is uniform in the insulating region, Eq. (1.54) leads to

$$j = j_c \theta ,\tag{1.133}$$

where j_c is given by

$$j_c = \frac{2\hbar e}{m^* d}|\Psi|^2 \tag{1.134}$$

with d denoting the thickness of the insulating layer. In Eq. (1.133) θ, which is the difference of the gauge-invariant phase of the two superconductors, is given by

$$\theta = \phi_1 - \phi_2 - \frac{2\pi}{\phi_0}\int_1^2 A_x \mathrm{d}x \tag{1.135}$$

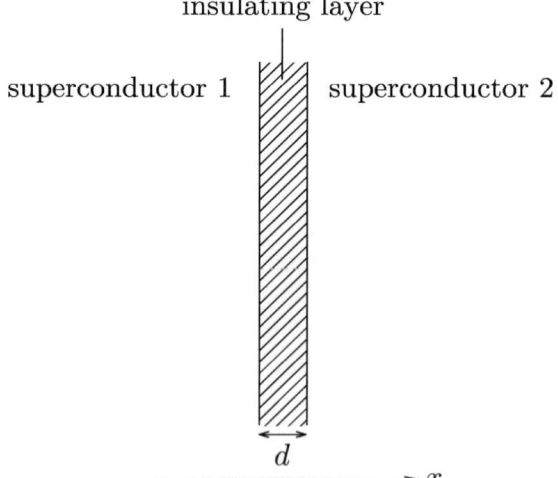

Fig. 1.10. Structure of Josephson junction

with ϕ_1 and ϕ_2 denoting the phases of superconductors 1 and 2, respectively. Equation (1.133) is correct when the phase difference θ is small. When θ becomes large, the relationship between the current density and θ starts to deviate from this equation. This can be understood from the physical requirement that the current should vary periodically with θ the period of the variation being 2π. Hence, a relationship of the form

$$j = j_c \sin \theta \tag{1.136}$$

is expected instead of Eq. (1.133). In fact, this relationship was derived by Josephson using BCS theory. Equation (1.136) can also be derived using G-L theory, if Eqs. (1.30) and (1.54) are solved simultaneously [13].

Since the phase difference θ contains the effect of the magnetic field in a gauge-invariant form, the critical current density, i.e., the maximum value of Eq. (1.136) averaged in the junction, varies with the magnetic field as

$$J_c = j_c \left| \frac{\sin(\pi \Phi / \phi_0)}{\pi \Phi / \phi_0} \right| \tag{1.137}$$

due to interference (see Fig. 1.11), where Φ is the magnetic flux inside the junction. This form is similar to the interference pattern due to Fraunhofer diffraction by a single slit. For example, when the magnetic flux just equal to one flux quantum penetrates the junction, the critical current density of the junction is zero. In this situation the phase inside the junction varies over 2π and the zero critical current density results from the interference of the positive and negative currents of the same magnitude. This influence of the magnetic field gives a direct proof of the DC Josephson effect. The SQUID (Superconducting Quantum Interference Device) in which a very small magnetic flux density can be measured is a device that relies on this property.

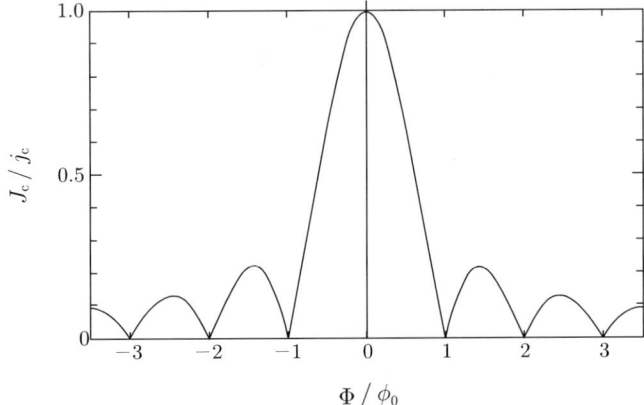

Fig. 1.11. Relation between the critical current density in a Josephson junction and the magnetic flux inside the junction

Another effect predicted by Josephson is the AC Josephson effect. In this phenomenon, when a voltage with V is applied to the junction, an AC superconducting current flows with angular frequency, ω, given by

$$\hbar\omega = 2eV . \tag{1.138}$$

In the voltage state the magnetic flux flows through the junction region and the phase of the order parameter varies in time. As will be shown in Sect. 2.2, the angular frequency given by Eq. (1.138) is the same as the rate of variation of the phase. When the junction is irradiated by microwave energy of this frequency, resonant absorption occurs and a DC step of the superconducting current, i.e., a "Shapiro step," appears. The AC Josephson effect was demonstrated by this kind of measurement. The present voltage standard is established by the AC Josephson effect expressed by Eq. (1.138) in association with an extremely exact frequency measuring technique.

1.8 Critical Current Density

The maximum superconducting current density that the superconductor can carry is a very important factor from an engineering standpoint. Some aspects of this property are mentioned in this section. According to the G-L theory, the superconducting current density may be transcribed from Eq. (1.54) into the form

$$\boldsymbol{j} = -2e|\Psi|^2 \boldsymbol{v}_s , \tag{1.139}$$

where

$$\boldsymbol{v}_s = \frac{1}{m^*}(\hbar\nabla\phi + 2e\boldsymbol{A}) \tag{1.140}$$

is the velocity of the superconducting electrons. If the size of superconductor is sufficiently small compared to the coherence length ξ, $|\Psi|$ can be probably regarded as approximately constant over the cross section of the superconductor. If we note that $\nabla\Psi \simeq i\Psi\nabla\phi$, the free energy density in Eq. (1.21) reduces to

$$F_s = F_n(0) + \alpha|\Psi|^2 + \frac{\beta}{2}|\Psi|^4 + \frac{1}{2}m^*|\Psi|^2 v_s^2 + \frac{B^2}{2\mu_0} . \tag{1.141}$$

Minimizing the free energy density with respect to $|\Psi|$, we have

$$|\Psi|^2 = |\Psi_\infty|^2 \left(1 - \frac{m^* v_s^2}{2|\alpha|}\right) . \tag{1.142}$$

From Eq. (1.139) the corresponding current density is given by

$$j = 2e|\Psi_\infty|^2 \left(1 - \frac{m^* v_s^2}{2|\alpha|}\right) v_s . \tag{1.143}$$

This becomes maximum when $m^* v_s^2 = (2/3)|\alpha|$, the maximum value being

$$j_{\mathrm{c}} = \left(\frac{2}{3}\right)^{3/2} \frac{H_{\mathrm{c}}}{\lambda}. \tag{1.144}$$

Under the condition that j is maximum, $|\Psi|$ takes a finite value, $(2/3)^{1/2}|\Psi_\infty|$, and the depairing of the superconducting electron pairs has not yet occurred. In fact, the velocity at which the depairing takes place resulting in zero $|\Psi|$ is $\sqrt{3}$ times as large as the velocity corresponding to j_{c}. However, according to the BCS theory the current density almost attains its maximum value when v_{s} is such that the energy gap diminishes to zero in the limit of $T=0$. Thus there is a clear relationship between the depairing velocity and the maximum current density. For this reason the current density given by Eq. (1.144) is sometimes called the depairing current density.

The Meissner current is another current associated with the superconducting phenomena. This current, which is localized near the surface according to Eq. (1.15), brings about the perfect diamagnetism. In type-2 superconductors its maximum value is

$$j_{\mathrm{c1}} = \frac{H_{\mathrm{c1}}}{\lambda}. \tag{1.145}$$

Here we shall investigate the above two critical current densities quantitatively. Take the practical superconducting material Nb_3Sn for example. From $\mu_0 H_{\mathrm{c}} \simeq 0.5$ T, $\mu_0 H_{\mathrm{c1}} \simeq 20$ mT and $\lambda \simeq 0.2$ µm, we have $j_{\mathrm{c}} \simeq 1.1 \times 10^{12}$ Am^{-2} and $j_{\mathrm{c1}} \simeq 8.0 \times 10^{10}$ Am^{-2} at 4.2 K. It is seen that these values are very high. However, the size of superconductor should be smaller than ξ to attain the depairing current density j_{c} over its entire cross section. Since ξ in Nb_3Sn is approximately 3.9 nm, the fabrication of superconducting wires sufficiently thinner than ξ is difficult. Furthermore, suppose that multifilamentary subdivision is adopted for keeping the current capacity at a sufficient level; i.e., suppose that a large number of fine superconducting filaments are embedded in a normal metal. In this case we have to confront an essential problem; viz. the proximity effect in which the superconducting electrons in the superconducting region soak into the surrounding normal metal matrix. Two consequences follow: (1) the superconducting property in the superconducting region becomes degraded. (2) Since superconductivity is induced in the normal metal, the superconducting filaments become coupled and the whole wire behaves as a single superconductor. This is contradictory to the premise that the size of superconductor is sufficiently smaller than the coherence length. Hence, it is necessary to embed the superconducting filaments in an insulating material to avoid the proximity effect. However, such a wire is hopelessly unstable. Application of the Meissner current j_{c1} is strongly restricted by the condition that the surface field should be lower than H_{c1}. In Nb_3Sn $\mu_0 H_{\mathrm{c1}}$ is as low as 20 mT. Hence j_{c1} cannot be practically used except some special uses.

Since the magnetic energy density is proportional to the second power of the magnetic field, superconducting materials are sometimes used as high-field magnets to store large amounts of energy. Therefore, the superconductivity is required to persist up to high magnetic fields. For such applications

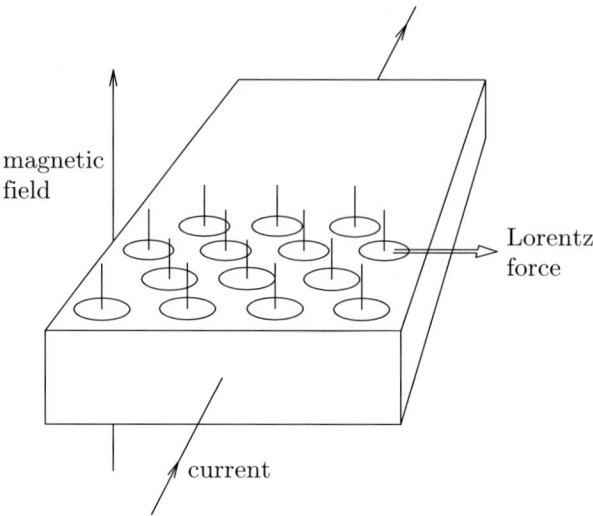

Fig. 1.12. Situation of a current-carrying superconductor in the magnetic field. The Lorentz force acts on the flux lines in the direction shown by the *arrow*

a type-2 superconductor with the short coherence length is required; the superconductor is then in the mixed state and is penetrated by flux lines. If the superconductor carries a transport current under this condition (suppose a superconducting wire composing a superconducting magnet under an operating condition), the relative direction of the magnetic field and the current is like the one shown in Fig. 1.12 and the flux lines in the superconductor experience a Lorentz force. The driving force on the flux lines will be described in more detail in Sect. 2.1. If the flux lines are driven by this Lorentz force with velocity $\boldsymbol{v}$, the electromotive force induced is:

$$\boldsymbol{E} = \boldsymbol{B} \times \boldsymbol{v} , \tag{1.146}$$

where $\boldsymbol{B}$ is the macroscopic magnetic flux density. When this state is maintained steadily, an energy dissipation, and hence an electrical resistance, should appear as in a normal metal. Microscopically, the central region of each flux line is almost in the normal state as shown in Fig. 1.6, and the normal electrons in this region are driven by the electromotive force, resulting in an ohmic loss. This phenomenon is inevitable as long as an electromotive force exists. Hence, it is necessary to stop the motion of flux lines ($\boldsymbol{v} = 0$) in order to prevent the electromotive force. This so-called flux pinning is provided by inhomogeneities and various defects such as dislocations, normal precipitates, voids and grain boundaries. These inhomogeneities and defects are therefore called pinning centers. Flux pinning is like a frictional force in that it prevents the motion of flux lines until the Lorentz force exceeds some critical value. In this state only the superconducting electrons are able to flow and energy dissipation does not occur. For the Lorentz force larger than the critical value

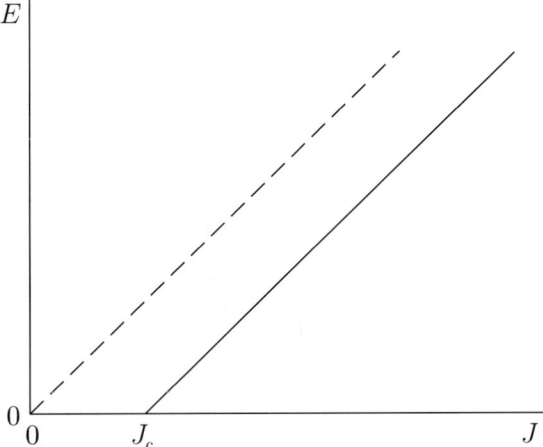

Fig. 1.13. Current-voltage characteristics in the presence of flux pinning interactions. The *broken line* shows the characteristics in the absence of pinning interactions

the motion of flux lines sets in and the electromotive force reappears, resulting in the current-voltage characteristics shown in Fig. 1.13. The total pinning force that all the elementary pinning centers in a unit volume can exert on the flux lines is called the pinning force density; it is denoted by F_p. At the critical current density J_c, at which the electromotive force starts to appear, the Lorentz force of $J_\mathrm{c}B$ acting on the flux lines in a unit volume is balanced by the pinning force density. Hence, we have the relation:

$$J_\mathrm{c} = \frac{F_\mathrm{p}}{B} \, . \tag{1.147}$$

The practical critical current density in commercial superconducting materials is determined by this flux pinning mechanism. This implies that this J_c is not an intrinsic property like the two critical current densities previously mentioned but is an acquired property determined by the macroscopic structure of introduced defects. That is, the critical current density depends on the density, type of, and arrangement of pinning centers. It is necessary to increase the flux pinning strength in order to increase the critical current density. In the above-mentioned Nb_3Sn, a critical current density of the order of $J_\mathrm{c} \simeq 1 \times 10^{10}$ Am^{-2} is obtained at $B = 5$ T.

As a matter of fact, the current-voltage characteristics are not the ideal ones shown in Fig. 1.13 and the electric field is not completely zero for $J \leq J_\mathrm{c}$. This comes from the motion of flux lines that have been depinned due to the thermal agitation. This phenomenon called the flux creep will be considered in detail in Sect. 3.8. However, in most cases at sufficiently low temperatures the critical current density J_c can be defined as in Fig. 1.13. Henceforth we will assume that the E-J relation depicted in Fig. 1.13 is approximately correct

1.9 Flux Pinning Effect

The practical critical current density in superconductors originates from the flux pinning interactions between the flux lines and defects. The flux line has spatially varying structures of order parameter Ψ and magnetic flux density b as shown in Fig. 1.6. The materials parameters such as T_c, H_c, ξ, etc., in the pinning center are different from those in the surrounding region. Hence, when the flux line is virtually displaced near the pinning center, the free energy given by Eq. (1.21) varies due to the interference between the spatial variation in Ψ or b and that of α or β. The rate of variation in the free energy, i.e., the gradient of the free energy gives the interaction force.

Each such individual pinning interaction is vectored in various directions depending on the relative location of the flux line and the pinning center. On the other hand, the resultant macroscopic pinning interaction force density is a force directed opposite to the motion of flux lines in the manner of a macroscopic frictional force. While the individual pinning force comes from the potential and is reversible, the macroscopic pinning force is irreversible. Furthermere, the macroscopic pinning force density is not generally equal to the sum all the elementary pinning forces, the maximum forces of individual interactions, in a unit volume; and the relationship between the macroscopic pinning force density and the elementary pinning force is not simple. The so-called pinning force summation problem will be considered in Chap. 7.

At first glance it might seem that the superconductor can carry some current of the density smaller than J_c without energy dissipation. However, this is correct only in the case of steady direct current. For an AC current or a varying current, loss occurs even when the current is smaller than the critical value. The loss is caused by the electromotive force given by Eq. (1.146) due to the motion of flux lines in the superconductor under the AC or varying condition. That is, the mechanism of the loss is the same as that of ohmic loss in normal metals. Hence, the resultant loss seems to be of the nature that the loss energy per cycle is proportional to the frequency, similarly to the eddy current loss in copper. However, it is the hysteresis loss independent of the frequency. What is the origin for such an apparent contradiction? This originates also from the fact that the flux pinning interaction comes from the potential. This will be mentioned in Chap. 2.

Exercises

1.1. Compare the energy treated in the London theory and that in the G-L theory.

1.2. With the use of the G-L equation (1.30), prove that the free energy density given by Eq. (1.21) is written as

$$F_{\rm s} = F_{\rm n}(0) + \frac{1}{2\mu_0}(\nabla \times \boldsymbol{A})^2 - \frac{\beta}{2}|\Psi|^4 + \frac{\hbar^2}{4m^*}\nabla^2|\Psi|^2.$$

1.3. Prove that the magnetic flux is quantized in a unit cell of the flux line lattice.

1.4. Calculate the contributions from the following matter to the energy of the flux line in the low field region:
   (1) the spatial variation in the order parameter inside the core and
   (2) the magnetic field inside the core.
   Use Eq. (1.74).

1.5. Derive Eq. (1.97). We can write as $C_n = C_0 \exp(i\pi n^2/2)$ so as to satisfy $C_{2m} = C_0$ and $C_{2m+1} = iC_0$.

1.6. It was shown by the approximate solution of Eq. (1.98) that $(x,y) = ((\sqrt{3}/4)a_{\rm f}, -a_{\rm f}/4)$ is one of the zero points of Ψ. Prove that Ψ given by Eq. (1.95) is exactly zero at this point.

1.7. Derive Eq. (1.111).

1.8. We calculate the magnetic flux of one flux line in the area shown in Fig. 1.14. The surface integral is given by the curvilinear integral of the vector potential $\boldsymbol{A}$. Since the current density $\boldsymbol{j}$ is perpendicular to the straight line L, the curvilinear integral of $\boldsymbol{A}$ is equal to the curvilinear integral of $-(\hbar/2e)\nabla\phi$ on L with ϕ denoting the phase of the order parameter. Equation (1.55) is valid also on the half circle R at sufficiently long distance. As a result the magnetic flux in the region shown in the figure should be an integral multiple of the flux quantum ϕ_0. This is clearly incorrect. Examine the reason why such an incorrect result was derived.

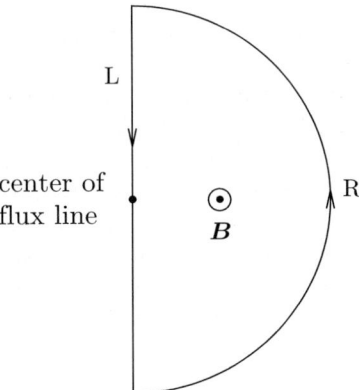

Fig. 1.14. Closed loop consisting of the *straight line* L passing through the center of the quantized magnetic flux and the *half circle* R at sufficiently long distance

1.9. Discuss the reason why the center of quantized flux line is in the normal state.

References

1. V. L. Ginzburg and L. D. Landau: Zh. Eksperim. i Teor. Fiz. **20** (1950) 1064.
2. P. G. de Gennes: *Superconductivity of Metals and Alloys* (W. A. Benjamin, New York, 1966) p. 227.
3. J. Bardeen, L. N. Cooper and J. R. Schrieffer: Phys. Rev. **108** (1957) 1175.
4. For example, see M. Tinkham: *Introduction to Superconductivity* (McGraw-Hill, New York, 1996) p. 119.
5. M. Tinkham: *Introduction to Superconductivity* (McGraw-Hill, New York, 1996) p. 151.
6. M. Nozue, K. Noda and T. Matsushita: *Adv. in Supercond. VIII* (Springer, Tokyo, 1996) p. 537.
7. A. A. Abrikosov: Zh. Eksperim. i Teor. Fiz. **32** (1957) 1442 (English translation: Sov. Phys.-JETP **5** (1957) 1174.
8. W. H. Kleiner, L. M. Roth and S. H. Autler: Phys. Rev. **133** (1964) A1226.
9. T. Matsushita, M. Iwakuma, K. Funami, K. Yamafuji, K. Matsumoto, O. Miura and Y. Tanaka: *Adv. Cryog. Eng. Mater.* (Plenum, New York, 1996) p. 1103.
10. S. Nakajima: *Introduction to Superconductivity* (Baihukan, Tokyo, 1971) p. 70 [in Japanese].
11. D. Saint-James and P. G. de Gennes: Phys. Lett. **7** (1963) 306.
12. B. D. Josephson: Phys. Lett. **1** (1962) 251.
13. D. A. Jacobson: Phys. Rev. **138** (1965) A1066.

2

Fundamental Electromagnetic Phenomena in Superconductors

2.1 Equations of Electromagnetism

Here we assume a sufficiently large superconductor. When a magnetic field is applied to the superconductor, flux lines penetrate it from the surface. Since the flux lines are expected to be pinned by pinning centers in the superconductor, those cannot penetrate deeply from the surface and the density of the flux lines will be higher near the surface and lower in the inner region, resulting in a nonuniform distribution in a macroscopic scale. When the external magnetic field is decreased, on the other hand, the flux lines go out of the superconductor and their density becomes lower near the surface. It is important to know correctly the magnetic flux distribution in the superconductor in such cases in order to understand or foresee its electromagnetic phenomenon exactly.

We assume a semi-macroscopic region which is sufficiently larger in size than the flux line spacing but sufficiently smaller than the superconductor. We designate r_n as the central position of this region; the mean magnetic flux density within it, given by the product of the density of flux lines and the flux quantum ϕ_0, is designated by $\boldsymbol{B}_n$. The superconductor is imagined to be divided into such small segments (see Fig. 2.1). If the differences in the magnetic flux density between adjacent segments are sufficiently small, the set $\{\boldsymbol{B}_n(\boldsymbol{r}_n)\}$ can be approximated by a continuous function of $\boldsymbol{B}(\boldsymbol{r})$, where $\boldsymbol{r}$ denotes the macroscopic coordinate in the superconductor. The macroscopic magnetic field $\boldsymbol{H}(\boldsymbol{r})$, current density $\boldsymbol{J}(\boldsymbol{r})$, and electric field $\boldsymbol{E}(\boldsymbol{r})$ can be defined in a similar manner.

The quantities $\boldsymbol{B}$, $\boldsymbol{H}$, $\boldsymbol{J}$ and $\boldsymbol{E}$ defined above, which are the semi-macroscopic averages of the local $\boldsymbol{b}$, $\boldsymbol{h}$, $\boldsymbol{j}$ and $\boldsymbol{e}$ (the word "macroscopic" is not used hereafter except special cases), satisfy the well-known Maxwell equations:

$$\boldsymbol{J} = \nabla \times \boldsymbol{H} , \qquad (2.1)$$

$$\nabla \times \boldsymbol{E} = -\frac{\partial \boldsymbol{B}}{\partial t} , \qquad (2.2)$$

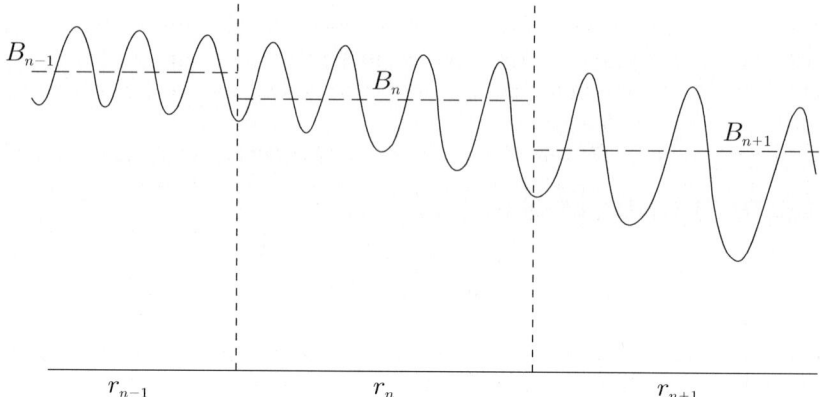

Fig. 2.1. Magnetic flux density averaged in a semi-macroscopic scale

$$\nabla \cdot \boldsymbol{B} = 0 \,, \tag{2.3}$$

$$\nabla \cdot \boldsymbol{E} = 0 \,. \tag{2.4}$$

Although the above equations are formally the same as the equations on the local quantities discussed in Chap. 1, these are the relationships for the macroscopic quantities. Equation (2.4) is based on the fact that the electric charge does not exist in the superconductor under the usual conditions and the displacement current is neglected in Eq. (2.1). In order to solve the Maxwell equations, two other equations for the superconductor are needed. One of them is the relationship between $\boldsymbol{B}$ and $\boldsymbol{H}$, which is simply given by

$$\boldsymbol{B} = \mu_0 \boldsymbol{H} \tag{2.5}$$

except for strongly paramagnetic superconductors, which can be confirmed by measurement in the normal state in a magnetic field greater than the upper critical field H_{c2}. The other is the relationship between $\boldsymbol{E}$ and $\boldsymbol{J}$, that gives the outstanding characteristic of the superconductor. These last two relationships together with Eqs. (2.1) and (2.2) yield the four quantities, $\boldsymbol{B}$, $\boldsymbol{H}$, $\boldsymbol{J}$ and $\boldsymbol{E}$.

The relationship between $\boldsymbol{E}$ and $\boldsymbol{J}$ describes the fundamental properties of the superconductor, such as the ability of DC steady transport current to flow within it without appearance of the electrical resistance or its irreversible magnetic behavior in a varying magnetic field. For example, the electrical properties of a material can be obtained by solving the equation of motion of its electrons. Since the most electromagnetic phenomena in the superconductor are concerned with the magnetic flux distribution within it, the motion of flux lines must be dealt with. Consider the case mentioned in the beginning of this section, where the flux lines penetrate the superconductor in an applied magnetic field. On one hand, an equilibrium may be attained under the balance between the driving force and the restraining force due to flux pinning on the flux lines. On the other hand where the equilibrium is not attained due to the driving force which exceeds the pinning force, the situation can be described by an equation of motion that contains a new term, the viscous force, as in the usual equation of mechanical motion. In what follows the various forces that appear in this equation are discussed.

If, as mentioned above, $\boldsymbol{B}$ is the mean magnetic flux density within a small region and $F(\boldsymbol{B})$ is the corresponding free energy density, the intensive variable, $\mathcal{H}$, corresponding to $\boldsymbol{B}$ is given by

$$\mathcal{H} = \frac{\partial F(\boldsymbol{B})}{\partial \boldsymbol{B}} . \tag{2.6}$$

This quantity, which has the dimension of magnetic field, is called the thermodynamic magnetic field. If the $\mathcal{H}$ that is in equilibrium with $\boldsymbol{B}$ is uniform in space, a driving force does not act on the flux lines. If, however, there is a distortion or an eddy in $\mathcal{H}$, a driving force does act on the flux lines. It should be noted that the driving force does not necessarily originate from a nonuniformity in $\boldsymbol{B}$. In Sect. 2.6 it will be shown that a driving force may not appear even when $\boldsymbol{B}$ varies spatially, provided that $\mathcal{H}$ is uniform.

In many cases, especially in superconductors with large G-L parameters κ, such as commercial superconductors, magnetic energy dominates the G-L energy given by Eq. (1.21). The other components are the condensation energy and the kinetic energy, that are of the order of $\mu_0 H_c^2/2$ at most. Hence, the ratio of this energy to the magnetic energy is of the order of $(H_c/H)^2$ and amounts only to $8/\kappa^2$ even at magnetic fields as high as a quarter of the upper critical field. In Nb-Ti which has a κ of approximately 70 this ratio is negligible. In such cases Eq. (2.6) reduces to

$$\mathcal{H} \simeq \frac{\partial}{\partial \boldsymbol{B}} \cdot \frac{\boldsymbol{B}^2}{2\mu_0} = \frac{\boldsymbol{B}}{\mu_0} . \tag{2.7}$$

This result is reasonable, since the energy associated with the diamagnetism is neglected in the above treatment. When $\mathcal{H}$ varies in space, the driving force on the flux lines in a unit volume is generally given by

$$\boldsymbol{F}_{\mathrm{d}} = (\nabla \times \mathcal{H}) \times \boldsymbol{B} . \tag{2.8}$$

From Eq. (2.7) which disregards the effect of diamagnetism, only an electromagnetic contribution to the force appears and we have

$$\boldsymbol{F}_{\mathrm{d}} \simeq \left(\nabla \times \frac{\boldsymbol{B}}{\mu_0}\right) \times \boldsymbol{B} = \boldsymbol{J} \times \boldsymbol{B} \equiv \boldsymbol{F}_{\mathrm{L}} . \tag{2.9}$$

In the above, Eqs. (2.1) and (2.5) were used. The driving force $\boldsymbol{F}_{\mathrm{L}}$ is known as the Lorentz force. This is the force felt by moving electrons, and hence the current, in the magnetic field. In the present case, the vortex current which forms the flux line experiences this force. The distortion of flux lines, such as a gradient of their density or a bending deformation, results in a transport current as shown in Eq. (2.1). That is, the transport current originates from superposition of vortex currents. Hence the Lorentz force may be regarded as acting on the flux lines themselves. In fact, the force acting on two flux lines is derived from their magnetic energy, and the expression of the Lorentz force is deduced for the general case from this result in [1]. A force of this type works not only on the quantized flux lines in superconductors but generally on magnetic flux lines. The Lorentz force can also be generally expressed in the form of the restoring force on the distorted magnetic structure. This will be discussed in Sect. 7.2. The driving force on an isolated flux line in a thin film by a transport current is discussed in [2]. The effect of the diamagnetism on the driving force is larger for the case of superconductor with a small G-L parameter. This effect is also pronounced for superconductors with weak pinning forces or small superconductors. This will be discussed in Sect. 2.6.

When flux lines are under the influence of a driving force, hereafter called the Lorentz force for simplicity, they are acted on by restraining forces. These are the pinning force and the viscous force. The pinning force comes from the potential energy that the flux line feels depending on its position and the viscous force originates from the mechanism of ohmic energy dissipation inside and outside of the normal core of the flux line due to its motion. The balance of these forces is described by

$$\boldsymbol{F}_\mathrm{L} + \boldsymbol{F}_\mathrm{p} + \boldsymbol{F}_\mathrm{v} = 0 , \tag{2.10}$$

where $\boldsymbol{F}_\mathrm{p}$ and $\boldsymbol{F}_\mathrm{v}$ are the pinning force density and the viscous force density, respectively. The mass of the flux line can usually be neglected [3] and the inertial force does not need to be introduced. Under the condition of Eq. (2.10) the superconductor is said to be in its "critical state" and the model in which this state is assumed is called the critical state model. Following Josephson [4] who treated the quasistatic case in which the viscous force can be neglected (see Appendix A.1), it is also possible to derive Eq. (2.10) from the requirement that the work done by the external source should be equal to the variation in the free energy in the superconductor. In Eq. (2.10) $\boldsymbol{F}_\mathrm{p}$ does not depend on the velocity of flux lines $\boldsymbol{v}$, while $\boldsymbol{F}_\mathrm{v}$ does. These force densities are written

$$\boldsymbol{F}_\mathrm{p} = -\boldsymbol{\delta} F_\mathrm{p}(|\boldsymbol{B}|, T) , \tag{2.11}$$

$$\boldsymbol{F}_\mathrm{v} = -\eta \frac{|\boldsymbol{B}|}{\phi_0} \boldsymbol{v} , \tag{2.12}$$

where $\boldsymbol{\delta} = \boldsymbol{v}/|\boldsymbol{v}|$ is a unit vector in the direction of flux line motion and F_p represents the magnitude of the pinning force density which depends on the magnetic flux density $|\boldsymbol{B}|$ and the temperature T. The quantity, η, is the viscous coefficient. As will be mentioned in the next section η is related to the flux flow resistivity; it is also a function of $|\boldsymbol{B}|$ and T. Sometimes electromagnetic phenomena are treated in an isothermal condition. In such a case the T is dropped in relationships deriving from Eq. (2.11). Substitution of Eqs. (2.9), (2.11) and (2.12) into Eq. (2.10) leads to

$$\frac{1}{\mu_0}(\nabla \times \boldsymbol{B}) \times \boldsymbol{B} - \delta F_{\mathrm{p}}(|\boldsymbol{B}|) - \eta \frac{|\boldsymbol{B}|}{\phi_0} \boldsymbol{v} = 0 \, . \tag{2.13}$$

To solve this equation using the Maxwell equations, a relationship between the electromagnetic quantities and the velocity $\boldsymbol{v}$ is needed. This is the subject of the next section.

2.2 Flux Flow

We assume that the superconductor is stationary in space and that the flux line lattice of flux density $\boldsymbol{B}$ moves with a velocity $\boldsymbol{v}$. We define two coordinate systems, or frames; a stationary one for the superconductor and another moving with the velocity $\boldsymbol{v}$ of the flux line lattice. If the electric fields measured in the stationary frame and the moving frame are represented by $\boldsymbol{E}$ and $\boldsymbol{E}_0$, respectively, Farady's law of induction becomes

$$\nabla \times (\boldsymbol{E}_0 - \boldsymbol{v} \times \boldsymbol{B}) = -\frac{\partial \boldsymbol{B}}{\partial t} \tag{2.14}$$

as well known in the theory of electromagnetism [5]. Since the magnetic structure does not change at all with time from the view of the moving frame, we have $\boldsymbol{E}_0 = 0$. Thus, Eq. (2.14) reduces to

$$\nabla \times (\boldsymbol{B} \times \boldsymbol{v}) = -\frac{\partial \boldsymbol{B}}{\partial t} \, . \tag{2.15}$$

This is called the continuity equation for flux lines [6]. This equation can also be derived directly by equating the magnetic flux coming in a small loop in unit time with the rate of variation of the magnetic flux in the loop. This derivation is offered as an exercise at the end of this chapter. Comparing Eq. (2.15) with one of the Maxwell equations, Eq. (2.2), we have generally

$$\boldsymbol{E} = \boldsymbol{B} \times \boldsymbol{v} - \nabla \Psi \, . \tag{2.16}$$

In the above, the scalar function, Ψ, represents the electrostatic potential for the usual geometry in which the magnetic field and the current are perpendicular to each other. It is zero in superconductors [7, 8] in which case

$$\boldsymbol{E} = \boldsymbol{B} \times \boldsymbol{v} \, . \tag{2.17}$$

With the so-called longitudinal magnetic field geometry in which the magnetic field and the current are parallel to each other, the additional term, $-\nabla \Psi$, is needed. This condition will be discussed in detail in Chap. 4. In this case Ψ is not the electrostatic potential, since all the electric field comes from the electromagnetic induction due to the motion of flux lines.

One more word is added here in order to avoid confusion. It might be considered that the electric field given by Eq. (2.17) is not the induced one in the steady flux flow state, since the macroscopic flux distribution does not change with time. In fact, the substitution of $\partial \boldsymbol{B}/\partial t = 0$ into Eq. (2.15) leaves $\boldsymbol{B} \times \boldsymbol{v}$ as the gradient of a scalar function. As mentioned above, energy is dissipated in the flux flow state, leading to the appearance of electrical resistance as in normal conductors. In this sense, if we confine ourselves within the framework of pure macroscopic electromagnetism, it is possible to interpret this scalar function as an electrostatic potential, as for a normal conductor, without any contradiction. It is not possible, however, to go on and explain from such a theoretical background why an electric field does not appear when the motion of flux lines is stopped. As is emphasized in [9], the motion of flux lines is the essence. That is, the observed electric field is in fact an induced one.

The electric field given by Eq. (2.17) is equivalent to Eq. (1.138) which describes the AC Josephson effect. To prove this we begin by assuming for simplicity that the flux line lattice is a square lattice and that two points, A and B, are separated by L along the direction of the current, as shown in Fig. 2.2. If the flux line spacing is represented by a_f, L/a_f rows of flux lines are moving with the velocity v in the direction perpendicular to the current between these two points. The flux lines move a distance a_f during a time interval of $\Delta t = a_\mathrm{f}/v$. The amount of magnetic flux that crosses the line AB during the interval Δt is given by $(L/a_\mathrm{f})\phi_0$. Since the change in the phase of the order parameter when circulating around one quantum ϕ_0 of magnetic flux is 2π, the variation in the phase difference between A and B is $\Delta\Theta = 2\pi(L/a_\mathrm{f})$. With the aid of $B = \phi_0/a_\mathrm{f}^2$, Eq. (2.17) reduces to

$$V = EL = \frac{\phi_0}{2\pi} \cdot \frac{\Delta\Theta}{\Delta t} = \frac{\hbar\omega}{2e} \tag{2.18}$$

and agrees with Eq. (1.138). In the above we replace $\Delta\Theta/\Delta t$ by an angular frequency, ω.

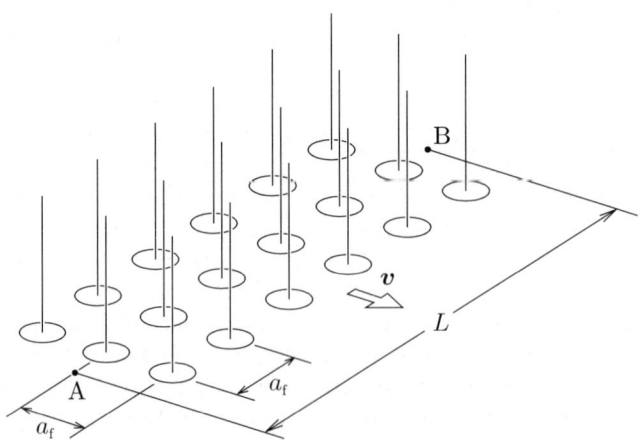

Fig. 2.2. Motion of flux line lattice

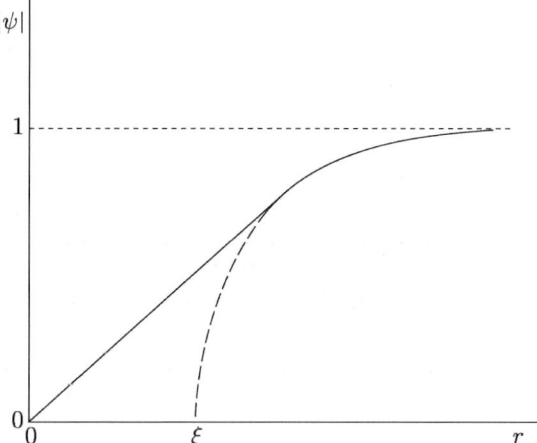

Fig. 2.3. Spatial variation in the order parameter in the vicinity of the center of normal core. The *broken line* represents the approximation of the local model

As a variation in the magnetic flux density is related to the velocity of flux lines through Eq. (2.15), the solution to $\boldsymbol{B}$ can be obtained in principle from this equation and Eq. (2.13). Before we proceed to solve these equations, some important phenomena related to flux motion will be described in this section. First we describe the various energy loss mechanisms such as a pinning energy loss in superconductors, and second we derive the flow resistivity corresponding to the energy loss and go on to clarify the relationship between the flow resistivity and the viscous coefficient η given by Eq. (2.12).

A description of the structure of the flux line is needed in order to explain the phenomenon plainly. For this we adopt the local model of Bardeen and Stephen [10]. Commentary on the rigorousness of this model is provided in detail in [9], and so is not repeated here. It is known that the results of this theoretical model are generally correct in spite of various assumptions for simplicity. We assume that the G-L parameter κ of the superconductor is sufficiently large. Bardeen and Stephen assumed the structure of the order parameter around the center of flux line to be as shown in Fig. 2.3 where the region inside a circle of radius ξ is in the normal state. In the original paper this radius was initially treated as an unknown quantity and then shown to coincide with ξ from its relationship to the upper critical field H_{c2}. In this section, we begin with this result. Outside the normal core of radius ξ the order parameter is approximately constant and the London equation can be used. In this region the superconducting current flows circularly around the normal core. Cylindrical coordinates are introduced with the z-axis along the flux line and with r denoting the distance from this axis. From Eqs. (1.4) and (1.64) the momentum of the superconducting electron in the circulating current in the region $\xi < r < \lambda$ is given by

$$\boldsymbol{p}_{\mathrm{s}} = m^{*}\boldsymbol{v}_{\mathrm{s}\theta}\boldsymbol{i}_{\theta} \simeq -\frac{\hbar}{r}\boldsymbol{i}_{\theta} \equiv \boldsymbol{p}_{\mathrm{s}0} \,, \tag{2.19}$$

where $\boldsymbol{i}_{\theta}$ is a unit vector in the azimuthal direction.

Cartesian coordinates are also introduced with the x- and y-axes in the plane normal to the flux line to express the flux motion. We assume that the current flows along the y-axis. If the Hall effect is disregarded for simplicity, the flux motion occurs along the x-axis and the mean macroscopic electric field is directed to the y-axis. The equation of motion of the superconducting electron is described as [10]

$$\frac{\mathrm{d}\boldsymbol{v}_{\mathrm{s}}}{\mathrm{d}t} = \frac{\boldsymbol{f}_{\mathrm{e}}}{m^{*}} \,, \tag{2.20}$$

where $\boldsymbol{f}_{\mathrm{e}}$ is the force on the electron. When the Lorentz force acts on the charge $-2e$, the relationship

$$m^{*}\boldsymbol{v}_{\mathrm{s}} = \boldsymbol{p}_{\mathrm{s}} + 2e\boldsymbol{A} \tag{2.21}$$

is known to hold with the vector potential $\boldsymbol{A}$. If we assume that the flux lines move uniformly along the x-axis with a small velocity $\boldsymbol{v}$, the secondary effects due to this motion can be expected to be sufficiently small and the variation with time in Eq. (2.20) can be approximated as $\mathrm{d}/\mathrm{d}t \simeq -(\boldsymbol{v} \cdot \nabla)$. Thus, we have

$$\boldsymbol{f}_{\mathrm{e}} = -(\boldsymbol{v}\cdot\nabla)(\boldsymbol{p}_{\mathrm{s}}+2e\boldsymbol{A}) = -v\frac{\partial}{\partial x}(\boldsymbol{p}_{\mathrm{s}}+2e\boldsymbol{A}) \,. \tag{2.22}$$

Within an accuracy of the order in v, we can approximately use $\boldsymbol{p}_{\mathrm{s}0}$ given by Eq. (2.19) for $\boldsymbol{p}_{\mathrm{s}}$. At the same time $\boldsymbol{A}$ is approximately given by $(Br/2)\boldsymbol{i}_{\theta}$ assuming that the magnetic field is nearly constant in the vicinity of the normal core. Substitution of these into Eq. (2.22) leads to

$$\begin{aligned}\boldsymbol{f}_{\mathrm{e}} &= v\frac{\partial}{\partial x}\left(\frac{\hbar}{r}-eBr\right)\boldsymbol{i}_{\theta}\\ &= \frac{v\hbar}{r^{2}}(-\boldsymbol{i}_{\theta}\cos\theta+\boldsymbol{i}_{r}\sin\theta)-eBv\boldsymbol{i}_{y} \,.\end{aligned} \tag{2.23}$$

This force originates from the local electric field $\boldsymbol{e}$, expressed as

$$\begin{aligned}\boldsymbol{e} &= -\frac{\boldsymbol{f}_{\mathrm{e}}}{2e} = \frac{\phi_{0}v}{2\pi r^{2}}(\boldsymbol{i}_{\theta}\cos\theta-\boldsymbol{i}_{r}\sin\theta)+\frac{Bv}{2}\boldsymbol{i}_{y}\\ &= \boldsymbol{e}_{1}+\frac{1}{2}(\boldsymbol{B}\times\boldsymbol{v}) \,,\end{aligned} \tag{2.24}$$

where $\boldsymbol{e}_{1}$ represents the nonuniform component of the electric field.

The electric field inside the normal core can be obtained from the boundary condition at $r=\xi$ that its tangential component is continuous to the outside. Since this component of the nonuniform component $\boldsymbol{e}_{1}$ given by Eq. (2.24) is $(\phi_{0}v/2\pi\xi^{2})\cos\theta$, the electric field inside the core is given by

$$\boldsymbol{e} = \frac{\phi_{0}v}{2\pi\xi^{2}}\boldsymbol{i}_{y}+\frac{1}{2}(\boldsymbol{B}\times\boldsymbol{v}) \tag{2.25}$$

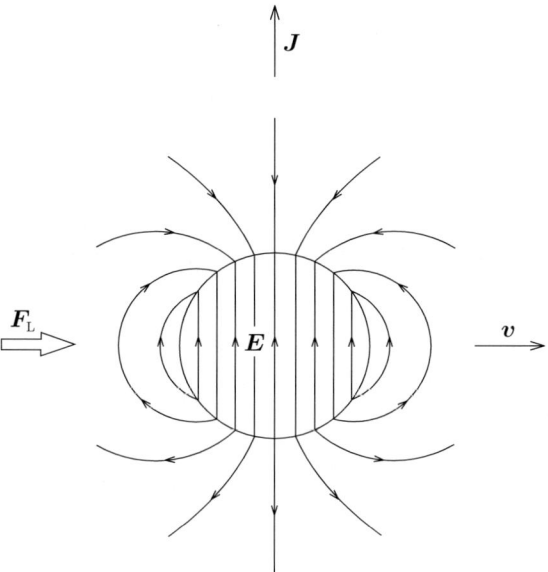

Fig. 2.4. Lines of electric force due to nonuniform component of the electric field inside and outside the normal core

and is found to be uniform and directed along the y-axis. The nonuniform component of the electric field around the normal core is schematically depicted in Fig. 2.4. It can be seen from the above result and the figure that the normal component of the electric field is discontinuous at the boundary, indicating that electric charge is distributed around the boundary according to $\sigma = -(\phi_0 v \epsilon / \pi \xi^2) \sin\theta$ with ϵ denoting the dielectric constant. This seems to contradict Eq. (2.4). However, this comes about through the use of the local model; the result does not contradict Eq. (2.4) on a macroscopic scale, since the total electric charge inside the core is zero.

Let us investigate the relationship between the local and macroscopic electric fields. We assume that the distance between the nearest flux lines is sufficiently large. If the unit cell of the flux line lattice is approximated by a circle of the radius R_0, then $B = \phi_0 / \pi R_0^2$. It is easily shown that the average of the nonuniform electric field $\boldsymbol{e}_1$ given by Eq. (2.24) in the region $\xi < r < R_0$ is zero. Thus, the second term of Eq. (2.24) and Eq. (2.25) contribute to the macroscopic electric field:

$$\boldsymbol{E} = \frac{\boldsymbol{i}_y}{\pi R_0^2} \int_0^\xi \frac{\phi_0 v}{2\pi \xi^2} 2\pi r \mathrm{d}r + \frac{1}{2}(\boldsymbol{B} \times \boldsymbol{v}) = \boldsymbol{B} \times \boldsymbol{v} \ . \qquad (2.26)$$

This result agrees with Eq. (2.17).

The electric field shown above causes the flow of normal electrons inside and outside the core, resulting in ohmic energy dissipation. This is the origin of the energy loss in superconductors and a corresponding electrical resistance is observed. Here we shall derive an expression for the flow resistivity from the energy loss due to the above electric field. The power loss inside the normal core per unit length of the flux line is given by

$$W_1 = \pi \xi^2 \frac{\mu_0^2 H_{c2}^2 v^2}{\rho_n} \left(1 + \frac{B}{2\mu_0 H_{c2}}\right)^2, \qquad (2.27)$$

where ρ_n is the normal resistivity and Eq. (1.52) is used. We assume that the resistivity outside the normal core is also approximately given by ρ_n. Thus, the power loss in this region per unit length of flux line is calculated as [10]

$$\begin{aligned}W_2 &= 2\pi \int_\xi^{R_0} \frac{r dr}{\rho_n} \left[\left(\frac{\phi_0 v}{2\pi}\right)^2 \frac{1}{r^4} + \left(\frac{Bv}{2}\right)^2\right] \\ &= \frac{\pi R_0^2 \mu_0 H_{c2} B}{2\rho_n} \left(1 - \frac{B^2}{4\mu_0^2 H_{c2}^2}\right) v^2 .\end{aligned} \qquad (2.28)$$

Hence, if the total power loss $W_1 + W_2$ is equated to the power loss, $\pi R_0^2 B^2 v^2/\rho_f$, in an equivalent uniform material with an effective resistivity, i.e., the flow resistivity ρ_f, we obtain

$$\rho_f = \frac{B}{\mu_0 H_{c2}} \left(1 + \frac{B}{2\mu_0 H_{c2}}\right)^{-1} \rho_n . \qquad (2.29)$$

In the limit $B \ll \mu_0 H_{c2}$ where the flux line spacing is sufficiently large, the above result reduces to

$$\rho_f = \frac{B}{\mu_0 H_{c2}} \rho_n . \qquad (2.30)$$

This agrees with the experimental result [11].

We go on to treat the case in which the magnetic field is applied in the direction of the z-axis and the current is applied along the y-axis. The velocity vector $\boldsymbol{v}$ lies along the x-axis. When Eq. (2.13) is rewritten in terms of Eqs. (1.147) and (2.17), we have

$$E = \rho_f (J - J_c) \qquad (2.31)$$

for $J \geq J_c$. This relationship between E and J gives the current-voltage characteristic in the flux flow state as shown in Fig. 1.13 and represents the characteristic feature of the superconductor. Thus, the flow resistivity is expressed as

$$\rho_f = \frac{\phi_0 B}{\eta} \qquad (2.32)$$

in terms of the viscous coefficient η.

2.3 Mechanism of Hysteresis Loss

It was shown in the last section that the energy loss in a superconductor is ohmic in nature due to the motion of normal electrons driven by an electric field induced by the motion of flux lines. It is well known that ohmic loss under AC conditions is proportional to the square of the frequency. Corresponding to this the power losses given by Eqs. (2.27) and (2.28) are proportional to the square of the flux line velocity v. The total power loss density is written as

$$P = \frac{1}{\pi R_0^2}(W_1 + W_2) = \frac{B^2 v^2}{\rho_{\rm f}} = -\boldsymbol{F}_v \cdot \boldsymbol{v} \tag{2.33}$$

and is the *viscous* power loss density. On the other hand, the *pinning* power loss density is given by $-\boldsymbol{F}_{\rm p} \cdot \boldsymbol{v}$, thus is proportional to the first power of flux line velocity, i.e., proportional to the frequency (here we note that $F_{\rm p}$ is independent of v). Thus, this loss is not of the ohmic type, a feature which is associated with the fact that the current-voltage characteristic shown in Fig. 1.13 is not ohmic. The above result seems to be in conflict with the notion that any kind of loss originates from the ohmic loss of normal electrons. It is necessary to understand the motion of flux lines in the pinning potential in order to resolve this contradiction.

The macroscopic electromagnetic phenomena in superconductors due to the motion of flux lines can be theoretically treated in an analogous way to the motion of a mechanical system. An example can be seen in Eq. (2.33). In terms of Eqs. (2.9) and (2.17) the input power density $\boldsymbol{J} \cdot \boldsymbol{E}$ to the superconductor may also be expressed as

$$\boldsymbol{J} \cdot \boldsymbol{E} = \boldsymbol{J} \cdot (\boldsymbol{B} \times \boldsymbol{v}) = \boldsymbol{F}_{\rm L} \cdot \boldsymbol{v} , \tag{2.34}$$

which can be regarded as the power given by the Lorentz force. At a more microscopic level such a correspondence to the mechanical system can also be expected to hold for the motion of individual flux lines. In that case, however, it should be noted that the pinning interaction does not give rise to an irreversible frictional force in a macroscopic sense but a reversible force originated from a pinning potential.

Here we look at one flux line in the lattice moving in the field of a pinning potential. We assume that the center of the lattice is moving with a constant velocity v. Because of the pinning interaction, the position u of a given flux line deviates from the equilibrium position u_0 determined by the elastic interaction between it and the surrounding flux lines. Consequently its velocity, $\dot{u} = \partial u/\partial t$, differs from the mean velocity v. As a result, the flux line experiences an elastic restoring force proportional to the displacement, $u - u_0$. According to Yamafuji and Irie [12] the equation describing such flux line motion is given by

$$\eta^* v - k_{\rm f}(u - u_0) + f(u) - \eta^* \dot{u} = 0 , \tag{2.35}$$

in which

$$\eta^* = \frac{B\eta}{\phi_0 N_{\rm p}} \ . \qquad (2.36)$$

Here η^* is the effective viscous coefficient per pinning center of number density $N_{\rm p}$, $k_{\rm f}$ is the spring constant of the elastic restoring force of the flux line lattice and $f(u)$ is the force due to the pinning potential. The fourth term in Eq. (2.35) is the viscous force and the first and the second terms give the driving force on the flux line. That is, the first term is the component for the case where the velocity is not disturbed by the pinning potential and the second term is the additional component due to the disturbance of the velocity. From the condition of continuity of the steady flow of flux lines, we have

$$\langle \dot{u} \rangle_t = \dot{u}_0 = v \ , \qquad (2.37)$$

where $\langle \ \rangle_t$ represents the average with respect to time.

The input power in this case is given by

$$\langle [\eta^* v - k_{\rm f}(u - u_0)] \dot{u} \rangle_t \ . \qquad (2.38)$$

From the mechanism of the energy dissipation, this should be equal to the viscous power loss, $\langle \eta^* \dot{u}^2 \rangle_t$. The proof of this equality is Exercise 2.2. On the other hand, the apparent viscous power loss is $\eta^* v^2$. Hence, the difference between these two quantities is the loss due to the pinning interaction. Thus, the pinning power loss density $P_{\rm p}$ is given by this difference multiplied by $N_{\rm p}$; i.e.

$$P_{\rm p} = \frac{B\eta}{\phi_0}(\langle \dot{u}^2 \rangle_t - v^2) \ . \qquad (2.39)$$

The pinning power loss is the additional power loss due to the fluctuation of the flow velocity of flux lines caused by the pinning potential. It should be noted that the pinning potential itself does not apparently influence this power loss. The question is whether or not this pinning power loss density is proportional to the mean velocity v.

Yamafuji and Irie [12] showed that the velocity of a flux line becomes very large when it drops into the pinning potential well and then jumps out again under the elastic interaction with the surrounding flux lines. Strictly speaking, in order to realize this feature, the pinning potential must be sufficiently steep to fulfill the condition $|\partial f / \partial u| \equiv k_{\rm p} > k_{\rm f}$, as will be shown later in Sect. 7.3. We assume that this condition is fulfilled. If the flux line reaches the edge of the pinning potential, $u = 0$, at $t = 0$, and if the pinning force varies as $f(u) \simeq k_{\rm p} u$, from Eq. (2.35) we have

$$u(t) \simeq -\frac{k_{\rm f} v t}{k_{\rm p} - k_{\rm f}} + \frac{k_{\rm p} \eta^* v}{(k_{\rm p} - k_{\rm f})^2}\left[\exp\left(\frac{t}{\tau}\right) - 1\right]; \quad t > 0 \ , \qquad (2.40)$$

where

$$\tau = \frac{\eta^*}{k_{\rm p} - k_{\rm f}} \qquad (2.41)$$

is a time constant. The details of this analysis are given in [13]. In the above we assumed that $u_0 = vt$, since $u = u_0$ for $t < 0$. It may be seen from the above result that the flux line motion becomes unstable and its velocity becomes very large when it drops into the pinning potential well. For simplicity, we assume that the mean velocity v is sufficiently small. If the instability continues from $t = 0(u = 0)$ to $t = \Delta t(u = d)$ when the flux line reaches the center of the pinning potential well (with $2d$ denoting the size of the pinning potential well), the contribution of this term to the integral of $\dot{u}^2$ with respect to time is given by

$$\int_0^{\Delta t} \dot{u}^2 dt = \frac{d^2(k_\mathrm{p} - k_\mathrm{f})}{2\eta^*} + O(v), \qquad (2.42)$$

where the second term on the right-hand side is a small quantity of the order in v. Strictly speaking, the period during which the flux line motion becomes unstable in the limit $v \to 0$ is from $t = 0(u = 0)$ to $t = \Delta t'(u = d')$ as shown in Fig. 2.5. The contribution from the instability when the flux line jumps out of the pinning potential well is also approximately given by $d^2(k_\mathrm{p} - k_\mathrm{f})/2\eta^*$. Thus, we have

$$\langle \dot{u}^2 \rangle_t = \frac{d^2(k_\mathrm{p} - k_\mathrm{f})}{T_0 \eta^*} + O(v^2), \qquad (2.43)$$

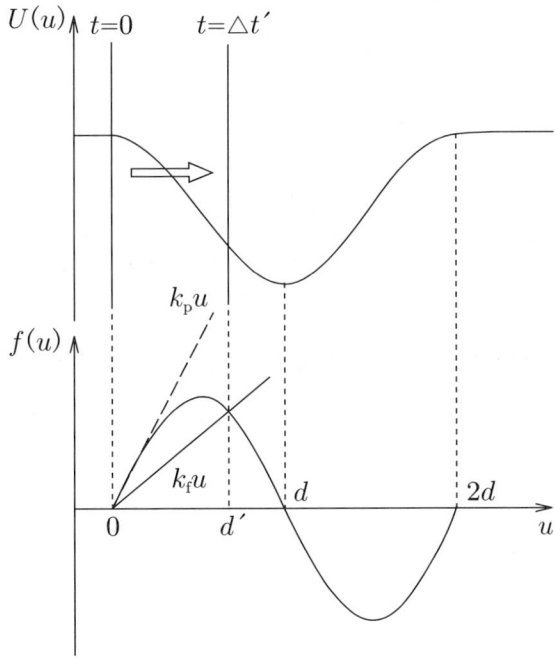

Fig. 2.5. Motion of the flux line in the pinning potential. *Upper* and *lower* figures show the pinning potential and the pinning force, respectively

where $1/T_0$ is the frequency at which one flux line meets the pinning potential wells in a second. If we represent the mean separation of the pinning potential wells by D_p, we have $T_0 = D_\mathrm{p}/v$, and the pinning power loss density is approximately given by

$$P_\mathrm{p} = \frac{N_\mathrm{p}}{D_\mathrm{p}}(k_\mathrm{p} - k_\mathrm{f})d^2 v \ . \tag{2.44}$$

This result also implicitly requires that $k_\mathrm{p} > k_\mathrm{f}$. Although the above theoretical treatment gives only a rough estimate (the detailed calculation will be given in Sect. 7.3), it tells us that, if the motion of the flux line is unstable inside the pinning potential well, the pinning power loss density is of the hysteretic type which is proportional to v (hence to frequency under AC conditions) as shown in Eq. (2.44). A nonzero critical current density can exist only under the condition $k_\mathrm{p} > k_\mathrm{f}$ (see Sect. 7.3). Therefore, the fact that the pinning loss is of the hysteretic type is identical with the fact that the current-voltage characteristic is non-ohmic.

The above result can be simply explained as follows. When the motion of a flux line is unstable, its velocity is approximately given by $[k_\mathrm{p}v/(k_\mathrm{p} - k_\mathrm{f})]\exp(\Delta t/\tau) \simeq d/\tau$ and takes a large value independent of the mean velocity v. Hence, the energy loss of the flux line during its interaction with one pinning center is a constant and the power loss is proportional to the number of pinning centers which the flux line meets in a second, i.e., to v.

2.4 Characteristic of the Critical State Model and its Applicable Range

It is possible to obtain the solutions of the magnetic flux density $\boldsymbol{B}$ and the velocity of flux lines $\boldsymbol{v}$ from the critical state model, i.e., Eqs. (2.13) and (2.15) describing the force balance and the continuity of flux lines, respectively. It is also possible to obtain $\boldsymbol{B}$ and $\boldsymbol{E}$ from Eqs. (2.2), (2.13) and (2.17). In the both cases the equation to be solved contains a second spatial derivative term and a first time derivative term. This equation is difficult to solve because of the existence of the coefficient, $\boldsymbol{\delta}$, that indicates the direction of the pinning force. A simple example of an approximate solution of this equation will be shown in Sect. 3.2.

In this section the characteristics of the critical state model are briefly mentioned and some phenomena which cannot be described by this model are discussed. One of the characteristics of this model is that the local current density does not take on smaller values than the critical current density as shown in Eq. (2.31). It means that the pinning interaction is expected to exert its effect as much as possible like the maximum static friction. Especially in the static case $\boldsymbol{E} = 0$, we have $|\boldsymbol{J}| = J_\mathrm{c}$ and the current density in the superconductor is equal to the critical value. This is called the critical state in a narrow sense. This corresponds to the case where the static magnetic flux distribution is determined by the balance between the Lorentz force and the pinning force.

The second characteristic is that the phenomena associated with a variation in the magnetic flux distribution are assumed to be completely irreversible. That is, the restraining forces always work in the opposite direction to the flux motion as given in Eqs. (2.11) and (2.12). The power from the external source such as $-\bm{F}_\mathrm{p} \cdot \bm{v}$ is always positive. Therefore, stored energy is not included at all and energy is always dissipated. It means that $\bm{E} \cdot \bm{J}$ is just the power loss density. In this case, the local power loss density can be obtained if we know the local $\bm{E}$ and $\bm{J}$. It should be noted that complete irreversibility is assumed even for the pinning force which comes originally from the potential. The pinning potential is reversible in nature at a microscopic level. It was shown in the last section that the irreversibility originates from the unstable flux motion associated with flux lines dropping into and jumping out of the pinning potential wells. Therefore, if the variation in the external magnetic field etc. is so small that the motion of flux lines is restricted mostly to the region inside the pinning potential wells, the phenomenon is regarded as almost reversible without appreciable energy dissipation. In this case the critical state model cannot be applied. The input power, $\bm{E} \cdot \bm{J}$, includes the stored power and sometimes takes on a negative value. In general, therefore, it is not possible to estimate the instantaneous power loss. Only in the case of periodically varying conditions can the energy loss per cycle be estimated from the integral of $\bm{E} \cdot \bm{J}$ with respect to time or from the area of closed hysteresis curve. Such a reversible phenomenon will be considered in Sect. 3.7.

2.5 Irreversible Phenomena

As mentioned in the last section, the dynamic force balance equation can be solved only approximately because of the direction coefficient δ. In addition, the force balance equation itself is sometimes a nonlinear differential equation. In presenting an example of the approximate solution, in this section we focus on the magnetic flux distribution and the magnetization in a quasistatic condition. The use of the term "quasistatic process" in this book is different from that used in thermodynamics and refers merely to processes in which the external magnetic field is varied slowly. That is, the quasistatic state is that obtained by a linear extrapolation of sweep rate of the external field to zero. Hence, such a state is in most cases a nonequilibrium state in the thermodynamic sense. In the context of this book the velocity of flux lines v has only to be so small to enable the viscous force to be neglected. We should note that this condition differs largely from material to material. In commercial Nb$_3$Sn wires, for example, the quasistatic state is attained over a wide range of sweep rates, since the pinning force is very large. We assume $J_\mathrm{c} = 5 \times 10^9$ Am^{-2} at $B = 5$ T as a typical case. The flux flow resistivity estimated from Eq. (2.30) is $\rho_\mathrm{f} \simeq 8 \times 10^{-8}$ Ωm, where we have used $\mu_0 H_{c2} = 20$ T and $\rho_\mathrm{n} = 3 \times 10^{-7}$ Ωm. Hence, the condition that the viscous force is as large as 1 percent of the pinning force is given by $E = \rho_\mathrm{f} J_\mathrm{c} \times 10^{-2} \simeq 0.4$ Vm^{-1}. Superconductors do not generally experience such high levels of electric field, hence they usually operate in a range where the viscous force can be neglected. We shall here clarify the corresponding sweep rate of the external magnetic field. We assume for simplicity that the external magnetic field completely penetrates the superconductor. The induced electric field is $E \simeq d\partial B/\partial t$, where d is a half-radius of a superconducting wire or a half-thickness of a superconducting slab. Hence, we find that the sweep rate of the field at which $E = 0.4$ Vm^{-1} is reached is 8×10^3 Ts^{-1} for $d = 50$ μm. Clearly the maximum sweep rate depends on the size of superconductor. Furthermore, the range within which the process can be regarded as quasistatic becomes narrower as the pinning force (hence J_c) becomes weaker.

Here we treat the case where the magnetic field is applied parallel to a sufficiently large superconducting slab. We assume that the slab occupies $0 \leq x \leq 2d$ and the magnetic field H_e is applied along the z-axis. From symmetry we have to consider only the half-slab, $0 \leq x \leq d$. All the electromagnetic quantities are uniform in the y-z plane and expected to vary only along the x-axis. The flux lines move along the x-axis, hence Eq. (2.13) reduces to

$$-\frac{\widehat{B}}{\mu_0} \cdot \frac{\partial \widehat{B}}{\partial x} = \delta F_\mathrm{p}(\widehat{B}) , \qquad (2.45)$$

where $\widehat{B} = |\boldsymbol{B}|$ and $\delta = \pm 1$ is the sign factor representing the direction of flux motion. That is, $\boldsymbol{\delta} = \delta \boldsymbol{i}_x$, where $\boldsymbol{i}_x$ is a unit vector along the x-axis. In this case the current flows along the y-axis. Equation (2.45) can be solved, if the functional form of $F_\mathrm{p}(\widehat{B})$ is given.

Many models have been proposed for the functional form of $F_\mathrm{p}(\widehat{B})$. Here we use the Irie-Yamafuji model [6] which can be applied over a relatively wide magnetic field range except in the high field region near the upper critical field:

$$F_\mathrm{p}(\widehat{B}) = \alpha_\mathrm{c} \widehat{B}^\gamma , \qquad (2.46)$$

where α_c and γ are the pinning parameters; usually $0 \leq \gamma \leq 1$. If we assume $\gamma = 1$, the above model reduces to the Bean-London model [14, 15]. This model is applicable to the case where J_c can be regarded as approximately field independent. Equation (2.46) reduces to the Yasukochi model [16] when $\gamma = 1/2$. This model is useful for practical superconductors in which grain boundaries or large normal precipitates are effective as pinning centers. The Silcox-Rollins model [17] is obtained when $\gamma = 0$ is used. As for other pinning models, the Kim model [18] is also known to express well the magnetic field dependence of the pinning force density within a certain range of magnetic field, although its functional form is different from that in Eq. (2.46). A correction is needed for these models at high fields so that the pinning force density $F_\mathrm{p}(\widehat{B})$ decreases with increasing $\widehat{B}$ (see Sect. 7.1).

2.5 Irreversible Phenomena 57

When Eq. (2.46) is substituted into Eq. (2.45) and it is integrated, we have

$$\delta \widehat{B}^{2-\gamma} = \delta_0 \widehat{B}_0^{2-\gamma} - (2-\gamma)\mu_0 \alpha_c x , \tag{2.47}$$

where δ_0 and $\widehat{B}_0$ are the values of δ and $\widehat{B}$ at the surface ($x = 0$), respectively. $\widehat{B}_0$ is sometimes different from the magnetic flux density $|\mu_0 H_e|$ corresponding to the external magnetic field H_e. This is attributed to the diamagnetic surface current or the surface irreversibility which will be discussed later in Sects. 2.6 and 3.5, respectively. In superconductors with large G-L parameters κ, however, the diamagnetic surface current is small. Hence, if the flux pinning strength or the size of the superconductor, i.e., the thickness $2d$, in this case, is sufficiently large, the effect of diamagnetism can be disregarded. On the other hand, the effect of surface irreversibility does not appear for a uniform superconductor. Thus, we assume

$$\widehat{B}_0 = \mu_0 \widehat{H}_e , \tag{2.48}$$

where $\widehat{H}_e = |H_e|$. Equation (2.47) expresses the magnetic flux distribution in the region from the surface to a certain depth, i.e., the point at which $\widehat{B}$ reduces to zero or the breaking point of the magnetic flux distribution, as will be shown later. The magnetic flux distribution is linear in case $\gamma = 1$ and is parabolic in case $\gamma = 0$.

The magnetic flux distribution in response to an increase in the external magnetic field from zero as the initial state is given by Eq. (2.47) with $\delta = \delta_0 = 1$ and is schematically shown in Fig. 2.6. Figures 2.6(a) and (b) represent cases in which the magnetic flux does not and does, respectively, penetrate to the center of the superconductor. We call the external magnetic field at which the flux front reaches the center of the superconducting slab as the "penetration field," H_p. It is given by

$$H_\mathrm{p} = \frac{1}{\mu_0}[(2-\gamma)\mu_0 \alpha_c d]^{1/(2-\gamma)} . \tag{2.49}$$

We assume that $\boldsymbol{B}$ is in the positive z-axis direction ($B > 0$) such that $\widehat{B} = B$. The current distribution is given by

$$J_y = -\frac{1}{\mu_0} \cdot \frac{\partial \widehat{B}}{\partial x} = \alpha_c [\widehat{B}_0^{2-\gamma} - (2-\gamma)\mu_0 \alpha_c x]^{(\gamma-1)/(2-\gamma)} . \tag{2.50}$$

This result is also directly obtained from $J = J_c = F_\mathrm{p}(\widehat{B})/\widehat{B} = \alpha_c \widehat{B}^{\gamma-1}$. That is, as was already mentioned, the local current density is always equal to the critical current density, $\pm J_c$, at the magnetic field strength at this point according to the critical state model. In a more exact expression, the local current density never takes on a smaller value than J_c including the contribution from the viscous force. The current distributions inside the superconducting slab corresponding to the magnetic flux distributions in the initial magnetization process are shown in the lower figures in Fig. 2.6. Although it is possible

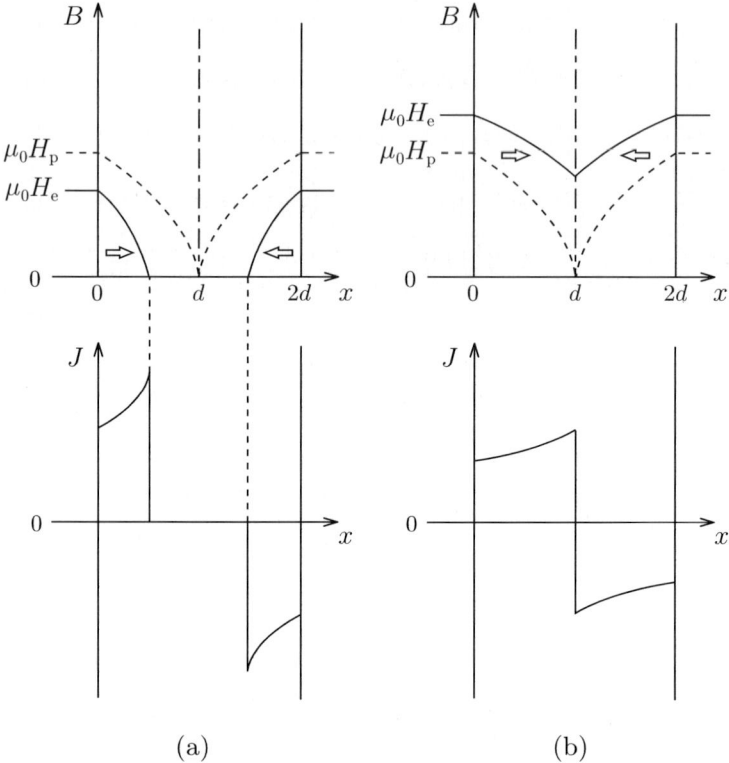

Fig. 2.6. Distributions of the magnetic flux (*upper figure*) and the current (*lower one*) in the superconductor in an initial increasing field (**a**) in case where the external field H_e is smaller than the penetration field H_p and (**b**) in case where H_e is larger than H_p. The *arrows* show the direction of flux motion

to generally describe the current density J_y including δ, δ_0 and the sign of B, it is complicated. According to Eq. (2.50) the current density J diverges at the point $\widehat{B} = 0$ in case $\gamma \neq 1$. This divergence, which results from the approximation of the magnetic field dependence of J_c over a relatively wide range of magnetic fields by a relatively simple function, does not represent the real situation. In spite of such unphysical limit, there is no anomaly in the averaged quantities such as the magnetization or the energy loss. Hence, it is not necessary to consider this problem.

We next consider the case where the external magnetic field is decreased after being increased up to H_m. The flux lines go out of the superconductor. In this case flux lines near the surface leave first and hence the variation in the magnetic flux distribution starts at the surface. In this region the pinning force prevents the flux lines from going out and the direction of the current is opposite to the direction in the initial state. Near the surface in the half-region of the slab, $0 \leq x \leq d$, we have $\delta = \delta_0 = -1$. Substituting these values into

Eq. (2.47), we have the magnetic flux distribution near the surface:

$$\widehat{B}^{2-\gamma} = (\mu_0\widehat{H}_e)^{2-\gamma} + (2-\gamma)\mu_0\alpha_c x \,. \tag{2.51}$$

On the other hand, the former flux distribution

$$\widehat{B}^{2-\gamma} = (\mu_0 H_m)^{2-\gamma} - (2-\gamma)\mu_0\alpha_c x \tag{2.52}$$

remains unchanged in the inner region of the superconductor. The breaking point, $x = x_b$, at which the crossover of the two distribution equations occurs is given by

$$x_b = \frac{-(\mu_0\widehat{H}_e)^{2-\gamma} + (\mu_0 H_m)^{2-\gamma}}{2(2-\gamma)\mu_0\alpha_c} = \frac{d}{2}\left[-\left(\frac{\widehat{H}_e}{H_p}\right)^{2-\gamma} + \left(\frac{H_m}{H_p}\right)^{2-\gamma}\right]. \tag{2.53}$$

The magnetic flux and current distributions associated with this process are depicted in Fig. 2.7. There are three cases depending on the value of H_m;

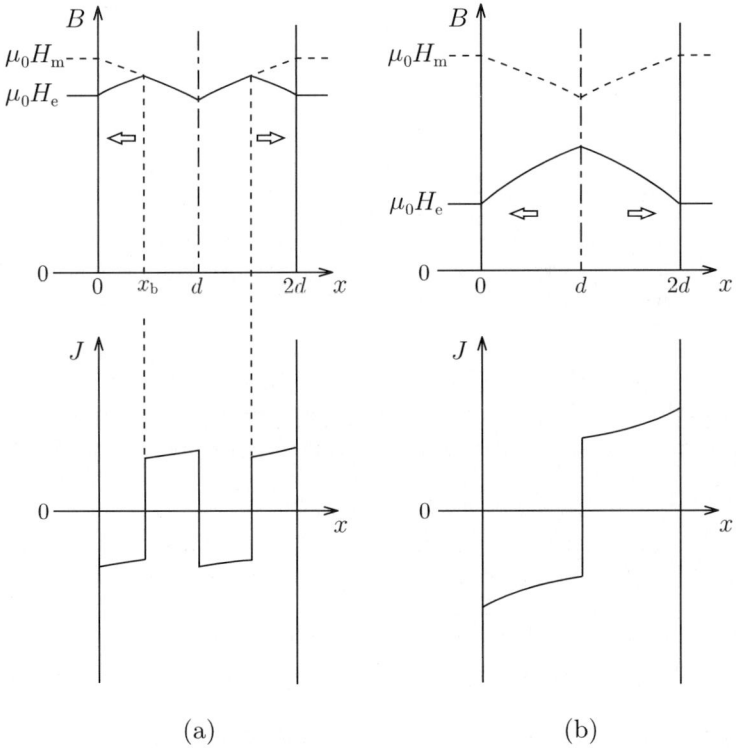

Fig. 2.7. Distributions of the magnetic flux (*upper figure*) and the current (*lower one*) in the superconductor in a decreasing field. The *arrows* show the direction of flux motion

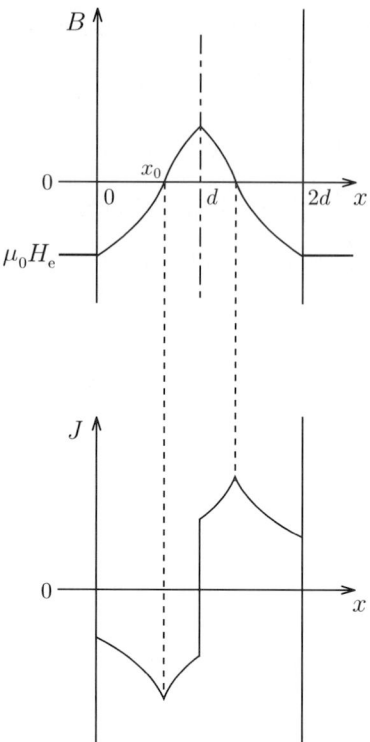

Fig. 2.8. Distributions of the magnetic flux (*upper figure*) and the current (*lower one*) when the external magnetic field is reversed

i.e., depending on whether $H_m < H_p$, $H_p < H_m < 2^{1/(2-\gamma)} H_p$ or $H_m > 2^{1/(2-\gamma)} H_p$. In case $H_m > 2^{1/(2-\gamma)} H_p$, the breaking point of the magnetic flux distribution disappears before $\widehat{H}_e$ reaches zero (see Fig. 2.7(b)) and the magnetization at $\widehat{H}_e = 0$ is on the major magnetization curve as will be shown later.

When the external magnetic field is further decreased to a negative value, the distributions of magnetic flux and current vary as shown in Fig. 2.8. In this case $\widehat{B} = -B$ for $0 < x < x_0$, i.e., in the region from the surface to the point at which $\widehat{B} = 0$.

We shall calculate the magnetization of the superconducting slab. The magnetization of a superconductor is defined by

$$M = \frac{1}{\mu_0 d} \int_0^d B(x) \mathrm{d}x - H_e \ . \tag{2.54}$$

The first term corresponds to the magnetic flux density averaged inside the superconductor. Hence, Eq. (2.54) is neither equivalent to the local relationship, $\boldsymbol{m} = \boldsymbol{b}/\mu_0 - \boldsymbol{h}$, for magnetic substances nor its spatial average. It is in fact the average of $\boldsymbol{b}/\mu_0 - \boldsymbol{H}_\mathrm{e}$ over the superconductor. For the case $H_\mathrm{m} > 2^{1/(2-\gamma)} H_\mathrm{p}$, after a simple calculation we have

$$\frac{M}{H_\mathrm{p}} = \frac{2-\gamma}{3-\gamma} h_\mathrm{e}^{3-\gamma} - h_\mathrm{e}; \qquad 0 < H_\mathrm{e} < H_\mathrm{p}, \quad (2.55\mathrm{a})$$

$$= \frac{2-\gamma}{3-\gamma}[h_\mathrm{e}^{3-\gamma} - (h_\mathrm{e}^{2-\gamma} - 1)^{(3-\gamma)/(2-\gamma)}] - h_\mathrm{e}; \quad H_\mathrm{p} < H_\mathrm{e} < H_\mathrm{m}, \quad (2.55\mathrm{b})$$

$$= \frac{2-\gamma}{3-\gamma}[2^{-1/(2-\gamma)}(h_\mathrm{m}^{2-\gamma} + h_\mathrm{e}^{2-\gamma})^{(3-\gamma)/(2-\gamma)}$$
$$- (h_\mathrm{m}^{2-\gamma} - 1)^{(3-\gamma)/(2-\gamma)} - h_\mathrm{e}^{3-\gamma}] - h_\mathrm{e}; \quad H_\mathrm{m} > H_\mathrm{e} > H_\mathrm{a}, \quad (2.55\mathrm{c})$$

$$= \frac{2-\gamma}{3-\gamma}[(h_\mathrm{e}^{2-\gamma} + 1)^{(3-\gamma)/(2-\gamma)} - h_\mathrm{e}^{3-\gamma}] - h_\mathrm{e}; \quad H_\mathrm{a} > H_\mathrm{e} > 0, \quad (2.55\mathrm{d})$$

$$= \frac{2-\gamma}{3-\gamma}\{[1 - (-h_\mathrm{e})^{2-\gamma}]^{(3-\gamma)/(2-\gamma)} - (-h_\mathrm{e})^{3-\gamma}\} - h_\mathrm{e};$$
$$0 > H_\mathrm{e} > -H_\mathrm{p}, \quad (2.55\mathrm{e})$$

where h_e and h_m are defined by

$$h_\mathrm{e} = \frac{H_\mathrm{e}}{H_\mathrm{p}}, \qquad h_\mathrm{m} = \frac{H_\mathrm{m}}{H_\mathrm{p}}, \qquad (2.56)$$

respectively, and

$$H_\mathrm{a}^{2-\gamma} = H_\mathrm{m}^{2-\gamma} - 2H_\mathrm{p}^{2-\gamma}. \qquad (2.57)$$

Each magnetic flux distribution in Fig. 2.9 has a corresponding description in Eq. (2.55). That is, in both cases 'a' and 'b' describe the increasing field processes, 'c' and 'd' are those in a decreasing field and 'e' corresponds to that when the magnetic field is reversed. The point 'e''' in the magnetization curve is just opposite to the point 'e'. The points b, d and e are on the major magnetization curve and correspond to the full critical state. The initial magnetization given by Eq. (2.55a) reaches the major curve at $H_\mathrm{e} = H_\mathrm{p}$. Magnetization curves for various values of γ are given in Fig. 2.10.

We have argued the case where only a magnetic field is applied. Now we consider the case where the transport current is also applied. It is assumed that the current with a mean density J_t is applied along the y-axis of the above superconducting slab and a magnetic field H_e is applied along the z-axis. Then, a self field according to Ampère's law leads to the boundary conditions at the surfaces of the slab:

$$B(x=0) = \mu_0 H_\mathrm{e} + \mu_0 J_\mathrm{t} d,$$
$$B(x=2d) = \mu_0 H_\mathrm{e} - \mu_0 J_\mathrm{t} d. \qquad (2.58)$$

Hence, the magnetic flux distribution is not symmetric with respect to the center, $x = d$, and we have to consider both halves of the slab. In addition, it should be noted that the magnetic flux distribution is different depending on

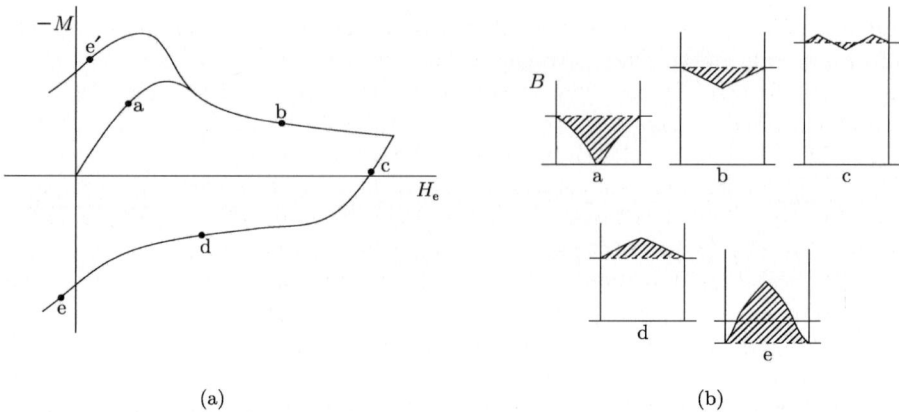

Fig. 2.9. (a) Magnetization and (b) magnetic flux distribution at each point. The letter at each point corresponds to the letter in equation number in Eq. (2.55)

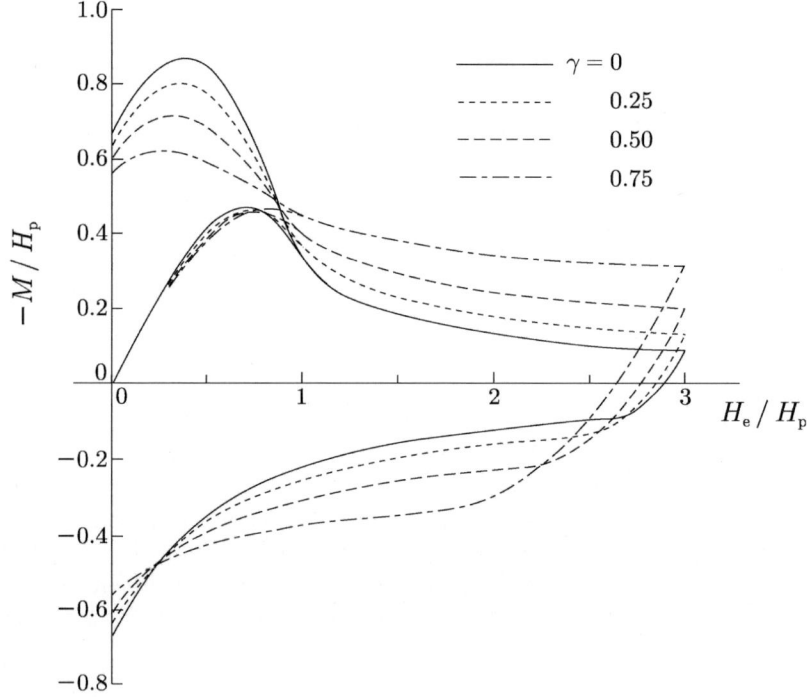

Fig. 2.10. Reduced magnetization curves for various values of γ

the history of application of magnetic field and current even for the same final boundary condition. For example, the upper part of Fig. 2.11(a) represents the magnetic flux distribution for the case where the magnetic field is applied (broken line) and then the current is applied (solid line). The upper part of Fig. 2.11(b) shows the distribution when the order of application is reversed. The corresponding current distributions are given in the lower figures. Thus, the magnetic flux distribution changes generally depending on the order of application of the magnetic field and the current. In any case the resultant magnetic flux distribution can be easily obtained from the critical state model with taking into account the history dependent boundary condition.

In the full critical state where the current flows in the same direction throughout the superconductor, the magnetic flux distribution is schematically shown in Fig. 2.12 and does not depend on the order of application of the magnetic field and the current. In this case from the boundary condition

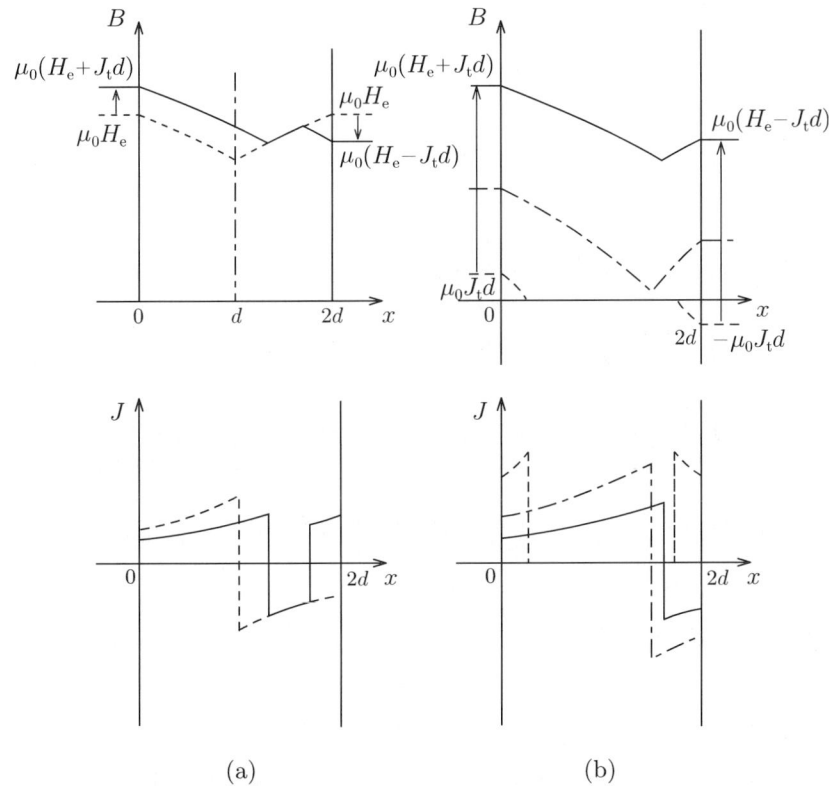

(a)   (b)

Fig. 2.11. Distributions of the magnetic flux (*upper figure*) and the current (*lower one*) in the superconductor when the magnetic field and the current are applied simultaneously: (**a**) in case where the current is applied after the magnetic field and (**b**) in the opposite case

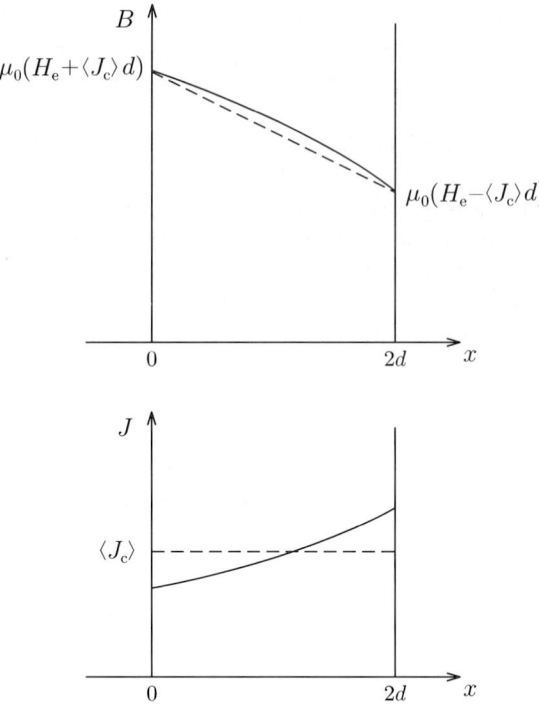

Fig. 2.12. Distributions of the magnetic flux (*upper figure*) and the current (*lower one*) in the complete critical state where the critical current flows. The *broken lines* show the distributions when the mean critical current density $\langle J_c \rangle$ flows

we have
$$B^{2-\gamma}(2d) = B^{2-\gamma}(0) - 2(\mu_0 H_p)^{2-\gamma} . \tag{2.59}$$
Hence, the average critical current density obtained from measurements, which corresponds to the slope of the broken line in the upper part of Fig. 2.12, is calculated from
$$(H_e + \langle J_c \rangle d)^{2-\gamma} - (H_e - \langle J_c \rangle d)^{2-\gamma} = 2H_p^{2-\gamma} . \tag{2.60}$$
In general H_e is replaced by $\widehat{H}_e$ and if the self field, $\langle J_c \rangle d$, is sufficiently smaller than $\widehat{H}_e$, Eq. (2.60) can be expanded in a series. Then, from an iteration approximation we have
$$\langle J_c \rangle = \frac{\alpha_c}{(\mu_0 \widehat{H}_e)^{1-\gamma}} \left[1 + \frac{(1-\gamma)\gamma}{6(2-\gamma)^2} \left(\frac{H_p}{\widehat{H}_e} \right)^{4-2\gamma} \right] . \tag{2.61}$$
In the above equation the first term is equal to the local critical current density at which the local magnetic field is equal to the external field H_e. The iteration approximation is correct when the ratio, $H_p/\widehat{H}_e = \epsilon$, is sufficiently

smaller than 1. The correction by the second term is relatively on the order of $\epsilon^{4-2\gamma}$.

For the transport current over this density, there is no solution of stable magnetic flux distribution. That is, the flux flow state sets in and we have to solve Eq. (2.13) including the viscous force.

In the above we have treated only the case where the magnetic field is applied parallel to the superconducting slab. Now we consider the case where the magnetic field is applied normal to the superconducting slab. It is assumed that the magnetic field and the current are applied along the z- and y-axes, respectively, of a wide superconducting slab parallel to the x-y plane. In this case the magnetic flux density inside the superconductor contains a uniform z-component $\mu_0 H_e$ and an x-component due to the current varying in the direction of the thickness, i.e., the z-axis. The distributions of current and magnetic flux due to the self field are shown in Fig. 2.13(a) and (b), respectively, in case of a total current smaller than the critical value. We have assumed the Bean-London model for simplicity. Figure 2.13(c) shows the magnetic flux structure and the direction of the Lorentz force as indicated by the arrows. The Lorentz force in this configuration can be regarded as the restoring force against the curvature of the flux lines and is called the "line tension." On the other hand, the Lorentz force which appears in the case shown in Fig. 2.6 originates from the gradient of the density of flux lines and is called the "magnetic pressure."

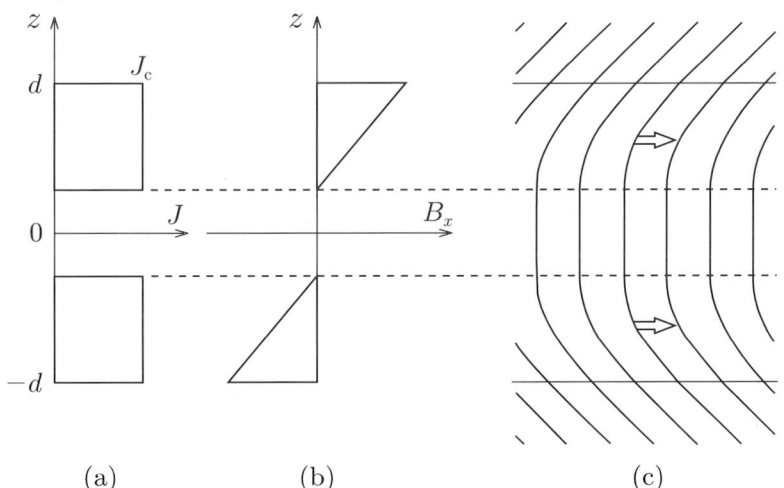

Fig. 2.13. Distributions of (a) the current and (b) the magnetic flux component parallel to the surface of the superconducting slab when the magnetic field normal to the slab and the current smaller than the critical value are applied. (c) represents the magnetic structure and the *arrows* show the direction of the Lorentz force

2.6 Effect of Diamagnetism

The magnetization described by the above critical state model is diamagnetic in an increasing field and paramagnetic in decreasing one, as shown in Fig. 2.9. Especially at high fields, the increasing and decreasing branches of the magnetization curve are almost symmetrical about the field axis. On the other hand, the magnetization curves shown in Fig. 2.14(a) and (b) are asymmetric and biased to the diamagnetic side. This comes from the essential diamagnetism of the superconductor. The magnetization curve of Fig. 2.9 corresponds to the case where the diamagnetic magnetization is much smaller than the magnetization due to the pinning effect.

The magnetic flux distribution inside the superconductor in an increasing field is schematically shown in Fig. 2.15. In the figure B_0 is the magnetic flux density in equilibrium with the external magnetic field H_e. That is, the difference between B_0/μ_0 and H_e gives the diamagnetic magnetization. The observed magnetization is proportional to the area of the hatched regions in the figure: region 'a' represents the contribution from the diamagnetism and region 'b' represents that from the flux pinning effect. The magnetization due to the pinning is relatively large in the case where the diamagnetism is small, the critical current density is large, and the superconductor is large in size. The last point is due to the fact that the pinning current is distributed throughout the whole region of the superconductor, while the diamagnetic shielding current is localized in the surface region. The magnitude of magnetization is proportional to H_p, and hence, is proportional to the sample size. Therefore, with reference to Fig. 2.14, the effect of diamagnetism cannot be neglected in the following cases:

(1) Superconductors in which the diamagnetism is strong, and hence, the lower critical field H_{c1} is large. These are mostly superconductors with small G-L parameters κ as shown in Fig. 2.14(a).
(2) Superconductors in which the pinning force is weak. In Bi-based oxide superconductors especially at high temperatures, for example, although the diamagnetism is small as characterized by low H_{c1}, the contribution of pinning is smaller, resulting in a conspicuous diamagnetism.
(3) Superconductors that are small in size. Figure 2.14(b) shows the magnetization of fine particles of V_3Ga. Although V_3Ga is used as a commercial superconductor and its pinning is strong, the magnetization due to the pinning is small because of small sample size. Similar results are observed in sintered Y-based oxide superconductors in which the coupling between fine grains is very weak.

The diamagnetic effect on the magnetic flux distribution in a superconductor is most simply treated by taking into account only the difference between $\mu_0 H_e$ and B_0 as a boundary condition and proceeding the manner of Sect. 2.5 to solve $B(x)$. However, this method is not correct in a strict sense, since the driving force on the flux lines is slightly different from the Lorentz force

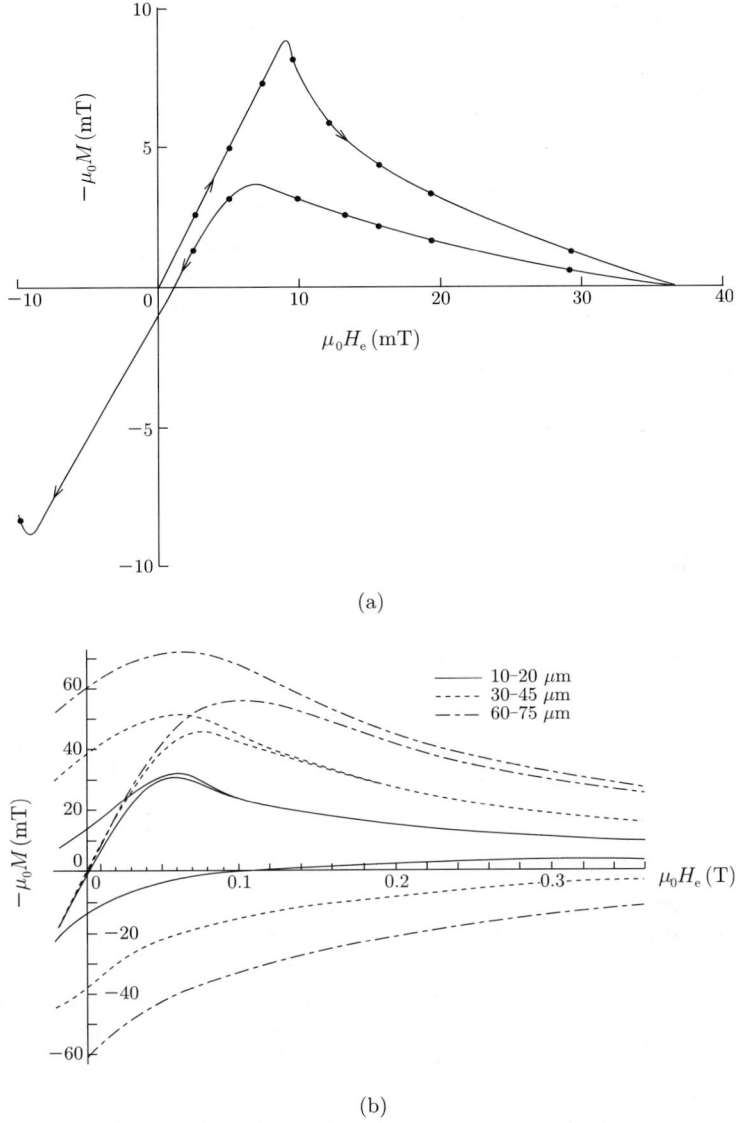

Fig. 2.14. Magnetization of (a) Ta with low κ-value at 3.72 K [19] and (b) V_3Ga powders at 4.2 K [20]. In (b) the effect of diamagnetism is relatively large because of the small grain size in spite of strong pinning

for superconductors in which the diamagnetic effect cannot be disregarded as mentioned in Sect. 2.1.

The thermodynamic magnetic field $\mathcal{H}$, which is the external variable corresponding to the internal variable $\boldsymbol{B}$, is defined by Eq. (2.6) and the flux lines are driven by the distortion of the thermodynamic magnetic field. We assume

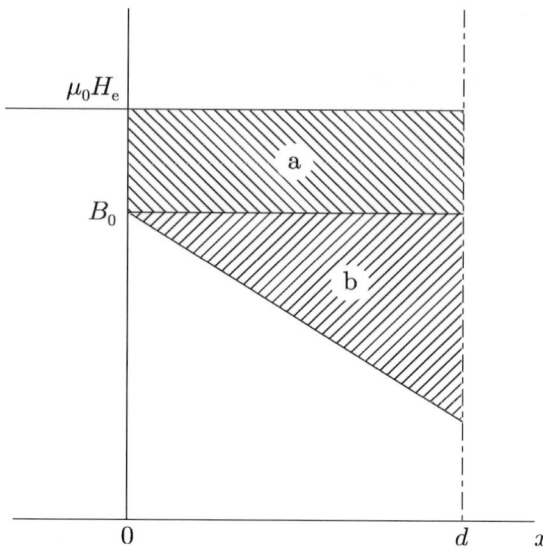

Fig. 2.15. Magnetic flux distribution in increasing field. 'a' and 'b' correspond to the contributions from the diamagnetism and the pinning to the magnetization, respectively

for example that there are no pinning centers in a region near the surface of the superconductor. The free energy in this region is given by Eq. (1.112) and it can be seen from Eqs. (2.6) and (1.113) that the thermodynamic magnetic field is equal to the external magnetic field. That is, if the thermodynamic magnetic field is different from the external magnetic field, a force proportional to the difference acts on the flux lines near the surface. The same also occurs inside the superconductor. If there exists a distortion in $\mathcal{H}$, a driving force acts so as to reduce the distortion. As mentioned above, the driving force comes from the gradient of $\mathcal{H}$ (the rotation of $\mathcal{H}$ in a strict sense) but not from the gradient of B which gives rise to the Lorentz force. Here we assume the case where two pin-free superconductors with different diamagnetism are in contact with each other as shown in Fig. 2.16 and in equilibrium with the external magnetic field. Since each superconductor is in equilibrium with the external field, an equilibrium state is also attained between the two superconductors. Hence, the flux lines do not move. In this case, a net current flows near the boundary of the two superconductors because of the difference in the magnetic flux density B due to the difference in the diamagnetism. The Lorentz force due to this current drives the flux lines from superconductor I to superconductor II. On the other hand, since the diamagnetism is stronger in superconductor II, the diamagnetic force pushes the flux lines to superconductor I. These forces cancel out and a net driving force does not act on the flux lines. The situation is the same also in the case where B and $\mathcal{H}$ vary continuously.

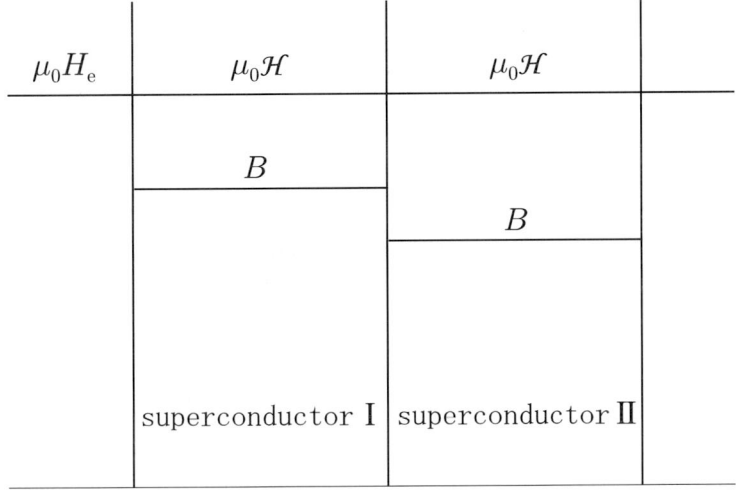

Fig. 2.16. Magnetic flux distribution in two pin-free superconductors, which are in contact with each other and in equilibrium with external magnetic field

It can be seen from the above argument that the relationship between $\mathcal{H}$ and B is the same as the one between the external magnetic field H_e and B in the case where the pinning energy is not contained in the free energy, i.e., $F = F_s$. Hence, this result in the ideal case seems to be approximately applicable to weakly pinned superconductors. When the variation with time is not too fast, the magnetic flux distribution can be obtained from the force balance equation

$$(\nabla \times \mathcal{H}) \times \mathbf{B} - \delta F_p(\widehat{B}) = 0 \tag{2.62}$$

and

$$\mathcal{H} = f(\mathbf{B}) \quad \text{or} \quad \mathbf{B} = f^{-1}(\mathcal{H}), \tag{2.63}$$

where f is a function derived from Eq. (2.6).

In practical cases it is not easy to derive Eq. (2.63) theoretically from Eq. (2.6). Especially it is not possible to express the feature over a wide range of magnetic field strengths by a single expression. However, an approximate expression which fits well with experimental results throughout the entire field range can be easily found for low κ superconductors. For example, Kes et al. [21] proposed the relationship

$$B = \mu_0 \mathcal{H} - \mu_0 H_{c1} \left[1 - \left(\frac{\mathcal{H} - H_{c1}}{H_{c2} - H_{c1}} \right)^n \right]; \quad H_{c1} \leq \mathcal{H} \leq H_{c2} \tag{2.64}$$

with

$$n = \frac{H_{c2} - H_{c1}}{1.16(2\kappa^2 - 1)H_{c1}}. \tag{2.65}$$

The expression for the pinning force density, Eq. (2.46), should also be modified so that it reduces to zero at the upper critical field in order to analyze the magnetization over the entire field range. We use the expression

$$F_\mathrm{p}(\widehat{B}) = \alpha_\mathrm{c}\widehat{B}^\gamma \left(1 - \frac{\widehat{B}}{\mu_0 H_{\mathrm{c}2}}\right)^\beta. \qquad (2.66)$$

Here, as in the last section, we consider an external magnetic field H_e applied parallel to the superconducting slab occupying $0 \leq x \leq 2d$. We treat the half-slab, $0 \leq x \leq d$, in an increasing magnetic field, for which $\delta = 1$. In this one-dimensional case, Eq. (2.62) leads to

$$\alpha_\mathrm{c}\mu_0^{\gamma-1}(x - x_\mathrm{c}) = -\int_0^{\widehat{\mathcal{H}}-H_{\mathrm{c}1}} (\Theta + c\Theta^n)^{1-\gamma}\left(1 - \frac{\Theta + c\Theta^n}{H_{\mathrm{c}2}}\right)^{-\beta} d\Theta \qquad (2.67)$$

with

$$c = \frac{H_{\mathrm{c}1}}{(H_{\mathrm{c}2} - H_{\mathrm{c}1})^n}. \qquad (2.68)$$

In the above $\widehat{\mathcal{H}} = |\mathcal{H}|$ and $x = x_\mathrm{c}$ represents the position where $\widehat{\mathcal{H}} = H_{\mathrm{c}1}$. Equation (2.67) can be solved only numerically. The relationship between $\widehat{\mathcal{H}}$ and x is obtained from this solution and the magnetic flux distribution, the relationship between $\widehat{B}$ and x, is obtained from Eq. (2.64). The distributions of $\widehat{B}$ and $\mu_0\widehat{\mathcal{H}}$ inside the superconducting slab are schematically shown in Fig. 2.17.

The magnetic flux distributions obtained from Eq. (2.67) are only those in the region where $\widehat{B} > 0$, i.e., $\widehat{\mathcal{H}} > H_{\mathrm{c}1}$. Here we shall discuss the distributions in the other region. For example, during the initial magnetization, i.e. during $0 \leq H_\mathrm{e} < H_{\mathrm{c}1}$, flux lines do not exist within the superconductor and $B = 0$. In this case, $\mathcal{H}$ cannot be defined by Eq. (2.6) and the definition itself is meaningless. As the external field continues to increase and H_e slightly exceeds $H_{\mathrm{c}1}$, the point, $x = x_\mathrm{c}$, exists inside the superconductor and the distributions of B and $\mathcal{H}$ are as shown in Fig. 2.18. In the region $x_\mathrm{c} \leq x \leq d$, flux lines do not exist and the definition of $\mathcal{H}$ is again meaningless. When H_e is further increased, the distributions again change as shown in Fig. 2.17.

Now we consider the case of decreasing field. For $H_\mathrm{e} > H_{\mathrm{c}1}$ distributions like the inverse of those given in Fig. 2.17 are expected. A question arises when H_e is reduced below $H_{\mathrm{c}1}$. Since the magnetic flux density B reduces to zero at the surface, $x = 0$, when $H_\mathrm{e} = H_{\mathrm{c}1}$, how do the distributions change when the external field is further decreased? Walmsley [22] speculated that the inner flux distribution remained unchanged as shown by the solid line in Fig. 2.19 when H_e is decreased below $H_{\mathrm{c}1}$, since $B = 0$ is in equilibrium with H_e for $0 \leq H_\mathrm{e} \leq H_{\mathrm{c}1}$. This is identical with the assumption that only the diamagnetic shielding current at the surface, i.e., the Meissner current changes with decreasing magnetic field. According to this model the magnetization

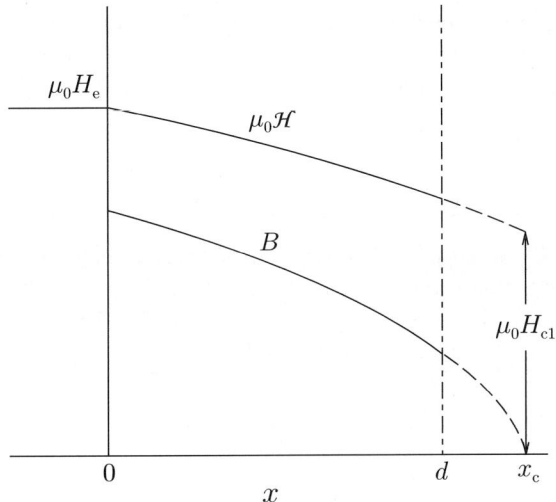

Fig. 2.17. Distributions of B and $\mu_0 \mathcal{H}$ in the superconductor in an increasing field when the magnetic flux penetrates the entire region of the superconductor

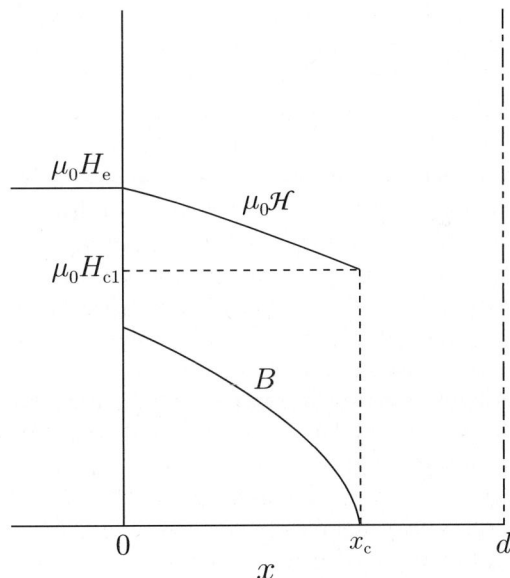

Fig. 2.18. Distributions of B and $\mu_0 \mathcal{H}$ when the external magnetic field slightly exceeds H_{c1}

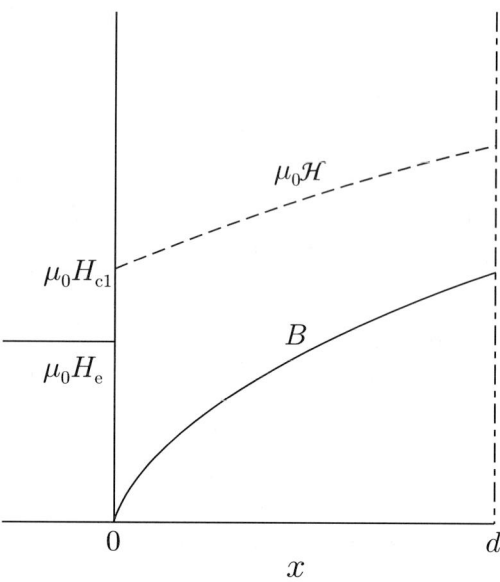

Fig. 2.19. Distributions of B and $\mu_0\mathcal{H}$ assumed by Walmsley [22] when the external field is decreased below H_{c1}. These distributions were assumed to remain unchanged even when the external magnetic field reduces to zero

varies, with a slope of -1, parallel to the initial magnetization curve with decreasing field from H_{c1}. However, as shown in Fig. 2.14, the slope of the magnetization is actually much gentler than this, suggesting a continuous discharge of flux lines. Frequently less than a half of the trapped flux at $H_e = H_{c1}$ remains inside the superconductor at $H_e = 0$. According to the expression of Campbell and Evetts, [23] the magnetization varies in this field range as if the boundary condition is given by $B(0) = \mu_0 H_e$. Practically the more flux lines than this speculation are discharged from the superconductor by the diamagnetism.

Why are the flux lines discharged from the superconductor? If the magnetic flux distribution shown in Fig. 2.19 is realized, the corresponding distribution of thermodynamic field $\mathcal{H}$ is speculated to be that indicated by the broken line in the figure, which is very different from the external field H_e at the surface. As a result of this difference the flux lines near the surface are expected to be driven to outside the superconductor. Experiments suggest that the magnetic flux distribution is as shown by the solid line in Fig. 2.20. Here it is necessary to define the thermodynamic magnetic field in the region below H_{c1}. If we remember the fact that the relationship therein between B and $\mathcal{H}$ is the same as that between B and H_e, it seems to be reasonable to assume

$$B = 0 \; ; \qquad 0 \leq \mathcal{H} \leq H_{c1} \qquad (2.69)$$

2.6 Effect of Diamagnetism   73

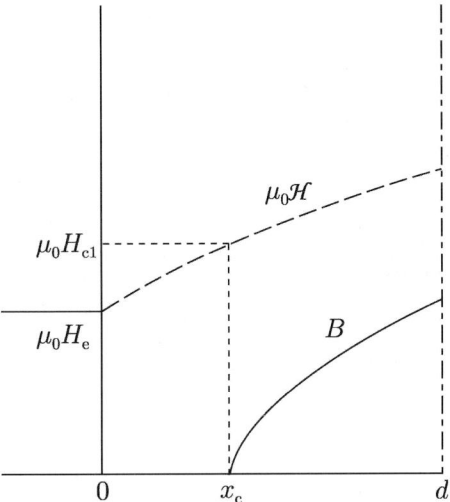

Fig. 2.20. Distributions of B and $\mu_0\mathcal{H}$ speculated from experimental results when the external field is decreased below H_{c1}

independently of the definition in Eq. (2.6). However, we should be careful about such an oversimplified treatment. Even if $\mathcal{H}$ can be defined, why is it possible for $\mathcal{H}$ to possess a gradient in the region where the flux lines do not exist? In such a region the magnetic property should be just the same as that in pin-free superconductors.

In practice it seems that some flux lines are trapped by "pinning layers" even in the region where B is macroscopically regarded as zero. This is a fundamental phenomenon that cannot be described by the macroscopic critical state model; hence a more microscopic discussion is necessary. For simplicity, we assume that the flux lines move across a multilayered structure composed of ideal superconducting layers and pinning layers [24] as shown in Fig. 2.21. In

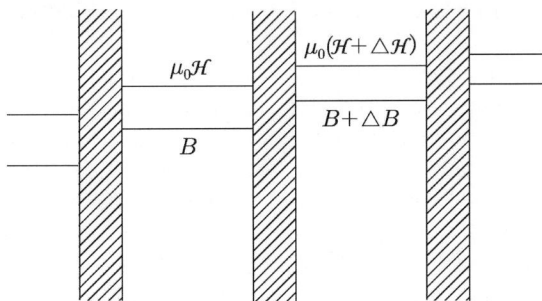

Fig. 2.21. Idealized one-dimensional multilayer structure composed of superconducting and pinning layers

each superconducting layer B and $\mathcal{H}$ are considered to be uniform. Because of the interaction between the flux lines and the pinning layer, differences appear in B and $\mathcal{H}$, respectively, in the pair of superconducting layers on either side of the pinning layer. The continuous distributions given in Fig. 2.17 are macroscopic representations of this step-wise variation of B and $\mathcal{H}$.

Consider the region near the surface and assume the situation where the external magnetic field has been reduced to H_{c1} and that $B = 0$ and $\mathcal{H} = H_{c1}$ are attained in the superconducting layer just inside the surface. The flux lines still remain in the next superconducting layer as shown in Fig. 2.22(a). When the external field is further decreased, the difference in $\mathcal{H}$ between the first and second superconducting layers exceeds the value determined by the flux pinning strength and the flux lines in the second superconducting layer cross the pinning layer and go out of the superconductor. Thus, B and $\mathcal{H}$ in the second superconducting layer decrease with decreasing external field. We denote the value of the external field by $H_e = H_{c1} - \Delta H_p$ at which $B = 0$ and $\mathcal{H} = H_{c1}$ are attained in the second superconducting layer. It should be noted that the magnetic flux density B is zero in the both superconducting layers, while the thermodynamic field $\mathcal{H}$ differs by ΔH_p. This can be explained by the fact that the flux lines are trapped in the pinning layer. In this situation a net current does not flow because $B = 0$ in the two superconducting layers. The diamagnetic force proportional to the difference in $\mathcal{H}$ on the trapped flux lines in the pinning layer is balanced by the pinning force. When the external magnetic field is further decreased, the diamagnetic force is enhanced and the flux lines leave the superconductor. Thus, the variation in $\mathcal{H}$ penetrates the superconductor and other flux lines come from the inner region and are trapped in the pinning layer. It is considered that flux lines always exist in the pinning

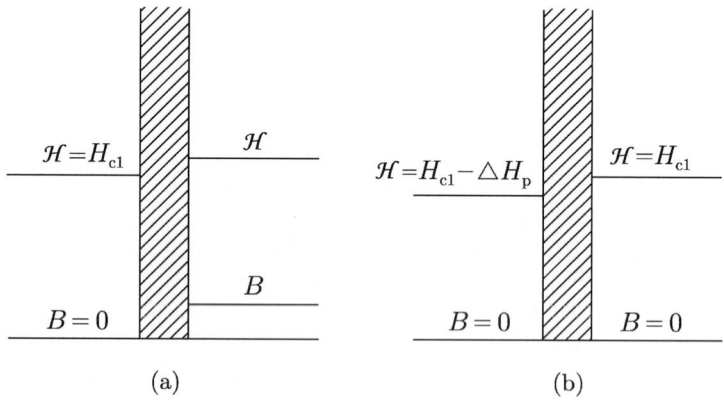

Fig. 2.22. Magnetic flux density and thermodynamic magnetic field in the superconducting layer just inside the surface (*left-hand side*) and the next superconducting layer (*right-hand side*); (**a**) when H_e is reduced to H_{c1} and (**b**) when H_e is slightly reduced from H_{c1} so that B becomes zero in the next superconducting layer. For simplicity μ_0 is omitted for $\mathcal{H}$

layer and that the difference in $\mathcal{H}$ across the pinning layer does not disappear. If the thickness of each superconducting layer is not very much larger than λ, such a penetration of B and $\mathcal{H}$ variation into the superconductor is expected to occur. As a result, it is proved that a gradient of $\mathcal{H}$ exists in the region where $B = 0$ macroscopically as drawn in Fig. 2.20.

Here we assume that $\mathcal{H}$ is described by

$$\left|\frac{d\mathcal{H}}{dx}\right| = a - b\mathcal{H} ; \qquad 0 \leq \mathcal{H} \leq H_{c1} , \tag{2.70}$$

where a and b are positive parameters. In a decreasing field, $d\mathcal{H}/dx$ is positive and from the boundary condition that $\mathcal{H} = H_e$ at $x = 0$, we have

$$\mathcal{H} = \frac{a}{b} - \left(\frac{a}{b} - H_e\right)\exp(-bx). \tag{2.71}$$

The position at which $\mathcal{H} = H_{c1}$ is given by

$$x_c = \frac{1}{b}\log\left(\frac{a - bH_e}{a - bH_{c1}}\right). \tag{2.72}$$

The distributions of $\mathcal{H}$ and B in the region $x > x_c$ are given by Eq. (2.67) (with a change of the sign of the right-hand side) and Eq. (2.64), respectively. The magnetization is calculated by inserting this result for B into Eq. (2.54). The result of numerical calculation [24] based on the above model is compared with an experimental result for Nb foil by Kes et al. [21] in Fig. 2.23. The good agreement indicates that the above model describes the phenomenon correctly.

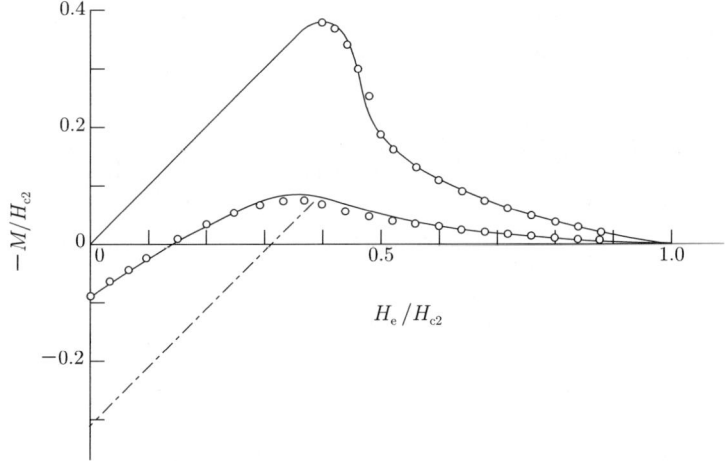

Fig. 2.23. Observed magnetization (*solid line*) on a Nb foil at 3.51 K [21] and corresponding theoretical result (*open circles*). The *chained line* shows the speculation from Walmsley's model

2.7 AC Losses

When the magnetic field or the current applied to the superconductor changes, the magnetic flux distribution in it also changes. This change induces an electromotive force and results in the energy loss. The input power density under the motion of flux lines driven by the Lorentz force is given by Eq. (2.34). According to the critical state model in which the phenomenon is assumed to be completely irreversible, the power loss density is written as

$$\boldsymbol{J} \cdot \boldsymbol{E} = \widehat{v} F_\mathrm{p}(\widehat{B}) + \eta \frac{\widehat{B}}{\phi_0} \widehat{v}^2 > 0 \qquad (2.73)$$

with the aid of Eq. (2.13), where $\widehat{v} = |\boldsymbol{v}|$. The first and second terms on the right-hand side in Eq. (2.73) are the pinning power loss density and the viscous power loss density, respectively, and both quantities are positive. Since the viscous force is generally much smaller than the pinning force, the viscous loss is not treated in this section. In what follows we consider an AC magnetic field to be applied parallel to a wide superconducting slab of thickness $2d$ as in Sect. 2.5.

If we again use the Irie-Yamafuji model [6] given by Eq. (2.46) for the pinning force density $F_\mathrm{p}(\widehat{B})$, the pinning power loss density is given by

$$P(x,t) = \alpha_\mathrm{c} \widehat{B}^\gamma \widehat{v} . \qquad (2.74)$$

Equation (2.15) is used for eliminating $\widehat{v}$. From the fact that $\widehat{v} = 0$ at the breaking point of the distribution, $x = x_\mathrm{b}$, Eq. (2.74) can be written as

$$P(x,t) = -\alpha_\mathrm{c} \widehat{B}^{\gamma-1} \int_{x_\mathrm{b}}^{x} \delta \frac{\partial \widehat{B}}{\partial t} \mathrm{d}x . \qquad (2.75)$$

The variation in magnetic flux distribution penetrates always from the surface of the superconductor and ends at the breaking point of the distribution. In the region beyond this breaking point the magnetic flux distribution does not vary and $\widehat{v}$ and E are zero. When the viscous force can be neglected, the time variation of the magnetic flux density $\widehat{B}$ in the superconductor comes only from that of its value at the surface, $\widehat{B}_0$. If the diamagnetic surface current and the surface irreversibility are neglected as assumed in Sect. 2.5, $\widehat{B}_0$ is approximately given by $\mu_0 \widehat{H}_\mathrm{e}$ as in Eq. (2.48), and the kernel in the integral in Eq. (2.75) may be written as

$$\delta \frac{\partial \widehat{B}}{\partial t} = \delta \frac{\partial \widehat{B}}{\partial \widehat{H}_\mathrm{e}} \cdot \frac{\partial \widehat{H}_\mathrm{e}}{\partial t} = \delta_0 \mu_0^{2-\gamma} \left(\frac{\widehat{H}_\mathrm{e}}{\widehat{B}} \right)^{1-\gamma} \frac{\partial \widehat{H}_\mathrm{e}}{\partial t} . \qquad (2.76)$$

Hence, Eq. (2.75) reduces to

$$P(x,t) = \frac{\partial \widehat{H}_e}{\partial t} \delta_0 \left(\frac{\mu_0 \widehat{H}_e}{\widehat{B}} \right)^{1-\gamma} (\delta \widehat{B} - \delta_b \widehat{B}_b) \,, \qquad (2.77)$$

where δ_b and $\widehat{B}_b$ are the values of δ and $\widehat{B}$ at the breaking point, $x = x_b$. Equation (2.77) gives the instantaneous pinning power loss density $P(x)$ at an arbitrary point in the superconductor. Examples of $P(x)$ distributions are shown in Fig. 2.24. The value of $P(x)$ diverges at the annihilation point of flux lines, i.e. at $\widehat{B} = 0$ in case $\gamma \neq 1$. This divergence in $P(x)$ originates from the divergence of the critical current density J_c at $\widehat{B} = 0$ which in turn is the inevitable result of the use of Eq. (2.46) as a simple approximation for the magnetic field dependence of J_c. However, even if the local power loss density diverges, the average value is finite as will be shown later. This is similar to the fact that the average critical current density is finite, even if the local value diverges. Hence, such a divergence in the power loss density is not a

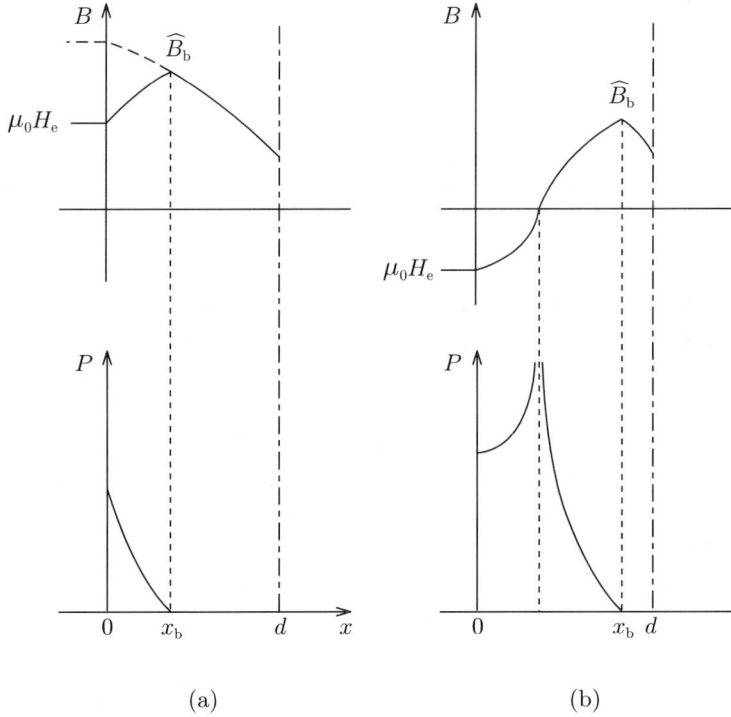

Fig. 2.24. Distributions of magnetic flux (*upper figure*) and pinning power loss density (*lower figure*) in a decreasing field for (**a**) $H_e > 0$ and (**b**) $H_e < 0$. The pinning power loss density diverges at the point $B = 0$ according to Irie-Yamafuji model [6]

serious problem. This result approximates rather well the tendency for the power loss density to be large when $\widehat{B} \sim 0$ and J_c is also large.

After averaging Eq. (2.77) in space and integrating it with time during a period of one cycle of the AC magnetic field, the pinning energy loss density is obtained:

$$W = \frac{1}{d} \int dt \int_0^{x_b} P(x,t) dx$$
$$= \frac{\mu_0^{1-\gamma}}{d} \int d\widehat{H}_e \delta_0 \widehat{H}_e^{1-\gamma} \int_0^{x_b} \widehat{B}^{\gamma-1}(\delta \widehat{B} - \delta_b \widehat{B}_b) dx \ . \quad (2.78)$$

In the above we used the fact that the loss occurs only in the region from the surface ($x = 0$) to the breaking point of the distribution ($x = x_b$). The integral with respect to $\widehat{H}_e$ is taken for a period of one cycle of the AC field. After integrating with respect to space we have

$$W = \frac{1}{2\alpha_c \mu_0^\gamma d} \int \delta_0 \widehat{H}_e^{1-\gamma}(\delta_0 \mu_0 \widehat{H}_e - \delta_b \widehat{B}_b)^2 d\widehat{H}_e \ . \quad (2.79)$$

In determining W from Eq. (2.79) there are three cases depending on the magnitude of AC field amplitude, H_m, for the same reason that the calculation of magnetization in Sect. 2.5 must be subdivided into three regions, i.e. for which: $H_m < H_p$, $H_p < H_m < 2^{1/(2-\gamma)} H_p$ and $H_m > 2^{1/(2-\gamma)} H_p$. In this section we treat the simplest case, $H_m < H_p$. Calculation of the pinning energy loss density in the other cases is given as an exercise. We divide the integral in Eq. (2.79) into (i) the region where $H_e = \widehat{H}_e$ varies from H_m to zero ($\delta_0 = \delta_b = -1$) and (ii) the region where H_e varies from zero to $-H_m$ ($\widehat{H}_e$ varies from zero to H_m, $\delta = 1$ and $\delta_b = -1$). By symmetry the pinning energy loss density is double the sum of the two contributions. Thus

$$W = \frac{2(2-\gamma)}{3} K(\gamma) \frac{\mu_0 H_m^{4-\gamma}}{H_p^{2-\gamma}} \ , \quad (2.80)$$

where $K(\gamma)$ is a function of γ defined by

$$K(\gamma) = 3 \left\{ \frac{2}{4-\gamma} - 2^{-1/(2-\gamma)} \int_0^1 \zeta^{2-\gamma}[(1+\zeta^{2-\gamma})^{1/(2-\gamma)} \right.$$
$$\left. - (1-\zeta^{2-\gamma})^{1/(2-\gamma)}] d\zeta \right\} \ . \quad (2.81)$$

As shown in Fig. 2.25, $K(\gamma)$ depends only weakly on γ and is 1.02 ± 0.02. It is seen from the above result that the energy loss density is finite even for $\gamma \neq 1$, although the local pinning power loss density diverges at the point where $\widehat{B} = 0$. Equation (2.80) shows that the pinning energy loss density increases with the AC field amplitude H_m in proportion to its $(4-\gamma)$-th power for

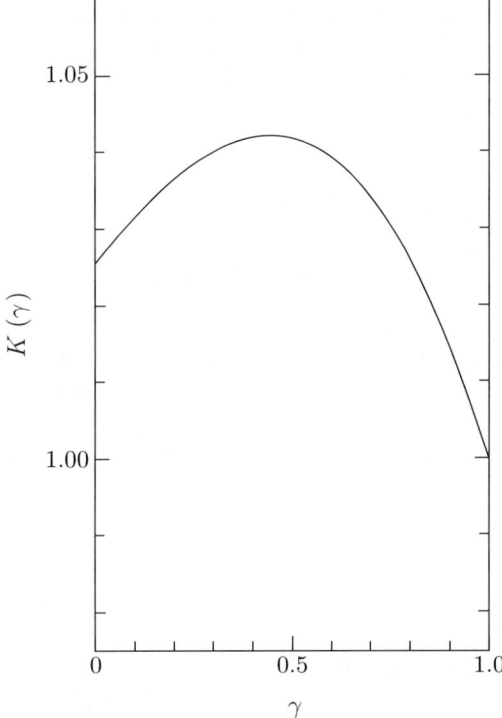

Fig. 2.25. Function $K(\gamma)$ defined by Eq. (2.81)

$H_{\mathrm{m}} < H_{\mathrm{p}}$. It is inversely proportional to α_{c} and hence assumes smaller values for more strongly pinned specimens.

On the other hand, the pinning energy loss density W increases more weakly than linearly with increasing H_{m} for $H_{\mathrm{m}} \gg H_{\mathrm{p}}$. In case $\gamma = 1$, W is approximately proportional to H_{m}. Its value is larger for more strongly pinned specimens. The AC field amplitude dependence of the energy loss density is shown in Fig. 2.26 for various values of γ.

Here we shall demonstrate an approximate method for calculating the pinning energy loss density. The pinning energy loss density is given by a product of the pinning force per unit length of the flux line, the distance of motion of flux lines and the density of moving flux lines. The pinning force on each flux line per unit length is $\phi_0 J_{\mathrm{c}}$, the mean distance by which the flux lines move during a half cycle is of the order of the maximum penetration depth, $H_{\mathrm{m}}/J_{\mathrm{c}}$, and the flux lines move twice in one cycle. The density of the moving flux lines is approximately given by the mean density of flux lines, $\mu_0 H_{\mathrm{m}}/2\phi_0$, multiplied by the fraction of the region where the flux distribution changes, viz. $H_{\mathrm{m}}/J_{\mathrm{c}}d$. Thus, we have

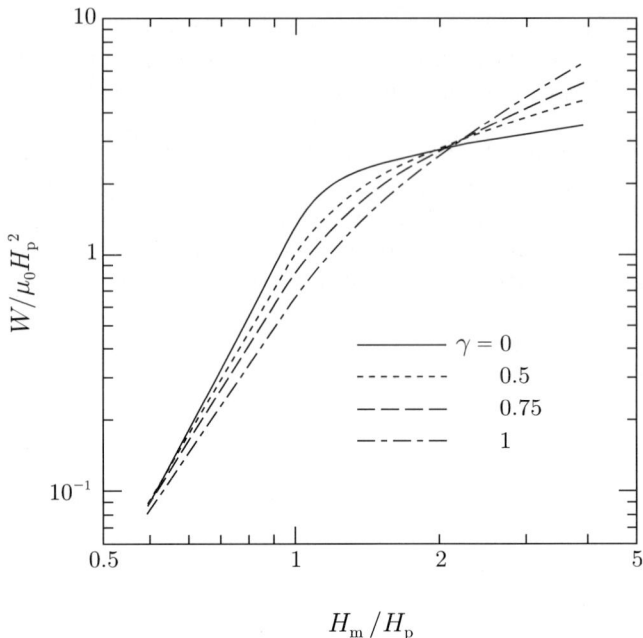

Fig. 2.26. AC field amplitude dependence of the pinning energy loss density for various values of γ

$$W \sim \phi_0 J_c \cdot \frac{H_m}{J_c} \cdot 2 \cdot \frac{\mu_0 H_m^2}{2\phi_0 J_c d} = \frac{\mu_0 H_m^3}{J_c d}. \tag{2.82}$$

This result is 3/2 times as large as the result in Eq. (2.80) in case $\gamma = 1$ (the Bean-London model) and agrees fairly well the exact calculation. The reason why the stronger pinning brings about the smaller energy loss is that the stronger pinning causes the density of moving flux lines and the distance of their motion to be smaller, although it makes the force larger. On the other hand, for $H_m \gg H_p$, the mean distance of the flux motion in a quarter cycle is approximately $d/2$ and the density of moving flux lines is about $\mu_0 H_m/\phi_0$. Hence, a calculation similar to the above yields

$$W \sim 2\mu_0 H_m J_c d = 2\mu_0 H_m H_p. \tag{2.83}$$

This result agrees approximately with the theoretical result in case $\gamma = 1$,

$$W = 2\mu_0 H_m H_p \left(1 - \frac{2H_p}{3H_m}\right). \tag{2.84}$$

As is seen in the above, the energy loss density can be estimated rather correctly even by a rough calculation. This result comes from the fact that the energy loss density is a quantity that is averaged with respect to space and time. The result that the energy loss density does not largely depend on the pinning property represented by γ also stems from the same reason.

In this section the pinning energy loss density is derived from Eq. (2.74) based on the critical state model. The merit of this method is that the instantaneous power loss density at an arbitrary point can be obtained. On the other hand, the areas of the hysteretic M-H loops or B-H loops are sometimes used to obtain the total energy loss in the specimen or its mean density. Although the latter method is much simpler than the former one, it can be used only in the steady repetitional case. For example, the energy loss in a superconducting magnet in the initial energizing process cannot be obtained by the simple method and we have to use the former method starting from Eq. (2.74). At the same time only the energy loss averaged over space and time can be obtained from the method using the area of hysteresis curves. But only for calculating the pinning energy loss density as in Eq. (2.80), the simpler method seems to be more useful.

It should be noted that Eq. (2.74) is correct only when the pinning is completely irreversible. As briefly mentioned in Sect. 2.4, when the flux motion is more or less confined within the pinning potential, the phenomenon is almost reversible and the critical state model cannot be used. In this case the magnetic energy stored in the superconductor is contained in $\boldsymbol{J}\cdot\boldsymbol{E}$. Therefore, the energy loss is generally obtainable only from the area of hysteresis curves and only in the steady repetitional, e.g. AC, case. However, if the irreversible component can be separated from the pinning force density as will be shown in Sect. 3.7, the energy loss can be calculated in terms of the work done by this component in the same manner as in Eq. (2.74).

Exercises

2.1. Derive the continuity equation (2.15) for flux lines by obtaining directly the magnetic flux which penetrates a loop C. (*Hint*: Express the magnetic flux coming in the loop by a curvilinear integral on the loop and transform it into the surface integral using Stokes' theorem.)

2.2. Prove that the input power given by Eq. (2.38) is equal to the viscous power loss, $\langle \eta^* \dot{u}^2 \rangle_t$.

2.3. Yamafuji and Irie used the expression

$$\langle \eta^* v - k_{\mathrm{f}}(u - u_0) \rangle_t v$$

for the input power on the flux line. Discuss the condition in which this coincides with the general expression in Eq. (2.38).

2.4. Discuss the motion of flux lines in the resistive state in a superconducting cylinder when the transport current only is applied without external magnetic field.

2.5. Discuss the distributions of B and $\mathcal{H}$ when the external magnetic field is reversed, following the discussion in Sect. 2.6.

2.6. Derive the pinning energy loss density in case where the AC magnetic field of amplitude H_m is applied parallel to a wide superconducting slab of thickness $2d$. Assume the Irie-Yamafuji model for the pinning force density and calculate in cases of $H_p < H_m < 2^{1/(2-\gamma)} H_p$ and $H_m > 2^{1/(2-\gamma)} H_p$.

2.7. Calculate the magnetization of a wide superconducting slab of thickness $2d$ in a parallel magnetic field. Use the model of Kim et al. [18]

$$J_c = \frac{\alpha_0}{|B| + \beta}$$

for the critical current density, where α_0 and β are constants.

2.8. Calculate the pinning energy loss density from the area of a loop of the magnetization curve using the Bean–London model ($\gamma = 1$) for $H_m < H_p$.

2.9. The pinning energy loss density in the case where the AC magnetic field of amplitude H_m is superposed to a sufficiently large DC field H_0 is given by Eq. (2.80) with $\gamma = 1$. Derive this result approximately from a similar estimate as in Eq. (2.82). (*Hint*: Estimate the displacement of flux lines using the continuity equation (2.15).)

2.10. Calculate the AC energy loss density when the AC magnetic field is applied parallel to a wide superconducting slab of thickness $2d$ in which the diamagnetism cannot be disregarded. Use the Bean–London model ($\gamma = 1$). It is assumed that the AC field amplitude is smaller than the penetration field and the magnetic flux density just inside the surface is given by

$$\begin{aligned} B_0 &= \mu_0(H_e - H_{c1}); & H_e &> H_{c1}, \\ &= 0; & H_{c1} &> H_e > -H_{c1}, \\ &= \mu_0(H_e + H_{c1}); & -H_{c1} &> H_e. \end{aligned}$$

References

1. M. Tinkham: *Introduction to Superconductivity* (McGraw-Hill, New York, 1996) p. 154.
2. A. G. van Vijfeijken: Phil. Res. Rep. Suppl. No. **8** (1968) 1.
3. J. I. Gittleman and B. Rosenblum: Phys. Rev. Lett. **16** (1966) 734.
4. B. D. Josephson: Phys. Rev. **152** (1966) 211.
5. W. K. H. Panofsky and M. Phillips: *Classical Electricity and Magnetism* (Addison-Wesley Pub., Massachusetts, 1964) p. 162.
6. F. Irie and K. Yamafuji: J. Phys. Soc. Jpn. **23** (1967) 255.
7. B. D. Josephson: Phys. Lett. **16** (1965) 242.
8. T. Matsushita, Y. Hasegawa and J. Miyake: J. Appl. Phys. **54** (1983) 5277.
9. M. Tinkham: *Introduction to Superconductivity* (McGraw-Hill, New York, 1996) p. 166.
10. J. Bardeen and M. J. Stephen: Phys. Rev. **140** (1965) A1197.
11. Y. B. Kim, C. F. Hempstead and A. R. Strnad: Phys. Rev. Lett. **13** (1964) 794.

12. K. Yamafuji and F. Irie: Phys. Lett. **25A** (1967) 387.
13. T. Matsushita, E. Kusayanagi and K. Yamafuji: J. Phys. Soc. Jpn. **46** (1979) 1101.
14. C. P. Bean: Phys. Rev. Lett. **8** (1962) 250.
15. H. London: Phys. Lett. **6** (1963) 162.
16. K. Yasukoch, T. Ogasawara and N. Ushino: J. Phys. Soc. Jpn. **19** (1964) 1649.
17. J. Silcox and R. W. Rollins: Rev. Mod. Phys. **36** (1964) 52.
18. Y. B. Kim, C. F. Hempstead and A. R. Strnad: Phys. Rev. **129** (1963) 528.
19. G. J. C. Bots, J. A. Pals, B. S. Blaisse, L. N. J. de Jong and P. P. J. van Engelen: Physica **31** (1965) 1113.
20. P. S. Swartz: Phys. Rev. Lett. **9** (1962) 448.
21. P. H. Kes, C. A. M. van der Klein and D. de Klerk: J. Low Tem. Phys. **10** (1973) 759.
22. D. G. Walmsley: J. Appl. Phys. **43** (1972) 615.
23. A. M. Campbell and J. E. Evetts: Adv. Phys. **21** (1972) 279.
24. T. Matsushita and K. Yamafuji: J. Phys. Soc. Jpn. **46** (1979) 764.

3

Various Electromagnetic Phenomena

3.1 Geometrical Effect

In the last chapter the magnetization and AC loss in a wide superconducting slab were calculated. In this section we discuss the electromagnetic phenomena in a superconductor with other geometries. The cases are treated where the current, the transverse AC field or the transverse rotating field is applied to a cylindrical superconductor.

3.1.1 Loss in Superconducting Wire due to AC Current

We assume that AC current is applied to a straight superconducting cylinder of radius R without external magnetic field. In this case only the self field in the azimuthal direction exists. If the magnitude of the AC current is denoted by $I(t)$, the value of the self field at the surface, $r = R$, is given by

$$H_\mathrm{I} = \frac{I}{2\pi R} \,. \tag{3.1}$$

The penetration of the azimuthal flux lines due to the self field is also described by the critical state model as in Sect. 2.5. We assume again the Irie-Yamafuji model [1] given by Eq. (2.46) for the magnetic field dependence of the pinning force density. The azimuthal magnetic flux density and its magnitude are represented by B and $\widehat{B}$, respectively. The force balance equation in the quasistatic process is described as

$$-\frac{\widehat{B}}{\mu_0 r} \cdot \frac{\mathrm{d}}{\mathrm{d}r}(r\widehat{B}) = \delta \alpha_\mathrm{c} \widehat{B}^\gamma \,, \tag{3.2}$$

where δ is a sign factor indicating the direction of the flux motion, e.g. $\delta = 1$ indicates that flux lines move in the radial direction. Equation (3.2) can be easily solved yielding for the magnetic flux distribution:

$$\delta(r\widehat{B})^{2-\gamma} = \delta_R(R\mu_0\widehat{H}_{\rm I})^{2-\gamma} + \frac{2-\gamma}{3-\gamma}\alpha_c\mu_0(R^{3-\gamma} - r^{3-\gamma}), \tag{3.3}$$

where $\widehat{H}_{\rm I} = |H_{\rm I}|$, and δ_R is the value of δ at the surface ($r = R$), and the boundary condition

$$B(r = R) = \mu_0 H_{\rm I} \tag{3.4}$$

was used.

The energy loss can be calculated from Eq. (2.74) as was done previously. But we calculate it more easily in terms of Poynting's vector. Since the induced electric field $\boldsymbol{E}$ and the magnetic flux density $\boldsymbol{B}$ are expressed as $(0, 0, E)$ and $(0, B, 0)$ from symmetry, Poynting's vector, $(\boldsymbol{E} \times \boldsymbol{B})/\mu_0$, at the surface is directed negative radially, and hence, towards the inside of the superconductor. Then, the energy loss density per cycle of the AC current is written as

$$\begin{aligned} W &= \frac{2}{R\mu_0} \int {\rm d}t E(R,\ t) B(R,\ t) \\ &= \frac{2}{R\mu_0} \int {\rm d}t B(R,\ t) \int_0^R \frac{\partial}{\partial t} B(r,\ t) {\rm d}r, \end{aligned} \tag{3.5}$$

where the integral with respect to time is carried out for the period of one cycle. From symmetry we have only to double the contribution from the period in which the current varies from the maximum value, $I_{\rm m}$, to $-I_{\rm m}$. If the maximum self field is denoted by $H_{\rm m} = I_{\rm m}/2\pi R$, this half cycle is divided into the periods (i) and (ii) in which $H_{\rm I}$ changes from $H_{\rm m}$ to 0 and from 0 to $-H_{\rm m}$ as shown in Fig. 3.1(a) and 3.1(b), respectively. In period (i), $B > 0$ and $\delta = 1(\delta_R = 1)$ in the entire area, $r_{\rm b1} \leq r \leq R$, in which the magnetic flux distribution changes. On the other hand, in period (ii), we have $\delta_R = -1$ and $B > 0$ and $\delta = 1$ for $r_{\rm b2} \leq r \leq r_0$, while $B < 0$ and $\delta = -1$ for $r_0 < r \leq R$. From Eq. (3.3) the critical current is given by

$$I_c = 2\pi \left(\frac{2-\gamma}{3-\gamma}\alpha_c\mu_0^{\gamma-1}R^{3-\gamma}\right)^{1/(2-\gamma)}. \tag{3.6}$$

If the corresponding self field is denoted by

$$H_{\rm Ip} = \frac{I_c}{2\pi R}, \tag{3.7}$$

$r_{\rm b1}$, $r_{\rm b2}$ and r_0 are respectively given by

$$1 - \left(\frac{r_{\rm b1}}{R}\right)^{3-\gamma} = \frac{1}{2H_{\rm Ip}^{2-\gamma}}(H_{\rm m}^{2-\gamma} - \widehat{H}_{\rm I}^{2-\gamma}), \tag{3.8a}$$

$$1 - \left(\frac{r_{\rm b2}}{R}\right)^{3-\gamma} = \frac{1}{2H_{\rm Ip}^{2-\gamma}}(H_{\rm m}^{2-\gamma} + \widehat{H}_{\rm I}^{2-\gamma}), \tag{3.8b}$$

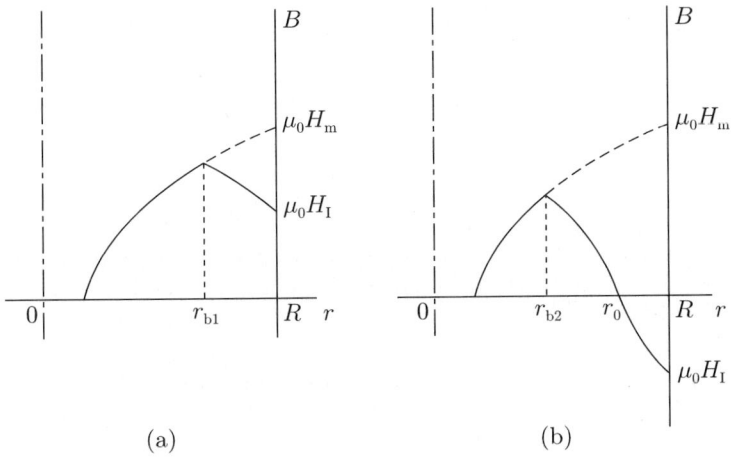

Fig. 3.1. Distribution of azimuthal magnetic flux in a superconducting cylinder while the self field due to the AC current changes from H_m to $-H_m$. (**a**) and (**b**) correspond to cases where the current flows in the positive and negative z-axis directions, respectively

$$1 - \left(\frac{r_0}{R}\right)^{3-\gamma} = \left(\frac{\widehat{H}_I}{H_{Ip}}\right)^{2-\gamma}. \tag{3.8c}$$

Since the variation in the magnetic flux distribution with respect to time comes only from the variation in H_I, Eq. (3.5) reduces to

$$W = 4\mu_0 H_{Ip}^2 \int_0^{h_m} dh_I h_I^{2-\gamma} \left[-\int_{x_1}^1 \frac{dx}{x}(1 + h_I^{2-\gamma} - x^{3-\gamma})^{(\gamma-1)/(2-\gamma)} \right.$$
$$+ \int_{x_2}^{x_0} \frac{dx}{x}(1 - h_I^{2-\gamma} - x^{3-\gamma})^{(\gamma-1)/(2-\gamma)}$$
$$\left. + \int_{x_0}^1 \frac{dx}{x}(x^{3-\gamma} - 1 + h_I^{2-\gamma})^{(\gamma-1)/(2-\gamma)} \right], \tag{3.9}$$

where

$$h_m = \frac{H_m}{H_{Ip}}, \qquad h_I = \frac{\widehat{H}_I}{H_{Ip}}, \tag{3.10}$$

$$x_0 = \frac{r_0}{R}, \qquad x_1 = \frac{r_{b1}}{R}, \qquad x_2 = \frac{r_{b2}}{R}. \tag{3.11}$$

An analytic calculation can be carried out only for $\gamma = 1$, yielding [2]

$$W = 4\mu_0 H_{Ip}^2 \left[h_m \left(1 - \frac{h_m}{2}\right) + (1 - h_m)\log(1 - h_m) \right]. \tag{3.12}$$

This value reduces to $W \simeq 4\mu_0 H_m^3/\alpha_c R$ for $h_m \ll 1$ and amounts to double that value for the superconducting slab, i.e., the value given by Eq. (2.80)

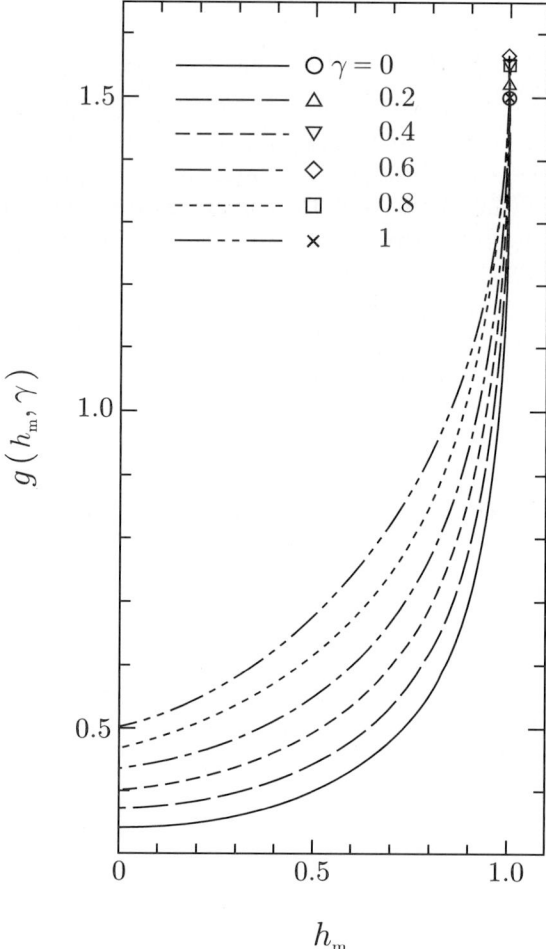

Fig. 3.2. Function $g(h_{\mathrm{m}}, \gamma)$

with $\gamma = 1$, where d is approximately replaced by R. This comes from the fact that the surface region where the energy dissipation occurs is relatively wider for the case of superconducting cylinder. The energy loss density is expressed as [3]

$$W = \frac{4}{3}(2 - \gamma)\mu_0 g(h_{\mathrm{m}}, \gamma)\frac{H_{\mathrm{m}}^{4-\gamma}}{H_{\mathrm{Ip}}^{2-\gamma}} \tag{3.13}$$

analogously to Eq. (2.80), where g is given by a double integral and is a function of h_{m} and γ as shown in Fig. 3.2. When γ is a rational number, g can be expressed in the form of a single integral.

3.1.2 Loss in Superconducting Wire of Ellipsoidal Cross Section and Thin Strip due to AC Current

Norris [4] calculated the loss in a superconducting wire with an ellipsoidal cross section and a thin superconducting strip due to AC current using the Bean-London model ($\gamma = 1$) [5]. According to the calculated result, the loss in the ellipsoidal wire of the cross sectional area S is essentially the same as that in a cylindrical wire given by Eq. (3.12). In terms of the current, it leads to

$$W = \frac{\mu_0 I_c^2}{\pi S} \left[i_m \left(1 - \frac{i_m}{2}\right) + (1 - i_m) \log(1 - i_m) \right] \tag{3.14}$$

with the normalized current amplitude:

$$i_m = \frac{I_m}{I_c} . \tag{3.15}$$

In the case of a thin superconducting strip of the cross sectional area S, the loss is given by

$$W = \frac{\mu_0 I_c^2}{\pi S} [(1 - i_m) \log(1 - i_m) + (1 + i_m) \log(1 + i_m) - i_m^2] . \tag{3.16}$$

The AC losses in the superconducting ellipsoidal wire and the thin strip are shown in Fig. 3.3. The loss in the ellipsoidal wire approaches that of the equivalent slab at small current amplitudes, while the current amplitude

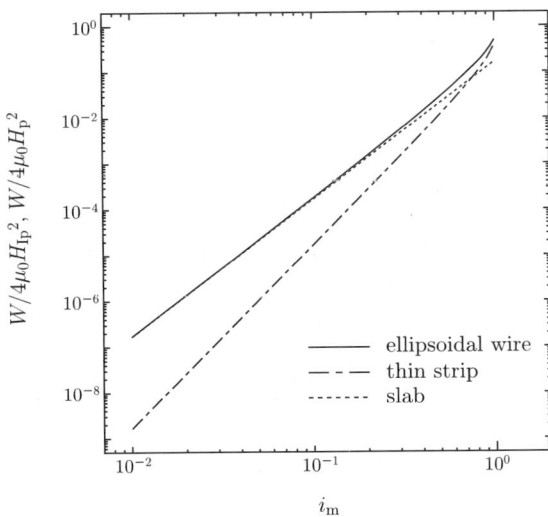

Fig. 3.3. Calculated AC current loss in superconducting ellipsoidal wire and thin strip [4] for the case of $\gamma = 1$. The *broken line* shows the loss in a superconducting slab

dependence is significantly different for the thin strip with a small loss at small current amplitudes.

3.1.3 Transverse Magnetic Field

We have treated the cases where the physical quantities depend only on one coordinate axis without being influenced by the geometrical factor such as a demagnetization coefficient of superconductor. In this subsection we treat the case where a transverse magnetic field is applied to a cylindrical superconductor. Because of the break in symmetry the physical quantities depend on two coordinate axes and we have to solve a two-dimensional problem.

A very small transverse magnetic field is supposed to be applied to a cylindrical superconductor. The initial state is assumed and the diamagnetism at the surface is disregarded for simplicity. The inner part of the superconductor is completely shielded and the magnetic flux density there is zero. The shielding current flows only in the vicinity of the surface. If the thickness of the region in which the shielding current flows is sufficiently small, an approximate solution can be obtained. We define the coordinates as shown in Fig. 3.4, where H_e is the uniform external magnetic field and R is the radius of the superconductor. After applying a method well known in electromagnetism we obtain the solution:

$$\left. \begin{array}{l} B_r = \mu_0 H_e \left(1 - \dfrac{R^2}{r^2}\right) \cos\theta \\[2mm] B_\theta = -\mu_0 H_e \left(1 + \dfrac{R^2}{r^2}\right) \sin\theta \end{array} \right\} ; \qquad r > R \qquad (3.17)$$

and

$$B_r = B_\theta = 0; \qquad 0 < r < R . \qquad (3.18)$$

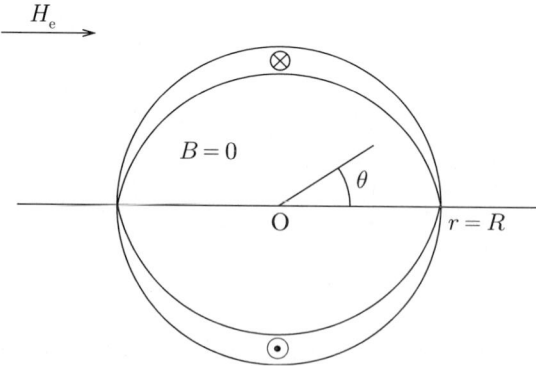

Fig. 3.4. Surface layer of shielding current in a superconducting cylinder in a small transverse magnetic field

The azimuthal magnetic flux density B_θ is not continuous at $r = R$ and the current corresponding to the difference flows along the z-axis on the surface of the superconductor. If we represent this surface current density by $\tilde{J}$ (Am^{-1}), we have

$$\tilde{J}(\theta) = -2H_e \sin\theta\ . \tag{3.19}$$

In practice the critical current density originating from flux pinning is finite and the thickness of the shielding-current region is also finite. If we use the Bean-London model [5] in which the critical current density is independent of the magnetic field, the thickness is $(2H_e/J_c)\,|\sin\theta\,|$.

When the magnetic field becomes much larger, the shielding-current region becomes wider and the completely shielded region becomes narrower as shown in Fig. 3.5. According to the critical state concept the distribution of the shielding current shown in Fig. 3.5(a) is determined so as to minimize the variation in the magnetic flux distribution inside the superconductor, i.e., to minimize the invasion of the magnetic flux. However, the analytic exact solution has not yet been obtained even for the simple Bean-London model. The detailed discussion on the magnetic flux distribution is given in [6–8]. Now the approximate schemes are used in which the region of shielding current is assumed to be of simple shape and determined by the condition that $B = 0$ is satisfied at some special points within the shielded region. In the simplest case a shielding-current region of a circular shape is assumed as shown by the broken line in Fig. 3.5(b) and its radius is determined by the condition that $B = 0$ at the center of the cylindrical superconductor. Even such a simplified approximation [9] with the Bean-London model leads to magnetization and loss due to the transverse AC magnetic field which are rather close to the results [8] of numerical analysis. In [10] the magnetization and the loss are analyzed using the Irie-Yamafuji model [1] for the magnetic field dependence

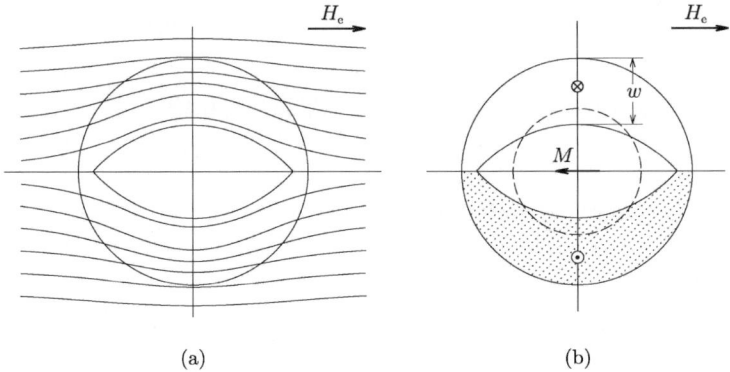

Fig. 3.5. (a) Magnetic structure and (b) shielding current distribution in a superconducting cylinder in large AC transverse magnetic field. Circular current layer shown by the *broken line* is sometimes assumed simply. M denotes the magnetization due to the shielding current

of the pinning force density, and the magnetic flux distribution is determined on the assumption that the magnitude of the shielding current density is a function only of the distance from the center of the cylinder. These calculated results are compared with experimental results in detail. According to the calculated result the AC loss in the range of small field amplitude is four times as large as Eq. (2.80) for a superconducting slab in a parallel field. This is caused by the fact that the amount of shielding current is enhanced due to the effect of demagnetization. That is, from an approximate estimate as in Eq. (2.82) the enhancement factor is calculated as the average of $(2\sin\theta)^3$ in the angular region $0 \leq \theta \leq \pi$, which is equal to $32/3\pi \simeq 3.4$. This is close to the analytical result, 4

When the transverse AC magnetic field becomes larger than the penetration field given by

$$H_{\mathrm{p}\perp} = \frac{1}{\mu_0}\left[\frac{2}{\pi}(2-\gamma)\mu_0\alpha_{\mathrm{c}}R\right]^{1/(2-\gamma)}, \tag{3.20}$$

the shielding current extends to the entire region of the cylinder and currents of the opposite directions flow in the upper and lower halves. In this case the exact solution has not yet been obtained except for $\gamma = 1$ where the magnetic flux distribution is uniquely determined. In case $\gamma \neq 1$ an approximate solution is obtained assuming that the magnitude of the shielding current depends only on the distance from the center of cylinder. The error in the hysteresis loss obtained from this result in comparison with the numerically calculated loss is within 10% [10] even in the vicinity of the penetration field where the error is largest. The losses obtained for various γ values are shown in Fig. 3.6.

3.1.4 Rotating Magnetic Field

We shall next consider the case where a transverse magnetic field is applied to a cylindrical superconductor and then rotated. Provided that the rotating angle is small, the rotation is almost identical to a superposition of a small magnetic field in the direction normal to the initial field. Hence, a new shielding current is induced by the superposed field. The net current distribution is obtained by superposition of the newly induced distribution upon the initial one. When the rotating angle becomes much larger, the current distribution must be obtained in a different way. Extrapolating from the distribution under the small rotating angle, the current distribution inside the superconductor in the steady state is deduced to be that shown in Fig. 3.7. Although this distribution is to be determined under the condition that $B = 0$ is satisfied throughout the entire shielded region, it cannot be generally determined correctly. In case $\gamma = 1$ where the magnitude of shielding current density is constant, a solution which satisfies $B = 0$ approximately in the shielded region can be obtained [11] only when the magnetic field is so small that the thickness of the shielding current layer is small. In case $\gamma \neq 1$ an approximate

3.1 Geometrical Effect   93

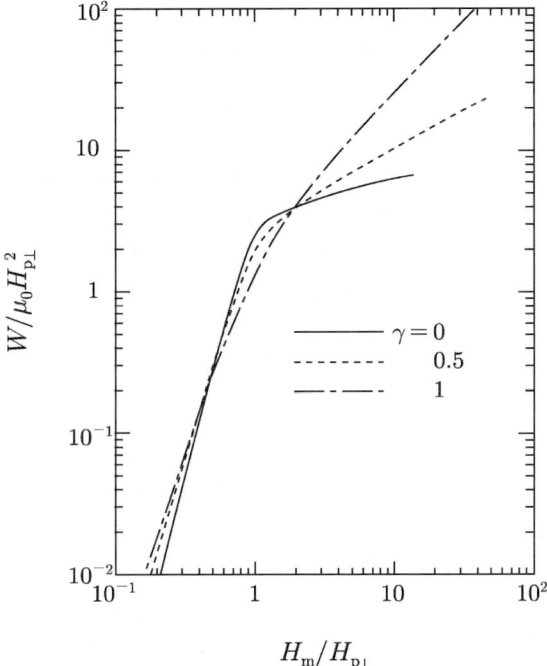

Fig. 3.6. Energy loss density in a superconducting cylinder due to AC transverse magnetic field [10]

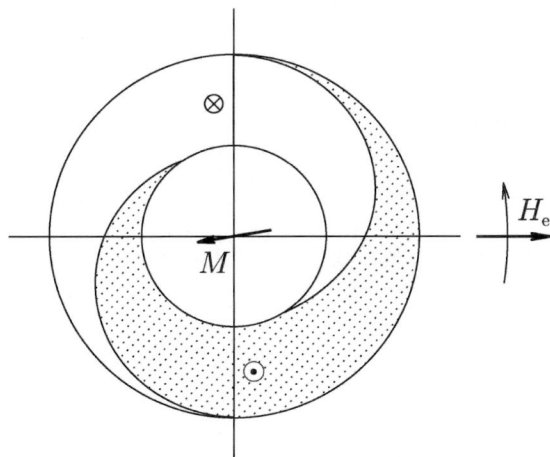

Fig. 3.7. Steady distribution of shielding current in a superconducting cylinder in a small rotating transverse magnetic field

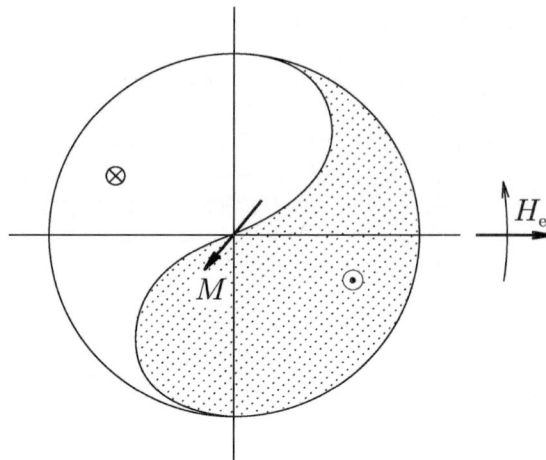

Fig. 3.8. Steady distribution of shielding current in a superconducting cylinder in a rotating transverse magnetic field of magnitude larger than the penetration field

solution is obtained [11] from the requirement that $B = 0$ at the center of cylinder with the assumption that the shielding current density depends only on the distance from the center. From the shielding current distribution the magnetic flux distribution is obtained. The induced electric field $\boldsymbol{E}$ can be calculated from the variation in the magnetic flux distribution and the loss is estimated from $\boldsymbol{J} \cdot \boldsymbol{E}$. The energy loss density [11] so obtained is $8/\pi$ times as large as the loss due to the transverse AC magnetic field with the same amplitude discussed in Subsect. 3.1.3, and hence, $32/\pi$ times as large as the value given by Eq. (2.80).

On the other hand, the current distribution in the steady state is shown in Fig. 3.8 where the magnetic flux penetrates up to the center in a transverse field greater than the penetration field $H_{\mathrm{p}\perp}$. In this case there is no region where the magnetic flux is completely shielded and the current distribution is determined using the condition that the electric field E is zero on the boundary of the two regions where the current flows are opposite to each other. This condition is based on the irreversibility in the critical state model which requires that the current and the electric field are in the same direction, i.e., $\boldsymbol{J} \cdot \boldsymbol{E} > 0$. The current distribution and the loss are calculated using the above-mentioned assumption that the current density depends only on the distance from the center.

In the intermediate region where the magnetic field is comparable to the penetration field, the approximate expression of the energy loss density is derived by interpolating the result in each region [11]. Agreement between this expression and the numerically calculated result [8] is obtained for case $\gamma = 1$.

3.2 Dynamic Phenomena

We have treated only the quasistatic process in which the variation in the magnetic flux distribution is very slow. This process is not the one used in thermodynamics but the one in which the variation in the inner magnetic flux distribution with time depends only on the time variation of the external sources such as the magnetic field. In this case the viscous force can be neglected and the magnetic flux distribution is determined from the balance between the Lorentz force and the pinning force. In this section we shall discuss the case where the external variable varies so quickly that the viscous force cannot be neglected.

For simplicity the one-dimensional problem is again treated where the external magnetic field is applied along the z-axis of a semi-infinite superconductor occupying $x \geq 0$. Since the force balance equation obtained by substitution of Eq. (2.46) into Eq. (2.13) is nonlinear, an analytic solution is not easily obtained [1]. We suppose that a small varying field, $h(t)$, is superposed on a large external magnetic field, H_e. The internal magnetic flux density is expressed as

$$B(x, t) = \mu_0 H_e + b(x, t) . \qquad (3.21)$$

In the above $b(x,t)$ is considered to be much smaller than $\mu_0 H_e$. If we assume that H_e is positive, B is also positive. The continuity equation for flux lines (2.15) is approximately rewritten as

$$\frac{\partial b}{\partial t} = -\mu_0 H_e \frac{\partial v}{\partial x} . \qquad (3.22)$$

The force balance equation (2.13) approximately reduces to

$$H_e \frac{\partial b}{\partial x} + \delta F_p(\mu_0 H_e) + \eta \frac{\mu_0 H_e}{\phi_0} v = 0 , \qquad (3.23)$$

where $v = \delta \hat{v}$ is used. Derivation of this equation with respect to x and elimination of v lead to a diffusion equation for b. The breaking point, x_b, is used as one of the boundary conditions to determine the magnetic flux distribution. However, this equation cannot be easily solved, since this boundary inside the superconductor varies with time.

In this section we treat the case where the viscous force is sufficiently small that the magnetic flux distribution can be approximately obtained by an iterative calculation from a quasistatic one. For example we assume that a slowly varying sinusoidal AC magnetic field of amplitude h_0 and frequency $\omega/2\pi$ is superposed on the DC field H_e. The condition required for the frequency will be discussed later. The boundary condition at the surface is given by

$$b(0, t) = \mu_0 h_0 \cos \omega t . \qquad (3.24)$$

When the viscous force can be neglected, the magnetic flux distribution is obtained from Eq. (3.23) as

$$b(x,t) = \mu_0(h_0 \cos\omega t - \delta J_c x)$$
$$\equiv b_0(x,t); \qquad 0 < x < x_{b0}, \tag{3.25}$$

where $F_p(\mu_0 H_e) = \mu_0 H_e J_c$ with J_c denoting the constant critical current density, $\delta = -\mathrm{sign}(\sin\omega t)$ and

$$x_{b0} = \frac{h_0}{2J_c}(1 + \delta\cos\omega t). \tag{3.26}$$

From Eqs. (3.22) and (3.25) we have

$$v = -\frac{1}{\mu_0 H_e}\int_{x_{b0}}^{x} \frac{\partial b_0}{\partial t}\,dx = \frac{h_0}{H_e}\omega\sin\omega t(x - x_{b0}). \tag{3.27}$$

Substitution of Eq. (3.27) into the third term in Eq. (3.23) leads to

$$b(x,t) = b_0(x,t) - \frac{\eta\mu_0 h_0\omega}{2\phi_0 H_e}\sin\omega t(x^2 - 2x_{b0}x). \tag{3.28}$$

This solution holds in the region from the surface to the breaking point of the magnetic flux distribution x_b, which is slightly different from x_{b0} in Eq. (3.26). The new breaking point is obtained as a crossing point between the distribution given by Eq. (3.28) and the "previous" distribution. Since the distribution given by Eq. (3.28) agrees with the quasistatic distribution at $\omega t = n\pi$, with n denoting an integer at which the sign factor δ changes, the "previous" distribution is the quasistatic one. Hence, after a simple calculation we have approximately

$$x_b = x_{b0} - \frac{\eta h_0\omega}{4\phi_0 H_e J_c}|\sin\omega t|x_{b0}^2 \tag{3.29}$$

up to the first order in ω. The second term in Eq. (3.29) should be smaller than the first so that the iterative approximation holds true. Since x_{b0} becomes as large as h_0/J_c, the condition for the frequency is written as

$$\omega \ll \frac{4\phi_0 H_e J_c^2}{\eta h_0^2} \equiv \omega_0. \tag{3.30}$$

The AC component of the magnetic flux density averaged over the superconductor in the period $0 \leq \omega t \leq \pi$ is to first order in ω given by

$$\langle b \rangle = \frac{\mu_0 h_0^2}{4J_c d}\left[\sin^2\omega t + 2\cos\omega t + \frac{2\omega}{3\omega_0}\sin\omega t(1 - \cos\omega t)^3\right]. \tag{3.31}$$

From symmetry the energy loss density per cycle of the AC field is

$$W = 2\int_{-h_0}^{h_0}\langle b\rangle\,d(h_0\cos\omega t) = \frac{2\mu_0 h_0^3}{3J_c d}\left(1 + \frac{7\pi\omega}{16\omega_0}\right). \tag{3.32}$$

The first term is the pinning energy loss density in the quasistatic process; it agrees with the result of Eq. (2.80) after substituting $\gamma = 1$. The second term is the viscous energy loss density. The reason why the pinning energy loss density is not different from that in the quasistatic case is that the magnetic flux distributions at $\omega t = 0$ and π are the same as those in the quasistatic case, i.e., the amount of magnetic flux which contributes to the pinning loss during one cycle is unchanged. The second term in Eq. (3.32) can also be calculated directly from the second term in Eq. (2.73) as the viscous energy loss density (see Exercise 3.2).

When the frequency of the AC magnetic field becomes higher, it is necessary to take into account terms to higher order in ω. In this case the "previous" distribution in the region where $x_b < x$ varies with time, and hence, the calculation becomes extremely complicated. According to the theoretical analysis of Kawashima et al. [12] the energy loss density in this case is predicted to be

$$W = \frac{2\mu_0 h_0^3}{3 J_c d} \left[1 + \frac{7\pi\omega}{16\omega_0} - \frac{512}{105} \left(\frac{\omega}{\omega_0} \right)^2 \right]. \qquad (3.33)$$

The decrease in the energy loss density at high frequencies is caused by the fact that the amount of moving flux decreases due to the stronger shielding action of the viscous force.

3.3 Superposition of AC Magnetic Field

3.3.1 Rectifying Effect

When a small AC magnetic field is applied to a current-carrying superconducting wire or tape in a transverse DC field, the current-voltage characteristics vary with a decrease in the critical current density [13, 14] as shown in Fig. 3.9. Sometimes the critical current density reduces to zero. This is commonly observed independently of whether the AC field is parallel or normal to the DC field. Here we shall first argue the case of parallel AC field. The current-voltage characteristics in this case can also be analyzed in terms of a magnetic flux distribution in the superconductor as predicted by the critical state model. The magnetic flux distribution is predicted to vary with the surface field during one cycle of AC field as shown in Fig. 3.10. The arrows in the figure represent the direction of flux motion. It is seen that the flux motion is not symmetric. That is, the amount of flux that moves from the left to the right is larger than that in the opposite direction and a DC component of electric field appears due to the rectifying effect of flux flow [13]. Strictly speaking, Eq. (2.13) should be used, since a resistive state is being treated. But here, for simplicity the viscous force is disregarded. When the pinning force is strong as in a commercial superconductor, this approximation is valid within the practical range of the electric field.

We assume a superconducting slab of width $2d$ carrying a transport current of density J_t. The Bean-London model ($\gamma = 1$ and $J_c = $ const.) is used for

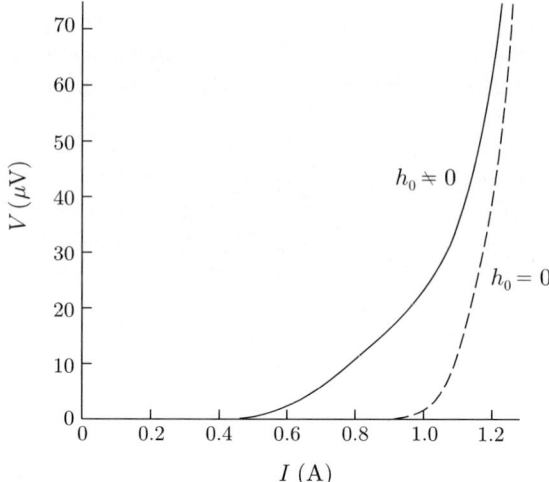

Fig. 3.9. Current-voltage characteristics in a superconducting Pb-Bi foil with (*solid line*) and without (*broken line*) superposed small AC magnetic field perpendicular to both the normal DC magnetic field and the current [14]

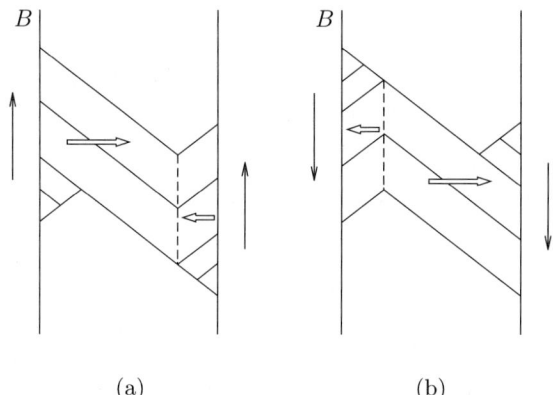

Fig. 3.10. Explanation of rectifying effect by the Kaiho model [13] in case where DC and AC magnetic fields are parallel to a superconducting slab. (**a**) and (**b**) show magnetic flux distributions in the phases of increasing and decreasing AC magnetic field, respectively

the pinning force density. The net magnetic flux Φ that flows from the left to the right during one cycle of the AC field of amplitude h_0 corresponds to the area of hatched region in Fig. 3.11 and can be calculated as

$$\Phi = 4\mu_0 j[h_0 - H_p(1-j)]d, \quad (3.34)$$

where $H_p = J_c d$ is the penetration field and $j = J_t/J_c$. Hence, the average value of the electric field is given by

3.3 Superposition of AC Magnetic Field

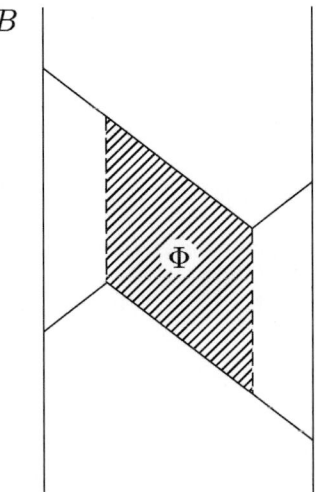

Fig. 3.11. *Shadowed region* corresponds to the magnetic flux passing through the superconducting slab during one cycle of AC magnetic field

$$\overline{E} = \Phi f , \qquad (3.35)$$

where f is the frequency of the AC field. The apparent critical current density J_c^* is obtained from the condition $\overline{E} = 0$ as

$$J_c^* = J_c \left(1 - \frac{h_0}{H_p}\right) . \qquad (3.36)$$

Hence, $J_c^* = 0$ for $h_0 > H_p$.

Now we shall estimate the energy loss in the resistive state. One part of the dissipated energy is supplied by the DC current source and is given by $W_c = J_t \overline{E}/f = J_t \Phi$ per unit volume. The other part is supplied by the AC magnet and its value per unit volume is given by

$$W_f = \int \langle B \rangle \mathrm{d}H(t) , \qquad (3.37)$$

where $\langle B \rangle$ is the magnetic flux density averaged over the superconducting slab and $H(t)$ is the instantaneous value of the AC magnetic field. After a simple calculation we have [15]

$$W_f = 2\mu_0 H_p h_0 (1 - j^2) - \frac{4}{3}\mu_0 H_p^2 (1 - 3j^2 + 2j^3) . \qquad (3.38)$$

Thus, the total energy loss density is

$$W = W_c + W_f = 2\mu_0 H_p h_0 (1 + j^2) - \frac{4}{3}\mu_0 H_p^2 (1 - j^3) . \qquad (3.39)$$

This result can also be directly obtained from the method shown in Eq. (2.74) (verify that the two methods derive the same result, Exercise 3.3).

Secondly we shall discuss the case where the AC and DC magnetic fields are perpendicular to each other. For example, we assume that the wide superconducting slab parallel to the y-z plane carries a DC transport current along the y-axis in a DC magnetic field along the x-axis and an AC field along the z-axis. In this case the assumption that $\partial/\partial y = \partial/\partial z = 0$ seems to be allowed. The magnetic flux density has only the x- and z-components, B_x and B_z. The condition of $\nabla \cdot \boldsymbol{B} = 0$ leads to a B_x that is uniform and equal to $\mu_0 H_e$ with H_e denoting the DC magnetic field. Hence, only the component B_z varies along the x-axis and the current density along the y-axis is given by

$$J = -\frac{1}{\mu_0} \cdot \frac{\partial B_z}{\partial x} . \tag{3.40}$$

Hence, the mathematical expression is similar to the case of parallel DC and AC fields discussed above and hence the same analysis can be repeated. Thus, a similar rectifying effect and reduction of the apparent critical current density can be explained. From the viewpoint of the flux motion, since the electric field, $\boldsymbol{E} = \boldsymbol{B} \times \boldsymbol{v}$, is along the y-axis and the magnetic flux density $\boldsymbol{B}$ is almost parallel to the x-axis, the velocity of flux lines is approximately directed along the negative z-axis. That is, the flux lines flow in the negative z-axis direction with an oscillating motion in the x-z plane. The details of this flux motion are described in [14]. In this reference the more general theoretical analysis of the force balance equation including the viscous force was carried out and an approximate solution expressed in a power series in frequency was obtained as in the last section. The obtained current-voltage characteristics were compared with experimental results.

3.3.2 Reversible Magnetization

Even for a superconductor with a hysteretic magnetization due to flux pinning, it is known [16] that the superposition of small parallel AC and DC magnetic fields results in a reduction of the hysteresis of the DC magnetization or sometimes even in reversible magnetization (see Fig. 3.12). Figure 3.13(a) shows the variation of the magnetic flux distribution in a superconductor during one cycle of the AC field in the presence of an increasing DC field. For simplicity the diamagnetism is disregarded and the Bean-London model is assumed for the pinning force density. The magnetic flux distribution averaged over one cycle is shown in Fig. 3.13(b); it is flatter than that in the absence of the AC field itself (represented by the broken line). Thus, the reduction in magnetization hysteresis can be explained. The magnitude of the hysteresis is predicted to be

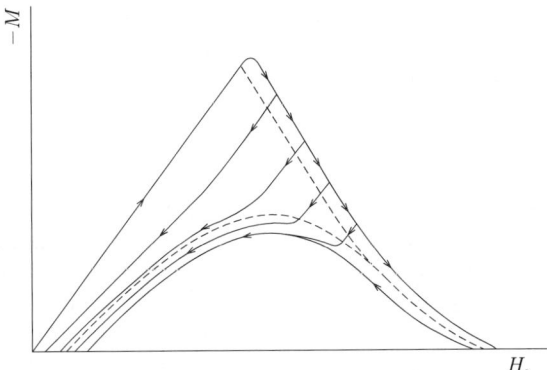

Fig. 3.12. Magnetization of a superconducting Pb-1.92at%Tl cylinder [16]. *Broken* and *solid lines* correspond to the cases with and without the superposition of small AC magnetic field, respectively

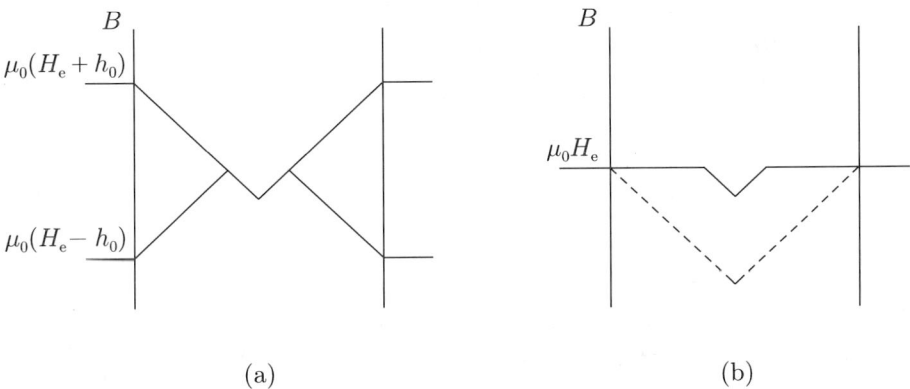

Fig. 3.13. (a) Maximum flux density (*upper line*) and minimum one (*lower line*) during one period of AC magnetic field of amplitude h_0 in an increasing DC magnetic field H_e. (b) The *solid line* shows averaged magnetic flux density in one period and the *broken line* corresponds to the case without the AC magnetic field

$$\Delta M = \Delta M_0 \left(1 - \frac{h_0}{H_p}\right)^2, \tag{3.41}$$

where ΔM_0 is the hysteretic magnetization in the absence of the AC field. Hence, when the AC field amplitude h_0 exceeds the penetration field H_p, the hysteresis disappears and the magnetization becomes reversible. This method is useful for investigation of the diamagnetism in superconductors.

3.3.3 Abnormal Transverse Magnetic Field Effect

When the AC magnetic field is superposed normal to the transverse DC field applied to a superconducting cylinder or tape as shown in Fig. 3.14, the magnetization due to the transverse DC field decreases gradually. Such a phenomenon is called the "abnormal transverse magnetic field effect" [17–19]. An example is shown in Fig. 3.15 where the AC field is applied parallel to a superconducting cylinder: (a) and (c) correspond to the processes of increasing and decreasing DC field, respectively. (b) depicts field cooled process wherein the DC field is applied at a temperature higher than the critical value T_c, and then the temperature is decreased below T_c. The initial magnetic flux distribution due to the application of the DC field in each case is shown in the right side of the figure. The magnetization decreases with application of the AC field and reduces approximately to zero in the steady state.

When the AC field is superposed in a different direction from the DC field as above-mentioned, it is necessary to obtain the distribution of the shielding current. If we assume that the current flows so as to shield the penetration of AC field as much as possible, the current that has shielded the DC field now has to change completely to shield the AC field, resulting in a complete penetration of the DC field. In this case the total amount of penetrating flux seems to be very large. Hence, the shielding current is predicted to flow in such a way that the total amount of penetrating DC and AC flux is minimum. Then, a part of the current that has shielded only the DC field changes so as to shield

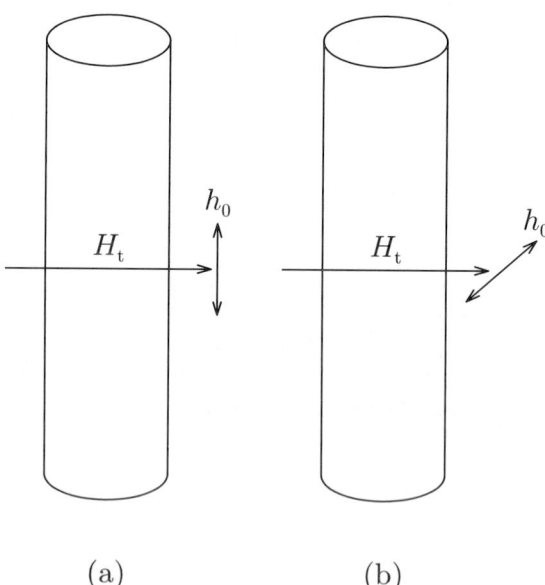

(a)　　　　　　　　(b)

Fig. 3.14. Application of transverse magnetic field H_t and normal small AC magnetic field h_0 to a superconducting cylinder

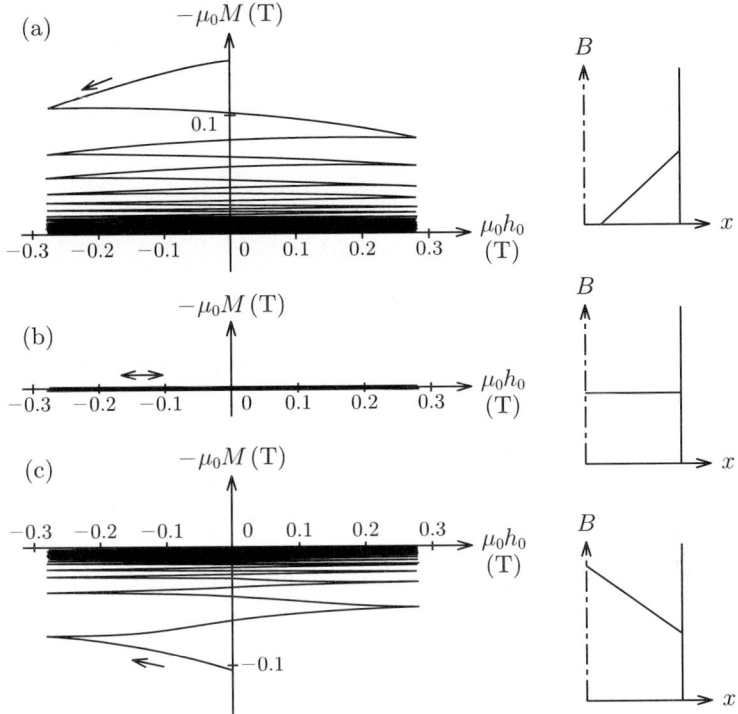

Fig. 3.15. Relaxation of longitudinal magnetization [17] due to superposition of AC magnetic field shown in Fig. 3.14(a) in the processes of (**a**) increasing field, (**b**) field cooling and (**c**) decreasing field. *Right figures* show the initial distributions of DC magnetic flux in each process

the AC field. In other words, a flowing pattern of the current changes gradually from one that shields the DC field to one that shields the AC field during each successive half-cycle of the AC field. Figure 3.16 represents the varying states of distribution of the shielding current when the AC field is applied normal to the superconducting cylinder. Therefore, the DC field penetrates gradually one cycle after another until complete penetration is finally reached. In this final state the current shields only the AC field. The abnormal transverse magnetic field effect is a kind of relaxation process in which the direction of the magnetic moment due to the shielding current changes gradually. In the field cooled process shown in Fig. 3.15(b) the DC field has already penetrated hence the current flows so as to fully shield the AC field.

3.4 Flux Jump

The magnetization in a superconductor sometimes varies discontinuously during the sweeping of a magnetic field as shown in Fig. 3.17. This phenomenon

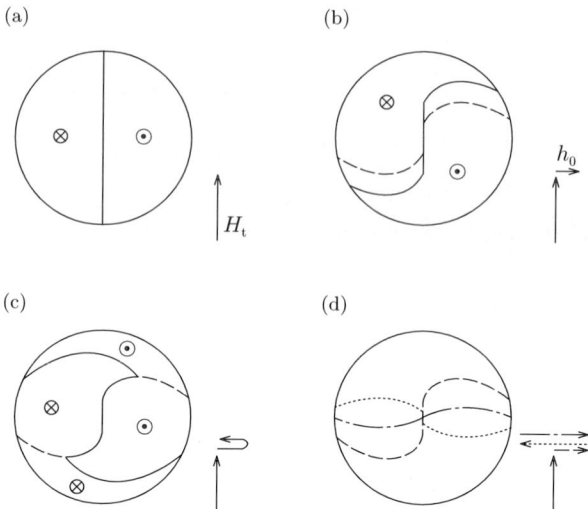

Fig. 3.16. Variation of the distribution of shielding current in the order from (**a**) to (**d**) due to the superposition of AC magnetic field shown in Fig. 3.14(b)

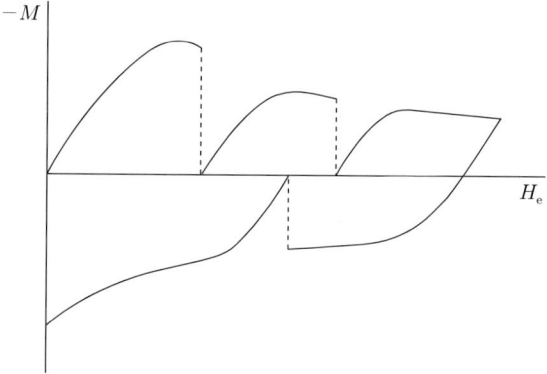

Fig. 3.17. Discontinuous variation of the magnetization due to flux jumping

is called the flux jump. An example of the observed magnetic flux distribution inside a flux-jumping superconductor [20] is shown in Fig. 3.18. It is seen from this observation that the disappearance of shielding current in the superconductor at the moment of the flux jump is accompanied by a sudden invasion of the magnetic flux. Such an instability originates from the irreversible nature of flux pinning. For instance, we assume that a local flux motion occurs for some reason. This will lead to some energy dissipation and a slight temperature rise. This temperature rise will reduce the pinning force that prevents the flux motion and more flux lines than the initial group will move. This will cause a further energy dissipation and temperature rise. The phenomenon will

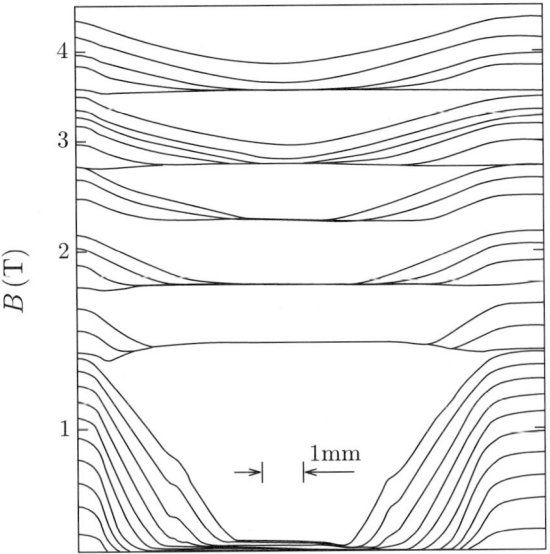

Fig. 3.18. Variation of the magnetic flux distribution in a Nb-Ti measured by scanning a Hall probe in a gap between two pieces of specimen [20]

continue until such positive feedback destroys the superconductivity and the flux motion is completely stopped. This is an outline of the mechanism of flux jumping.

In metallic superconductors the diffusion velocity of the heat is usually much faster than that of the flux lines. Hence, the isothermal approximation is adequate for a superconductor. Thus we assume that the temperature T is uniform throughout the superconductor. The heat produced in the superconductor by the flux motion is absorbed into a coolant such as liquid helium; the equation of heat flow is

$$P = C\frac{\mathrm{d}T}{\mathrm{d}t} + \Phi_\mathrm{h} , \qquad (3.42)$$

where P is the power loss density in the superconductor, C is the heat capacity of a unit volume of the superconductor and Φ_h is the heat flux absorbed by the coolant. When the temperature of the superconductor is not much higher than the temperature of the coolant T_0, the heat flux to the coolant is given by $\Phi_\mathrm{h} = K(T - T_0)$, where K is the heat transfer coefficient per unit volume of the superconductor. Viscous loss and related quantities are disregarded. P contains not only the pinning power loss density P_0 at constant temperature but also an additional component, $C_\mathrm{p}(\mathrm{d}T/\mathrm{d}t)$, due to the temperature rise. The temperature rise causes a variation in the parameter α_c, the flux pinning strength. The resultant variation in the magnetic flux distribution is obtained from Eq. (2.47) as

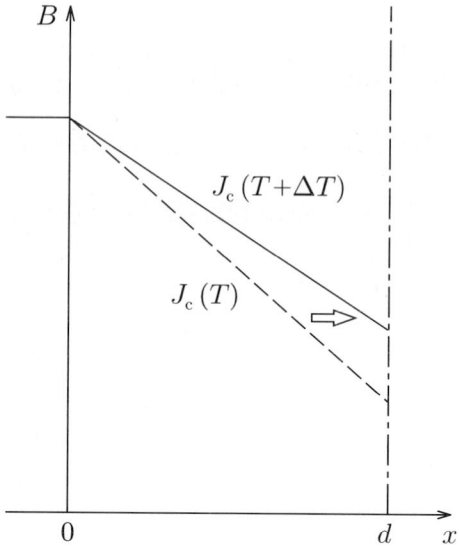

Fig. 3.19. Variation of the magnetic flux distribution in a superconducting slab when the temperature is changed by ΔT

$$\delta \widehat{B}^{1-\gamma} \frac{\partial \widehat{B}}{\partial T} = -\mu_0 \frac{d\alpha_c}{dT} x \ . \tag{3.43}$$

From Eq. (2.75) the additional power loss density is

$$P_1 = -\alpha_c \widehat{B}^{\gamma-1} \frac{dT}{dt} \int_{x_b}^{x} \delta \frac{\partial \widehat{B}}{\partial T} dx \ . \tag{3.44}$$

Equations (2.47) and (3.43) are substituted into Eq. (3.44) and after some calculation we obtain the mean power loss density:

$$\langle P_1 \rangle = \frac{1}{d} \int_0^{x_b} P_1 dx \equiv C_p \frac{dT}{dt} \ . \tag{3.45}$$

Here we assume the Bean-London model ($\gamma = 1$). In case the magnetic flux penetrates to the center of a superconducting slab of thickness $2d$ as in Fig. 3.19, we have $\delta_0 = \delta_b = 1$ in the region $0 \leq x \leq d$ and $x_b = d$, and C_p is given by

$$C_p = \frac{1}{3} \mu_0 H_p \left(-\frac{dH_p}{dT} \right) \ . \tag{3.46}$$

Because of this term Eq. (3.42) is rewritten as

$$(C - C_p) \frac{dT}{dt} = P_0 - \Phi_h \ . \tag{3.47}$$

According to this equation the rate of temperature rise, dT/dt, diverges when

$$C - C_{\rm p} = 0 \qquad (3.48)$$

is satisfied. This is the condition of the rapid temperature rise, i.e., the flux jump.

Yamafuji et al. [21] discussed the temperature rise in detail, taking account a higher order term $(dT/dt)^2$ which originated from the viscous loss. According to their argument the condition that dT/dt becomes indefinite must be satisfied for a flux jump to occur. This means that the condition

$$P_0 - \Phi_{\rm h} = 0 \qquad (3.49)$$

must be satisfied simultaneously with the heat-capacity condition of Eq. (3.48). However, the validity of Eq. (3.49) has not yet been clarified. Anyhow Eq. (3.48) is the condition for the flux jump.

Since flux jumping reduces the critical current density to zero, it must be avoided in practical superconducting wires. Hence, the inequality that $C > C_{\rm p}$ is required so that Eq. (3.48) cannot be satisfied. Since $H_{\rm p} = J_{\rm c} d$, this inequality is equivalent to

$$d < \left[\frac{\mu_0}{3C} \left(-\frac{dJ_{\rm c}}{dT} \right) J_{\rm c} \right]^{-1/2} \equiv d_{\rm c} . \qquad (3.50)$$

This implies that the thickness of superconductor $2d$ should be less than $2d_{\rm c}$. This is the principle of stabilization of superconducting wire by reduction of the superconducting filament diameter. In practical superconducting wires, many fine superconducting filaments are embedded in matrix materials such as copper and are stabilized by the above principle. At the same time the high thermal conductivity of the matrix material ensures rapid dissipation of generated heat. In the case of Nb$_3$Sn for example, if we assume that $J_{\rm c} = 1 \times 10^{10}$ Am^{-2}, $-dJ_{\rm c}/dT = 7 \times 10^8$ Am^{-2}K^{-1} and $C = 6 \times 10^3$ Jm^{-3}K^{-1}, we have $2d < 90$ μm from Eq. (3.50). In practical multifilamentary Nb$_3$Sn wires the diameter of superconducting filaments is smaller than several 10 μm. The filament diameter in multifilamentary wires for AC use is sometimes reduced below 1 μm to reduce the hysteresis loss drastically.

The condition of stabilization (3.50) can also be derived from the following simple argument. Again consider the magnetic flux distribution shown in Fig. 3.19 and assume that the temperature in the superconductor rises from T to $T + \Delta T$ within a short period of time, Δt. The resultant change in critical current density is then $\Delta J_{\rm c} = (dJ_{\rm c}/dT)\Delta T$, where of course $\Delta J_{\rm c} < 0$. Hence, the magnetic flux distribution changes as shown in Fig. 3.19 and the induced electric field due to this change is

$$E(x) = \int_d^x \mu_0 \frac{\Delta J_{\rm c}}{\Delta t} x {\rm d}x = \frac{\mu_0}{2} \left(-\frac{\Delta J_{\rm c}}{\Delta t} \right) (d^2 - x^2) . \qquad (3.51)$$

The resultant energy loss density is given by

$$W = \frac{1}{d}\int_0^{\Delta t} \mathrm{d}t \int_0^d J_c E(x)\mathrm{d}x = \frac{\mu_0}{3}\left(-\frac{\mathrm{d}J_c}{\mathrm{d}T}\right)J_c d^2 \Delta T. \tag{3.52}$$

If Δt is sufficiently small, the above variation is supposed to occur adiabatically. In this case the additional temperature rise in the superconductor is estimated to be $\Delta T' = W/C$. Provided that this temperature rise $\Delta T'$ is smaller than the initial temperature rise ΔT, the initial disturbance will not develop into a flux jump by positive feedback. This condition agrees with Eq. (3.50).

3.5 Surface Irreversibility

During measurement of the DC magnetization of a superconductor, when the sweep of the external magnetic field changes from increasing to decreasing, the magnetization curve is sometimes linear with slope -1 over a certain range of field variation denoted by ΔH, as shown in Fig. 3.20. This is similar to the variation of magnetization in the Meissner state. That is, the magnetic flux distribution is macroscopically unchanged during the variation of the external field. If the external field is increased again within this range, the magnetization reverses. It should be noted, however, that, although in the usual magnetization measurement the magnetization seems to behave reversible as in Fig. 3.20, sensitive B-H_e measurements in fact reveal it to be irreversible. Such behavior is also observed when the sweep of external field changes from decreasing to increasing. This phenomenon insists that an irreversible current with a very high density flows in the surface region and shields the variation of the external field. The magnitude of the magnetization ΔH due to the surface current is sample dependent and varies with the magnetic field; ΔH usually decreases with increasing field.

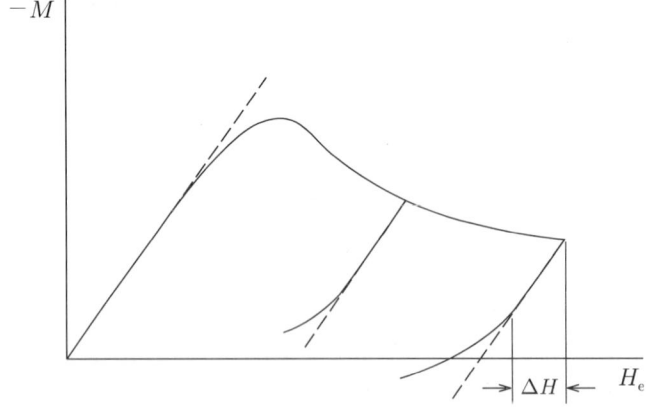

Fig. 3.20. Macroscopic magnetization due to the surface irreversibility

Three mechanisms, viz. surface sheath, surface barrier, and surface pinning have been proposed to explain the irreversible surface current.

The surface superconductivity treated in Sect. 1.6 is associated with a special property of the surface which permits the superconducting order parameter to take on a nonzero value even when the applied magnetic field is above H_{c2}. Fink [22] speculated that a similar two-dimensional surface superconductivity exists even below H_{c2} independently of the three-dimensional flux line structure inside the superconductor. This surface superconductivity is called the surface sheath and considered to cause an irreversible surface current.

The idea of a "surface barrier" was proposed by Bean and Livingston [23] who suggested that the surface itself provided a barrier against the invasion and elimination of flux lines. The surface barrier was originally proposed during investigations of the first entry field of flux lines into a practical superconductor in comparison with H_{c1}, the theoretical result of Abrikosov for an infinitely large superconductor. We begin by assuming that a flux line has entered the superconductor from the surface. In addition to the external field that decays within a characteristic distance of λ from the surface we consider the flux line, and a postulated image flux line directed opposite to it as in Fig. 3.21. The image is necessary to fulfill the boundary condition that the current around the flux line should not flow across to the surface. The total

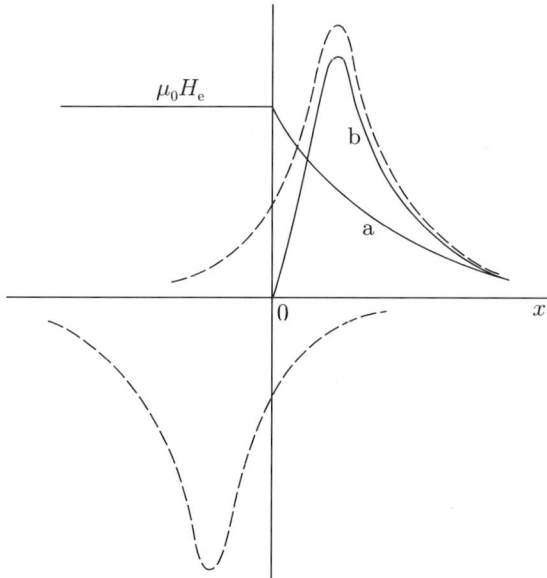

Fig. 3.21. Surface barrier model by Bean and Livingston [23]. 'a' represents the penetrating magnetic flux from the surface given by Eq. (1.14) and 'b' is the sum of the penetrating flux line and its image

magnetic flux of the flux line (the component of b in Fig. 3.21) is smaller than the flux quantum ϕ_0 and hence, flux quantization is not fulfilled. This is because the current is not zero on the surface, which is a part of the loop enclosing the magnetic flux.

Bean and Livingston treated the case where the G-L parameter κ is large, in which case the modified London equation is valid and the magnetic flux density of the flux line penetrating sufficiently deeply from the surface is given by Eq. (1.61). We assume that the superconductor occupies $x \geq 0$ and the external magnetic field H_e is applied parallel to the z-axis. The position of the flux line in the x-y plane is represented as $(x_0, 0)$, where $x_0 > 0$. Its image is located at $(-x_0, 0)$ and the total magnetic flux density in the superconductor is given by

$$b = \mu_0 H_e \exp\left(-\frac{x}{\lambda}\right) + \frac{\phi_0}{2\pi\lambda^2}\left[K_0\left(\frac{\sqrt{(x-x_0)^2 + y^2}}{\lambda}\right) - K_0\left(\frac{\sqrt{(x+x_0)^2 + y^2}}{\lambda}\right)\right] \quad (3.53)$$

except the region of the normal core. The Gibbs free energy is given by

$$G = \int \left\{\frac{1}{2\mu_0}[b^2 + \lambda^2(\nabla \times b)^2] - H_e \cdot b\right\} dV, \quad (3.54)$$

where the volume integral is over the superconductor ($x \geq 0$). The first term in Eq. (3.53) may be symbolized by b_0 and the sum of the second and third terms which represent the penetrating flux line and its image may be symbolized by b_1. After substituting these into Eq. (3.54), partially integrating and using the modified London equation, the Gibbs free energy becomes

$$G = \lambda^2 \int_S (\nabla \times b_1) \times H_e \cdot dS$$
$$+ \frac{\lambda^2}{2\mu_0} \int_{S_c} [b_1 \times (\nabla \times b_1) + 2b_1 \times (\nabla \times b_0) + 2\mu_0(\nabla \times b_1) \times H_e] \cdot dS$$
$$+ \frac{1}{2\mu_0} \int_{\Delta V} [b_1^2 + \lambda^2(\nabla \times b_1)^2 + 2b_0 \cdot b_1 + 2\lambda^2(\nabla \times b_0) \cdot (\nabla \times b_1)$$
$$- 2\mu_0 b_1 \cdot H_e] dV. \quad (3.55)$$

In the above the first integral is carried out on the surface of superconductor, $S(x = 0)$, the second one on the surface of normal core of the flux line, S_c (dS is directed inward the surface), and the third one inside the normal core. The constant terms which are functions only of b_0 are omitted. In the first integral in Eq. (3.55), since H_e is equal to b_0/μ_0 on the surface, it can be replaced by b_0/μ_0. If we replace the integral on S to an integral on S and S_c minus an integral on S_c, we have

$$\frac{\lambda^2}{\mu_0} \int_{S+S_c} (\nabla \times \boldsymbol{b}_1) \times \boldsymbol{b}_0 \cdot \mathrm{d}\boldsymbol{S} = \frac{\lambda^2}{\mu_0} \int_{S+S_c} (\nabla \times \boldsymbol{b}_0) \times \boldsymbol{b}_1 \cdot \mathrm{d}\boldsymbol{S}$$

$$= \frac{\lambda^2}{\mu_0} \int_{S_c} (\nabla \times \boldsymbol{b}_0) \times \boldsymbol{b}_1 \cdot \mathrm{d}\boldsymbol{S}, \quad (3.56)$$

where partial integration is done and the modified London equation and the boundary condition of $\boldsymbol{b}_1 = 0$ on the superconductor surface are used. Thus, Eq. (3.55) reduces to

$$G = \frac{\lambda^2}{2\mu_0} \int_{S_c} [\boldsymbol{b}_1 \times (\nabla \times \boldsymbol{b}_1) - 2\mu_0 \boldsymbol{H}_\mathrm{e} \times (\nabla \times \boldsymbol{b}_1) + 2\boldsymbol{b}_0 \times (\nabla \times \boldsymbol{b}_1)] \cdot \mathrm{d}\boldsymbol{S}$$
$$+ \frac{1}{2\mu_0} \int_{\Delta V} [\boldsymbol{b}_1^2 + \lambda^2(\nabla \times \boldsymbol{b}_1)^2 + 2\boldsymbol{b}_0 \cdot \boldsymbol{b}_1 + 2\lambda^2(\nabla \times \boldsymbol{b}_0) \cdot (\nabla \times \boldsymbol{b}_1)$$
$$- 2\mu_0 \boldsymbol{b}_1 \cdot \boldsymbol{H}_\mathrm{e}] \mathrm{d}V. \quad (3.57)$$

Now we write $\boldsymbol{b}_1 = \boldsymbol{b}_\mathrm{f} + \boldsymbol{b}_\mathrm{i}$ to indicate the sum of the flux line and its image, where the full expressions for these components are just the second and third terms in Eq. (3.53). After a simple calculation we have

$$G = \frac{\lambda^2}{2\mu_0} \int_{S_c} [\boldsymbol{b}_\mathrm{f} \times (\nabla \times \boldsymbol{b}_\mathrm{f}) + \boldsymbol{b}_\mathrm{i} \times (\nabla \times \boldsymbol{b}_\mathrm{f}) - \boldsymbol{b}_\mathrm{f} \times (\nabla \times \boldsymbol{b}_\mathrm{i})$$
$$- 2\mu_0 \boldsymbol{H}_\mathrm{e} \times (\nabla \times \boldsymbol{b}_\mathrm{f}) + 2\boldsymbol{b}_0 \times (\nabla \times \boldsymbol{b}_\mathrm{f}) - 2\boldsymbol{b}_\mathrm{f} \times (\nabla \times \boldsymbol{b}_0)] \cdot \mathrm{d}\boldsymbol{S}$$
$$+ \frac{1}{2\mu_0} \int_{\Delta V} [\boldsymbol{b}_\mathrm{f}^2 + \lambda^2(\nabla \times \boldsymbol{b}_\mathrm{f})^2 - 2\mu_0 \boldsymbol{b}_\mathrm{f} \cdot \boldsymbol{H}_\mathrm{e}] \mathrm{d}V. \quad (3.58)$$

With the aid of Eq. (1.78) it turns out that the sum of the first term in the first integral and the first and second terms in the second integral gives the self energy of the flux line, $\epsilon = \phi_0 H_{\mathrm{c}1}$, and the sum of the fourth term in the first integral and the third term in the second integral gives a constant term, $-\phi_0 H_\mathrm{e}$. The fifth and second terms in the first integral represent the interactions of the flux line with the Lorentz force due to the surface current and with the image, respectively. It can be easily shown that the third and sixth terms in the first integral are sufficiently small and can be disregarded. Thus, we have

$$G = \phi_0 \left[H_\mathrm{e} \exp\left(-\frac{x_0}{\lambda}\right) - \frac{\phi_0}{4\pi\mu_0\lambda^2} K_0\left(\frac{2x_0}{\lambda}\right) + H_{\mathrm{c}1} - H_\mathrm{e} \right] \quad (3.59)$$

(per unit length of the flux line), which is identical with the result obtained by Bean and Livingston [23] and by de Gennes [24]. This equation is valid for the case where the normal core completely penetrates the superconductor, i.e., $x_0 > \xi$. Since the constant term that depends only on $\boldsymbol{b}_0$ is omitted, we have $G = 0$ when the flux line does not penetrate the superconductor, i.e., $x_0 = 0$. When $H_\mathrm{e} = H_{\mathrm{c}1}$, G goes to zero in the limit $x_0 \to \infty$. That is, the condition of a bulk superconductor is naturally satisfied.

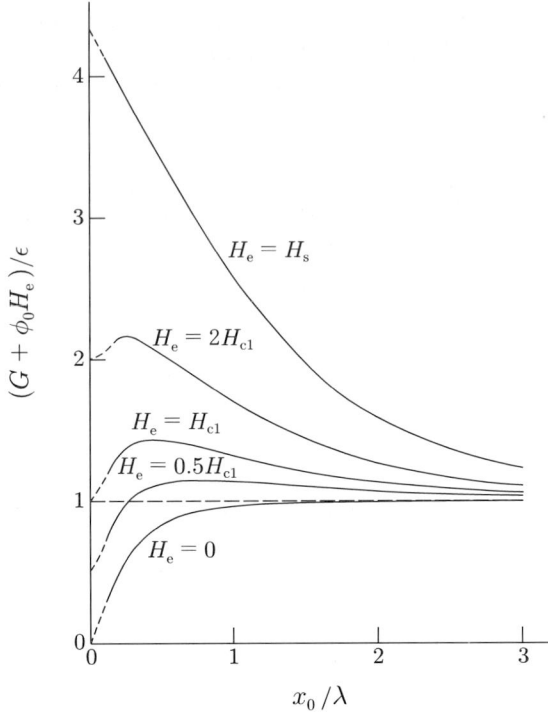

Fig. 3.22. Variation of the energy G vs the position, x_0, of the flux line in case $\kappa = 10$ [23]. The ordinate is normalized by the self energy, ϵ, of flux line per unit length

Figure 3.22 shows the variation in the free energy G with the position of the flux line x_0. It means that the energy barrier exists even when H_e exceeds H_{c1} and the flux line cannot penetrate the superconductor. Hence, the magnetization remains perfectly diamagnetic until the external magnetic field reaches H_s sufficiently greater than H_{c1}, and then the flux line first penetrates. Conversely, the surface barrier prevents the flux lines from exiting the superconductor as the field decreases. The flux lines are predicted to be trapped in the superconductor until the external field is reduced to zero. Also in this case the magnetization curve is a line parallel to the Meissner line, suggesting that the internal magnetic flux distribution remains unchanged even under variation of the external field. Such feature agrees qualitatively with the surface irreversibility phenomenon observed experimentally. For this reason the surface barrier model seems to be applicable not only to the estimation of the first penetration field, its initial purpose, but also to the general phenomena of surface irreversibility.

3.5 Surface Irreversibility

Here, we shall estimate the first penetration field H_s of the flux line from the above result of the Bean-Livingston model. We treat the case in which the flux line exists near the surface ($x_0 \sim \xi$). The corresponding magnetic flux density is approximately given by Eq. (1.62a). Thus, the Gibbs free energy reduces to

$$G = \phi_0 \left[H_e \exp\left(-\frac{x_0}{\lambda}\right) + \frac{\phi_0}{4\pi\mu_0\lambda^2} \log 2x_0 \right] + \text{const}. \tag{3.60}$$

Since the penetration of flux line occurs when $\partial G / \partial x_0 = 0$ is attained at $x_0 \sim \xi$, we have

$$H_s \simeq \frac{\phi_0}{4\pi\mu_0\lambda\xi} = \frac{H_c}{\sqrt{2}}. \tag{3.61}$$

The calculation in terms of the modified London equation in the above is not exact when x_0 is close to ξ. Then, de Gennes [25] argued the first penetration field H_s using the G-L equations. We assume again that the superconductor occupies $x \geq 0$. de Gennes treated this problem as an extrapolation of the Meissner state above H_{c1}, i.e., the superheated state and assumed that the order parameter and the vector potential vary one-dimensionally only along the x-axis. In this case, the order parameter can be chosen as a real number as known well. If we normalize Ψ by $|\Psi_\infty|$ as in Eq. (1.38) and the vector potential $\boldsymbol{A}$ and the coordinate x as

$$\boldsymbol{a} = \frac{\boldsymbol{A}}{\sqrt{2}\mu_0 H_c \lambda}, \tag{3.62}$$

$$\widetilde{x} = \frac{x}{\lambda}, \tag{3.63}$$

the G-L equations (1.30) and (1.31) reduce to

$$\frac{1}{\kappa^2} \cdot \frac{\mathrm{d}^2 \psi}{\mathrm{d}\widetilde{x}^2} = \psi(-1 + \psi^2 + a^2), \tag{3.64}$$

$$\frac{\mathrm{d}^2 a}{\mathrm{d}\widetilde{x}^2} = \psi^2 a. \tag{3.65}$$

In the above, a is the y-axis component, if the magnetic field is applied along the z-axis. As will be shown later, ψ and a vary gradually with the distance of the order of 1 in the $\widetilde{x}$-coordinate (λ in real space). Hence, the left-hand side of Eq. (3.64) can be approximately replaced by zero for a superconductor with the large G-L parameter κ. Then, Eq. (3.64) reduces to

$$\psi^2 = 1 - a^2. \tag{3.66}$$

Substitution of this into Eq. (3.65) leads to

$$\frac{\mathrm{d}^2 a}{\mathrm{d}\widetilde{x}^2} = a - a^3. \tag{3.67}$$

Multiplying both sides by $da/d\tilde{x}$ and integrating, we have

$$\left(\frac{da}{d\tilde{x}}\right)^2 - a^2 + \frac{a^4}{2} = \text{const} . \tag{3.68}$$

Since a and $da/d\tilde{x}$ are expected to drop to zero at deep inside the superconductor, $\tilde{x} \to \infty$, the constant term on the right-hand side of Eq. (3.68) must be zero. Thus,

$$a = -\frac{\sqrt{2}}{\cosh(\tilde{x}+c)} , \tag{3.69}$$

where c is a constant determined by the boundary condition at the surface, $\tilde{x} = 0$. This solution satisfies the above-mentioned requirement that the physical quantities vary gradually with the distance of the order of λ in real space. From Eq. (3.69) the magnetic flux density is

$$B = \sqrt{2}\mu_0 H_c \frac{da}{d\tilde{x}} = \frac{2\mu_0 H_c \sinh(\tilde{x}+c)}{\cosh^2(\tilde{x}+c)} . \tag{3.70}$$

From the boundary condition that the magnetic flux density is $\mu_0 H_e$ at $\tilde{x} = 0$, c can be evaluated from

$$\frac{H_e}{H_c} = \frac{2\sinh c}{\cosh^2 c} . \tag{3.71}$$

The maximum value of H_e, i.e., the first penetration field, H_s, corresponds to $c = \sinh^{-1} 1$, and hence [25]

$$H_s = H_c . \tag{3.72}$$

If we neglect the term proportional to $(d\psi/d\tilde{x})^2$, the free energy density is given by

$$F = \mu_0 H_c^2 \left[-\psi^2 + \frac{1}{2}\psi^4 + \left(\frac{da}{d\tilde{x}}\right)^2 + a^2\psi^2 \right]$$

$$= \mu_0 H_c^2 \left[-\frac{1}{2} + \frac{4\sinh^2(\tilde{x}+c)}{\cosh^4(\tilde{x}+c)} \right] \tag{3.73}$$

or

$$F = -\frac{1}{2}\mu_0 H_c^2 + \frac{B^2}{\mu_0} = F_n - \frac{1}{2}\mu_0 H_c^2 + \frac{B^2}{2\mu_0} , \tag{3.74}$$

where $F_n = B^2/2\mu_0$ is the energy density of magnetic field, i.e., the free energy density in the normal state. Figure 3.23 shows the magnetic flux density, the normalized order parameter and the free energy density in the critical state at $H_e = H_c$. At the surface where the magnetic flux density reaches $\mu_0 H_c$, the order parameter ψ is zero and the free energy density F is equal to F_n, its normal-state value.

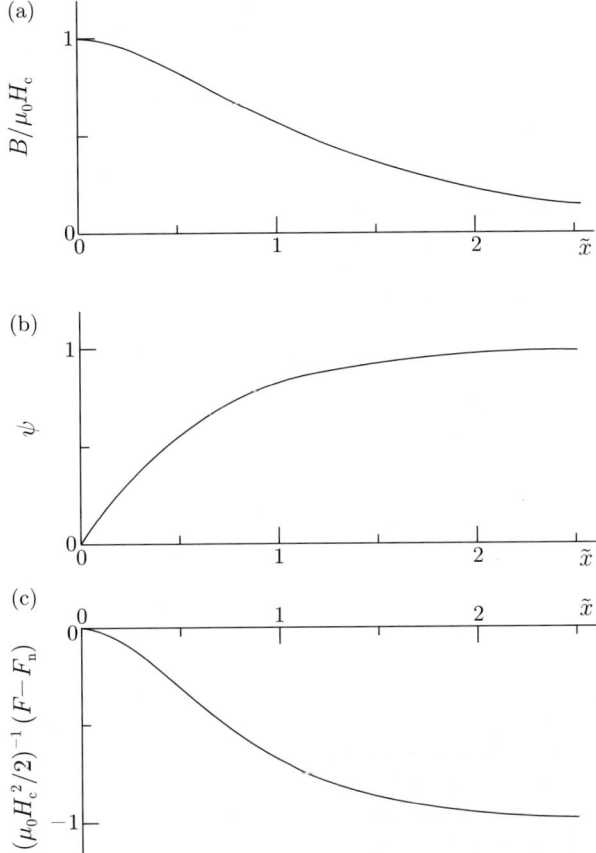

Fig. 3.23. (a) Magnetic flux density, (b) normalized order parameter and (c) free energy density in the vicinity of the surface of a superconductor in the critical superheated state ($H_e = H_c$)

As shown above the first penetration fields, as obtained by Bean and Livingston and by de Gennes, are of the order of H_c, although these differ by a factor of $\sqrt{2}$. In the case of high-κ superconductors, which is the required condition for the approximations, the predicted values are much greater than the bulk value, H_{c1}. Such large penetration fields have not yet been observed.

Here the relationship between the two theories will be discussed. As mentioned above the superheated state has been treated by de Gennes. In this case, the assumption that the order parameter gradually varies one-dimensionally does not hold any more as the field decreases after the penetration of flux lines, and hence, the superheated state cannot be re-established. It follows that the magnetization curve is predicted to be reversible after the penetration of flux lines, as shown in Fig. 3.24. Hence, in oder to discuss the surface irreversibility, the interaction between the flux line and the surface should be

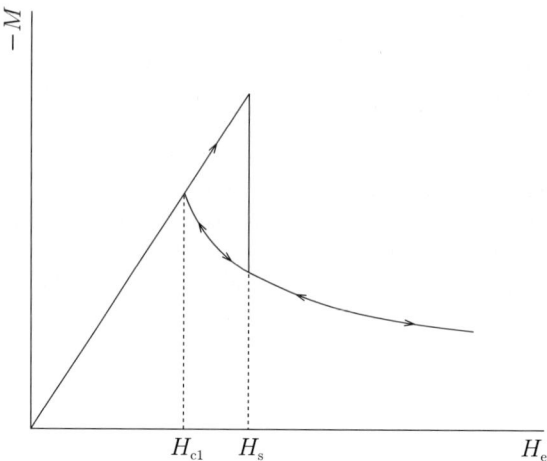

Fig. 3.24. Magnetization after the superheated state is destroyed

treated as it was done in the surface barrier model. It is also important to investigate the effect of surface roughness. If the surface roughness is of the order of ξ, it is accompanied by a steep spatial variation of Ψ, in which case the assumption of a gradual one-dimensional variation is no longer valid. In high-κ superconductors, ξ is small and it seems to be quite difficult to make the surface roughness of bulk specimens smaller than ξ. Hence, it seems unphysical to image that the superheated state could be maintained up to high fields. On the other hand, the penetration of flux line results in a two-dimensional spatial variation of Ψ and the surface barrier appears. In this case the image of flux line is considered to become dim due to the surface roughness resulting in a weakening of the interaction between the flux line and its image. However, the surface barrier should remain. Thus, it can be seen that there is a difference between the surface barrier mechanism and the superheating mechanisms proposed by de Gennes, and furthermore that the former provides a more practical explanation of surface irreversibility.

When the surface is roughened, it is speculated that the effects of surface sheath and surface barrier are reduced. However, the surface irreversibility is enhanced in most cases. It has been shown that neither bulk irreversibility nor surface irreversibility is observed in materials with few defects and clean surfaces [26], indicating that the surface sheath and the surface barrier are not the main causes of surface irreversibility. On the other hand, Hart and Swartz [27] speculated that the pinning by surface roughness and defects near the surface causes the surface irreversibility based on the correlation between the surface roughness and the irreversibility. This mechanism is called "surface pinning."

Experimentally it has been shown that surface pinning is a dominant mechanism. Matsushita et al. [28] showed that residual pinning centers could be removed from several Nb-50at%Ta tape specimens by heat treatment at very high temperatures under very high vacuum. After this heat treatment, dislocations with different densities were introduced to the specimens by different rates of rolling deformation. The initial thickness of each specimen was changed so that the final thicknesses of all the specimens were the same. Since the superconducting properties such as T_c and H_c and the condition of the surface were almost the same in each one, ΔH should have been approximately the same, if either the surface sheath or the surface barrier was the origin of the surface irreversibility. Figure 3.25 shows the bulk critical cur-

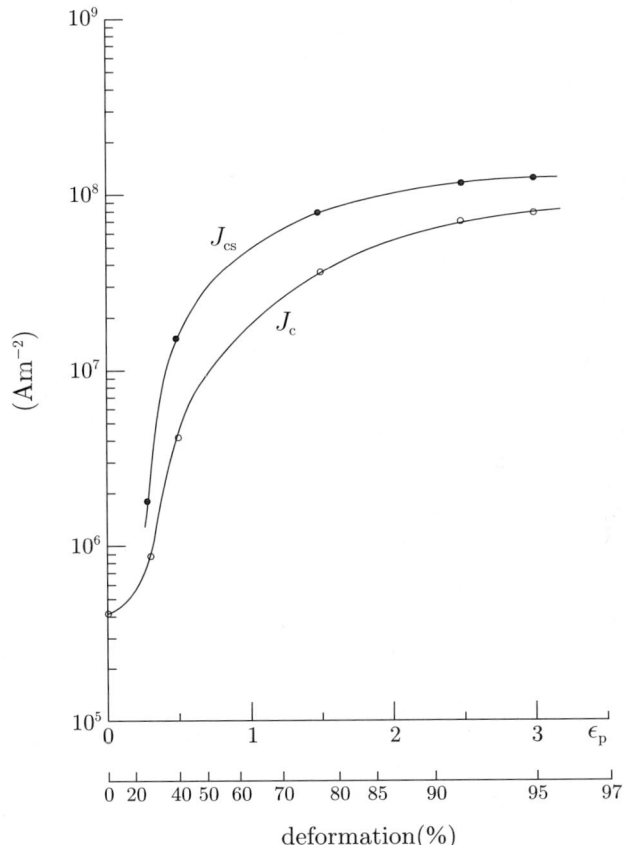

Fig. 3.25. Bulk critical current density J_c and surface one J_{cs} vs deformation by the cold rolling in Nb-Ta specimens [28]. These critical current densities increase with increasing density of pinning dislocations. The deformation is defined as $1 - (A_0/A)$ and we have $\epsilon_p = \log(A/A_0)$, where A and A_0 represent the surface area of superconductor before and after the rolling, respectively

rent density J_c and the one near the surface J_{cs} estimated using Campbell's method described in Sect. 5.3, where J_{cs} is approximately proportional to ΔH. This result shows that not only the bulk critical current density, J_c, but also the surface one J_{cs} increases significantly with increasing density of pinning dislocations and that the two critical current densities are saturated to almost the the same value in the strong pinning limit. From the fact that the surface irreversibility is enhanced by two orders of magnitude by introduction of dislocations, it can be concluded that the dominant cause of the surface irreversibility is surface pinning and that the effects of a surface barrier etc. can be neglected. In addition, a saturation behavior of critical current density in the strong pinning regime and its special magnetic field dependence [28] are known to be characteristic features of the saturation phenomenon for the bulk flux pinning, which will be described in Sect. 7.5.

It has been concluded that since with rolling deformation the defects are nucleated with the higher density in the surface region, the surface critical current density J_{cs} increases faster than the bulk value.

It follows that the surface irreversibility is not an intrinsic surface effect as such but rather a secondary phenomenon caused by defects that are likely to be concentrated in the surface region. It is generally known [28] that the magnitude of surface irreversibility ΔH decreases with increasing magnetic field and disappears at high fields. This is caused by the nonlocal nature of the flux pinning. That is, the critical current density is a value averaged over the range of the pinning correlation length of the flux line lattice (Campbell's AC penetration depth that will be described in the next section). At high fields, this correlation length increases, and the region of the average is no longer limited to the surface region with strong pinning forces but extends into the inner region. Thus, the surface irreversibility is diluted quickly with increasing magnetic field strength.

As discussed above, surface pinning rather than the surface barrier effect is the dominant mechanism of surface irreversibility. As indicated in Fig. 3.22 the energy barrier itself is not particularly large, and in any case its effectiveness reduces in the presence of the usual surface roughness which by dimming the image of flux line weakens the attraction between the flux line and the image. In addition, the penetration of flux lines through the surface barrier can be facilitated by flux creep, to be discussed in Sect. 3.8.

The surface pinning force itself can be reduced by various kinds of surface treatments such as metallic coating [29] or oxidation [30] (see Fig. 3.26). In the former case, a proximity effect between the normal metal coating and the superconductor reduces the order parameter at the surface and hence the strength of surface pinning.

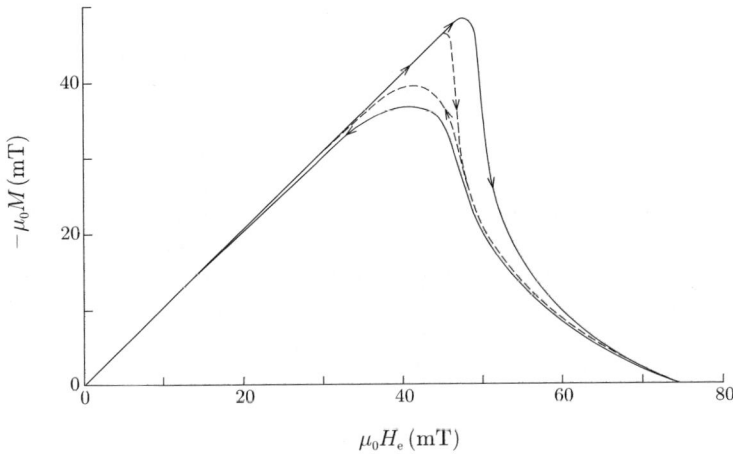

Fig. 3.26. Magnetization of V specimen at 4.2 K [30]. *Solid* and *broken lines* show results on the specimen before and after the oxygenation, respectively

3.6 DC Susceptibility

Measurement of DC susceptibility in the field cooled process was carried out for evaluating the superconducting volume fraction of a specimen just after the discovery of high-temperature superconductors. A constant susceptibility at sufficiently low temperatures was regarded as related to the volume fraction of a superconducting phase. However, this is correct only for pin-free superconductors. As the temperature decreases, the superconductor becomes diamagnetic and the flux lines are expelled from the superconductor, resulting in a negative susceptibility. If the pinning interactions in the superconductor become effective as the temperature decreases, flux lines will be prevented from leaving the superconductor, and the susceptibility will be influenced by the pinning. That is, the susceptibility is proposed to be small for a strongly pinned superconductor. Thus, the result does not reflect correctly the volume fraction of superconducting material.

For a description by the critical state model, it is assumed that a magnetic field H_e is applied parallel to a very wide superconducting slab ($0 \leq x \leq 2d$). From symmetry we need to treat only half of the slab, $0 \leq x \leq d$. The critical temperature in the magnetic field H_e is denoted by T_c'. When the temperature T is higher than T_c', the magnetic flux density in the superconductor is uniform and given by $B = \mu_0 H_e$. When the temperature is slightly decreased from T_c' to $T_1 = T_c' - \Delta T$, the superconductor becomes diamagnetic. If T_1 is higher than the irreversibility temperature, $T_i(H_e)$, the pinning does not yet work, and the internal magnetic flux distribution is as schematically shown in Fig. 3.27(a), where $M(<0)$ is the magnetization. When the temperature is further reduced to $T_n = T_i(H_e) - \Delta T$, the pinning interaction becomes effective. If the critical current density at this stage is denoted by ΔJ_c, the

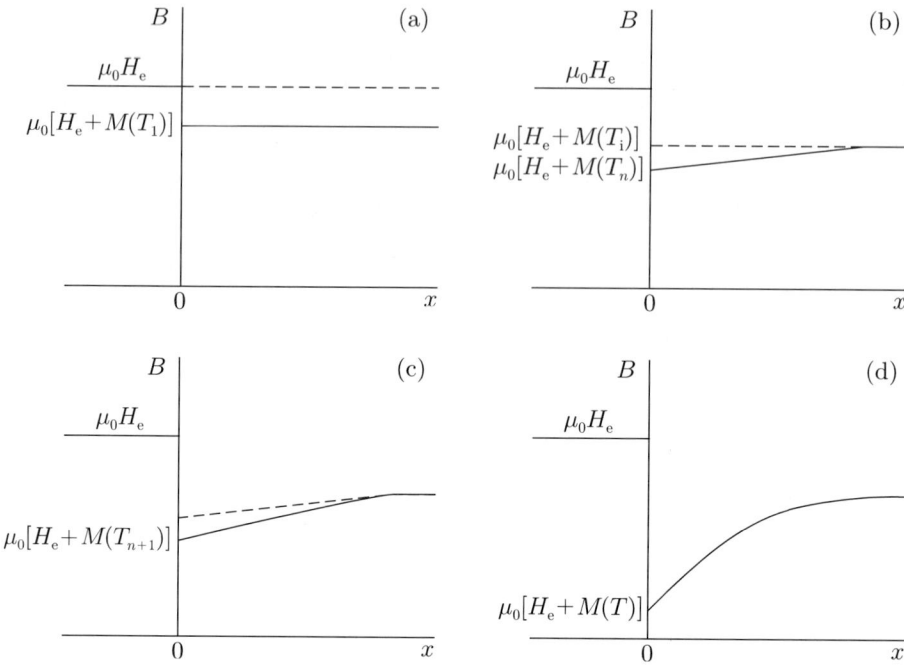

Fig. 3.27. Magnetic flux distribution in a superconductor in the field cooled process (a) for $T_1 \leq T \leq T_c'$, (b) when the temperature is decreased slightly by ΔT from T_1, (c) when the temperature is further decreased and (d) at sufficiently low temperatures

magnetic flux distribution inside the superconducting slab is expected to be like the one shown in Fig. 3.27(b), where the slope of the magnetic flux distribution near the surface is equal to $\mu_0 \Delta J_c$. When the temperature is further decreased, the diamagnetism of the superconductor becomes stronger and the flux lines near the surface are driven to the outside of the superconductor. At the same time the pinning also becomes stronger, and the flux distribution shown in Fig. 3.27(c) results. Thus, the magnetic flux distribution at a sufficiently low temperature is expected to be like that in Fig. 3.27(d).

Here the magnetic flux distribution is calculated analytically. For simplicity the diamagnetic property of the superconductor is approximated as shown in Fig. 3.28. That is, if the temperature at which H_e is equal to the lower critical field H_{c1} is denoted by T_{c1}, the magnetization is given by

$$M(T) = -\epsilon[H_{c2}(T) - H_e] \tag{3.75}$$

for temperatures higher than T_{c1} and by

$$M(T) = -H_e \tag{3.76}$$

for temperatures lower than T_{c1}. In the above the parameter ϵ is given by

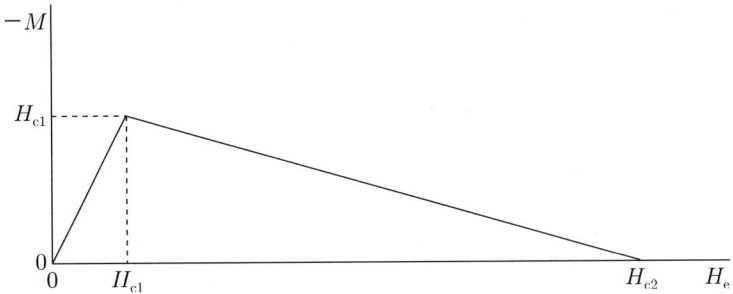

Fig. 3.28. Approximate diamagnetism of a superconductor

$$\epsilon = \frac{H_{c1}}{H_{c2} - H_{c1}} . \tag{3.77}$$

If the temperature dependence of κ is neglected, this parameter does not depend on the temperature. Correctly speaking, this parameter should be given by $\epsilon = 1/1.16(2\kappa^2 - 1)$ in the vicinity of H_{c2}, but the above approximation is used to simplify the analysis. The temperature dependence of H_{c2} is also approximated by

$$H_{c2}(T) = H_{c2}(0)\left(1 - \frac{T}{T_c}\right) . \tag{3.78}$$

In addition, the critical current density is assumed to be a function only of the temperature as

$$J_c(T) = A\left(1 - \frac{T}{T_i}\right)^{m'} \tag{3.79}$$

for sufficiently low H_e. If the irreversibility temperature T_i is approximately given by the critical temperature T_c', we have

$$T_i = (1 - \delta)T_c \tag{3.80}$$

with $\delta = H_e/H_{c2}(0)$.

The magnetic flux distribution near the surface of the superconductor is determined only by M and J_c at a given temperature as

$$B(x) = \mu_0 H_e + \mu_0 M(T) + \mu_0 J_c(T)x . \tag{3.81}$$

In general m' is larger than 1. Thus, the history of magnetic flux distribution at higher temperatures remains in the superconductor as shown in Fig. 3.27(d). Namely, the internal flux distribution is equal to the envelope of Eq. (3.81) at higher temperatures in the past. If the region in which the magnetic flux distribution is expressed by Eq. (3.81) is $0 \leq x \leq x_0$, x_0 is obtained from

$$\frac{\partial B(x_0)}{\partial T} = 0 \tag{3.82}$$

at temperatures above T_{c1}. Under the assumptions of Eqs. (3.75) and (3.78)–(3.80) we have

$$x_0 = \frac{\epsilon[H_{c2}(0) - H_e]}{Am'}\left[1 - \frac{T}{(1-\delta)T_c}\right]^{1-m'} . \tag{3.83}$$

This depth is $x_0 = \infty$ at $T = T_1$ and decreases with decreasing temperature. The envelope of the flux distribution in the internal region can be derived by substituting the obtained x_0 into Eq. (3.81). We have only to eliminate T in Eq. (3.81) in terms of $x_0(T)$ in Eq. (3.83). Then, replacing x_0 by x, the flux distribution in the envelope region is given by

$$B(x) = \mu_0 H_e - (m'-1)\mu_0 \left[\frac{\epsilon H_{c2}(0)(1-\delta)}{m'}\right]^{m'/(m'-1)} (Ax)^{-1/(m'-1)} . \tag{3.84}$$

This holds within the region $x_0 \leq x \leq d$.

If the external magnetic field H_e is sufficiently small, the temperature T_{c1} at which H_{c1} is equal to H_e exists. Below this temperature M is given by Eq. (3.76), and B just inside the surface is zero. Hence, the magnetic flux distribution at temperatures lower than this remains unchanged. Strictly speaking, the flux lines near the surface are continuously expelled from the superconductor with the strengthened diamagnetism from decreasing temperature as discussed in Sect. 2.6, and hence, the flux distribution does not remain completely unchanged. However, the remaining flux distribution is the "heritage" of distributions at higher temperatures, and hence, its gradient is small and the resultant driving force to expel flux lines from the superconductor is relatively smaller than the pinning force at the ambient temperature. Thus, although this effect increases the diamagnetism slightly, its influence is considered not to be large. The effect of the reversible motion of flux lines, to be discussed in Sect. 3.7, is rather larger than this, since it is considered that the flux lines are likely to be nucleated in the bottom of pinning potentials where the energy is lowest in the field cooled process.

The magnetic flux density and the DC susceptibility can be calculated from the above results. The temperature T_0, at which x_0 is equal to d, is given by

$$T_0 = T_c(1-\delta)\left\{1 - \left[\frac{\epsilon H_{c2}(0)(1-\delta)}{Am'd}\right]^{1/(m'-1)}\right\} . \tag{3.85}$$

After a simple but long calculation we have [31]

3.6 DC Susceptibility

$$\chi = \epsilon - \frac{\epsilon}{\delta}\left(1 - \frac{T}{T_c}\right) + \frac{Ad}{2H_e}\left[1 - \frac{T}{(1-\delta)T_c}\right]^{m'} \quad ; \quad T_i \geq T > T_0 \, , \quad (3.86a)$$

$$= -\frac{[\epsilon H_{c2}(0)(1-\delta)]^2}{2m'(2-m')dAH_e}\left[1 - \frac{T}{(1-\delta)T_c}\right]^{2-m'}$$

$$+ \frac{(m'-1)^2}{(2-m')(Ad)^{1/(m'-1)}H_e}\left[\frac{\epsilon H_{c2}(0)(1-\delta)}{m'}\right]^{m'/(m'-1)} ;$$

$$T_0 \geq T > T_{c1} \, , \quad (3.86b)$$

$$= -\frac{[\epsilon H_{c2}(0)(1-\delta)]^{m'}}{2m'(2-m')dAH_e^{m'-1}}$$

$$+ \frac{(m'-1)^2}{(2-m')(Ad)^{1/(m'-1)}H_e}\left[\frac{\epsilon H_{c2}(0)(1-\delta)}{m'}\right]^{m'/(m'-1)} \equiv \chi_s;$$

$$T_{c1} \geq T \, , \quad (3.86c)$$

where χ_s is the saturated susceptibility at sufficiently low temperatures. The above results are useful for $m' \neq 2$. Calculate the susceptibility also for the case of $m' = 2$ (Exercise 3.5).

Calculated results [31] of DC susceptibility in the field cooled process for various values of A are shown in Fig. 3.29. The DC susceptibility when the temperature is increased in a fixed magnetic field is also shown for comparison. With increasing A, i.e., strengthening pinning force, the susceptibility in the field cooled process takes a smaller negative value, but a larger negative value in the process of increasing temperature in a fixed magnetic field. This can be understood, since the motion of flux lines is more restricted by the stronger pinning force. Figure 3.30 shows the relation between the saturated susceptibility and the size of the superconducting specimen [31]. It turns out that the diamagnetism becomes stronger with decreasing specimen size. This is because the internal flux lines can more easily leave the superconductor when the superconductor is smaller.

The above various results can be qualitatively explained from the magnetic flux distribution in Fig. 3.27. Figure 3.31 shows the dependence of the saturated susceptibility on the external magnetic field for a single crystal specimen of La-based superconductor [31], and the experimental results agree with the above theoretical predictions of the critical state model.

It is assumed here that the superconducting volume fraction is 100%. However, the obtained saturated susceptibility differs greatly depending on the conditions as shown in Figs. 3.29 and 3.30. Hence, this measurement technique is not suitable for evaluation of the superconducting volume fraction.

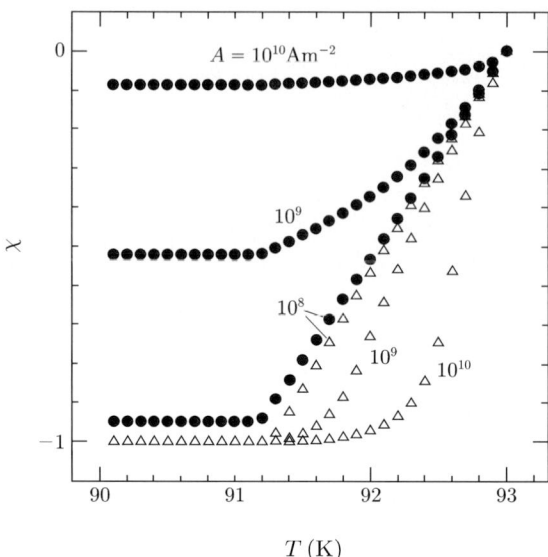

Fig. 3.29. Results of calculated DC susceptibilities for various values of A representing the flux pinning strength in the field cooled process (••) and when the temperature is increased in a constant magnetic field (△△) [31]. Assumed parameters are $T_c = 93$ K, $\mu_0 H_{c2}(0) = 100$ T, $\epsilon = 5.13 \times 10^{-4}$, $m' = 1.8$, $d = 1$ mm and $\mu_0 H_e = 1$ mT

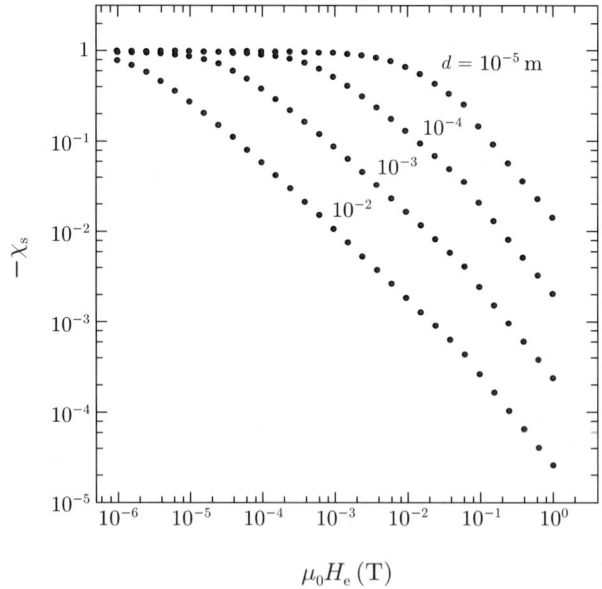

Fig. 3.30. Dependence of saturated DC susceptibility on external magnetic field for various sizes of superconductor [31]. Assumed parameters are $A = 1.0 \times 10^{10}$ Am^{-2} and the same values of T_c, $\mu_0 H_{c2}(0)$, ϵ and m' as in Fig. 3.29

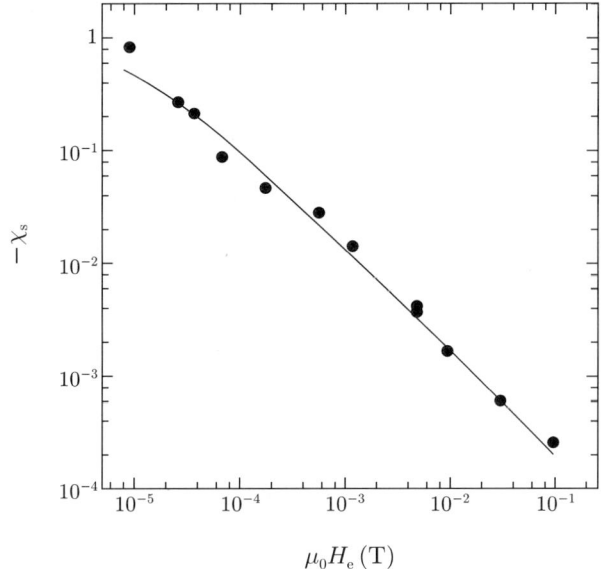

Fig. 3.31. Magnetic field dependence of saturated DC susceptibility of a La-based superconducting specimen [31]. The *solid line* is the theoretical result for $T_c = 35$ K, $\mu_0 H_{c2}(0) = 27.3$ T, $\epsilon = 5.1 \times 10^{-4}$, $A = 8.0 \times 10^{10}$ Am^{-2} and $m' = 1.8$

3.7 Reversible Flux Motion

Most electromagnetic properties of superconductors are irreversible and can be well described in terms of the critical state model. The irreversibility stems from the interaction of flux lines and the pins, i.e., the instability of flux lines as they drop into and jump out of the pinning potential, as discussed in Sect. 2.3. However, if the displacement of flux lines is so small that the flux motion is restricted to the interior of the pinning potential, the corresponding electromagnetic phenomena are expected to become reversible and hence deviate from the critical state description.

Here we assume that the flux lines in some region are in an equilibrium state inside an averaged pinning potential. When the flux lines are displaced by a distance u from the equilibrium position in response to a change of the external magnetic field etc., the pinning potential felt by the flux lines within a unit volume is of the form $\alpha_L u^2/2$, where α_L, a constant, is called the Labusch parameter. Hence, the force on the flux lines per unit volume is

$$F = -\alpha_L u \,, \tag{3.87}$$

which depends only on the position of flux lines u and is reversible. Note the difference between this force and that based on the critical state model according to which the force takes on only one of two values, $\pm J_c B$, depending on the direction of the flux motion. If now b represents the variation in the magnetic flux density due to the movement of flux lines from their equilibrium positions (we assume that the flux movement occurs along the x-axis), using the continuity equation (2.15) we have

$$\frac{\mathrm{d}u}{\mathrm{d}x} = -\frac{b}{B}, \quad (3.88)$$

where B is an equilibrium value of the magnetic flux density. The Lorentz force that arises from this variation is

$$F_\mathrm{L} = -\frac{B}{\mu_0} \cdot \frac{\mathrm{d}b}{\mathrm{d}x}. \quad (3.89)$$

Solutions for u and b can be obtained from the balance between the Lorentz force and the pinning force given by Eq. (3.87), $F_\mathrm{L} + F = 0$, under given boundary conditions. That is, eliminating u from Eqs. (3.87)–(3.89), we have

$$\frac{\mathrm{d}^2 b}{\mathrm{d}x^2} = \frac{\mu_0 \alpha_\mathrm{L}}{B^2} b \quad (3.90)$$

and hence,

$$b(x) = b(0) \exp\left(-\frac{x}{\lambda'_0}\right). \quad (3.91)$$

In the above the superconductor is assumed to occupy $x \geq 0$, $b(0)$ is a value of b at the surface, $x = 0$, and

$$\lambda'_0 = \frac{B}{(\mu_0 \alpha_\mathrm{L})^{1/2}}. \quad (3.92)$$

λ'_0 is a length called Campbell's AC penetration depth [32]. A solution of the same form can also be obtained for the displacement, $u(x)$. The variation in the magnetic flux density given by Eq. (3.91) is similar to Eq. (1.14) representing the Meissner effect. This is the reason for referring to λ'_0 as a "penetration depth." According to the above solution, the depth to which the variation penetrates is given by λ'_0 and is independent of the variation in the magnetic flux density at the surface $b(0)$, for $b(0)$ below some value.

The reversible phenomenon appears for example when the applied field changes from decreasing to increasing. Hence, the initial condition just before the appearance of the reversible phenomenon is mostly the critical state. The variation in the magnetic flux distribution after the field changes from decreasing to increasing is schematically shown in Fig. 3.32(a). On the other hand, based on the critical state model, the distribution would change according to $b(x) = b(0) - 2\mu_0 J_\mathrm{c} x$. In this case the depth to which the variation penetrates is $b(0)/2\mu_0 J_\mathrm{c}$ and increases in proportion to $b(0)$ (see Fig. 3.32(b)). Thus, the variation in the magnetic flux distribution is different between the reversible and completely irreversible states.

According to experiments, the pinning force density changes from $J_\mathrm{c} B$ to $-J_\mathrm{c} B$ or inversely (see Fig. 3.33), as the magnetic flux distribution changes as

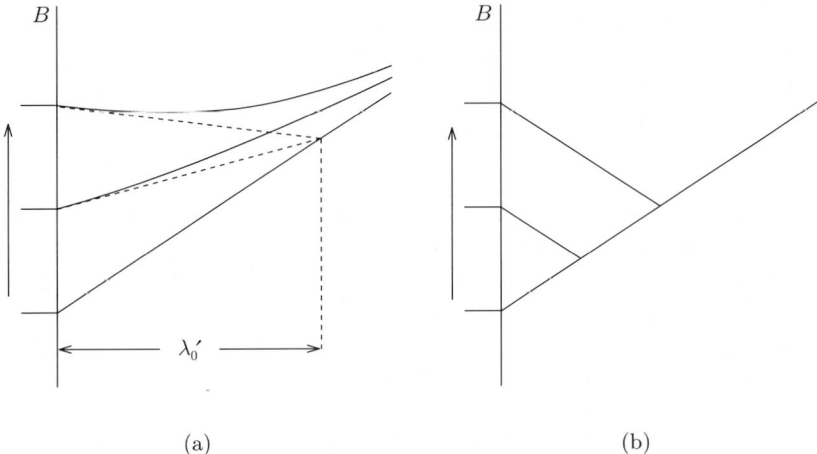

Fig. 3.32. Variations in the magnetic flux distribution in a superconductor when the external magnetic field is increased from the critical state in a decreasing field: (**a**) the case of noticeable reversible motion of flux lines and (**b**) the prediction of the critical state model

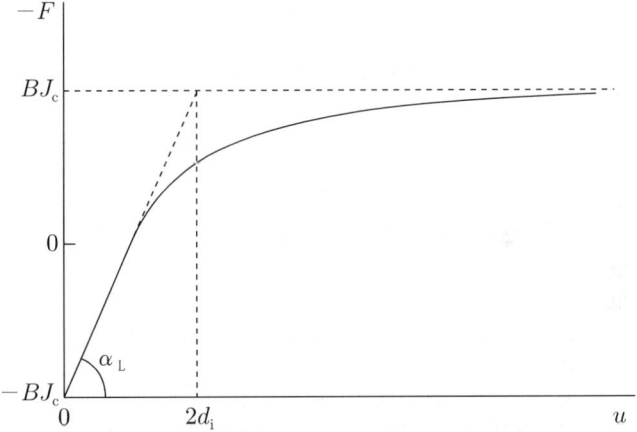

Fig. 3.33. Variation of the pinning force density vs the displacement of flux lines. Origin is the critical state and the figure shows the characteristics when the flux lines are displaced reversely

shown in Fig. 3.32(a). That is, the pinning force density varies linearly with u and the phenomenon is reversible as described above, while the displacement u from the initial condition is small. As the mean displacement of the flux lines increases, some flux lines jump out of individual pinning potentials locally and the characteristics of pinning force density vs displacement vary gradually from reversible to irreversible. When the displacement increases further, the pinning force density approaches asymptotically $-J_c B$ and the phenomenon

128   3 Various Electromagnetic Phenomena

becomes describable by the irreversible critical state model. Measurements of AC penetration depth λ_0' and the characteristics of pinning force density vs displacement shown in Fig. 3.33 may be carried out using Campbell's method described in Sect. 5.3.

In practice (see Figs. 3.32(a) and 3.33), the absolute value of the pinning force density is actually given by $F = -\alpha_L u + J_c B$ rather than Eq. (3.87) in the reversible region where u is sufficiently small. On the other hand, b is a variation of the magnetic flux density from the initial condition, and hence the Lorentz force is given by $F_L = -(B/\mu_0)(db/dx) - J_c B$. From the balance between the two forces the solution of Eq. (3.91) is again obtained.

Here we shall estimate the AC loss in the vicinity of the reversible region. We assume that a DC magnetic field H_e and an AC one of amplitude h_0 are applied parallel to an infinite superconducting slab of thickness $2D$ ($0 \leq x \leq 2D$). From symmetry we treat only the half-region, $0 \leq x \leq D$. If h_0 is sufficiently small, the critical current density J_c can be regarded as a constant. We assume that the initial magnetic flux distribution at the surface field of $H_e - h_0$ is in the critical state and that the variation from this distribution is as shown in Fig. 3.34. The variation in the magnetic flux density from the initial state is again denoted by $b(x)$. Campbell [32] expressed the variation of pinning force density with displacement in Fig. 3.33 as

$$F = -J_c B \left[1 - 2 \exp\left(-\frac{u}{2d_i}\right) \right] \tag{3.93}$$

and we go on to make the approximation as $B \simeq \mu_0 H_e$. d_i is half of the displacement when the linear extrapolation of the pinning force density in the reversible regime reaches $J_c B$ in the opposite critical state. That is, d_i

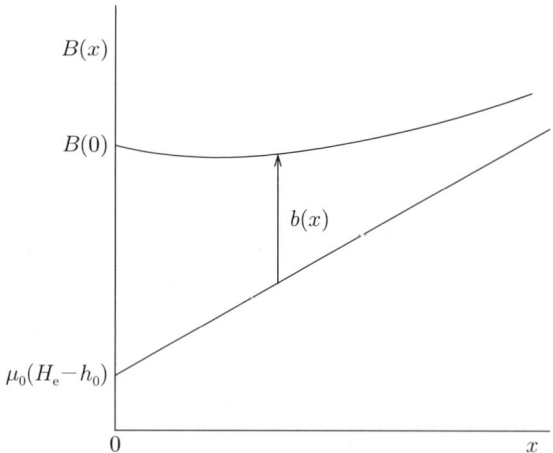

Fig. 3.34. Variation of the magnetic flux distribution in a superconductor when the surface field is increased from $H_e - h_0$ in the critical state

represents a radius of the averaged pinning potential and for this reason is referred as the "interaction distance." From the relation

$$J_c B = \alpha_L d_i \tag{3.94}$$

with Eq. (3.92) we have

$$d_i = \frac{\mu_0 J_c \lambda_0'^2}{B}. \tag{3.95}$$

Elimination of u from the force balance equation using Eq. (3.88) leads to

$$\frac{d^2 b}{dx^2} - \frac{b}{\lambda_0'^2}\left(1 + \frac{1}{2\mu_0 J_c} \cdot \frac{db}{dx}\right) = 0. \tag{3.96}$$

From symmetry the condition $u(D) = 0$ should be satisfied. This is written as $F(D) = J_c B$ or

$$\left.\frac{db}{dx}\right|_{x=D} = 0. \tag{3.97}$$

Equation (3.96) is only numerically solved under this condition and the boundary condition of $b(0)$ at the surface. When the external magnetic field is increased to $H_e + h_0$ and then decreased to $H_e - h_0$, the magnetic flux distribution does not go back to the initial condition. Hence, strictly speaking, it is necessary to obtain the distribution in the steady state after many periods of AC field to estimate the AC loss observed usually. However, this is not easily done and we shall approximate for simplicity that the curve of averaged magnetic flux density vs external field is symmetric between the increasing and decreasing field processes. In which case, after one period, the last point of the $\langle B \rangle$-H curve meets the initial point and the loop closes. In this way, the AC loss can be estimated approximately. Here it should be noted that the AC loss can be obtained not only from the area of the $\langle B \rangle$-H curve but also from the area of the closed F-u curve as shown in Fig. 3.35. In the latter case, the F-u curve is believed to be approximately symmetric between O → A and A → O [32].

Figure 3.36 shows the AC losses observed for a bulk Nb-Ta specimen [33], along with the result of theoretical analysis using the Campbell model and with a critical state prediction based on Eq. (2.80) and the assumption $\gamma = 1$. According to these results, the difference between the Campbell model and the critical state model is small even when the AC field amplitude is small and hence when the flux motion should be almost reversible; hence it is not clear which model better explains the experimental result. It is, however, possible to distinguish between the two models in terms of the "power factor," which is generally given by

$$\eta_p = \left[1 + \left(\frac{\mu'}{\mu''}\right)^2\right]^{-1/2}, \tag{3.98}$$

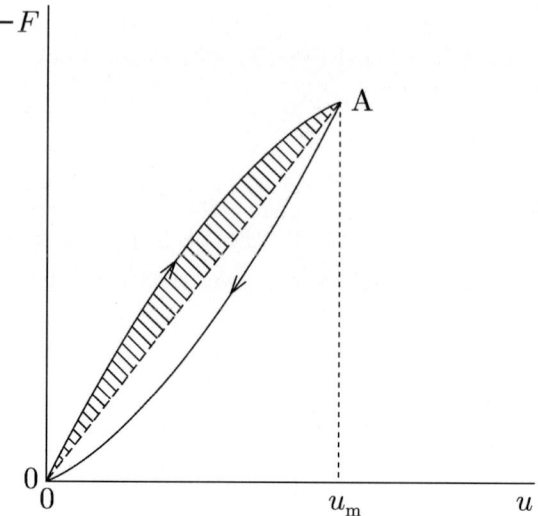

Fig. 3.35. Hysteresis loop of the pinning force density vs displacement of the flux lines in one cycle of the AC magnetic field

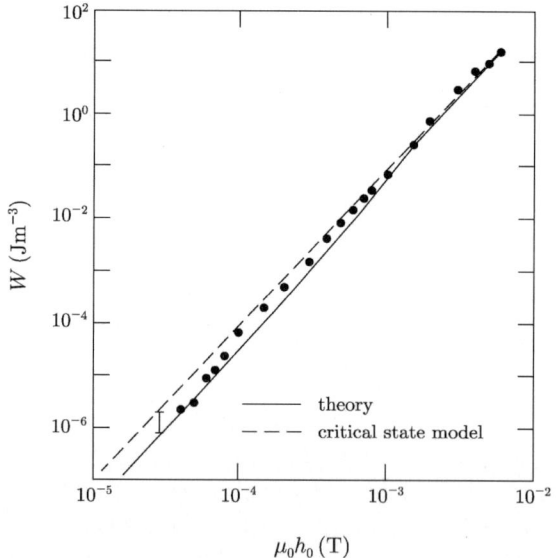

Fig. 3.36. AC energy loss density in a bulk Nb-Ta specimen [33] in a DC bias field $\mu_0 H_e = 0.357$ T. The *solid* and *broken lines* represent the theoretical predictions of the Campbell model and the critical state model, respectively

where μ' and μ'' are the real and imaginary parts of the fundamental AC permeability, respectively. If we express the time variation of external magnetic field as $h_0 \cos \omega t$, these are written as

$$\mu' = \frac{1}{\pi h_0} \int_{-\pi}^{\pi} \langle B \rangle \cos \omega t \, d\omega t \,, \tag{3.99}$$

$$\mu'' = \frac{1}{\pi h_0} \int_{-\pi}^{\pi} \langle B \rangle \sin \omega t \, d\omega t \,. \tag{3.100}$$

The AC energy loss density, W, is related to the imaginary AC permeability μ'' through

$$W = \pi \mu'' h_0^2 \,. \tag{3.101}$$

According to the critical state model with $\gamma = 1$ (the Bean-London model) Eq. (3.98) reduces to

$$\eta_\mathrm{p} = \left[1 + \left(\frac{3\pi}{4} \right)^2 \right]^{-1/2} \simeq 0.391 \,, \tag{3.102}$$

which is independent of h_0. The observed power factor for the Nb-Ta specimen and the predictions of the two models are compared in Fig. 3.37. It is found from this figure that the phenomenon is well explained by the Campbell model in which the effect of reversible flux motion is taken into account, while the prediction of the critical state model deviates from the experiment. η_p is

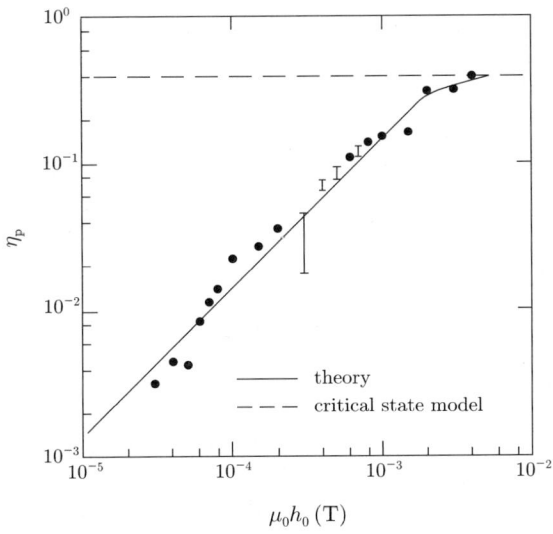

Fig. 3.37. Power factor of AC energy loss density in a bulk Nb-Ta specimen [33] shown in Fig. 3.36. The *solid* and *broken lines* represent the theoretical predictions of the Campbell model and the critical state model, respectively

proportional to h_0 in the region of small h_0. The derivation of η_p predicted by the Campbell model is Exercise 3.8.

As is seen from Fig. 3.36, the AC loss itself in a bulk superconductor is close to the prediction of the critical state model. This is the reason why the reversible effect has not been noticed. Why is the AC loss close to the prediction of the irreversible critical state model? In the regime of almost reversible flux motion, the displacement u is sufficiently small and the pinning force density in Eq. (3.93) can be approximately expanded in a power series in u:

$$F = J_\text{c} B \left[1 - \frac{u}{d_\text{i}} + \left(\frac{u}{2d_\text{i}}\right)^2 \right]. \tag{3.103}$$

Here F increases with increasing u as in the upper curve of Fig. 3.35 and the area of the hatched region gives a half of the energy loss density in one period of AC field. Hence, the irreversible component in F is a deviation from the linear line connecting the origin O and the point A. After a simple calculation we have

$$F_\text{irr} = -\frac{J_\text{c} B}{4 d_\text{i}^2} (u_\text{m} u - u^2), \tag{3.104}$$

where $u_\text{m}(x)$ is the maximum displacement. Here we shall estimate the displacement. Since the flux motion is almost reversible, the penetration of the AC flux can be approximated by Eq. (3.91). From Eq. (3.88) we have

$$u(x) = \frac{b(0) \lambda_0'}{B} \exp\left(-\frac{x}{\lambda_0'}\right), \tag{3.105}$$

where $b(0)$ is the variation in the magnetic flux density at the surface from the initial condition. $u_\text{m}(x)$ is given by this equation with a replacement of $b(0)$ by $2\mu_0 h_0$. Hence, the energy loss density during the variation in $b(0)$ by $db(0)$ is given by $-F_\text{irr} du$ with du denoting the variation in u in this period and is written as

$$dw = \frac{J_\text{c} \lambda_0'^3}{4 d_\text{i}^2 B^2} \exp\left(-\frac{3x}{\lambda_0'}\right) [2\mu_0 h_0 b(0) - b^2(0)] db(0). \tag{3.106}$$

Since $b(0)$ varies from 0 to $2\mu_0 h_0$ in a half period, the energy loss density is calculated to be

$$W = \frac{2}{D} \int_0^{2\mu_0 h_0} db(0) \int_0^D dx \cdot \frac{dw}{db(0)} = \frac{2\mu_0 h_0^3}{9 J_\text{c} D}. \tag{3.107}$$

In the above we assumed that $D \gg \lambda_0'$. This value is 1/3 of the prediction of the critical state model. The reason for the relatively small difference is that the displacement and the region in which the loss occurs are enhanced, while the irreversible force density $|-F_\text{irr}|$ is decreased, resulting in an approximate offset.

The reversible phenomenon does not affect the electromagnetic property appreciably for a bulk superconductor sufficiently thicker than λ'_0. However, it is considered from a comparison between (a) and (b) in Fig. 3.32 that reversibility will become noticeable for a superconductor with the size comparable to or smaller than λ'_0. λ'_0 is of the order of 0.5 µm at 1 T in a superconductor with the flux pinning strength comparable to a commercial Nb-Ti wire; accordingly the reversible phenomenon is really noticeable [34] in multifilamentary wires for AC use which have superconducting filaments thinner than the above value. As a result the dependence of the AC loss on the filament diameter shows a departure from the critical state prediction. That is, the critical state model predicts that the breaking point of the loss curve should shift to smaller AC field amplitudes accompanied by increasing loss in the lower amplitude region with decreasing filament diameter (see inset

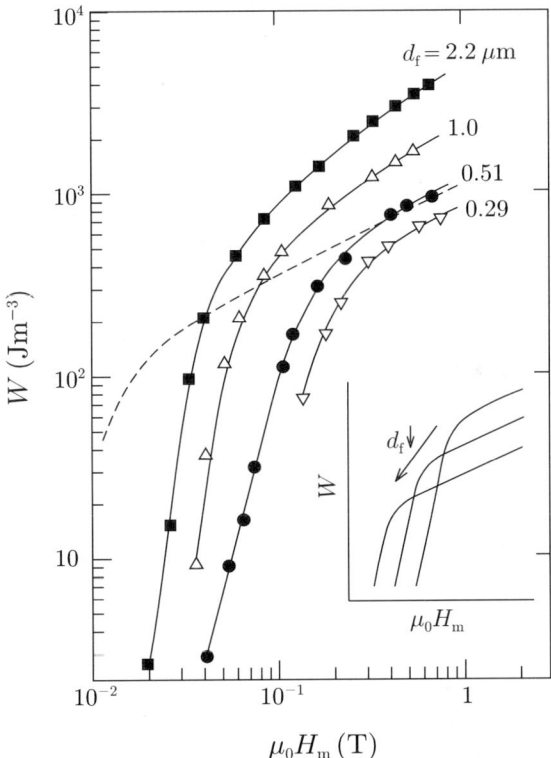

Fig. 3.38. AC energy loss density in a multifilamentary Nb-Ti wires with very fine filaments [34]. H_m is an AC magnetic field amplitude and a DC bias field is not applied. The *broken line* shows the prediction of the critical state model with the observed critical current density for filament diameter 0.51 µm. Inset represents the prediction of the critical state model on the variation of AC energy loss density with the filament diameter

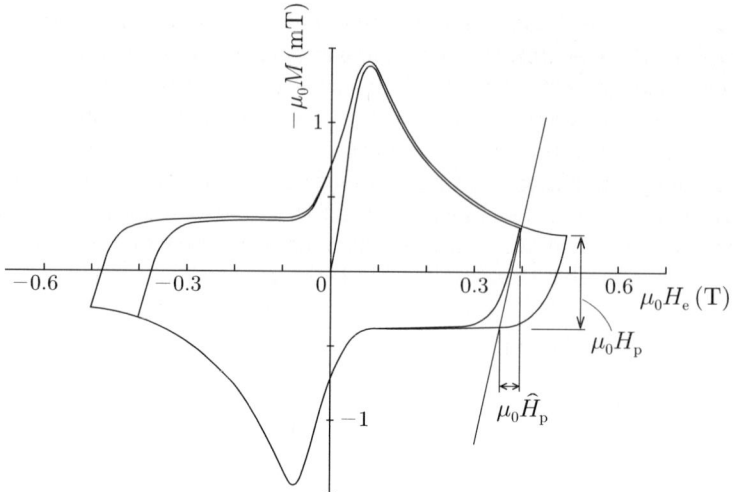

Fig. 3.39. Magnetization curve for a Nb-Ti multifilamentary wire [34] of filament diameter 0.51 μm

of Fig. 3.38) in complete conflict with the experimental result. On the other hand, in the region of large AC field amplitudes, the loss agrees with the prediction of the critical state model. Another result is that the slope of the minor magnetization curve when the sweep of magnetic field is changed from increasing to decreasing takes on a much smaller value than theoretically predicted, i.e., it is 1 for a large slab in a parallel field and 2 for a cylinder in a perpendicular field, as shown in Fig. 3.39 (note a difference in scales between the ordinate and the abscissa). This slope becomes smaller with increasing magnetic field.

Such an abnormal phenomenon originates from the reversible flux motion. Usually the filament diameter d_f is not sufficiently greater than the flux line spacing a_f. For instance, at $B = 1$ T a_f is 49 nm and hence only ten rows of flux lines exist in a filament of diameter 0.5 μm. Hence, the applicability of the semimacroscopic Campbell model to the macroscopic description of the spatial variation of magnetic flux distribution is in doubt. However, in the usual specimens the number of filaments is very large and the dimension along the length of the filament is also large. The magnetic quantity usually observed is the average within a large number of long filaments and hence the semimacroscopic description is considered to be possible only as an averaged flux distribution. The local flux distribution is expected to be different from such an averaged one. This is also the case in a bulk superconductor and even in the critical state. That is, it is not correct to postulate that the local flux distribution is of uniform slope equal to $\mu_0 J_\mathrm{c}$ in the critical state of a bulk superconductor. In fact, the slope may take on various values locally and $\mu_0 J_\mathrm{c}$ is nothing other than the average value. The fact that the critical state model holds for multifilamentary wires with many filaments and sufficient length has been validated by many experiments. This does not contradict the above speculation that the semimacroscopic description is possible for the averaged distribution. Hence, the Campbell model is considered to be applicable to multifilamentary superconducting wires even with very fine filaments.

Takács and Campbell [35] calculated the AC loss in a wire with very fine superconducting filaments in a small AC magnetic field of amplitude h_0 superposed to DC field. They assumed that the magnetic flux penetrates uniformly the very fine superconducting filaments. The filament of diameter d_f was approximated by a slab of thickness d_f. Here we shall calculate the AC loss using Eq. (3.103) in a different manner from [35]. Only the half, the region $0 \leq x \leq d_f/2$ is considered. The displacement of flux lines in this region is obtained from Eq. (3.88) as

$$u(x) = \frac{b(0)}{B}\left(\frac{d_f}{2} - x\right), \qquad (3.108)$$

where the symmetry condition, $u = 0$ at the center, $x = d_f/2$, was used. The loss in this case can also be estimated by substituting Eq. (3.108) into Eq. (3.104) as done previously; u_m is again given by Eq. (3.108) with $b(0)$ replaced by $2\mu_0 h_0$. D is replaced by $d_f/2$ in Eq. (3.107), and after some calculation we have

$$W = \frac{\mu_0 h_0^3}{3 J_c d_f}\left(\frac{d_f}{2\lambda_0'}\right)^4. \qquad (3.109)$$

This agrees with the result of Takács and Campbell [35] and is $(d_f/2\lambda_0')^4/4$ times as large as the prediction of the Bean-London model. Thus, the loss decreases rapidly with decreasing filament diameter. It is concluded that the reversible effect is very large in small superconductors for the following reason. Because of symmetry, the flux lines in the center of the filament do not move and are restrained around the origin of the pinning force vs displacement curve shown in Fig. 3.33. The average displacement of flux lines is approximately proportional to the filament diameter, and hence, most of the flux lines in the filament are in the reversible regime.

When the AC field amplitude becomes large, the loss approaches asymptotically $2\mu_0 H_p h_0$, where $H_p = J_c d_f/2$ is the penetration field (see Eq. (2.84)). From the intersecting point between this relationship and the extrapolation of Eq. (3.109), the breaking point of the loss curve shown in Fig. 3.38 is

$$\widetilde{H}_p = 2\sqrt{3}\left(\frac{2\lambda_0'}{d_f}\right)^2 H_p = 4\sqrt{3}\frac{J_c \lambda_0'^2}{d_f}. \qquad (3.110)$$

Hence, the breaking point shifts to higher AC field amplitudes with decreasing filament diameter. Thus, the dependence of the loss on the filament diameter obeys the Campbell description of reversible phenomenon. In the irreversible Bean-London model $\widetilde{H}_p = \sqrt{3} H_p$.

Suppose we extrapolate the tangent to the minor magnetization curve in Fig. 3.39, then the magnetic field variation needed to reach the opposite major curve is represented by $\widehat{H}_\mathrm{p}$. According to the Campbell model the slope of the minor magnetization curve, $H_\mathrm{p}/\widehat{H}_\mathrm{p}$, is a function only of $d_\mathrm{f}/2\lambda_0'$ and given by [34]

$$\frac{H_\mathrm{p}}{\widehat{H}_\mathrm{p}} = 1 - \frac{2\lambda_0'}{d_\mathrm{f}} \tanh\left(\frac{d_\mathrm{f}}{2\lambda_0'}\right), \tag{3.111}$$

where the slab approximation is used. In the extreme reversible case in which $d_\mathrm{f}/2\lambda_0' \ll 1$ is satisfied, we have $\widehat{H}_\mathrm{p} = (\sqrt{3}/2)\widetilde{H}_\mathrm{p} = 3(2\lambda_0'/d_\mathrm{f})^2 H_\mathrm{p}$. On the other hand, in the case where the Bean-London model holds and $d_\mathrm{f}/2\lambda_0' \gg 1$, $\widehat{H}_\mathrm{p}$ coincides with H_p. Because of demagnetization the slope of the minor magnetization curve takes on double the value given by Eq. (3.111) for multifilamentary wires in the transverse magnetic field. Figure 3.40 shows the dependences of H_p and $\widehat{H}_\mathrm{p}$ on filament diameter d_f for Nb-Ti multifilamentary wires and it is found that these are well described by the Campbell model.

Numerically calculated results of the energy loss density [33] for various filament diameters are shown in Fig. 3.41. In the case of very fine filaments in (a) the result is close to the analytic expression of Eq. (3.109) represented by the straight broken lines. The chained lines are the results of the irreversible

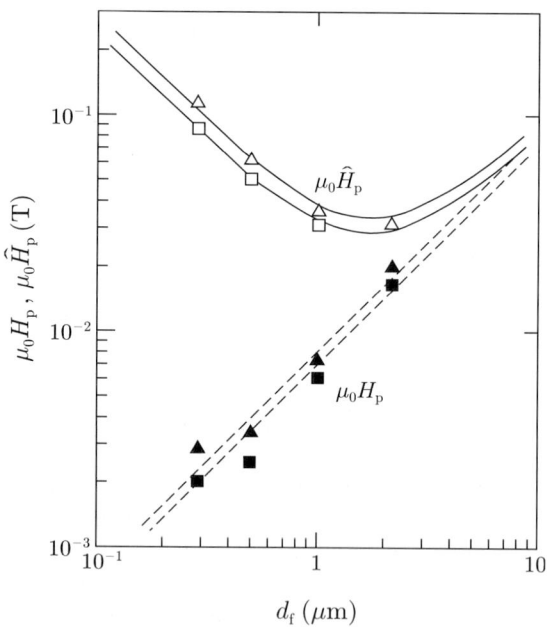

Fig. 3.40. Dependences of the characteristic fields, H_p and $\widehat{H}_\mathrm{p}$, on the filament diameter for Nb-Ti multifilamentary wires [34]. *Triangular* and *square* symbols show the values of the characteristic fields for $\mu_0 H_\mathrm{e} = 0.40$ T and 0.55 T, respectively. The *solid lines* are $\widehat{H}_\mathrm{p}$ estimated from Eq. (3.111) with H_p shown by the *broken lines* and the assumptions of $\lambda_0' = 0.56$ μm (0.40 T) and 0.54 μm (0.55 T)

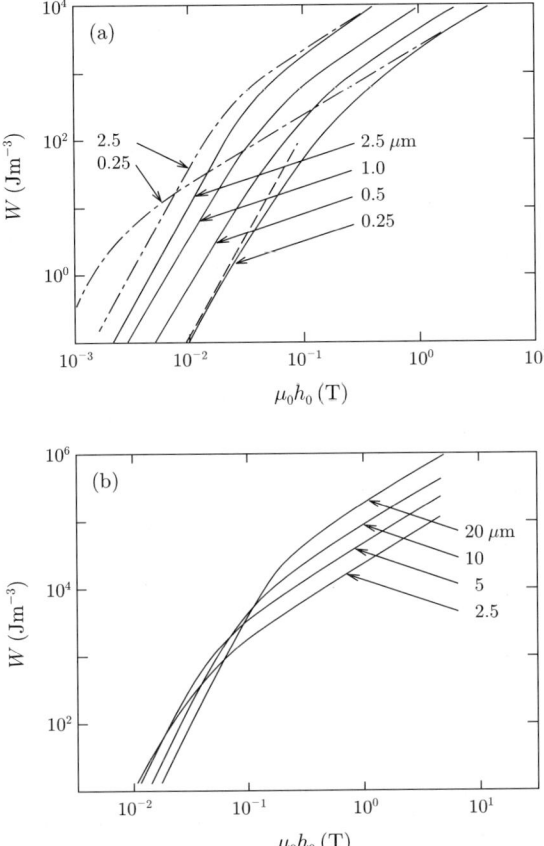

Fig. 3.41. AC energy loss density estimated using the Campbell model for various filament diameters [33]. The *broken* and *chained lines* show the results of Eq. (3.109) and the Bean-London model, respectively. Assumed parameters are $J_c = 1.0 \times 10^{10}\,\mathrm{Am}^{-2}$ and $\lambda'_0 = 0.63\,\mu\mathrm{m}$

Bean-London model. The numerical result tends to approach the prediction of this model as the filament diameter increases.

The AC energy loss density shown in Fig. 3.38 was observed in the absence of a large DC bias field, which is different from the above condition. In this case Eq. (3.88), the approximate formula based on the continuity equation for flux lines does not hold. In addition, not only J_c but also λ'_0 depends on the magnetic field strength. Therefore, a rigorous analysis is necessary. However, Eq. (3.109) is expected to be qualitatively correct. In practice the observed AC loss in recent multifilamentary wires with ultra fine filaments under small AC field amplitudes is even much smaller than the prediction of Eq. (3.109). In such wires the diameter of superconducting filaments is comparable to, or smaller than, the London penetration depth and the first penetration field is significantly enhanced as discussed in Appendix A.2. For example, if d_f is not much smaller than the London penetration depth λ, the effective lower critical field is predicted to be

$$H_{\mathrm{c}1}^* \simeq \left[1 - \frac{2\lambda}{d_\mathrm{f}} \tanh\left(\frac{d_\mathrm{f}}{2\lambda}\right)\right]^{-1} H_{\mathrm{c}1} . \tag{3.112}$$

From the result of Exercise 2.10, the corresponding energy loss density is given by

$$W = \frac{\mu_0}{3J_\mathrm{c} d_\mathrm{f}} \left(\frac{d_\mathrm{f}}{2\lambda_0'}\right)^4 (H_\mathrm{m} - H_{\mathrm{c}1}^*)^2 \left(H_\mathrm{m} + \frac{H_{\mathrm{c}1}^*}{2}\right) , \tag{3.113}$$

where h_0 is rewritten as H_m.

The AC penetration depth λ_0' defined by Campbell is an important quantity related to the AC loss in multifilamentary wires with very fine filaments. However, this length cannot be measured using Campbell's method in case the filament diameters are smaller than λ_0' (see Exercise 5.3). There are two methods for estimating λ_0' in this case; one is to compare the slope of minor magnetization curve with Eq. (3.111) and the other is to analyze the imaginary part of AC susceptibility as will be mentioned in Sect. 5.4.

3.8 Flux Creep

The superconducting current originated from the flux pinning mechanism has been assumed to be persistent in time so long as the external conditions are unchanged. However, if the DC magnetization of a superconducting specimen is measured for a long period, it is found to decrease slightly as shown in Fig. 3.42. That is, the superconducting current supported by the flux pinning is not a true persistent current but decreases with time. This results from the fact that the state in which the flux lines are restrained by the pinning potentials is only a quasistable one corresponding to a local minimum of the free energy in the state space and is not an actual equilibrium state. Therefore, a relaxation to the real equilibrium state, i.e., a decay of the shielding current takes place; it does so logarithmically with time as indicated in Fig. 3.42. The decay of the persistent current is accompanied by a decrease of the slope of the magnetic flux distribution caused by the motion of flux lines. Such flux motion is called "flux creep" which according to Anderson and Kim [36] is caused by thermal activation. It is supposed that thermally activated flux motion is not a macroscopic and continuous phenomenon like flux flow, but a partial and discontinuous one. The group of flux lines that move collectively is called the flux bundle.

We imagine one flux bundle to move under the influence of the transport current. When the flux bundle is virtually displaced in the direction of the Lorentz force, the variation in the energy of the flux bundle will be as shown schematically in Fig. 3.43(a). Point A corresponds to the state in which the

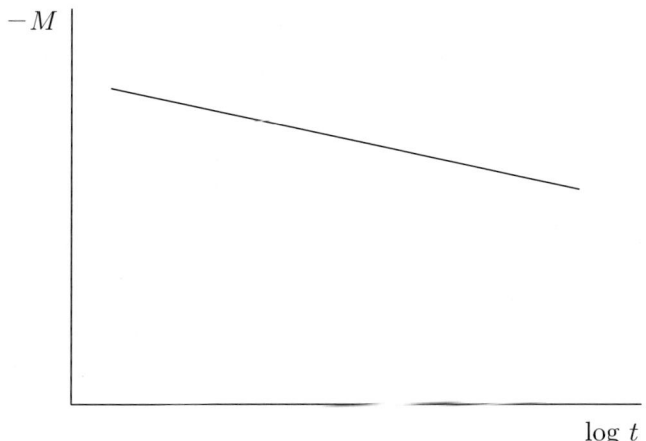

Fig. 3.42. Relaxation of magnetization due to flux creep

flux bundle is pinned, and the gradual decrease of energy that takes place when the flux bundle moves to the right represents the work done by the Lorentz force. It is necessary for the flux bundle to overcome the energy barrier at point B so as to be depinned. If there is no thermal activation, the state indicated in this figure is stable and the flux bundle does not move. In this virtual case, it is considered that the critical state is attained when the current density is increased until the peak and the bottom of the energy curve coincide with each other as shown in Fig. 3.43(b). At a higher current density continuous flux motion i.e., flux flow is expected to occur as in (c).

At a finite temperature T, thermal activation enables the flux bundle to overcome the energy barrier even in the state represented by Fig. 3.43(a). If the thermal energy, $k_B T$, is sufficiently small compared to the energy barrier, U, where k_B is the Boltzmann constant, the probability for the flux bundle to overcome the barrier for each attempt is given by the Arrhenius expression, $\exp(-U/k_B T)$. Hence, if the attempt frequency of the flux bundle is ν_0 and the distance by which the flux bundle moves during one hopping is a, the mean velocity of the flux lines to the right-hand side is given by $a\nu_0 \exp(-U/k_B T)$. The oscillation frequency ν_0 is expressed in terms of the Labusch parameter α_L and the viscous coefficient η as [37]

$$\nu_0 = \frac{\phi_0 \alpha_L}{2\pi B \eta} . \tag{3.114}$$

This is the frequency of damped oscillation within the averaged pinning potential. This is seen from the following argument: the relaxation time in the pinning potential is given as $\tau \sim \eta^*/k_p$ from Eq. (2.41), where k_p, the Labusch parameter per unit pinning center, is given by $k_p = \alpha_L/N_p$ where N_p is the pin concentration; thus, $\nu_0 \simeq 1/2\pi\tau$. In Sect. 2.3 the number of flux lines in the flux bundle is assumed to be 1.

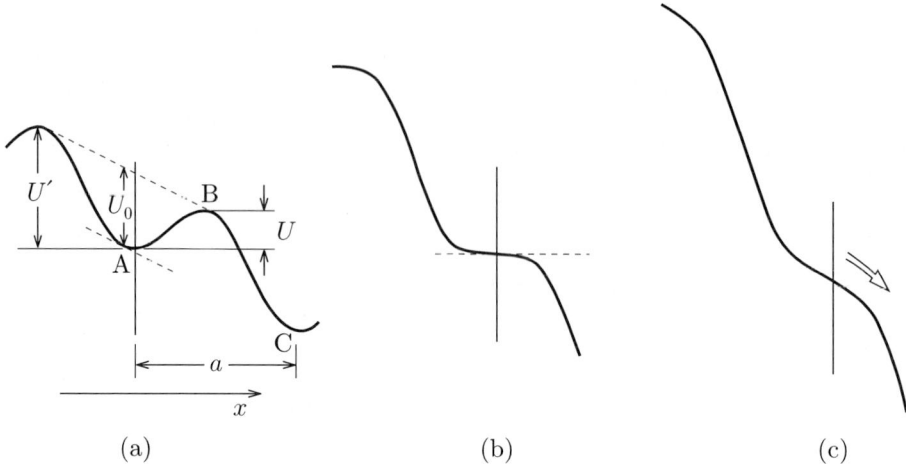

Fig. 3.43. Energy of flux bundle vs its position: (a) the case of transport current less than the virtual critical value. The flux bundle must overcome the barrier U so as to be depinned from the potential. (b) the virtual critical state and (c) the flux flow state

When the flux bundle is displaced by the flux line spacing a_f, its condition is expected to be approximately the same as before the displacement. In other words, the hopping distance a is supposed to be comparable to a_f. In general flux motion towards the left-hand side is also considered, hence the induced electric field according to Eq. (2.17) is given by

$$E = Ba_\mathrm{f}\nu_0 \left[\exp\left(-\frac{U}{k_\mathrm{B}T}\right) - \exp\left(-\frac{U'}{k_\mathrm{B}T}\right)\right], \qquad (3.115)$$

where U' is the energy barrier for the flux motion opposite to the Lorentz force (see Fig. 3.43(a)). Thus, the electrical field is generated by the motion of flux lines due to the flux creep. The mechanism responsible for the appearance of the electric field is essentially the same as that for flux flow in spite of the quantitative difference, and hence, a distinction in experiments between flux creep and flow is difficult. According to analysis of experimental results, most of the observed electric field at which the critical current density is determined by the usual four-terminal method comes from the mechanism of flux creep, as will be shown in Chap. 8. Hence, it is necessary to take account of the mechanisms of both flux creep and flow to analyze the practical E-J characteristics. The theoretical model of flux creep and flow used for the analysis of the E-J curves is described in Subsect. 8.5.2. The electromagnetic phenomena in high-temperature superconductors will be analyzed using this model, and the results will be discussed in Subsect. 8.5.3

Here we treat for simplicity the magnetic relaxation of a large superconducting slab ($0 \leq x \leq 2d$) in a magnetic field along the z-axis. From symmetry

3.8 Flux Creep

we need to treat only the half, $0 \leq x \leq d$. In an increasing field, the current flows along the positive y-axis and the motion of flux lines due to the flux creep occurs along the positive x-axis. If the average current density is denoted by J, the magnetic flux density is $B = \mu_0(H_e - Jx)$. In terms of its average value $\langle B \rangle$, the electric field at the surface, $x = 0$, is given by the Maxwell equation (2.1) as

$$E = \frac{\partial d \langle B \rangle}{\partial t} = -\frac{\mu_0 d^2}{2} \cdot \frac{\partial J}{\partial t}. \tag{3.116}$$

The relaxation of the superconducting current density with time can be obtained by substituting this equation into the left-hand side of Eq. (3.115) with U and U' expressed as functions of J.

Here we shall treat the case where the relaxation of the superconducting current is small in the vicinity of the virtual critical state. In this case $U \ll U'$ and the second term in Eq. (3.115) can be neglected. It is clear from Fig. 3.43(a) that U increases with decreasing J. Hence, it is reasonable to express U by expanding it in the form $U = U_0^* - sJ$, where U_0^* is the apparent pinning potential energy in the limit $J \to 0$ and s is a constant. As shown in Fig. 3.43(b) $U = 0$ is attained in the virtual critical state and the current density in this state is denoted by J_{c0}. Then, we have approximately $s = U_0^*/J_{c0}$ and

$$U = U_0^* \left(1 - \frac{J}{J_{c0}}\right). \tag{3.117}$$

Hence, the equation describing the time variation of the current density is given by

$$\frac{\partial J}{\partial t} = -\frac{2Ba_f\nu_0}{\mu_0 d^2} \exp\left[-\frac{U_0^*}{k_B T}\left(1 - \frac{J}{J_{c0}}\right)\right]. \tag{3.118}$$

This equation is easily solved and under the initial condition that $J = J_{c0}$ at $t = 0$ we obtain

$$\frac{J}{J_{c0}} = 1 - \frac{k_B T}{U_0^*} \log\left(\frac{2Ba_f\nu_0 U_0^* t}{\mu_0 d^2 J_{c0} k_B T} + 1\right). \tag{3.119}$$

After a sufficient time, the 1 in the argument of the logarithm can be neglected and the time variation of the current density shown in Fig. 3.42 can be derived. The apparent pinning potential energy U_0^* can be estimated from the logarithmic relaxation rate:

$$-\frac{d}{d \log t}\left(\frac{J}{J_{c0}}\right) = \frac{k_B T}{U_0^*}. \tag{3.120}$$

The energy barrier U is not generally a linear function of J, as in Eq. (3.117), over a wide range of J. The relaxation of the current for such a case will be discussed below: we simply approximate the relationship between the energy of the flux bundle and its central position, x, shown in Fig. 3.43(a) as

$$F(x) = \frac{U_0}{2} \sin kx - fx, \qquad (3.121)$$

where $k = 2\pi/a_{\rm f}$ and $f = JBV$ with V denoting the volume of the flux bundle. Differentiating Eq. (3.121) with respect to x, the quasiequilibrium position of the flux bundle is obtained:

$$x = -x_0 = -\frac{1}{k}\cos^{-1}\left(\frac{2f}{U_0 k}\right). \qquad (3.122)$$

On the other hand, $F(x)$ is locally maximum at $x = x_0$. Hence, the energy barrier is obtained as $U = F(x_0) - F(-x_0)$. That is,

$$\frac{U}{U_0} = \left[1 - \left(\frac{2f}{U_0 k}\right)^2\right]^{1/2} - \frac{2f}{U_0 k}\cos^{-1}\left(\frac{2f}{U_0 k}\right). \qquad (3.123)$$

If there is no thermal activation, the virtual critical state with $U = 0$ will be attained. In this case $x_0 = 0$ will be reached, and hence, $2f/U_0 k = 1$ will be satisfied. Since J in this case is equal to $J_{\rm c0}$, the general relation

$$\frac{2f}{U_0 k} = \frac{J}{J_{\rm c0}} \equiv j \qquad (3.124)$$

is derived. In terms of the normalized current density j, Eq. (3.123) can be written as

$$\frac{U}{U_0} = (1 - j^2)^{1/2} - j\cos^{-1} j. \qquad (3.125)$$

In case j is very close to 1 so that $1 - j \ll 1$, Eq. (3.125) reduces to $U/U_0 \simeq (2\sqrt{2}/3)(1-j)^{3/2}$. In this case j is described by

$$\frac{\partial j}{\partial t} = -c \exp\left[-\frac{U(j)}{k_{\rm B}T}\right], \qquad (3.126)$$

where $c = 2Ba_{\rm f}v_0/\mu_0 J_{\rm c0} d^2$. $U(j)$ is strictly a nonlinear function. If we expand $U(j)$ as in Eq. (3.117) within a narrow region, the variation of j as in Eq. (3.119) will be obtained. However, the value of U_0^* estimated from the relaxation is different from the real pinning potential energy, U_0. That is, U_0^* is usually smaller than U_0 as shown in Fig. 3.44. Hence, the measurement of magnetization relaxation leads to an underestimate of the pinning potential energy.

Here we shall show an example of the numerical analysis. We assume the temperature dependence of the virtual critical current density to be $J_{\rm c0} = A[1 - (T/T_{\rm c})^2]^2$. In case of strong pinning, the pinning potential energy U_0 is proportional to $J_{\rm c0}^{1/2}$ as will be shown in Sect. 7.7. Hence, the temperature dependence of U_0 is given by

$$U_0 = k_{\rm B}\beta\left[1 - \left(\frac{T}{T_{\rm c}}\right)^2\right], \qquad (3.127)$$

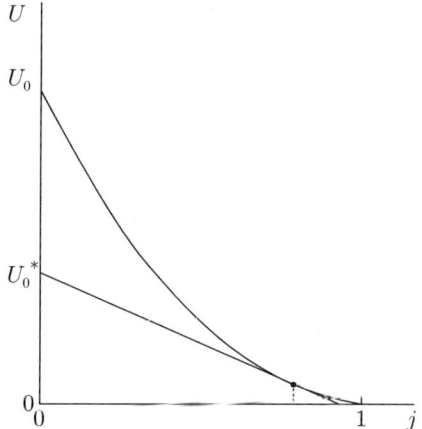

Fig. 3.44. Relationship between the energy barrier U and the normalized current density j. Extending the tangent at a given value of current density to $j = 0$, the intercept gives the apparent pinning potential energy U_0^*

where β is a constant dependent on the flux pinning strength. Here we assume an Y-based high-temperature superconductor with $T_c = 92$ K and other parameters: $B = 0.1$ T($a_f = 0.15$ μm), $\nu_0 = 1.0 \times 10^6$ Hz, $d = 1.0 \times 10^{-4}$ m and $A = 3.0 \times 10^9$ Am^{-2}. The results of numerical calculation [38] on the time dependence of j at various temperatures for $\beta = 3{,}000$ K are shown in Fig. 3.45. Figure 3.46(a) shows the apparent pinning potential energy U_0^* obtained from

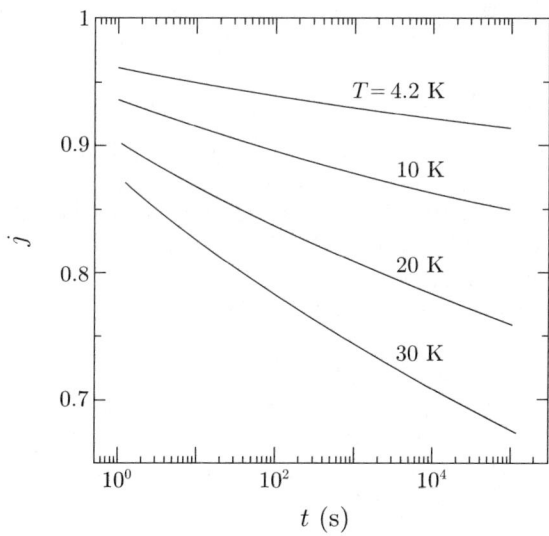

Fig. 3.45. Relaxation of normalized current density [38] obtained from Eq. (3.126) in case $\beta = 3{,}000$ K

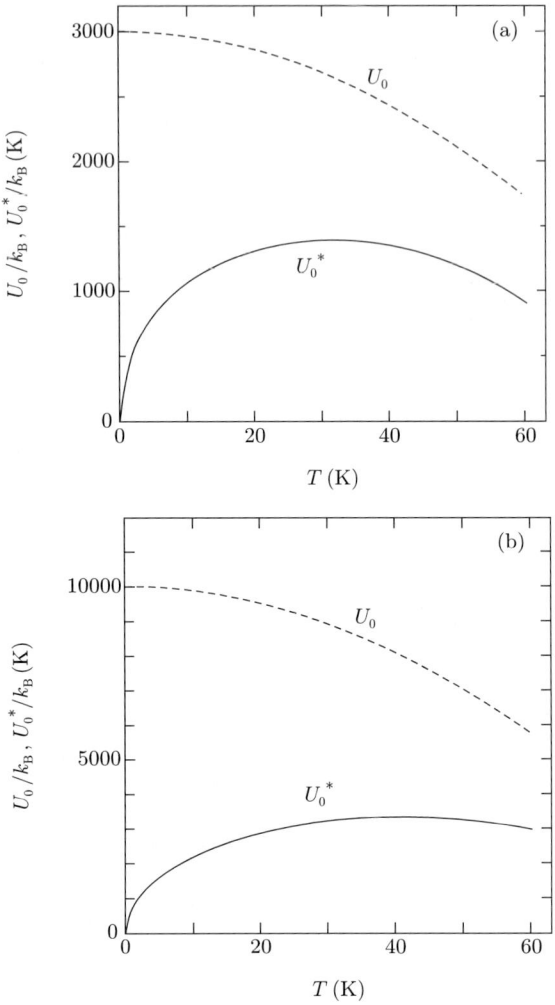

Fig. 3.46. Calculated apparent pinning potential energy U_0^* for given values of U_0 [38] for (**a**) $\beta = 3,000$ K and (**b**) $\beta = 10,000$ K

the average logarithmic relaxation rate in the range of $1 \leq t \leq 10^4$ s according to Eq. (3.120). In addition, (b) represents the relationship between U_0 and U_0^* for $\beta = 10,000$ K. It turns out that the U_0^* obtained is much smaller than the given U_0 and the difference becomes larger at lower temperatures, especially in the limit $T \to 0$, U_0^* approaches 0. Furthermore U_0^*/U_0 decreases as U_0 increases. In practice, according to the numerical calculation by Welch [39], if the current dependence of the activation energy is given by $U/U_0 \propto (1-j)^N$, the apparent pinning potential is expressed as (see Exercise 3.11)

$$U_0^* = c_N[(k_\mathrm{B}T)^{N-1}U_0]^{1/N} . \quad (3.128)$$

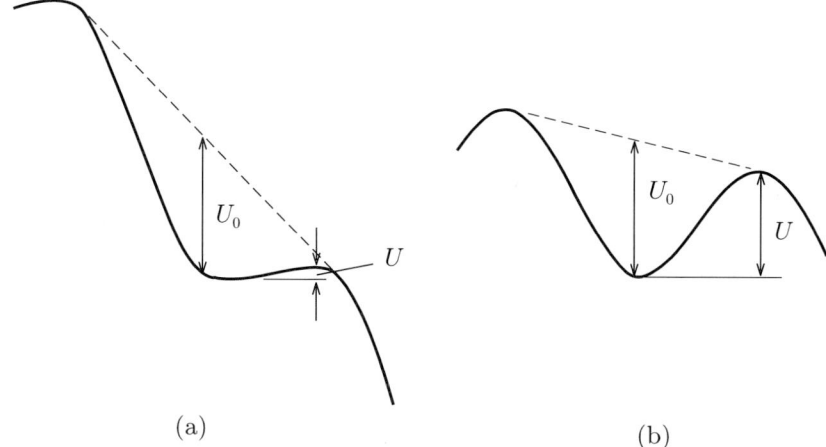

Fig. 3.47. (a) Relationship between the energy of flux bundle and its position at measurement of magnetization at low temperatures. Since the relaxation from the virtual critical state is not large, the energy barrier U is small. (b) Condition at measurement at high temperatures. Since the relaxation has already taken place to a considerable extent, U is large

In the case of sinusoidal washboard potential discussed above, $N = 3/2$ and $c_{3/2} = 1.65$. This result explains the above behavior exactly.

How can we understand this result? Assume that the initial state at $t = 0$ is the virtual critical state shown in Fig. 3.43(b).* The measurement of magnetic relaxation starts some time after the establishment of the initial state, at which time the variation in the energy of the flux bundle vs its position is shown in Fig. 3.47 for (a) low temperatures and (b) high temperatures. That is, at the low temperature in (a) little relaxation has taken place, and the energy barrier U is small; hence, the flux creep takes place easily, and the apparent pinning potential energy U_0^* is small. On the other hand, in (b) where the temperature is higher, the relaxation has already taken place revealing a large U; in this case, the flux creep is suppressed and the resultant U_0^* is large. This result can also be explained from Fig. 3.44. At low temperatures j at the time of measurement is close to 1 and U_0^* obtained from extrapolating the tangential line is much smaller than U_0. On the other hand, at high temperatures j is small and U_0^* is close to U_0. Furthermore the dependence of U_0^* on U_0 at a constant temperature has a similar explanation. The cases of large and small U_0 correspond qualitatively to Fig. 3.47(a) and (b), respectively.

Figure 3.48 shows some experimental U_0^* data for Y-Ba-Cu-O [39]: U_0^* takes on a small value at low temperatures and its temperature dependence

---

* In practice, even if we try to instantaneously establish an ideal external condition such as magnetic field before the flux creep starts, the relaxation due to the viscosity shown in Sect. 3.2 is added. Hence, the condition in Fig. 3.43(b) is not realized in a strict sense. However, the results after a sufficient long time do not

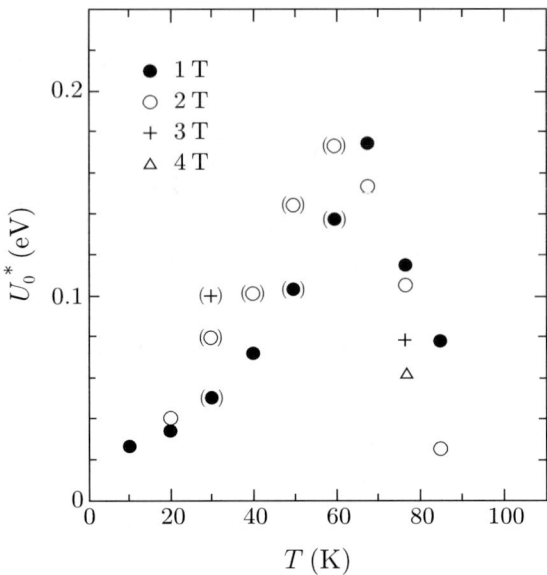

Fig. 3.48. Temperature dependence of apparent pinning potential energy U_0^* obtained from the magnetic relaxation for a melt-processed Y-Ba-Cu-O [39]

agrees qualitatively with the theoretical prediction as in Fig. 3.46. The observed flux creep phenomena can be approximately explained by the model assuming a simple sinusoidal variation in the energy given by Eq. (3.121). Flux creep in response to such a spatial variation in the potential was first pointed out by Beasley et al. [40] and later investigated in detail by Welch [39]. It should be noted that the shape of potential around the inflection point has a significant influence on the magnetic relaxation. However, any discussion has not yet been given on this problem in literature. Other mechanisms have also been proposed to explain the temperature dependence of U_0^* shown in Fig. 3.48, such as; the statistical distribution of the pinning potential energy [41], the nonlinear dependence of the energy barrier U on J [42] due to an enlargement of the pinning correlation length, that gives the flux bundle size, with decreasing J, etc. However, the width of distributed pinning potential energy necessary to explain the temperature dependence of U_0^* seems to be much larger than the observed distribution width of the critical current density. As for pinning correlation length enhancement, Campbell's AC penetration depth, as measured using Campbell's method described in Sect. 5.3, is not enhanced in the vicinity of $J = 0$. From these observations and from the fact that the shape of the pinning potential is necessarily involved when the flux bundle overcomes the barrier, it seems natural that the temperature

---

seem to depend sensitively on the initial condition as usually observed, and the above assumption will be admitted.

dependence of U_0^* originates mainly from the shape of the pinning potential. However, it is difficult to explain the temperature dependence of U_0^* at low temperature from only the simple effects of the pinning potential shape. This problem will be discussed in Appendix A.8.

It should be noted that, from the measurement of magnetic relaxation, the actual pinning potential energy that is important to the physics of flux pinning cannot be obtained, while the relaxation rate that is concerned with the lifetime of the persistent current and is an important engineering quantity can be obtained. In principle it is possible to obtain a value closer to the true pinning potential energy by inducing a current at slightly higher temperature and then measuring the relaxation of that current at some chosen lower temperature. However, if we wish to get a more exact value, an astronomically long time will be needed for the measurements.

When flux creep becomes pronounced at high temperatures, the flux motion occurs frequently, resulting in a steady electric field even for a small transport current. That is, the critical current density J_c is zero. In this regime magnetic hysteresis does not appear under a quasistatic variation of the applied field; i.e., the magnetization is reversible. The boundary between the reversible region with $J_c = 0$ and the irreversible one with $J_c \neq 0$ on the temperature vs magnetic field plane is called the irreversibility line (see Fig. 3.49). Figure 3.50 is a set of magnetization curves for Pb-In [43]. It can be seen that

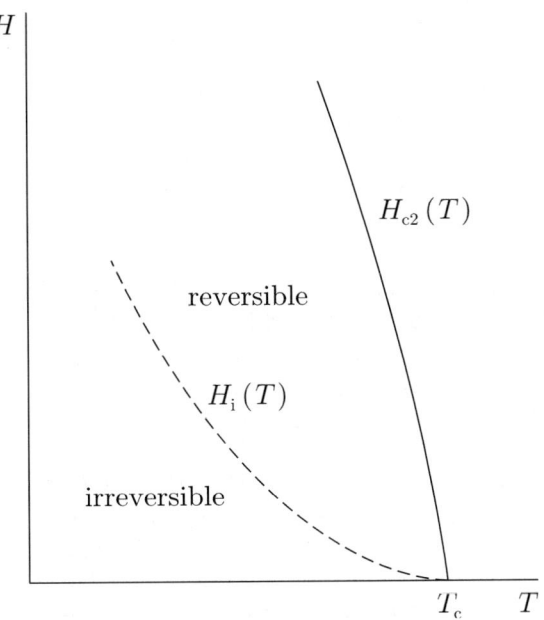

Fig. 3.49. Phase boundary $H_{c2}(T)$ and irreversibility line $H_i(T)$ on the temperature-magnetic field plane

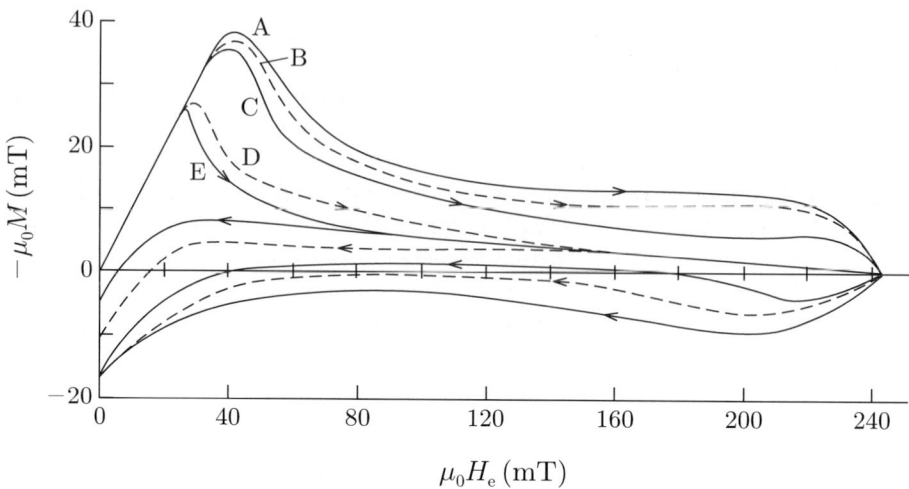

Fig. 3.50. Magnetization curves for Pb-8.23wt%In specimens with various flux pinning strengths [43]. 'A' is a specimen after cold working and 'B', 'C', 'D' and 'E' are specimens annealed at room temperature for 30 min., 1 day, 18 days and 46 days, respectively

the magnetization becomes reversible at high magnetic fields in specimens with weak pinning forces, and that the reversible region shrinks with increasing pinning strength. This shows that the irreversibility line depends on the pinning strength. At higher temperatures flux creep is stronger and hence, the above features are more noticeable in high-temperature superconductors. The melting of flux line lattice, the vortex glass-liquid transition, etc., were also proposed as for the origin of the irreversibility line. In this book we follow the mechanism of the flux creep. The detailed discussion on this point for high-temperature superconductors will be given in Sect. 8.2.

The irreversibility line at a given temperature T is defined as the magnetic field, $H_i(T)$, at which the critical current density determined in terms of the electric field criterion, $E = E_c$, for example, reduces to zero. That is, neglecting the second term in Eq. (3.115) again, from the requirement that $U = U_0$ in the limit $J = J_c = 0$ we have

$$U_0(H_i) = k_\mathrm{B} T \log\left(\frac{\mu_0 H_i a_\mathrm{f} \nu_0}{E_c}\right) . \tag{3.129}$$

As expected, U_0 depends on the flux pinning strength and is a function of magnetic field and temperature. Hence, the irreversibility line, $H_i(T)$, can be obtained from Eq. (3.129). The estimation of U_0 will take place in Sect. 7.9, and examples of the irreversibility line for high-temperature superconductors will be shown in Sect. 8.5. It was mentioned above that only the apparent pinning potential energy U_0^* can be obtained from the measurement of magnetic relaxation. On the other hand, the irreversibility line is directly related to the true pinning potential energy U_0. Hence, U_0 can be estimated from a measured value of the irreversibility field.

At higher temperatures and/or higher magnetic fields the flux lines tend to creep in the direction of the Lorentz force and a voltage appears. This mechanism is identical with that of flux flow. Based on this concept the voltage states in Fig. 3.43(a) and (c) are the creep state and the flow state, respectively. However, these are not easily discerned experimentally. In the regime of flux creep, we make the approximation

$$U' \simeq U + fa_{\rm f} = U + \pi U_0 \frac{J}{J_{\rm c0}} \quad (3.130)$$

for use in Eq. (3.115). If the second term is sufficiently less than $k_{\rm B}T$, the electric field may be written

$$E \simeq \frac{\pi B a_{\rm f} \nu_0 U_0 J}{J_{\rm c0} k_{\rm B} T} \exp\left(-\frac{U_0}{k_{\rm B}T}\right) . \quad (3.131)$$

This is an ohmic current-voltage characteristic, which takes into consideration the fact that U approaches U_0 in the range of sufficiently small J. The corresponding electrical resistivity is obtained as

$$\rho = \rho_0 \exp\left(-\frac{U_0}{k_{\rm B}T}\right), \quad (3.132)$$

where $\rho_0 = \pi B a_{\rm f} \nu_0 U_0 / J_{\rm c0} k_{\rm B} T$ can be approximately regarded as a constant within a narrow temperature range. This suggests that U_0 may be estimated from the slope of the $\log \rho$ vs $1/T$. However, as pointed out by Yeshurun and Malozemoff, [44] such an attempt would lead to error, since U_0 varies with temperature. If we write $U_0 = K(1 - T/T_{\rm c})^p$, for instance, at high temperatures, it is easy to derive

$$\frac{\partial \log \rho}{\partial (1/T)} = -\frac{U_0}{k_{\rm B}} \left(1 + \frac{pT}{T_{\rm c} - T}\right) . \quad (3.133)$$

This suggests that a simple plot of $\log \rho$ vs $1/T$ would lead to an overestimate of U_0, especially so in the vicinity of $T_{\rm c}$. Generally the value of p is unknown and U_0 cannot be obtained. This is due to the fact that Eq. (3.132) is correct in a fixed magnetic field only within a very narrow temperature range. It is possible to observe an electrical resistivity similar to that of Eq. (3.132) in the presence of a large transport current. In this case the condition of Fig. 3.47(a) holds and what is obtained is none other than U_0^*. The apparent pinning potential energy obtained in this method agrees well with that obtained from magnetic relaxation [45].

A very wide reversible region exists between the irreversibility line, $H_{\rm i}(T)$, and the phase boundary, $H_{\rm c2}(T)$, as shown in Fig. 3.49 for a superconductor with a weak pinning force; this leads to a wide resistive transition. Hence, it is

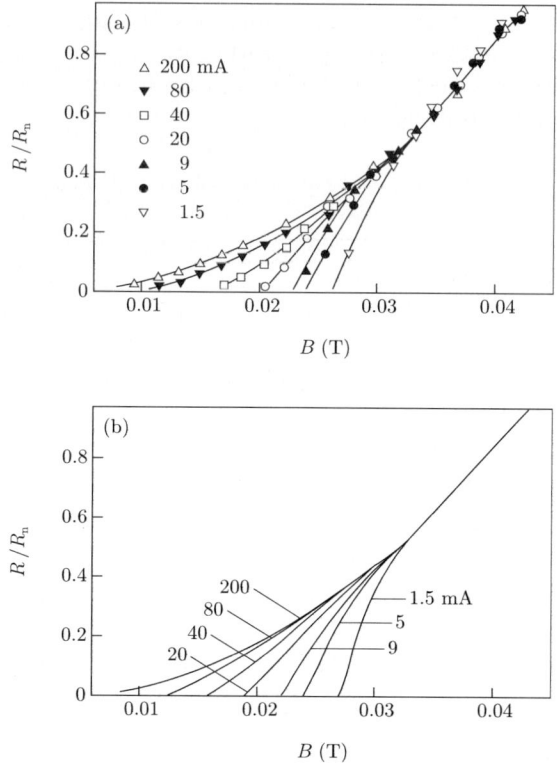

Fig. 3.51. (a) Resistance vs magnetic field at 4.2 K for a Nb-Ta specimen with weak pinning force [26]. (b) Theoretically calculated resistance vs magnetic field assuming a flux flow with observed critical current density and flux flow resistivity [46]

claimed that the width of the resistive transition is determined by the pinning strength. When the temperature is lowered slightly below the irreversibility line, J_c recovers suddenly. On the other hand, if the temperature is slightly increased, a variation from the flux creep state to the flux flow state takes place. Hence, the flux lines are considered to be in the flow state throughout most of the wide resistive transition. Figure 3.51(a) shows the broad magnetic field range of the resistive transition that has been measured for a Nb-Ta alloy [26] under various current densities; (b) shows the corresponding theoretical results [46] constructed from the observed critical current density and flux flow resistivity. These agree well with each other in the range where a small resistivity due to the flux creep can be disregarded. This supports the speculation that the dominant component of the usually observed resistive transition comes from flux flow. Such a broad resistive transition is also observed for high-temperature superconductors and the same discussion can be repeated in principle. However, the effect of flux creep is then much more noticeable

3.8 Flux Creep   151

and the region of low resistivity will be further widened. In addition the effect of the superconducting fluctuations is considered to be large around the phase boundary and the shape of the resistive transition itself seems to be strongly influenced by the fluctuations.

Exercises

3.1. Derive Eqs. (3.17), (3.18) and (3.19).

3.2. Derive the viscous energy loss density given by the second term of Eq. (3.32) directly from the second term of Eq. (2.73). (*Hint*: Since the viscous energy loss density is small, Eq. (3.27), the quasistatic value for the velocity of flux lines, can be used.)

3.3. Using the method of Eq. (2.74), derive Eq. (3.39), the energy loss density when an AC magnetic field is applied to a current-carrying superconductor.

3.4. Derive Eq. (3.59) from Eq. (3.58).

3.5. Calculate the DC susceptibility in the field cooled process when the parameter m' in Eq. (3.79), representing the temperature dependence of J_c, is equal to 2.

3.6. Calculate the DC susceptibility when a constant magnetic field is applied and then, the temperature is elevated after cooling down at zero field.

3.7. From the area of the $\langle B \rangle$-H loop, derive Eq. (3.109), the AC energy loss density in a superconducting slab thinner than the AC penetration depth λ'_0, in a parallel AC magnetic field.

3.8. Derive η_p in Fig. 3.37 for a bulk superconductor for a sufficiently small AC field amplitude h_0 using the Campbell model.

3.9. Derive Eq. (3.111), where the superconductor is a slab of thickness d_f.

3.10. Prove that the half size of a superconductor must be smaller than the AC penetration depth, λ'_0, for the effect of reversible flux motion to be significant. For simplicity it is assumed that the AC magnetic field is applied parallel to a wide superconducting slab of thickness $2d$. (*Hint*: Use the condition that the maximum displacement of flux lines in the superconductor in one period is less than the diameter of the pinning potential, $2d_i$.)

3.11. The current density dependence of the energy barrier, U, is assumed as $U(J) = U_0(1 - J/J_{c0})^N$, where $N > 1$. Discuss the dependences of the apparent pinning potential energy, U_0^*, on the temperature, T, and U_0, using Fig. 3.44.

3.12. When the resistivity criterion, $\rho = \rho_c$, is used for the definition of the critical current density, how is the expression of the irreversibility line different from Eq. (3.129)? (*Hint*: Use Eq. (3.131)).

References

1. F. Irie and K. Yamafuji: J. Phys. Soc. Jpn. **23** (1967) 255.
2. R. Hancox: Proc. IEE **113** (1966) 1221.
3. T. Matsushita, F. Sumiyoshi, M. Takeo and F. Irie: Tech. Rep. Kyushu Univ. **51** (1978) 47 [in Japanese].
4. W. T. Norris: J. Phys. D (Appl. Phys.) **3** (1970) 489.
5. C. P. Bean: Phys. Rev. Lett. **8** (1962) 250; H. London: Phys. Lett. **6** (1963) 162.
6. M. Askin: J. Appl. Phys. **50** (1979) 7060.
7. V. B. Zenkevitch, V. V. Zheltov and A. S. Romanyuk: Sov. Phys. Dokl. **25** (1980) 210.
8. C. Y. Pang, A. M. Campbell and P. G. McLaren: IEEE Trans. Magn. **17** (1981) 134.
9. Y. Kato, M. Noda and K. Yamafuji: Tech. Rep. Kyushu Univ. **53** (1980) 357 [in Japanese].
10. M. Noda, K. Funaki and K. Yamafuji: Mem. Faculty Eng. Kyushu Univ. **46** (1986) 63.
11. M. Noda, K. Funaki and K. Yamafuji: Tech. Rep. Kyushu Univ. **58** (1985) 533 [in Japanese].
12. T. Kawashima, T. Sueyoshi and K. Yamafuji: Jpn. J. Appl. Phys. **17** (1978) 699.
13. K. Kaiho, K. Koyama and I. Todoroki: Cryog. Eng. Jpn. **5** (1970) 242 [in Japanese].
14. N. Sakamoto and K. Yamafuji: Jpn. J. Appl. Phys. **16** (1977) 1663.
15. T. Ogasawara, Y. Takahashi, K. Kambara, Y. Kubota, K. Yasohama and K. Yasukochi: Cryogenics **19** (1979) 736.
16. F. Rothwarf, C. T. Rao and L. W. Dubeck: Solid State Commun. **11** (1972) 1123.
17. K. Funaki and K. Yamafuji: Jpn. J. Appl. Phys. **21** (1982) 299.
18. K. Funaki, T. Nidome and K. Yamafuji: Jpn. J. Appl. Phys. **21** (1982) 1121.
19. K. Funaki, M. Noda and K. Yamafuji: Jpn. J. Appl. Phys. **21** (1982) 1580.
20. H. T. Coffey: Cryogenics **7** (1967) 73.
21. K. Yamafuji, M. Takeo, J. Chikaba, N. Yano and F. Irie: J. Phys. Soc. Jpn. **26** (1969) 315.
22. H. J. Fink: Phys. Lett. **19** (1965) 364.
23. C. P. Bean and J. D. Livingston: Phys. Rev. Lett. **12** (1964) 14.
24. P. G. de Gennes: *Superconductivity of Metals and Alloys* (Translated by P. A. Pincus) (W. A. Benjamin, Inc., New York, 1966) section 3.2.
25. P. G. de Gennes: Solid State Commun. **3** (1965) 127.
26. R. A. French, J. Lowell and K. Mendelssohn: Cryogenics **7** (1967) 83.
27. H. R. Hart, Jr. and P. S. Swartz: Phys. Rev. **156** (1967) 403.
28. T. Matsushita, T. Honda, Y. Hasegawa and Y. Monju: J. Appl. Phys. **54** (1983) 6526. As to the magnetic field dependence of surface critical current density, T. Matsushita, T. Honda and K. Yamafuji: Memo. Faculty of Engineering, Kyushu University, **43** (1983) 233.
29. L. J. Barnes and H. J. Fink: Phys. Lett. **20** (1966) 583.
30. S. T. Sekula and R. H. Kernohan: Phys. Rev. B **5** (1972) 904.
31. T. Matsushita, E. S. Otabe, T. Matsuno, M. Murakami and K. Kitazawa: Physica C **170** (1990) 375.

32. A. M. Campbell: J. Phys. C **4** (1971) 3186.
33. T. Matsushita, N. Harada, K. Yamafuji and M. Noda: Jpn. J. Appl. Phys. **28** (1989) 356.
34. F. Sumiyoshi, M. Matsuyama, M. Noda, T. Matsushita, K. Funaki, M. Iwakuma and K. Yamafuji: Jpn. J. Appl. Phys. **25** (1986) L148.
35. S. Takács and A. M. Campbell: Supercond. Sci. Technol. **1** (1988) 53.
36. P. W. Anderson and Y. B. Kim: Rev. Mod. Phys. **36** (1964) 39.
37. K. Yamafuji, T. Fujiyoshi, K. Toko and T. Matsushita: Physica C **159** (1989) 743.
38. T. Matsushita and E. S. Otabe: Jpn. J. Appl. Phys. **31** (1992) L33.
39. D. O. Welch: IEEE Trans. Magn. **27** (1991) 1133.
40. M. R. Beasley, R. Labusch and W. W. Webb: Phys. Rev. **181** (1969) 682.
41. C. W. Hagen and R. Griessen: Phys. Rev. Lett. **62** (1989) 2857.
42. M. V. Feigel'man, V. B. Geshkenbein, A. I. Larkin and V. M. Vinokur: Phys. Rev. Lett. **63** (1989) 2303.
43. J. D. Livingston: Phys. Rev. **129** (1963) 1943.
44. Y. Yeshurun and A. P. Malozemoff: Phys. Rev. Lett. **60** (1988) 2202.
45. K. Yamafuji, Y. Mawatari, T. Fujiyoshi, K. Miyahara, K. Watanabe, S. Awaji and N. Kobayashi: Physica C **185–189** (1991) 2285.
46. T. Matsushita and B. Ni: Physica C **166** (1990) 423.

4

Longitudinal Magnetic Field Effect

4.1 Outline of Longitudinal Magnetic Field Effect

When a transport current is applied to a superconducting cylinder or tape in a longitudinal magnetic field, as shown in Fig. 4.1, various characteristic phenomena are observed and these are generically called the longitudinal field effect. The phenomena are as follows:

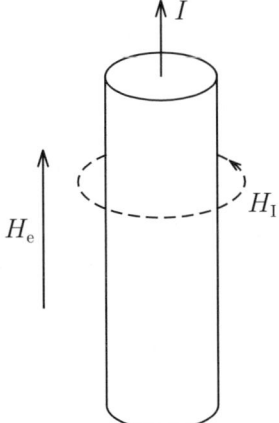

Fig. 4.1. Current and self field H_I in a longitudinal magnetic field

(1) A longitudinal paramagnetic magnetization is induced by the current. Figure 4.2 shows the variation in the longitudinal magnetization [1] when the transport current is applied after the longitudinal magnetic field is increased up to a certain value. It can be seen that the magnetization changes from diamagnetic to paramagnetic. This is called the paramagnetic effect.

156   4 Longitudinal Magnetic Field Effect

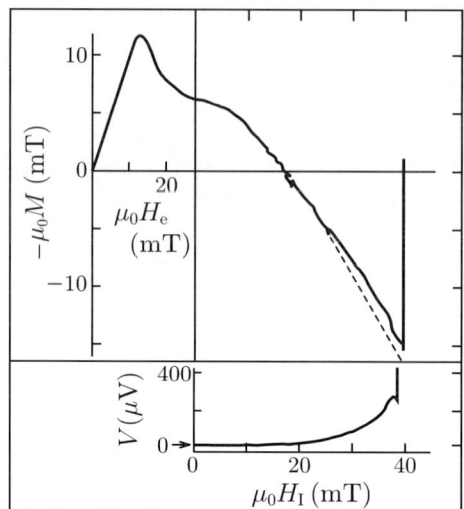

Fig. 4.2. Variation in magnetization (*upper figure*) and longitudinal voltage (*lower figure*) when a longitudinal magnetic field of 28 mT is applied and then a transport current is applied to a cylindrical Pb-Tl specimen [1]

(2) The critical current density is much larger than in the case in a transverse magnetic field. Figure 4.3 shows the data for Ti-Nb [2]. The enhancement factor sometimes exceeds 100.

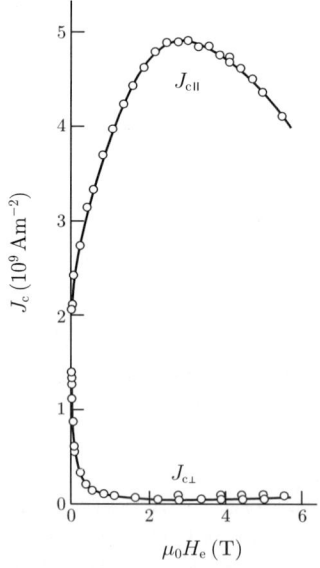

Fig. 4.3. Critical current densities of a cylindrical Ti-36%Nb specimen in transverse and longitudinal magnetic fields [2]

4.1 Outline of Longitudinal Magnetic Field Effect   157

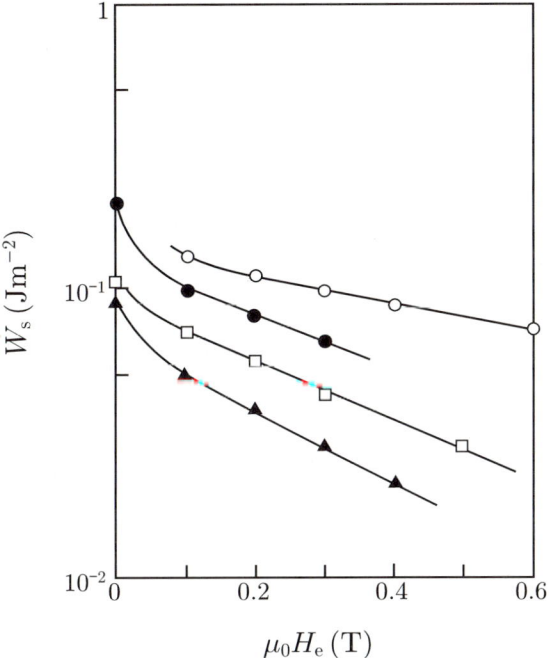

Fig. 4.4. Loss energy per unit surface area of four Nb-Ti wire specimens due to AC current vs. longitudinal magnetic field [3]. The self field amplitude of the AC current is kept constant at 0.14 T

(3) The loss due to an AC current decreases with increasing longitudinal magnetic field. Figure 4.4 represents the variations in the AC loss of four Nb-Ti specimens [3] with the same amplitude of self field due to the AC current.

(4) The fundamental equation, Eq. (1.146) derived by Josephson which relates the motion of flux lines to the electromagnetic phenomena is not obeyed [4]. That is,

$$\bm{E} \neq \bm{B} \times \bm{v} \,. \tag{4.1}$$

In the experiment by Cave et al. in [4], a small AC current was superimposed on a DC current as shown in Fig. 4.1, and the induced electric field due to the variation in the flux distribution was observed. It was found that $\bm{E}$ was almost parallel to $\bm{B}$, so the inequality, Eq. (4.1), was demonstrated.

(5) In the resistive state where the current density exceeds the critical value, it was found [5] that a region with negative electric field in the longitudinal direction existed, and the structure of the surface electric field shown in Fig. 4.5 was observed [5].

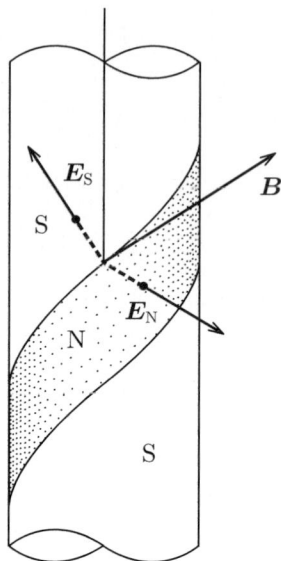

Fig. 4.5. Structure of surface electric field of a cylindrical superconductor in a longitudinal magnetic field in a resistive state

In a longitudinal magnetic field it can be supposed that the current flows parallel to the field or is close to this ideal condition. This fact has been clarified by various experimental and theoretical investigations. The force-free model [6] is one of the models proposed to explain the phenomena. In this model it is assumed that the local current flow is parallel to the magnetic field, so that a Lorentz force is not applied to flux lines. This is the origin of the name of the model. The force-free situation is expressed as

$$\boldsymbol{J} \times \boldsymbol{B} = 0 , \qquad (4.2)$$

where $\boldsymbol{J}$ and $\boldsymbol{B}$ are the current density and the magnetic flux density, respectively. This requires that

$$\boldsymbol{J} = \frac{\alpha_\mathrm{f} \boldsymbol{B}}{\mu_0} , \qquad (4.3)$$

where α_f is a scalar quantity. If the above equation is combined with the concept of a critical state model, α_f is a parameter determining the critical current density. Hence, α_f is in general a function of B. However, it is assumed that α_f is a constant for simplicity. This is not a bad approximation in the low field region where the critical current density increases with magnetic field as shown in Fig. 4.3.

Assume the case of a long superconducting cylinder of radius R in a parallel magnetic field H_e. Using the Maxwell equation, Eq. (4.3) is written in cylindrical coordinates as

4.1 Outline of Longitudinal Magnetic Field Effect

$$-\frac{\partial B_z}{\partial r} = \alpha_f B_\phi , \quad (4.4)$$

$$\frac{1}{r} \cdot \frac{\partial (rB_\phi)}{\partial r} = \alpha_f B_z . \quad (4.5)$$

In the equation above it is assumed from symmetry that the magnetic flux distribution is uniform in the azimuthal and longitudinal directions, i.e., $\partial/\partial\phi = \partial/\partial z = 0$. Eliminating B_z, the following is obtained:

$$\frac{\partial^2 B_\phi}{\partial r^2} + \frac{1}{r} \cdot \frac{\partial B_\phi}{\partial r} + \left(\alpha_f^2 - \frac{1}{r^2}\right) B_\phi = 0 . \quad (4.6)$$

As is well known, the solution of this equation is given by

$$B_\phi = A J_1(\alpha_f r) , \quad (4.7)$$

where J_1 is the Bessel function of first order and A is a constant. If the total current flowing through the superconducting cylinder is represented by I, the boundary condition is described as

$$B_\phi(r=R) = \frac{\mu_0 I}{2\pi R} . \quad (4.8)$$

The constant A can be determined from this condition. Then,

$$B_\phi = \frac{\mu_0 I}{2\pi R} \cdot \frac{J_1(\alpha_f r)}{J_1(\alpha_f R)} . \quad (4.9)$$

From Eqs. (4.5) and (4.9) it follows that [1]

$$B_z = \frac{\mu_0 I}{2\pi R} \cdot \frac{J_0(\alpha_f r)}{J_1(\alpha_f R)} , \quad (4.10)$$

where J_0 is the Bessel function of zeroth order. It should be noted that another boundary condition

$$B_z(r=R) = \mu_0 H_e \quad (4.11)$$

should also be satisfied. This means that α_f is determined by the boundary condition. However, this is contradictory to the previous statement that α_f is determined by the local critical condition. This fact clearly shows the applicable limit of the force-free model. However, this point will be ignored for now and the results from the model will be investigated.

The magnetic flux distribution obtained above is depicted in Fig. 4.6. In general, since $J_0(r')$ decreases monotonically with increasing r' up to the first zero point, the z-component of the magnetic flux density in the superconductor is maximum at the center and takes values larger than $\mu_0 H_e$ over the entire region. That is, the flux distribution is paramagnetic in the longitudinal direction. Thus, the paramagnetic effect can be explained. The longitudinal magnetization is

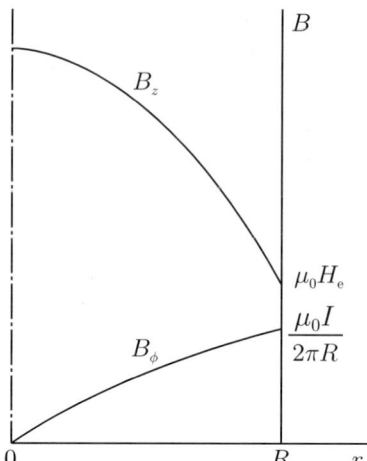

Fig. 4.6. Magnetic field distribution in a cylindrical superconductor in a force-free state

$$M_z = \frac{1}{\mu_0}\langle B_z \rangle - H_e = H_e \frac{J_2(\alpha_f R)}{J_0(\alpha_f R)}, \quad (4.12)$$

where J_2 is the Bessel function of second order. In the region being treated M_z is positive. The broken line in Fig. 4.2 shows this theoretical result [1]. Hence, it can be seen that the force-free model explains the experimental result well, although it contains the physical problem mentioned above. The paramagnetic effect can be easily derived for a superconducting slab also. This is Exercise 4.1. It should be noted that the force-free model holds for the distribution of the longitudinal component of magnetic flux, although the effect of flux pinning is present in ordinary superconductors. (Compare Eq. (4.2) and the force balance equation in Chap. 2.)

As shown above the magnetic flux distribution in a longitudinal magnetic field is in the force-free state described by Eq. (4.2) or in a state close to this one. This equation was theoretically derived by Josephson [7] as the one describing the equilibrium state in pin-free superconductors. According to this result it is believed that the force-free state is stable, and this equation does not constrain the upper limit of the density of transport current, i.e., the critical current density. It may be expected that the current can flow infinitely, since the Lorentz force does not work on flux lines. However, there exists a certain limit in practical cases. In particular, the critical current density depends on the flux pinning strength [8], which is approximately proportional to the critical current density in a transverse magnetic field, as shown in Fig. 4.7. The critical current density in a superconductor with a very weak pinning force is very small even in a longitudinal magnetic field. This fact suggests that the force-free state may be unstable without the influence of the flux-pinning interactions.

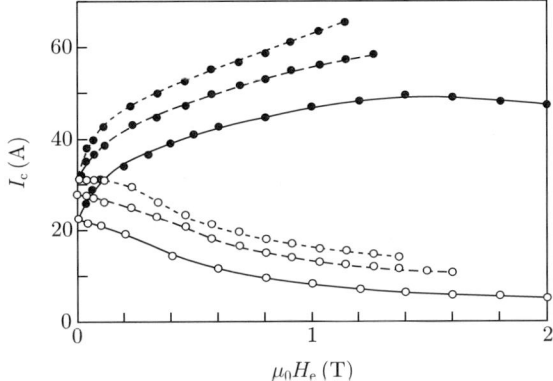

Fig. 4.7. Critical current densities of Nb_3Sn thin films in longitudinal (●●) and transverse (○○) magnetic fields [8]. Both critical current densities increase with increasing pinning strength due to neutron irradiation

4.2 Flux-Cutting Model

In Sect. 4.1 we learned that the magnetic flux distribution in a longitudinal magnetic field is in the force-free state or a state close to this ideal one. Then, the question arises about the motion of flux lines related to this distribution. As is done usually in experiments, for example, it is assumed that the longitudinal magnetic field is applied first so that the shielding current flows in the azimuthal direction perpendicular to the magnetic flux density and then the current is applied. What is the flux motion that occurs during this process, in which the flux distribution changes from the usual one to the force-free one? Campbell and Evetts [9] first considered that flux lines tilted by the angle determined by the external field and the self field are nucleated at the surface of the superconducting cylinder, and that then, these flux lines penetrate the superconductor, keeping their angle constant. If this is correct, Josephson's equation (1.146) for the induced electric field is satisfied. However, the experimental results of Cave et al. [4] show that this equation is not satisfied. This means that such a motion of flux lines does not occur. In this experiment the DC and superimposed small AC currents were applied to a superconducting cylinder in a longitudinal magnetic field, and the observed electric field which was induced was almost parallel to the longitudinal direction. This means that only the azimuthal component of the magnetic flux changes, while the longitudinal component is almost unchanged. From this experimental result Cave et al. proposed a kind of flux-cutting model, in which it was assumed that only the azimuthal flux caused by the AC current moves into and out of the superconductor by crossing the longitudinal flux (see Fig. 4.8).

The flux-cutting model was originally proposed for the solution of the contradiction between a constant longitudinal paramagnetic magnetization and a steady longitudinal electric field due to a constant transport current in the

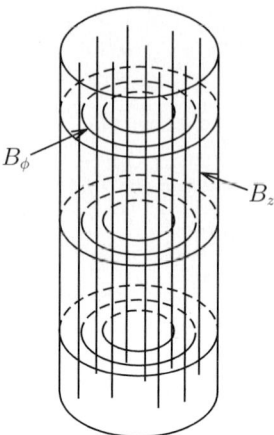

Fig. 4.8. Model of direct cutting between a longitudinal component of magnetic flux B_z and an azimuthal one B_ϕ proposed by Cave et al. [4]

resistive state [1]. That is, if the steady longitudinal electric field is attributed to the continuous penetration of the azimuthal component of the magnetic flux, as was treated in Exercise 2.4, and if the flux lines penetrate continuously into the superconductor, the longitudinal component of the magnetic flux should increase continuously with time. This contradicts the experimental result shown in Fig. 4.2. Since the longitudinal magnetization does not change with time, Walmsley [1] considered that only the azimuthal component penetrates the superconductor and proposed the flux-cutting model. In comparison to this, the flux-cutting mechanism of Cave et al. [4] was proposed to explain the phenomenon in the non-resistive state below the critical current.

The flux-cutting model is classified into the inter-cutting model, in which flux lines with different angles cut each other, and the intra-cutting model, in which flux lines are cut and reconnection between different flux lines occurs. The model of Cave et al. [4] belongs to the former category, and the intersection and cross-joining model by Clem [10] and Brandt [11] to the latter one. The process considered in the intersection and cross-joining model is shown in Fig. 4.9. The flux lines on opposite sides have different angles to each other in (a) and these are cut at the nearest positions to each other, and then these are reconnected as in (b). These flux lines become straight as in (c). If we compare the conditions between (a) and (c), i.e., before and after the cutting, it is found that the transverse component of the magnetic flux is changed, while the longitudinal component is unchanged. After this occurs, the flux lines again cut each other with adjacent rows. If such a variation occurs continuously, penetration only of the transverse or azimuthal component of the magnetic flux actually occurs. Later Brandt [11] clarified that the electric field strength in the resistive state which is derived from the mechanism of intersection and cross-joining is several orders of magnitude smaller than experimental results.

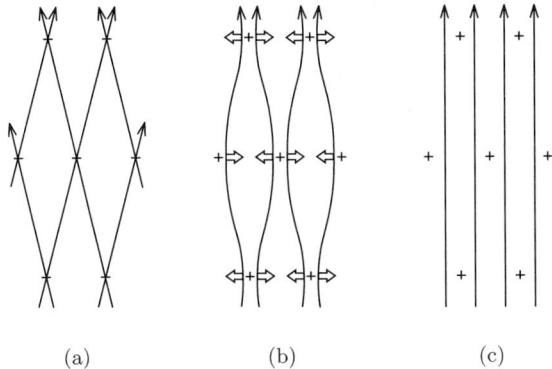

Fig. 4.9 Process of flux cutting in a model of intersection and cross-joining [10, 11]

To overcome this problem Clem [12] proposed a double cutting mechanism, in which the intersection and cross-joining shown in Fig. 4.9 is assumed to occur twice successively during the process of cutting. This is consequently a kind of direct inter-cutting between flux lines with different angles.

All the flux-cutting models are based on the hypothesis that only the transverse or azimuthal component of the magnetic flux penetrates the superconductor. Although this is acceptable for a dynamic state when the current is below the critical current, where the longitudinal electric field is induced by the variation in the distribution of transverse magnetic flux, it should be noted that there is no theoretical foundation for this idea in the case of the steady resistive state above the critical current. That is, Josephson's relation (1.146) which connects flux motion with the electric field is also not obeyed in a steady resistive state, as will be described in Sect. 4.7. Hence, there is no reason for the insistence that the longitudinal electric field in the resistive state should be attributed to the steady penetration of transverse magnetic flux. In other words, the flux-cutting model in the resistive state is based on Eq. (1.146), although this equation is not satisfied. It may be considered that the situation is also the same for the case of the non-resistive state below the critical current. However, the penetration of the transverse component of magnetic flux in the non-resistive state can be proved by experiments.

The flux-cutting models are based on the theoretical result of Josephson that the force-free state is essentially stable. It is assumed that the critical state is attained when the angle between adjacent flux lines, which is proportional to the current density, reaches the threshold value for cutting. However, as can be seen from the fact that many flux-cutting models have been proposed, the mechanism of the flux cutting has not yet been theoretically clarified, and no unified concept has been decided upon for the threshold value. The only common standpoint among the theoretical treatments is the calculation of the magnetic repulsive force between tilted flux lines. According to the calculation by Brandt et al. [13] the repulsive force between adjacent flux lines with tilt angle $\delta\theta$ is

$$f_c = \frac{\phi_0^2}{2\mu_0\lambda^2} \cot \delta\theta \tag{4.13}$$

per intersection. Hence, the flux cutting is believed to take place when flux lines are pushed toward each other by a force larger than this repulsive force. One of the methods to derive the cutting threshold is based on the assumption that the Lorentz force, which originates from local deviations from the perfect force-free state due to pinning centers, etc., contributes to the cutting, although this force does not necessarily push the tilted flux lines together: it can do the opposite sometimes. However, we shall admit this assumption tentatively. If the force-free current density is denoted by $J_\parallel$, the relationship $B\delta\theta = \mu_0 J_\parallel a_\mathrm{f}$ holds, where a_f is the flux line spacing. The Lorentz force on one intersection is about $\phi_0 a_\mathrm{f} J_\parallel / 2 \cos(\delta\theta/2)$. From the balance between the Lorentz force and the repulsive force, the cutting threshold $J_{c\parallel}$ at which $\delta\theta = \delta\theta_c$ is given as

$$J_{c\parallel} \simeq \frac{\phi_0}{2\mu_0 a_\mathrm{f} \lambda^2} \cdot \frac{\cos \delta\theta_c}{\sin(\delta\theta_c/2)} \simeq \frac{\phi_0}{\mu_0 a_\mathrm{f} \lambda^2 \delta\theta_c} \cdot \tag{4.14}$$

Using the relationship between $\delta\theta_c$ and $J_{c\parallel}$, the critical current density is

$$J_{c\parallel} \simeq \frac{(B\phi_0)^{1/2}}{\mu_0 \lambda a_\mathrm{f}} = \left(\frac{2}{\sqrt{3}}\right)^{1/2} \frac{B}{\mu_0 \lambda} \cdot \tag{4.15}$$

On the other hand, Clem and Yeh [14] treated the stability of a successive row of flux line planes as depicted in Fig. 4.10, and assumed that the above Lorentz force works between the planes. They investigated the dynamics when displacements shown by the arrows in the figure were introduced as perturbations, and they obtained the cutting threshold using a numerical calculation of the critical value of the tilted angle at which the displacements were enhanced unstably. The threshold value they obtained depends only on the normalized magnetic field $B/\mu_0 H_{c2}$ and the G-L parameter κ, and it is naturally independent of the flux-pinning strength. However, such a threshold value is obtained only for infinite number of planes of flux lines, for which the direction of the flux lines are rotated many turns in the superconductor. In practice, the rotation angle of flux lines should be less than π from one surface of the superconductor to the other. The cutting threshold in such a finite system has not yet been reported.

According to experimental results, the practical critical current density depends on the flux-pinning strength as shown in Fig. 4.7. Figure 4.11(a) and (b) are the experimental results for Nb-50at%Ta with dislocations and normal precipitates as pinning centers, respectively [15, 16]. These results show the same tendency and suggest that, in the limit of weak pinning where the critical current density in the transverse field $J_{c\perp}$ is very small, the critical

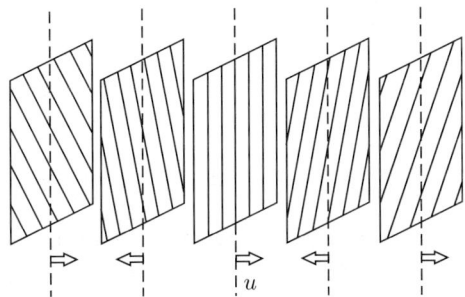

Fig. 4.10. Successive planes of flux lines with continuously rotating angle in a force-free state in a longitudinal magnetic field. Clem and Yeh [14] calculated a cutting threshold from an analysis of stability under a perturbation shown by the *arrows* in the figure

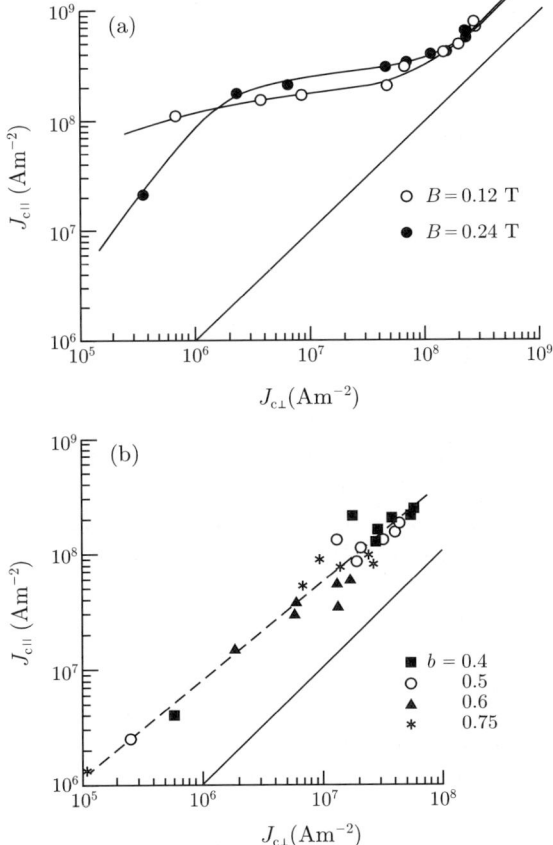

Fig. 4.11. Comparison of critical current density in a longitudinal magnetic field and that in a transverse magnetic field for Nb-50at%Ta slab specimens with pinning centers of (**a**) dislocations [15] and (**b**) normal precipitates [16]. $b = B/\mu_0 H_{c2}$ is the reduced magnetic field

current density in the longitudinal field $J_{c\parallel}$ is also very small. These results are compared with the critical current density estimated from the cutting threshold. In the case of Nb-Ta ($\kappa = 5.5$, $\mu_0 H_{c2} = 0.33$ T) shown in Fig. 4.11, $\delta\theta_c \simeq 3.5°$ is obtained at $B/\mu_0 H_{c2} = 0.7$ from the result of Clem and Yeh, [14] and the corresponding critical current density is $J_{c\parallel} \simeq 1.10 \times 10^{11}$ Am^{-2}. On the other hand, Eq. (4.15) gives $J_{c\parallel} \simeq 1.14 \times 10^{12}$ Am^{-2}. These values are extremely large when compared to observed results. Especially in comparison with the result $J_{c\parallel} \simeq 2.0 \times 10^7$ Am^{-2} for the specimen with weak pins, these theoretical values are larger by factors of 5×10^3 and 5×10^4, respectively. If a specimen with even weaker pins can be prepared, its $J_{c\parallel}$ value would be even smaller, and this would result in an even larger difference from the theories. The threshold values of flux cutting above exceed the depairing current density given by Eq. (1.144) multiplied by the correction factor of $(1 - B/\mu_0 H_{c2})$ at high fields, 3.16×10^{10} Am^{-2}. This means that the superconductor enters the normal state before flux cutting can be realized.

Since the threshold value of cutting is much larger than the practical critical current density, some corrections of the theory of Brandt et al. [13] were attempted. Wagenleithner [17] assumed that the flux lines are not straight but curved at the intersection point and calculated the repulsive force between them. He showed that the repulsive force becomes smaller. However, the flux-line shape was determined by considering that only the interaction energy of the two flux lines involved in the cutting has a minimum, and this is far from practical condition. In the usual condition, i.e. not in the vicinity of H_{c1}, the spacing of flux lines is smaller than λ and the elastic interaction of the flux-line lattice should be considered. This means that a flux line cannot be bent freely without consideration of the surrounding flux lines. The curvature by which the flux lines can be bent is determined by the tilt modulus C_{44} and the shear modulus C_{66}, and the amount of increase of the tilt angle at the intersection is [18] on the order of $(C_{66}/C_{44})^{1/2}$. Since this interaction is magnetic, the local limit should be applied to C_{44}, as will be discussed in Sect. 7.2. The increment of the crossing angle is about 1° for the case of Nb-Ta discussed above. Since this value is much smaller than the critical value, no remarkable effects are expected. In addition, even if the calculation by Wagenleithner is correct, the reduction of the repulsive force is about one order of magnitude, and hence the reduction of the critical current density in Eq. (4.15) is about a factor of 3. If this result is considered in the context of Clem and Yeh, since the repulsive force between the adjacent planes of flux lines and the attractive force between the second-nearest planes are changed by the same factor, no large change in the result takes place. Thus, the correction by Wagenleithner does not essentially solve the problem of the cutting threshold value. The low value of the practical cutting threshold might be ascribed to deviations from the perfect force-free state due to inhomogeneities such as pinning centers. However, experimental results show that, as the pinning becomes stronger (i.e., the number density of pinning centers increases in most cases), $J_{c\parallel}$ increases. Hence, this speculation is not correct.

The flux-cutting model gave explanations of the experimental results of Cave et al. and of Walmsley shown in Fig. 4.2. However, all the experimental results cannot be explained by this model. For example, if flux cutting really occurs, the electric field should be uniform on the surface of the superconductor in the resistive state, and the structure of the electric field depicted in Fig. 4.5 cannot be explained. As for the experimental results in the resistive state shown in Fig. 4.2, there is no theoretical basis which connects the longitudinal electric field with the penetration of the azimuthal magnetic flux as mentioned above, and hence, it is not necessary to invoke a flux-cutting model. In addition, the fact that the experimental values of the flux-cutting threshold are extremely large is the most serious problem. Although the mechanism of flux cutting is not clear, the repulsive force between the tilted flux lines is estimated from a calculation based on electromagnetic principles, and the error is not expected to be large. Thus, it seems reasonable to consider that flux cutting does not occur in practicality. It is known for a plasma that reconnection of flux lines similar to cutting occurs only in an energy-dissipative state such as the resistive state. In the case of a superconductor with a current density below the critical value, it is in a quasi-static state and if we suppose that the variation of external magnetic field, etc., is stopped, the variation in the flux distribution is stopped, and the energy dissipation does not occur. Hence, flux cutting is not expected to occur easily. It seems to be rather natural that flux cutting might be realized only when a current of huge density, such that superconductivity is destroyed, is applied to the superconductor, as predicted by the theories.

Some reports have been published which claimed to prove flux cutting by experiment. Fillion et al. [19] used the following procedure: first, azimuthal flux was trapped in a hollow cylinder of superconducting vanadium by applying a current to a lead through the cylinder and then reducing the current to zero. Then, an external magnetic field was applied in the direction of the long axis of the cylinder, as represented in Fig. 4.12. When the longitudinal magnetic field was increased, it was found that the trapped azimuthal flux did not change while the longitudinal flux penetrated the inside of the cylinder. From this result they argued that the longitudinal and azimuthal components of magnetic flux cut each other.

Blamire and Evetts [20] measured the magnetic field dependence of the critical current of a Pb-Tl thin film with thickness less than 1 µm in a longitudinal magnetic field, and found that the critical current increased in a stepwise manner with increasing magnetic field, as shown in Fig. 4.13. From the fact that the magnetic fields at which the steps occurred were independent of temperature and corresponded to the entry field of rows of flux lines into the thin film, Blamire and Evetts explained the experimental result by flux cutting. That is, they stated that the number of cutting positions, to which the critical current was proportional, increased with increasing magnetic field.

Other analysis of these experimental results will be described in Sects. 4.4 and 4.5.

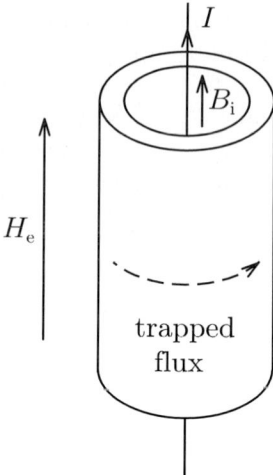

Fig. 4.12. Arrangement of experiment by Fillion et al. [19] Current is transported in a wire through the center of hollow cylindrical vanadium specimen and azimuthal flux lines are trapped in the specimen. Then, a parallel magnetic field H_e is applied and the internal magnetic flux density B_i and the trapped azimuthal magnetic flux are measured

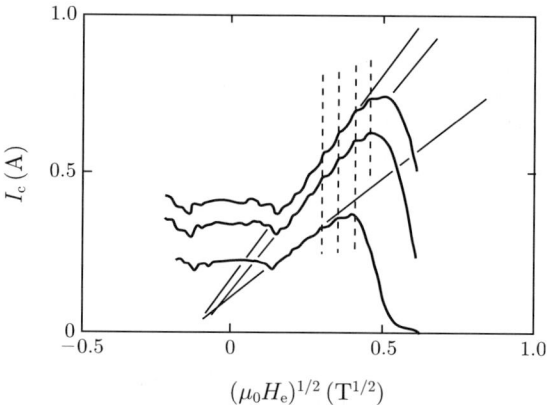

Fig. 4.13. Magnetic field dependence of critical current density in a Pb-Tl thin film in a longitudinal field geometry [20]. Results (from the *top*) are at 2.45 K, 3.1 K and 4.2 K. At magnetic fields shown by the *broken lines* the film thickness is equal to an integral multiple of the flux line spacing, and this is believed to show that penetration of new rows of flux lines occurs at these fields

4.3 Stability of the Force-Free State

Because the cutting threshold value is much larger than indicated by experiments, it is speculated that another instability occurs instead of flux cutting when the current density reaches the critical value. In addition, various

4.3 Stability of the Force-Free State

experimental results suggest that the critical condition is governed by the flux pinning strength. Hence, Josephson's theoretical result [7], that the force-free state is stable without stabilization by flux-pinning interactions, seems to be in doubt. That is, the force-free state might be unstable.

Here we shall investigate in general terms the structure of flux lines when a current flows. The magnetic flux density is expressed as

$$\boldsymbol{B} = B\boldsymbol{i}_B , \qquad (4.16)$$

where $\boldsymbol{i}_B$ represents a unit vector in the direction of $\boldsymbol{B}$. Then, the current density can be written as

$$\boldsymbol{J} = \frac{1}{\mu_0}\nabla \times \boldsymbol{B} = -\frac{1}{\mu_0}(\boldsymbol{i}_B \times \nabla)B + \frac{1}{\mu_0}B\nabla \times \boldsymbol{i}_B . \qquad (4.17)$$

The current has three components. These are a component caused by the gradient of B, one caused by the curvature of flux lines, and the force-free component. The first and second components contribute to the magnetic pressure and the line tension, respectively, and are connected to the Lorentz force. The component of the current responsible for the magnetic pressure is expressed by the first term, and the force-free component is expressed by the second term of Eq. (4.17). The component connected to the line tension comes from both terms.

In order to clarify the relationship between the magnetic structure and the current it is assumed for simplicity that the flux lines lie in planes parallel to the x-z plane and are uniformly distributed in each plane. Hence, the flux lines are straight in each plane and the current component related to the line tension is zero. If the angle of a flux line measured from the z-axis is denoted by θ, we have

$$\boldsymbol{i}_B = \boldsymbol{i}_x \sin\theta + \boldsymbol{i}_z \cos\theta . \qquad (4.18)$$

From the above assumption B and θ do not depend on x and z, and hence, Eq. (4.17) reduces to

$$\boldsymbol{J} = \frac{1}{\mu_0}\cdot\frac{\partial B}{\partial y}\boldsymbol{i}_L - \frac{B}{\mu_0}\cdot\frac{\partial \theta}{\partial y}\boldsymbol{i}_B , \qquad (4.19)$$

where

$$\boldsymbol{i}_L = \boldsymbol{i}_x \cos\theta - \boldsymbol{i}_z \sin\theta \qquad (4.20)$$

is a unit vector perpendicular to $\boldsymbol{i}_B$ and satisfies $\boldsymbol{i}_L \times \boldsymbol{i}_B = -\boldsymbol{i}_y$. It is clear that the first term in Eq. (4.19) is the current component related to the magnetic pressure and the second term is the force-free current. Thus, the flux-line lattice contains a rotational shear distortion, $\partial\theta/\partial y$, when the force-free current component exists.

As shown above the structure of the magnetic flux is distorted when current flows. Fundamental distortions are shown in Fig. 4.14: (a) is the distortion

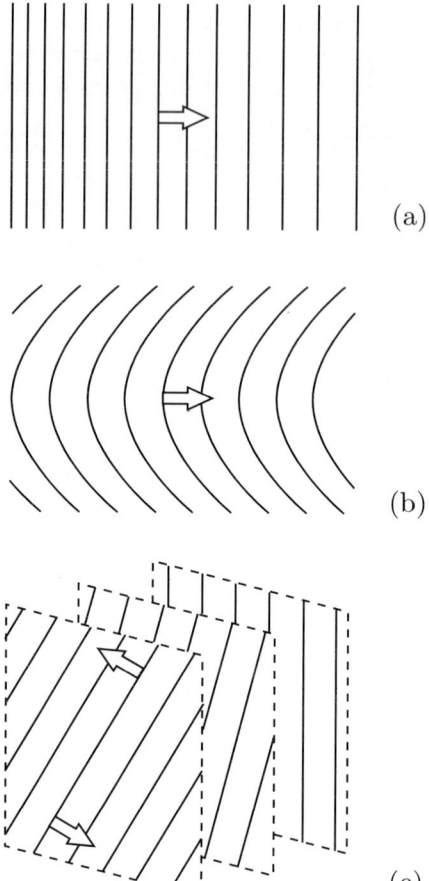

Fig. 4.14. Distortions of flux lines: (**a**) gradient of magnetic flux density and (**b**) bending of flux lines. Restoring forces of magnetic pressure and line tension work respectively corresponding to these distortions in the direction of the *arrows*. These forces are expressed as the Lorentz force. A rotational shear distortion of flux lines under a force-free current is shown in (**c**), and a torque is considered to work as shown by the *arrow*

due to the gradient of B, (b) is the bending distortion of flux lines. These contribute to the Lorentz force given by

$$\boldsymbol{F}_{\mathrm{L}} = \boldsymbol{J} \times \boldsymbol{B} = -\frac{1}{2\mu_0}\nabla B^2 + \frac{1}{\mu_0}(\boldsymbol{B}\cdot\nabla)\boldsymbol{B} \ . \tag{4.21}$$

The first term is the magnetic pressure caused by the gradient of B and the second term is the line tension caused by the bending strain of the flux lines. Thus, the Lorentz force can be described as the elastic restoring force against those respective strains. Now, consider the case of rotational shear distortion shown in Fig. 4.14(c) for the force-free current. By analogy to an elastic material the existence of a generalized force to reduce the distortion is expected. In this case the generalized force is a torque which rotates the flux lines so as to reduce the tilting angle. The existence of such a torque is compatible with the fact that the Lorentz force is zero. The above speculation clearly suggests that the force-free state is unstable.

In order to investigate practically the stability of the force-free state we have only to look at the variation in the energy when such a distortion is virtually introduced. Here we shall do this. We assume that the magnetic field is applied parallel to a wide superconducting slab of thickness $2d$ ($0 \leq y \leq 2d$). From symmetry only the half, $0 \leq y \leq d$, is treated. Assume that the external magnetic field is initially directed along the z-axis and the internal magnetic flux density is also directed along the z-axis and is uniform. This initial condition can be achieved by the field cooled process. Then, a force-free distortion is introduced. When the distortion $\alpha_\mathrm{f} = -\partial\theta/\partial y$ is uniform in space, the relationship between α_f and the force-free current density $J_\parallel$ is identical with Eq. (4.3), $\alpha_\mathrm{f} = \mu_0 J_\parallel / B$. The angle of the flux lines θ is assumed to be

$$\begin{aligned} \theta &= \alpha_\mathrm{f}(y_0 - y) ; & 0 \leq y \leq y_0 , \\ &= 0 ; & y_0 < y \leq d . \end{aligned} \quad (4.22)$$

This distortion can be attained by rotating the external magnetic field quasi-statically by $\theta_0 = \alpha_\mathrm{f} y_0$ (see Fig. 4.15). Here, y_0 is the depth of penetration

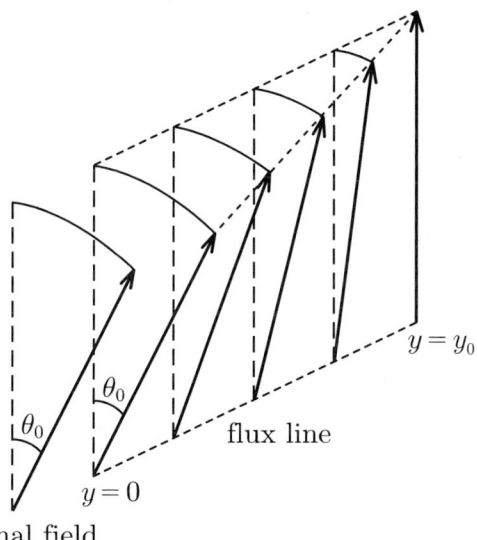

Fig. 4.15. Distortion of flux line structure introduced virtually by rotation of external magnetic field

4 Longitudinal Magnetic Field Effect

of the distortion and is assumed to be constant. Now we shall estimate the increment of the energy in the superconductor during the increase of α_f from 0 to $\Delta\alpha_f$. In terms of the Poynting vector the input power to the superconductor is given by

$$P = -\frac{1}{\mu_0} \int_S (\boldsymbol{E} \times \boldsymbol{B}) \cdot \mathrm{d}\boldsymbol{S} , \qquad (4.23)$$

where $\boldsymbol{E}$ is the electric field induced by the variation in the flux distribution and $\mathrm{d}\boldsymbol{S}$ is the surface element vector on the superconductor and is directed outward, i.e., in the negative direction of the y-axis in the present case. If the Maxwell equation is used, the input power can be rewritten as

$$P = \int \left(\frac{1}{2\mu_0} \cdot \frac{\partial \boldsymbol{B}^2}{\partial t} + \boldsymbol{E} \cdot \boldsymbol{J} \right) \mathrm{d}V . \qquad (4.24)$$

During the process of the virtual displacement treated here $\boldsymbol{B}^2$ is constant in time and only the second term remains. That is, the input energy is transformed into the energy of a current. Now we will derive an expression of the input energy for the displacement given by Eq. (4.22). After a simple calculation it is found that the induced electric field in the region of $0 \leq y \leq y_0$ is

$$\boldsymbol{E} = (E_x, 0, E_z) , \qquad (4.25a)$$

$$E_x = \frac{B}{\alpha_f^2} \cdot \frac{\partial \alpha_f}{\partial t} (\sin\theta - \theta \cos\theta) ,$$

$$E_z = \frac{B}{\alpha_f^2} \cdot \frac{\partial \alpha_f}{\partial t} (\theta \sin\theta + \cos\theta - 1) . \qquad (4.25b)$$

Substituting this expression into Eq. (4.24), the input power density in the region $0 \leq y \leq y_0$ is obtained:

$$p = \frac{B^2}{\mu_0 \alpha_f^2 y_0} \cdot \frac{\partial \alpha_f}{\partial t} [\alpha_f y_0 - \sin(\alpha_f y_0)] . \qquad (4.26)$$

Hence, the energy density which flows into the superconducting slab during the displacement is given by

$$w = \int p \, \mathrm{d}t = \frac{B^2}{\mu_0 y_0} \int_0^{\Delta\alpha_f} \frac{1}{\alpha_f^2} [\alpha_f y_0 - \sin(\alpha_f y_0)] \mathrm{d}\alpha_f . \qquad (4.27)$$

When the angle of the magnetic field at the surface $\theta_\mathrm{m} = \Delta\alpha_f y_0$ is sufficiently small, Eq. (4.27) reduces to

$$w = \frac{B^2}{12\mu_0} \theta_\mathrm{m}^2 . \qquad (4.28)$$

Thus, the energy is proportional to the second power of the rotation angle, i.e., the strain in analogy to the general case. The magnitude of the resultant restoring torque density is [21]

$$\Omega = \left|-\frac{\partial w}{\partial \theta_{\mathrm{m}}}\right| = \frac{B^2}{6\mu_0}\theta_{\mathrm{m}} = \frac{1}{6}BJ_{\parallel}y_0 \ . \qquad (4.29)$$

It can be seen from the above result that, when the rotational shear strain (the torsional strain in a cylindrical superconductor) is induced in the flux-line lattice by a force-free current, even if it is very small, the restoring torque acts on the flux-line lattice so that the strain is eliminated. This is similar to the fact that, when the strain associated with the gradient of the magnetic flux density or the bending strain is induced in the flux-line lattice by a transport current in a transverse magnetic field, the Lorentz force acts on the flux-line lattice to eliminate the strain. Therefore, it can be concluded that the force-free state is unstable as predicted previously. This means that the critical current density in a longitudinal magnetic field is also determined by the flux-pinning strength, similar to the case for a transverse magnetic field. This result is consistent with the experimental results in Figs. 4.7 and 4.11. The critical current density in a transverse magnetic field is determined by the balance between the Lorentz force and the pinning force, and that in a longitudinal magnetic field is determined by the balance between the restoring torque density given by Eq. (4.29) and the moment of pinning forces in a unit volume. We will later discuss y_0 in Eq. (4.29).

The second term in Eq. (4.24) is not the loss power in this case. This can be understood from the fact that, if the time is reversed in Eq. (4.25), $\boldsymbol{E}$ changes to $-\boldsymbol{E}$ and $\boldsymbol{E}\cdot\boldsymbol{J} < 0$ is obtained. That is, the energy in Eq. (4.28) is the energy stored by the inductance of the superconductor.

Although the torque on the distorted flux-line lattice in the force-free state has been predicted by Matsushita [21], it has not been proved experimentally. The driving force which has been verified is only the Lorentz force in a transverse magnetic field, i.e., when the magnetic field and the current are perpendicular to each other. Hereafter we shall call the torque mentioned above the force-free torque. As can be seen from the derivation, this torque is in principle not restricted to superconducting materials but is a general phenomenon. The torque on a magnetic needle in a tilted magnetic field seems to be similar. This comes from the magnetostatic interaction between the magnetic field and the magnetic moment of the needle. However, such a torque can also be described as the moment due to the Lorentz force on a circulating current, since the circulating current is equivalent to a magnetic moment as is well known in electromagnetism. There is no magnetic moment in the force-free state. In addition, only the force-free current is involved, and it does not give rise to a Lorentz force. That is, the force-free torque is completely different from the usual magnetostatic one. Note that the rotation of flux lines in the rotating transverse magnetic field treated in Subsect. 3.1.4 is caused by the Lorentz force. The difference in the rotational motion of flux lines between transverse and longitudinal magnetic fields will be discussed in Sect. 4.4.

Here we shall discuss why the force-free torque has not been noticed. When a current is applied to a wire in a magnetic field, the Lorentz force can

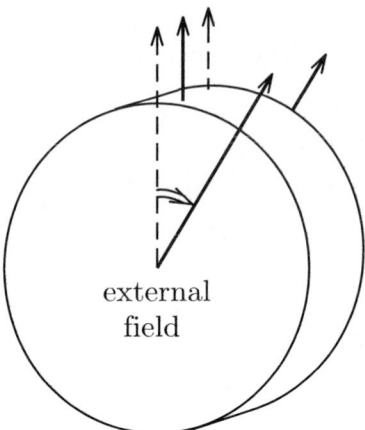

Fig. 4.16. Case where a magnetic field is applied to a metal disk and then rotated

easily be found. On the other hand, when a shielding current is induced, as in a magnetization measurement, it is not possible to detect the Lorentz force, i.e., the magnetic pressure. (Strictly speaking, this can be detected by a measurement of the strain of the material, i.e., expansion or contraction, although it seems to be difficult.) This is similar to a measurement of the pressure on a solid body in a gas, neglecting buoyancy. That is, the hydrostatic pressure of the gas reduces to zero after integrating on the surface. In the present case, there is no gradient of the magnetic pressure along the length of the wire, and hence a force analogous to buoyancy does not exist. When current is applied to a wire in a longitudinal magnetic field, the total torque on the flux lines, which is equal to the torque on the wire interacting with the flux lines, is zero because of a similar symmetry. One method to break the symmetry is to induce a shielding current by rotating the external magnetic field. Figure 4.16 shows an example. When an external magnetic field parallel to the flat surface of a metal disk is rotated as in the figure, the flux lines inside the metal disk lag behind the rotation angle, and a force-free torque proportional to the lag angle acts on the flux lines. When the flux lines are driven to rotate in such a way, the metal disk is affected by the torque because of the interaction between the flux lines and the metal. Therefore, if the torque on the disk can be measured, the existence of the force-free torque can be ascertained. This kind of measurement can also be done for a normal metal, where the quantized flux lines are now the magnetic flux as usual. However, the measurement is not easy, since the induced shielding current decays with time. Hence, the use of a superconducting disk is better because such a difficulty is avoided. In this case a static measurement is possible, since the current does not change with time. In addition, different magnitudes of the force-free torque can be measured by varying the flux-pinning strength, since the moment of the pinning force is balanced with the force-free torque. Note that the shielding current flows in the direction normal to the flux lines at the edge of the disk, and the Lorentz force acts on the flux lines there. Thus, there is some contribution from the Lorentz force to the measured torque.

As discussed above, the force-free state contains a distortion of the flux lines and is unstable. This result is consistent with the experimental result that the critical current density in a longitudinal magnetic field depends on the flux-pinning strength as it does in a transverse magnetic field. Then, a question arises: how does this relate to the theoretical result of Josephson [7] that the force-free state is stable? The details of this discussion are given in Appendix A.1. From that discussion, it can be concluded that the gauge which Josephson assumed in the derivation of Eq. (4.2) for a pin-free superconductor is not correct in a longitudinal magnetic field [22]. The gauge on the vector potential Josephson used is equivalent to Eq. (1.146), which, as has been frequently pointed out in this book, is not obeyed in a longitudinal magnetic field. Considering this point, it can be concluded that the equation describing the equilibrium state in a pin-free superconductor is

$$\boldsymbol{J} = 0 \,. \tag{4.30}$$

In a transverse magnetic field Eq. (1.146) is satisfied. Hence, Eq. (4.2), which then holds, is identical with Eq. (4.30). That is, Eq. (4.30) is the only general equation which applies to all cases. This discussion also leads to the same conclusion that the force-free state is unstable.

4.4 Motion of Flux Lines

The arrangement of flux lines in a longitudinal magnetic field is given by the force-free state, where the current flows parallel to the flux lines, or by a state close to this one, whose stabilization is provided by flux-pinning interactions. Since Eq. (4.2) describing the force-free state is the same as the force-balance equation describing the state in a pin-free superconductor in a transverse magnetic field, it gives rise to the question, why is the superconductor in such a force-free state even with flux pinning? The answer to this question will be given in Sect. 4.6. For now it will be assumed that the force-free state is established as observed by experiments. Consider a longitudinal magnetic field being applied to a superconductor first, where the shielding current flows normal to the field, and then a current is applied, such as is usually done in experiments as shown in Fig. 4.2. How do flux lines then penetrate the superconductor and move, and how is the force-free state established during this process? What is the critical state in the whole superconductor? What flux motion occurs in the resistive state? These questions will be treated in the following sections, and here the motion of flux lines in the force-free state below the critical current and the related electromagnetic phenomena, especially the connection with Eq. (4.1), are discussed.

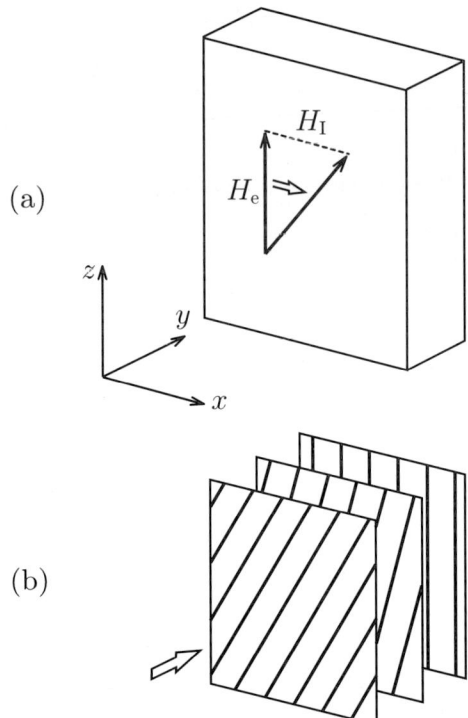

Fig. 4.17. (a) Variation in a surface magnetic field and (b) formation of a force-free state in internal flux lines when an external magnetic field is applied along the z-axis of a sufficiently wide superconducting slab and then a current is applied in the same direction. The *arrow* in (b) shows the direction of penetration of variation

The motion of flux lines can be investigated using the continuity equation for flux lines (2.15). Here it is assumed for simplicity that the initial magnetic flux distribution is uniform inside the superconductor at an external magnetic field H_e along the z-axis and then a current I to a sufficiently wide superconducting slab ($0 \leq y \leq 2d$) as shown in Fig. 4.17. The force-free state is assumed to be given by Eq. (4.3) with α_f being a constant. From symmetry we are allowed to assume that $\partial/\partial x = 0$ and $\partial/\partial z = 0$. The flux distribution in this case is sought in Exercise 4.1. According to this result, the magnetic flux density inside the superconductor lies in the x-z plane and is uniform:

$$B = \mu_0 (H_\mathrm{e}^2 + H_\mathrm{I}^2)^{1/2}, \qquad (4.31)$$

where H_I is the self field of the current given by

$$H_\mathrm{I} = \frac{I}{2L} \qquad (4.32)$$

with L denoting the width of the slab along the x-axis. Hence, the magnetic flux density is expressed as

$$\boldsymbol{B} = (B\sin\theta, 0, B\cos\theta) \,, \tag{4.33a}$$

$$\theta = \theta_0 - \alpha_f y \,; \quad 0 \leq y \leq \frac{\theta_0}{\alpha_f} \,,$$

$$= 0; \quad \frac{\theta_0}{\alpha_f} \leq y \leq d \,. \tag{4.33b}$$

In the above θ_0 is the angle between the external magnetic field and the z-axis and is given by

$$\theta_0 = \tan^{-1}\left(\frac{H_\mathrm{I}}{H_\mathrm{e}}\right) . \tag{4.34}$$

The magnetic flux distribution in the other half of the slab, $d \leq y \leq 2d$, is symmetric with the above one. Campbell and Evetts [9] assumed uniform flux motion along the y-axis. However, this cannot explain Eq. (4.1) and is incorrect. In fact, Yamafuji et al. [23] showed that a contradiction was found when such a restriction was imposed on the motion of flux lines (Exercise 4.4). Here the velocity of flux lines takes the form:

$$\boldsymbol{v} = (v_x, \ v_y, \ v_z) \,. \tag{4.35}$$

In the above only v_y is associated with the penetration of flux lines into the superconductor, i.e., the variation of B with time. Since this component is independent of x and z, it can be assumed that $\partial v_y/\partial x = \partial v_y/\partial z = 0$. In addition, since the flux motion in the x-z plane does not bring about a variation in the magnitude of B, there is no divergence of $\boldsymbol{v}$ in this plane and we have

$$\frac{\partial v_x}{\partial x} + \frac{\partial v_z}{\partial z} = 0 \,. \tag{4.36}$$

Usually $\boldsymbol{v}$ is defined so as to be normal to $\boldsymbol{B}$. This is written as

$$v_x \sin\theta + v_z \cos\theta = 0 \,. \tag{4.37}$$

Using Eqs. (4.36) and (4.37), the continuity equation (2.15) is reduced to

$$\frac{\partial B}{\partial t} = -B\frac{\partial v_y}{\partial y} \,, \tag{4.38}$$

$$\frac{\partial \theta}{\partial t} = \alpha_f v_y + \frac{1}{\sin\theta \cos\theta} \cdot \frac{\partial v_x}{\partial x} \,. \tag{4.39}$$

In the case of quasi-static variation we have

$$\frac{\partial B}{\partial t} = \mu_0 \sin\theta_0 \frac{\partial H_\mathrm{I}}{\partial t} \,, \tag{4.40}$$

$$\frac{\partial \theta}{\partial t} = \frac{\partial \theta_0}{\partial t} = \frac{\mu_0 \cos\theta_0}{B} \cdot \frac{\partial H_\mathrm{I}}{\partial t} \,. \tag{4.41}$$

Then, Eqs. (4.38) and (4.39) can be solved immediately in the region of $0 \leq y \leq \theta_0/\alpha_\mathrm{f}$ as [24]

$$v_x = \frac{\partial \theta_0}{\partial t} \cos\theta \left[1 - \frac{H_\mathrm{I}}{H_\mathrm{e}}\alpha_\mathrm{f}(d-y)\right] \left[x\sin\theta + z\cos\theta + g_\mathrm{r}\left(y - \frac{\theta_0}{\alpha_\mathrm{f}}\right)\right], \quad (4.42\mathrm{a})$$

$$v_y = \frac{\partial H_\mathrm{I}}{\partial t} \cdot \frac{\mu_0^2 H_\mathrm{I}}{B^2}(d-y), \quad (4.42\mathrm{b})$$

$$v_z = -\frac{\partial \theta_0}{\partial t} \sin\theta \left[1 - \frac{H_\mathrm{I}}{H_\mathrm{e}}\alpha_\mathrm{f}(d-y)\right] \left[x\sin\theta + z\cos\theta + g_\mathrm{r}\left(y - \frac{\theta_0}{\alpha_\mathrm{f}}\right)\right], \quad (4.42\mathrm{c})$$

where g_r is a function satisfying

$$g_\mathrm{r}(0) = 0. \quad (4.43)$$

v_x and v_z are zero on the line given by

$$x\sin\theta + z\cos\theta + g_\mathrm{r}\left(y - \frac{\theta_0}{\alpha_\mathrm{f}}\right) = 0 \quad (4.44)$$

in a plane where $y = $ const. Here we watch one flux line, and the position of the intersection point of the flux line with the line given by Eq. (4.44) is represented by $x = x_0$ and $z = z_0$. In terms of these coordinates, Eqs. (4.42a) and (4.42c) can be transformed respectively to

$$v_x = r\frac{\partial \theta_0}{\partial t} \cos\theta \left[1 - \frac{H_\mathrm{I}}{H_\mathrm{e}}\alpha_\mathrm{f}(d-y)\right], \quad (4.45\mathrm{a})$$

$$v_z = -r\frac{\partial \theta_0}{\partial t} \sin\theta \left[1 - \frac{H_\mathrm{I}}{H_\mathrm{e}}\alpha_\mathrm{f}(d-y)\right], \quad (4.45\mathrm{b})$$

where

$$r = (x - x_0)\sin\theta + (z - z_0)\cos\theta \quad (4.46)$$

represents the distance of the point on the flux line from the stationary point, (x_0, y, z_0), i.e., the radius of rotation. Hence, this stationary point is a rotation center, and Eq. (4.44) is the line passing through consecutive rotation centers (see Fig. 4.18). Thus, Eqs. (4.45a) and (4.45b) describe the rotation of flux lines. The factor $[1 - (H_\mathrm{I}/H_\mathrm{e})\alpha_\mathrm{f}(d-y)]$ in the equations above comes from the variation in B.

As shown above, the solution of the continuity equation for flux lines shows a rotation of flux lines. This can be interpreted as a result of flux motion driven by the force-free torque derived in the last section. If we watch the flux motion carefully, v_y varies uniformly and gradually along the y-axis, so that a given row of flux lines does not outstrip another one. At the same time the flux lines rotate uniformly in each plane. Thus, flux-cutting does not occur.

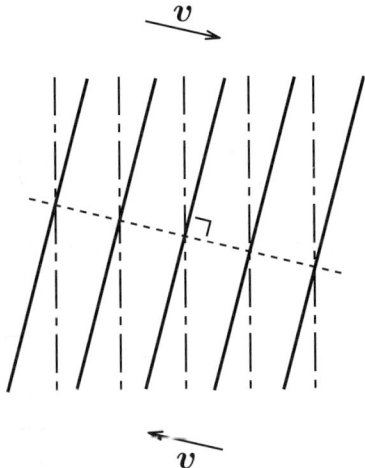

Fig. 4.18. Rotational motion of flux lines. The *broken line* shows a line connecting rotation centers given by Eq. (4.44)

Here the electric field induced by the rotational motion of flux lines will be calculated. For the variation in the magnetic flux distribution given by Eqs. (4.33a) and (4.33b), the electric field is directly obtained from the Maxwell equation (2.2) as [24]

$$E_x = -\frac{\mu_0}{\alpha_f} \cdot \frac{\partial H_I}{\partial t}[\cos(\theta_0 - \theta) - \cos\theta_0 + (\alpha_f d - \theta_0)\sin\theta_0]; \quad 0 \leq y \leq \frac{\theta_0}{\alpha_f},$$
$$= -\mu_0 \frac{\partial H_I}{\partial t}(d-y)\sin\theta_0; \quad \frac{\theta_0}{\alpha_f} < y \leq d, \quad (4.47a)$$

$$E_z = \frac{\mu_0}{\alpha_f} \cdot \frac{\partial H_I}{\partial t}[\sin\theta_0 - \sin(\theta_0 - \theta)]; \quad 0 \leq y \leq \frac{\theta_0}{\alpha_f},$$
$$= 0; \quad \frac{\theta_0}{\alpha_f} < y \leq d. \quad (4.47b)$$

By comparing the above results with Eqs. (4.42a)–(4.42c), it can be seen that Eq. (1.146) is not satisfied in the region $0 \leq y \leq \theta_0/\alpha_f$, where the rotation of flux lines occurs. From Eqs. (2.2) and (2.15) the following is obtained:

$$\boldsymbol{E} = \boldsymbol{B} \times \boldsymbol{v} - \nabla\Psi, \quad (4.48)$$

where Ψ is a scalar function. When the electric field is described in this form, Ψ is usually an electrostatic potential. However, Ψ is not the electrostatic potential in this case, since the electric field given by Eqs. (4.47a) and (4.47b) originates from the induction $-\partial \boldsymbol{A}/\partial t$. That is, when the sweep of the current is stopped and $\partial H_I/\partial t = 0$, $\boldsymbol{v} = 0$ and $\nabla\Psi = 0$ are simultaneously obtained. Because flux lines have nonzero velocity components v_x and v_z, $\boldsymbol{B} \times \boldsymbol{v}$ has a y-component and becomes larger at a point farther from the rotation center. On the other hand, the induced electric field has no y-component and is uniform in the x-z plane. Thus, the term arising from $\boldsymbol{B} \times \boldsymbol{v}$ is a quantity completely different from the practical electric field by itself and does not satisfy the symmetry. This can also be understood from the fact that the position of the rotation center in Eq. (4.45) is arbitrary and does not directly influence the electric field. If $-\nabla\Psi$ is added to $\boldsymbol{B}\times\boldsymbol{v}$, an electric field that has a physical meaning is obtained. In fact, the work done by the force-free torque comes from $-\nabla\Psi$. (The proof of this is Exercise 4.5.) This is because $\boldsymbol{v}$ is not directed along the flow of energy (the direction of the rotational motion corresponding to v_x and v_z is perpendicular to the flow of energy) and has only the meaning of the phase velocity of the variation in $\boldsymbol{B}$. As shown above, in the case of rotational shearing of flux lines in a longitudinal magnetic field, the flux motion cannot be understood by analogy with the motion of a mechanical system, as in the case of flux motion in a transverse magnetic field. This is associated with the fact that the force-free torque is independent of the Lorentz force. By contrast, in mechanical systems the mechanical torque originates from the existence of the force.

Here a rotation of flux lines driven by the Lorentz force is considered for comparison. It is assumed that a static magnetic field H_e is applied normal to a sufficiently wide superconducting slab of width $2d$, lying in the x-y plane ($|z|\leq d$), and then a magnetic field H_t is applied in the direction of the x-axis as shown in Fig. 4.19. In this case the magnetic field on the surface of the superconducting slab changes as in figure (a) and the flux lines rotate in the x-z plane as in figure (b) (compare with Fig. 4.17). Such a motion of flux lines is caused by the Lorentz force due to the shielding current flowing in the direction of the y-axis. Because of the demagnetization factor it is reasonable to assume that the field along the z-axis penetrates completely. Then, the magnetic flux distribution inside the slab is given by

$$\boldsymbol{B}=(B_x,\ 0,\ \mu_0 H_e)\,,\tag{4.49a}$$

$$\begin{aligned}B_x &= 0; & 0\leq z < d-\frac{H_t}{J_{c\perp}}\,,\\ &= \mu_0 H_t - \mu_0 J_{c\perp}(d-z); & d-\frac{H_t}{J_{c\perp}} \leq z \leq d\,.\end{aligned}\tag{4.49b}$$

From symmetry we have only to treat the half $0\leq z\leq d$. The velocity of flux lines that corresponds to the variation in the magnetic flux distribution is

$$\boldsymbol{v}=(v_x,\ 0,\ v_z)\,,\tag{4.50a}$$

$$v_x = v_z = 0;\qquad 0\leq z < d-\frac{H_t}{J_{c\perp}}\,,\tag{4.50b}$$

$$\left.\begin{aligned}v_x &= \frac{\partial H_t}{\partial t}\cdot\frac{H_e(z-d+H_t/J_{c\perp})}{H_e^2 + J_{c\perp}^2(z-d+H_t/J_{c\perp})^2}\,,\\ v_z &= -\frac{\partial H_t}{\partial t}\cdot\frac{J_{c\perp}(z-d+H_t/J_{c\perp})^2}{H_e^2 + J_{c\perp}^2(z-d+H_t/J_{c\perp})^2}\,,\end{aligned}\right\}\quad d-\frac{H_t}{J_{c\perp}}\leq z\leq d\,.\tag{4.50c}$$

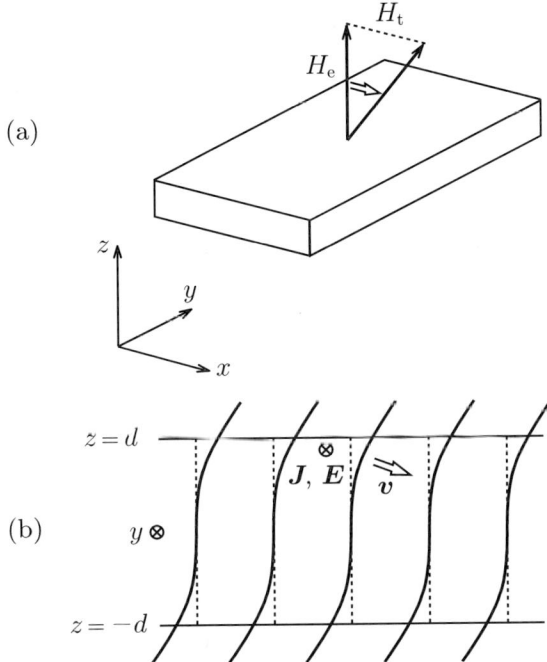

Fig. 4.19. (a) Variation in a surface magnetic field and (b) rotational motion of internal flux lines when a magnetic field is applied normal to a superconducting slab and then a magnetic field parallel to the slab is superimposed

On the other hand, the induced electric field is

$$\boldsymbol{E} = (0,\ E_y,\ 0), \tag{4.51a}$$

$$E_y = 0; \qquad 0 \leq z < d - \frac{H_{\rm t}}{J_{\rm c\perp}},$$

$$= \mu_0 \frac{\partial H_{\rm t}}{\partial t}\left(z - d - \frac{H_{\rm t}}{J_{\rm c\perp}}\right); \quad d - \frac{H_{\rm t}}{J_{\rm c\perp}} \leq z \leq d. \tag{4.51b}$$

Hence, the relationship $\boldsymbol{E} = \boldsymbol{B} \times \boldsymbol{v}$ is clearly satisfied in this case. In addition, it is easily shown that the Poynting vector $\boldsymbol{N} = \boldsymbol{E} \times \boldsymbol{B}/\mu_0$ is parallel to $\boldsymbol{v}$. That is, $\boldsymbol{v}$ is directed along the flow of energy and the motion of flux lines is analogous to the motion of mechanical systems. Thus, the phenomenon is different depending on whether the flux motion is caused by the Lorentz force or the force-free torque.

Here the features that characterize the rotation of flux lines by the force-free torque are discussed. Assume that the transport current is applied after the longitudinal magnetic field, which was discussed when Eq. (4.47) was derived. From this equation we have

$$\frac{E_x}{E_z} = -\tan\frac{\theta_0}{2} \tag{4.52}$$

at the surface ($\theta = \theta_0$). Thus, the angle of the electric field from the z-axis is $-\theta_0/2$, while the angle of the magnetic field is θ_0. Hence, E and B are approximately parallel to each other when θ_0 is small. This situation is close to the experiment of Cave et al. [4] Their experiment can thus be interpreted as evidence that the rotational shearing of flux lines really occurs. In fact, when the magnetic field is tilted slightly from the z-axis toward the x-axis, the variation in the z-component of the magnetic flux is much smaller than that of the x-component.

Now the rotation of flux lines is compared with the flux cutting model for the purpose of identification of the mechanism by experiments. It is postulated for simplicity that the flux lines originally directed along the z-axis are tilted toward the x-axis. Such a rotation of flux lines can be described by a combination of an x-component of magnetic flux that penetrates into the superconductor and a z-component that is eliminated from the superconductor. The flux-cutting models adopt this viewpoint. They assume independent motion of each component of magnetic flux (inter-cutting), or equivalent changes in these components by an intersection and cross-joining process (intra-cutting). Hence, flux rotation and flux cutting are equivalent descriptions of the variation in the magnetic flux distribution, and these mechanisms cannot be distinguished by measurements of the induced electric field or the magnetization. That is, not only the pure rotation of flux lines, but any variation in the macroscopic magnetic flux density can also be described in general either as coupled variations of the magnitude and the direction or as coupled variations in different components of magnetic flux. The former case corresponds to translational flux motion and rotation by the force-free torque, and the latter case corresponds to flux cutting. Therefore, an explanation from the opposite viewpoint to flux cutting is also possible.

The experiment by Fillion et al. [19] was explained from the viewpoint of flux cutting. Here the attempt is made to explain this experimental result from the opposite viewpoint. Kogan [25] insisted on the possibility that the flux lines, which nucleated near the surface of the specimen when a longitudinal field was applied, were not perfectly parallel to the specimen axis as assumed by Fillion et al., but might be helical and slightly tilted toward the inner flux lines. Even if the flux lines parallel to the axis are nucleated just inside the surface, the theoretical result predicts that the force-free torque acts between these flux lines and the inner flux lines, which suggests that the nucleated flux lines tilt. Hence, Kogan's opinion seems to be natural. According to Kogan, the fact that the azimuthal component of the magnetic flux is approximately constant with increasing external field results from a cancellation between the elimination of flux line rings at the inner surface of the hollow cylinder and the nucleation of helical flux lines at the outer surface. This is an alternative explanation of the experimental result of Fillion et al. For the reasons mentioned above, investigation of flux cutting should be based on measurements of critical current density or electric field in the resistive state.

However, thought experiments like a following one [26] seem to be useful for investigation of the practical flux motion. Assume that a magnetic field is applied parallel to a thin superconducting disk, and then is rotated within the plane parallel to the disk as shown in Fig. 4.16. The result should be the same between the case where the field is rotated as in the figure and the case where the disk is rotated in the opposite direction in a fixed magnetic field. In most experiments [27] the latter method has been chosen. When the magnetic field is rotated, each model assumes the corresponding characteristic flux motion, i.e., flux rotation or flux cutting.

Then, what results when the superconducting disk is rotated? Flux lines in the superconductor are driven to rotate by interactions with the superconductor, especially by pinning interactions. From the viewpoint of flux rotation (which hereafter shall be called the flux-rotation model), when the angle between the flux lines at the surface and the external magnetic field exceeds some critical value $\delta\theta_c$ determined by the flux pinning, the force-free torque exceeds the moment of pinning forces. Thus, flux lines cannot rotate with the superconductor any more and start to slip. When the superconductor is rotated further, the same thing occurs between the flux lines at the surface and the next set of flux lines. Thus, the critical state penetrates the superconductor in this manner (see Fig. 4.20).

What is the prediction of the flux-cutting model? The intersection and cross-joining model shown in Fig. 4.9 is assumed for an example. According to this model, when the angle between the external field and the flux lines at the surface reaches the cutting threshold $\delta\theta_c$, the flux cutting occurs and the flux lines are drawn back (see Fig. 4.21(b)). When the superconductor is

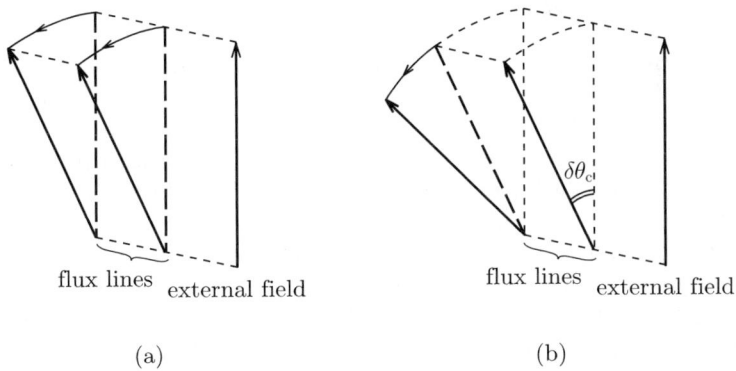

(a) (b)

Fig. 4.20. Flux motion and process of establishment of critical state expected from the flux-rotation model for the case where a superconductor is rotated in a fixed magnetic field: (**a**) in the beginning, internal flux lines rotate with the superconductor and (**b**) their angle with the external magnetic field reaches a critical value for pinning $\delta\theta_c$, so that they cannot rotate any more and slip, resulting in a rotation of only inner flux lines

184   4 Longitudinal Magnetic Field Effect

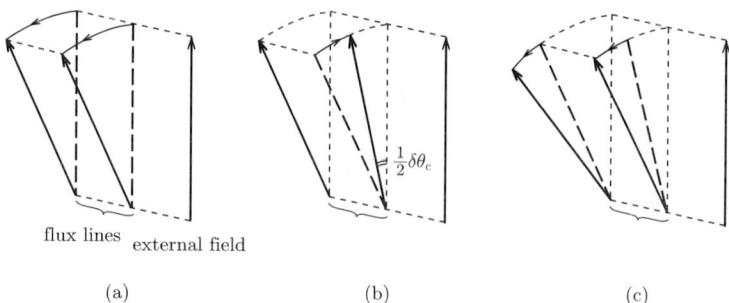

Fig. 4.21. Flux motion and process of establishment of the critical state expected from the flux-cutting model for the case where a superconductor is rotated in fixed magnetic field: (**a**) in the beginning internal flux lines rotate with the superconductor, (**b**) their angle reaches a critical value and the flux cutting occurs and (**c**) then, inner flux lines rotate

further rotated, flux lines in the superconductor are driven to rotate with the superconductor by flux pinning interactions (see Fig. 4.21(c)). For a further rotation, flux cutting occurs again between the external field and the flux lines at the surface, and similar flux cutting occurs between the flux lines at the surface and the inner flux lines. As a result, flux lines slip from the superconductor, and the critical state penetrates from the surface. Thus, flux lines excessively rotate with the superconductor, cut each other and then swing back, resulting in an unnatural oscillation. If we look at the flux motion carefully, it is found that such an unnatural oscillation occurs also when the external field is rotated. Figure 4.22(a) shows the variation in the flux-line angle while the critical state penetrates from the fourth row of flux lines from the surface to the fifth row. The flux lines oscillate as a → b → c. This result shows that the angle between adjacent rows of flux lines is close to $\delta\theta_c/2$. That is, the angle in the critical state is a half as large as the cutting threshold when the critical state is established over a wide area. The reason will be easily understood. Figure 4.22(b) shows the variation in the flux-line angle predicted by the flux-rotation model in the same condition. As discussed above, there is no difference in the macroscopic flux motion between the cases of rotating field and rotating superconductor for both models when the flux-pinning interactions exist. That is, these models are self-consistent.

Now the case of a pin-free superconductor is considered. Assume that the rotation is carried out quasi-statically so that an eddy current may not flow inside the superconductor. Hence, flux lines are expected not to interact at all with the superconductor. When the external field is rotated, flux lines are expected to rotate with the external field in the flux-rotation model, since the force-free torque acts on flux lines tilted from the external field. On the other hand, in the flux-cutting model the same force-free state as in the case of the superconductor with pins is derived, since the force-free state is believed to be stable. A discussion of the different results obtained from the two models

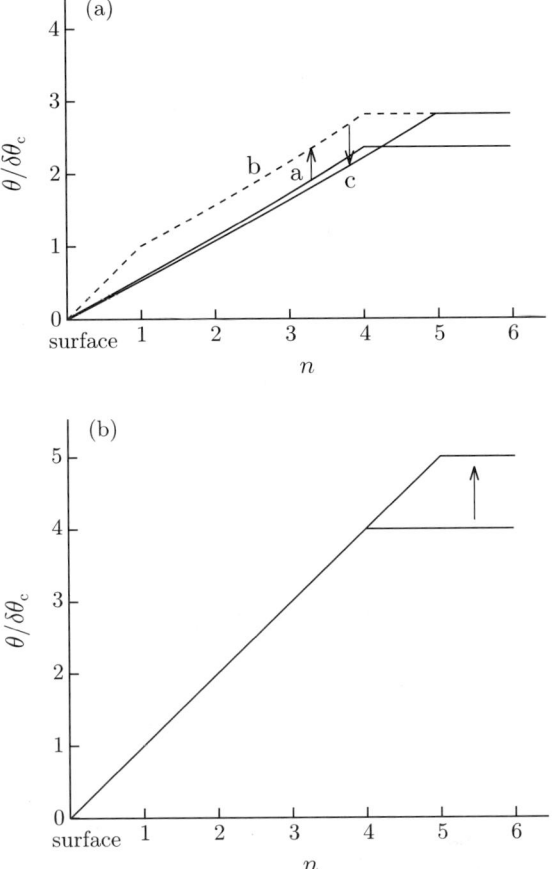

Fig. 4.22. Variation of the angle θ while a critical state penetrates from the fourth row of flux lines from the surface to the fifth row. n represents the number of rows from the surface. (**a**) is a prediction from the flux-cutting model, and the distribution varies as a$\to$ b $\to$ c. (**b**) is the variation predicted from the flux-rotation model

is identical with that on the critical current density, and such a discussion is not the purpose of the present thought experiment. The purpose of this thought experiment is a discussion of the relationship between the two cases for each model, and hence we raise no questions about the different results shown above from the two models. Next the case is considered where the superconductor is rotated in the opposite direction in a fixed magnetic field. Flux lines are considered to be fixed in the space in both models, since they do not interact with anything, including the superconductor. Hence, flux lines in the superconductor have the same angle as the external field. This result is equivalent to the case of the rotating external field in the flux-rotation model, while it is not in the flux-cutting model. That is, the flux-cutting model is not consistent between the cases of rotating field and rotating superconductor, and does not satisfy the equivalence [26]. The same conclusion can be obtained also for other flux-cutting models besides the intersection and cross-joining model. It might be thought that some eddy current effect exists in practical cases. Discuss the problem in this situation (Exercise 4.6)

4.5 Critical Current Density

As mentioned in Sect. 4.3, the flux pinning interaction is necessary to sustain the distortion of flux lines so that the force-free current can flow stably, in analogy to the case of the transverse magnetic field. The critical current density $J_{c\|}$ is determined from the balance between the force-free torque and the moment of pinning forces in the critical state:

$$\Omega_c = \Omega_p \,, \tag{4.53}$$

where Ω_c is given by Eq. (4.29) with replacement of $J_\|$ by $J_{c\|}$. The moment of pinning forces is formally given by

$$\Omega_p = \sum_i f_{pi} l_i \,, \tag{4.54}$$

where f_{pi} is the pinning force of the i-th pinning center, l_i is the effective rotation radius of the flux line interacting with this pinning center, and the summation is taken over a unit volume. When a local rotation of a flux line around a strong pinning center is considered, the effective rotation radius l_i will be given by the distance between the fulcrum and the point where the force is applied, i.e., the distance between one pinning center and the next (see Fig. 4.23). Hence, the mean value of l_i will be comparable to the mean spacing of pinning centers d_p. This expectation will be correct only when each pinning center is sufficiently strong and the concentration of pinning centers $N_p(=d_p^{-3})$ is not high. In such a case the moment of pinning forces will be proportional to the product of the pin concentration N_p and the average value of individual pinning forces f_p, obeying a linear summation, in analogy to the pinning force density in a transverse field. Hence, Eq. (4.54) will be described as

$$\Omega_p = \eta_\| N_p f_p d_p \,, \tag{4.55}$$

where $\eta_\|$ is a parameter representing the pinning efficiency and takes a value smaller than unity. When f_p is small and/or N_p is very large, collective pinning occurs, where many pinning centers work collectively. In this case, l_i takes a larger value than d_p, and f_{pi} is not the force of individual pinning centers but the force of pinning centers which work collectively.

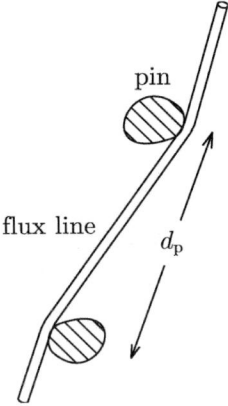

Fig. 4.23. Local rotation of flux line around strongly interacting pinning centers

It is necessary to evaluate y_0 in Eq. (4.29). Note that there is no method to determine y_0 analytically [21]. However, when individual pinning forces are sufficiently strong, it is expected that rows of flux lines shearing against each other behave independently. That is, when one row of flux lines is rotated, the adjacent row will not be rotated until the tilting angle between them reaches the critical value. In this case y_0 can be replaced by the flux line spacing a_f. The fact that y_0 takes on the minimum value means that the critical current density takes on a value as large as possible. The critical state model known to describe the phenomena in a transverse magnetic field demands that a variation in the internal magnetic flux distribution should be as small as possible when the external field is varied. This is analogous to the principle of minimum energy dissipation, i.e., the principle of irreversible thermodynamics for linear dissipative systems (see Appendix A.3). Hence, the hypothesis that y_0 takes on the minimum value can be understood as an extrapolation of the law of irreversible thermodynamics.

From the above argument, the critical current density in a longitudinal field is formally given as

$$J_{c\|} = 6\eta_\| \frac{N_p f_p}{B} \cdot \frac{d_p}{a_f} , \qquad (4.56)$$

when pinning centers are sufficiently strong and their concentration is not too high. On the other hand, the critical current density in a transverse field is written as

$$J_{c\perp} = \eta_\perp \frac{N_p f_p}{B} , \qquad (4.57)$$

where $\eta_\perp$ is the corresponding pinning efficiency (see Eq. (7.81)). Hence, the enhancement factor of the critical current density is given by

$$\frac{J_{c\|}}{J_{c\perp}} = 6 \frac{\eta_\|}{\eta_\perp} \cdot \frac{d_p}{a_f} . \qquad (4.58)$$

188   4 Longitudinal Magnetic Field Effect

If $\eta_\parallel$ and $\eta_\perp$ are comparable to each other, the enhancement factor amounts approximately to $6d_\mathrm{p}/a_\mathrm{f}$ and is considerably larger than 1. Hence, the enhancement of the critical current density from the transverse field geometry can be qualitatively explained. The above result shows that the enhancement factor decreases with increasing pin concentration, i.e., with decreasing pin spacing d_p.

Figure 4.24 shows the experimental results [28] for Pb-Bi specimens in which $\eta_\perp$ in Eq. (4.57) is known to be constant in a transverse magnetic field, so that the theoretically predicted dependence of the critical current density on pinning parameters can be checked. It is found from the figure that the critical current density is proportional to $N_\mathrm{p}f_\mathrm{p}d_\mathrm{p}$ as predicted in Eq. (4.56). Pinning centers in this series of specimens are normal precipitates several μm in size. It may be pointed out that flux cutting occurs more easily in normal precipitates. However, the critical current density generally increases with increasing amounts of normal precipitates as shown in Figs. 4.24 and 4.11(b). Hence, the proposal of easier flux cutting in normal precipitates is not correct. Flux cutting might take place in regions where flux lines are crossing each other. Note that crossing of flux lines does not take place in normal precipitates, since the superconducting current cannot flow there.

Here the experimental results of Blamire and Evetts [20] shown in Sect. 4.2 are discussed. Blamire and Evetts claimed that a stepwise increase in the

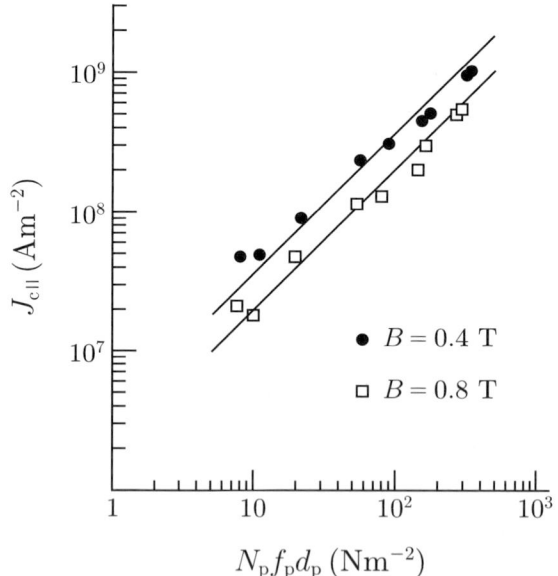

Fig. 4.24. Dependence of critical current density in a longitudinal magnetic field on pinning parameters for Pb-Bi slab specimens with precipitates of normal Bi phase [28]. The *solid line* represents the dependence given by Eq. (4.56)

critical current density with increasing magnetic field, as shown in Fig. 4.13, should be ascribed to the increase in cut portions owing to the penetration of new flux lines. However, note that the amount of pinned portions also increases, owing to the penetration of new flux lines as shown by Eq. (4.54). This suggests that the variation in the critical current density in Fig. 4.13 can also be explained by the mechanism of flux pinning. This result is obtained for a deposited thin film specimen. It is proposed that the grain size is sufficiently small, and hence, the density of grain boundaries that work as pinning centers is expected to be very high. In fact the critical current density in the vicinity of zero field, where the effect of longitudinal field is not appreciable, is very large (about 2×10^9 Am^{-2} at 4.2 K) in comparison with that for bulk specimens. Hence, flux lines that newly penetrate the superconducting film will be surely pinned by pinning centers. In this case the number of pinning points is given by the number of flux lines, and we have $\Omega_p \simeq (B/\phi_0)\langle f_p \rangle \langle l \rangle$, where $\langle \ \rangle$ denotes an average quantity. Hence, the critical current density behaves as

$$J_{c\parallel} \propto B^{1/2} \langle f_p \rangle . \tag{4.59}$$

Blamire and Evetts derived a similar equation $J_{c\parallel} \propto B^{1/2} f_c$, where f_c is an elementary cutting force, and explained that $J_{c\parallel}$ is proportional to $B^{1/2}$, as indicated by the straight lines shown in Fig. 4.13, with the assumption that f_c is a constant. This assumption is different from the theoretical result of Brandt et al. [13] given by Eq. (4.13), which claims that f_c is approximately inversely proportional to the tilt angle $\delta\theta$ (so that in this case $J_{c\parallel} \propto B$ is obtained from Eq. (4.15)). This claim by Blamire and Evetts is based on the theoretical result of Wagenleithner [17]. They estimated the cutting force f_c from the experimental result of $J_{c\parallel}$. Their result is shown in Fig. 4.25. In this case the ordinate has the units of 1×10^{-13} N. Here the temperature dependence of this force is discussed. If it is assumed that the cutting force does not depend on the tilt angle $\delta\theta$, as originally assumed by Blamire and Evetts, this leads to

$$f_c \propto \lambda^{-2} \propto 1 - \left(\frac{T}{T_c}\right)^4, \tag{4.60}$$

which gives the temperature dependence represented by the broken line in Fig. 4.25 [18]. In the above the temperature dependence of λ from the two fluid model is used.

Now their experimental result is discussed from the viewpoint of the flux pinning mechanism. In this case the pin concentration is proposed to be very high. Hence, $\langle f_p \rangle$ is not an elementary pinning force of individual pinning centers. It seems to be reasonable to regard it as a collective force by plural pinning centers. However, there is no information on pinning centers, which would be necessary to derive the dependencies of $\langle f_p \rangle$ on temperature and magnetic field. Thus, these dependencies are estimated from many experimental results in a transverse magnetic field. It is known that the pinning force density in a transverse field is expressed in the form of the

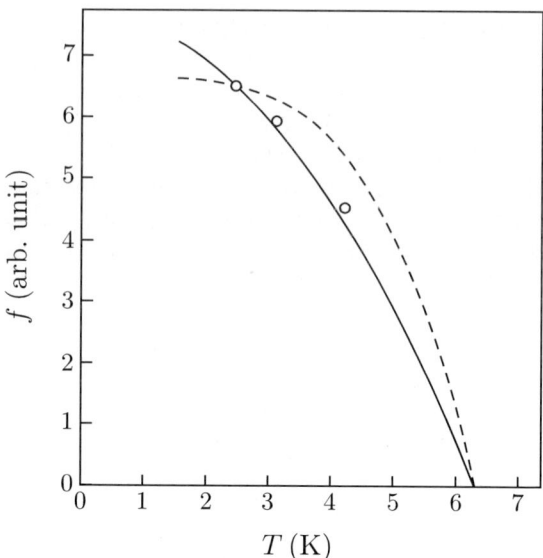

Fig. 4.25. Temperature dependence of the elementary cutting force f_c evaluated from the critical current density of the Pb-Tl thin film shown in Fig. 4.13 by Blamire and Evetts [20]. The *broken line* represents the prediction of Wagenleithner's theory [17]. On the other hand, the *solid line* shows the temperature dependence of the elementary pinning force f_p, which is estimated with the assumption that the critical current density comes from pinning interactions [18]

scaling law $F_p \propto H_{c2}^{m-\gamma}(T) B^\gamma$ except in the vicinity of the upper critical field (see Sect. 7.1), where m and γ are pinning parameters. As mentioned above, the flux pinning is expected to be sufficiently strong. In such a case $m = 2$ and $\gamma = 1$ are expected from experimental results for Nb-Ti, etc. In the case of strong pinning it is considered that individual flux lines are pinned almost independently of each other. This allows $F_p \propto (B/\phi_0)\langle f_p \rangle$ to be derived, and

$$\langle f_p \rangle \propto H_{c2}(T) \propto 1 - \left(\frac{T}{T_c}\right)^2 \tag{4.61}$$

can be obtained [18]. This result is shown by the solid line in Fig. 4.25, and the agreement with experimental results is better than with the broken line derived from the flux-cutting mechanism. In addition, $\langle f_p \rangle$ is a constant with respect to magnetic field, and hence, the magnetic field dependence of $J_{c\|} \propto B^{1/2}$ shown in Fig. 4.13 can also be explained from Eq. (4.59). As discussed above, although there are some uncertain factors, the experimental results of Blamire and Evetts are not exclusively explained by the flux-cutting mechanism, while their temperature dependence is explained better by the flux-pinning mechanism.

4.6 Generalized Critical State Model

A current generally has two components: a component normal to the flux lines and one parallel to the flux lines. For example, in the process of the usual experiments where a longitudinal magnetic field is first applied to a superconducting cylinder or slab and then a current is applied, the current flows normal to the flux lines in the beginning, and then turns gradually toward a parallel direction. How does the magnetic flux distribution vary during the process? In this section, the distribution in a wide superconducting slab is again considered for simplicity. Assume that the slab is parallel to the x-z plane and that the flux lines and current are parallel to this plane. Hence, these quantities vary only along the direction of slab thickness, i.e., along the y-axis. In this case the magnetic flux distribution is obtained when the magnitude of the magnetic flux density B and the angle of the flux lines θ from the z-axis are determined. As for the equations describing such a distribution, a set of equations of the following type has been proposed:

$$\frac{\partial B}{\partial y} = \mu_0 \delta_\perp J_{c\perp} f , \qquad (4.62a)$$

$$B\frac{\partial \theta}{\partial y} = \mu_0 \delta_\parallel J_{c\parallel} g , \qquad (4.62b)$$

where $\delta_\perp$ and $\delta_\parallel$ are sign factors related to the current direction, and f and g are factors which will be discussed below. Comparing these equations with Eq. (4.19), it is found that Eq. (4.62a) corresponds to the current component normal to the flux lines which gives rise to the Lorentz force, and that Eq. (4.62b) corresponds to the force-free current component. LeBlanc et al. [29] assumed that $f = 1$ and that g is a parameter depending on θ without any explanation of this assumption. On the other hand, Clem et al. [30] considered that Eqs. (4.62a) and (4.62b) originated from flux pinning and flux cutting, respectively, and that these were independent of each other. Hence, they assumed that $f = g = 1$.

However, the force-free current also originates from flux pinning interactions as shown above. Hence, the current components in Eqs. (4.62a) and (4.62b) cannot be independent of each other but come from the common pinning energy. Hence, the pinning energy should be shared between two components. Here, generalized coordinates (y, θ) which describe the position of flux lines are introduced. If the pinning potential for flux lines can be approximately expanded around an equilibrium position (y_e, θ_e) as

$$U = \frac{a}{2}(y - y_e)^2 + \frac{b}{2}(\theta - \theta_e)^2 , \qquad (4.63)$$

and if the critical state is attained when U reaches a certain threshold value U_p,

$$f = \sin\psi , \qquad g = \cos\psi . \qquad (4.64)$$

are obtained [31], where ψ is a parameter representing the sharing of pinning energy (see Appendix A.4). How is this parameter determined?

This parameter cannot be determined from the viewpoint of electromagnetism. Thus, the principle of minimum energy dissipation mentioned in Sect. 4.5, which is known as the principle of irreversible thermodynamics for linear dissipative systems, seems to be useful. If P denotes the loss power due to the variation in magnetic flux density in the superconductor, the pinning energy is expected to be shared so that P can be minimized. Hence, the parameter ψ is considered to be determined by the condition:

$$\frac{\partial P}{\partial \psi} = 0 \ . \tag{4.65}$$

The energy dissipation in the flux pinning process in a superconductor is different from that in the usual linear energy-dissipation processes, and there is no formal correspondence. However, the critical state model, which describes correctly the irreversible electromagnetic phenomena in a transverse magnetic field, satisfies the condition of minimum energy dissipation as discussed previously (see Appendix A.3). Hence, the above hypothesis seems to be reasonable.

Here an experimental result [32], which seems to be influenced strongly by the principle of irreversible thermodynamics, is introduced. When a longitudinal magnetic field and a transport current are applied simultaneously to a superconducting tape, and when their strengths are proportional to each other, the magnetic field at the surface keeps its angle constant and only its strength changes. Hence, it is simply expected that the penetration of flux lines from each surface of the tape is independent from the other, so long as the flux lines from each side do not meet at the center. That is, the penetration of flux lines seems to occur translationally without changing their angle, which is similar to the case where only a transverse field is applied and increased. However, the observed longitudinal magnetization was paramagnetic, and the critical current density was enhanced as in the usual cases involving the longitudinal field [32]. This result shows that flux lines are rotated when they penetrate the superconducting tape. In other words, the current does not flow in a perpendicular direction, but almost parallel to the flux lines. The current flow originally has a degree of freedom under the given conditions, and it is expected that the principle of irreversible thermodynamics determines the current flow. The loss due to the current generally takes on a smaller value for a larger critical current density as shown by Eq. (2.82). Therefore, the current flow is proposed to become force-free-like, so that the loss is minimized. Note that the current does not flow so that the path length may be the shortest as in the normal state. If the current flows parallel to the tape length, the paramagnetic moment does not appear in a longitudinal magnetic field.

As mentioned in Sect. 4.1, the force-free model is known to describe this phenomenon as shown in Fig. 4.2 even for specimens with flux-pinning effects. The principle of minimum energy dissipation mentioned above gives us an answer to this apparent contradiction. The pinning energy may be mostly allotted to the pinning interaction with larger critical current density so as to minimize the energy dissipation. As a result, when the field angle changes under the transport current, the pinning interaction mostly shields the rotation of flux lines, but is almost ineffective in shielding the translational penetration of flux lines driven by the Lorentz force ($\psi \simeq 0$). This is similar to the variation of the current distribution in the case of the abnormal transverse-field effect mentioned in Subsect. 3.3.3. The difference is only that the variation of the current distribution from the transverse flow to the longitudinal one is expected to be accomplished more quickly than in the case of the abnormal transverse-field effect, since the anisotropy of the critical current density is very large. In the flux-cutting model, the paramagnetic magnetization is predicted to be smaller than in the force-free model, since the flux pinning interaction is believed to shield the translational penetration of flux lines to the full extent ($f = 1$).

Assume that an AC current or small AC transverse field is applied to a long superconductor in a parallel DC field. The force-free state ($\psi = 0$) is expected to be approximately attained in the steady state because of the large difference in the critical current density. In this case Eq. (4.62b) predicts that the distribution of the transverse magnetic flux component is like the one described by the Bean-London model. In fact the experimental result [15] shown in Fig. 4.26 proves that this prediction is correct. The experimental result of Cave et al. [4] that the magnitude of the transverse magnetic flux

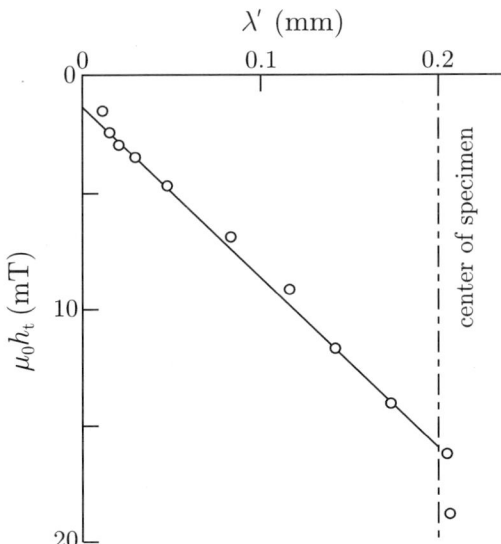

Fig. 4.26. Distribution of the transverse component of magnetic flux in a superconducting Nb-Ta slab [15]. h_t is the amplitude of an AC magnetic field superimposed perpendicularly on a DC longitudinal field, and λ' is the penetration depth of the AC field

component is proportional to the second power of the AC current amplitude also supports this prediction.

Here the experimental result of a simultaneous sweep of the longitudinal field and the current mentioned above is briefly discussed. In this experimental condition flux cutting is not believed to occur, since the direction of the magnetic field at the surface is unchanged and is parallel to the direction of the internal flux lines.

The magnetic flux distribution in a longitudinal field is expected to be described by Eqs. (4.62), (4.64) and (4.65). However, the questions of whether the approximation of Eq. (4.64) is sufficient and whether Eq. (4.65) is really satisfied have not yet been clarified, and these equations are now being compared with experiments. Hence, the model will be evaluated on the basis of investigations in the future.

4.7 Resistive State

When the transport current density in a longitudinal magnetic field exceeds the critical value $J_{c\|}$, some flux motion is induced and the superconductor becomes resistive. As mentioned previously, Walmsley [1] has proposed a flux-cutting model in which only the transverse component of magnetic flux was assumed to penetrate the superconductor to explain compatibly a steady longitudinal electric field and a constant longitudinal magnetization. Various flux-cutting models were proposed afterwards by Cave et al. [4], Clem [10, 12] and Brandt [11]. According to these models the flux cutting is believed to occur uniformly on the macroscopic scale because of symmetry and the uniformity of superconducting specimens. Hence, the resultant electric field is also believed to be uniform. However, the observed electric field in this state is not uniform and is significantly different from the case of a transverse field. That is, the observed electric field has a macroscopic structure on a scale comparable to the size of specimens. The most prominent feature of this structure is a negative electric field shown in Fig. 4.27. Namely, the direction of observed electric field between voltage terminals is opposite to the current direction given by the boundary conditions. (Note that this direction is not necessarily identical with the direction of practical current flow.) This means that there exists a region in which energy production seems to take place. Such a macroscopic structure in the electric field cannot be explained by the flux-cutting model in which all the difficulties are believed to be solved by the local flux-cutting mechanism.

The structure of the electric field in the resistive state was clarified in detail by Ezaki et al. [5, 34] They measured the current-voltage curves in the longitudinal and azimuthal directions of a superconducting cylinder specimen as shown in Fig. 4.28, and found the potential distribution represented in Fig. 4.29. Figure 4.5 shows this structure schematically. The structure is helical, and there are two regions, i.e., the regions of positive and negative electric

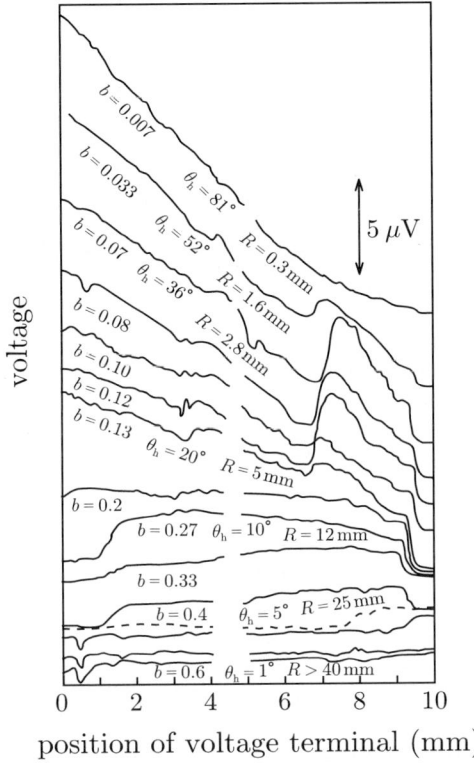

Fig. 4.27. Variation in longitudinal electric potential measured for a cylindrical Pb-In specimen in a longitudinal magnetic field [33]. b is the reduced magnetic field, and θ_h and R are the angle and the pitch of the magnetic field at the surface, respectively

fields. After their definition, the regions of positive and negative electric fields shall be called the S-region and the N-region, respectively.

In the critical state in a longitudinal magnetic field, the driving force-free torque is balanced with the moment of pinning forces. When the transport current is further increased, this balance breaks and some motion of flux lines is induced. What is this flux motion? Assume a distorted structure of flux lines as shown in Fig. 4.30. When the tilt angle $\delta\theta$ exceeds the critical value $\delta\theta_\mathrm{c}$, the structure becomes unstable, and the net torque appears to reduce $\delta\theta$. This torque is an internal one, and hence it cannot be predicted which flux line is practically rotated. This is similar to the example that, although it is known that a standing egg is unstable, it is unknown in which direction the egg falls down. It is assumed that the left flux line in Fig. 4.30 is driven towards the direction of $\boldsymbol{v}_1$ by the force-free torque, which exceeds the moment of pinning forces. However, such a motion varies the direction of flux lines in the superconductor and does not result in a steady state. Not only the rotational motion $\boldsymbol{v}_1$ but also the translational motion $\boldsymbol{v}_2$ is necessary to satisfy the

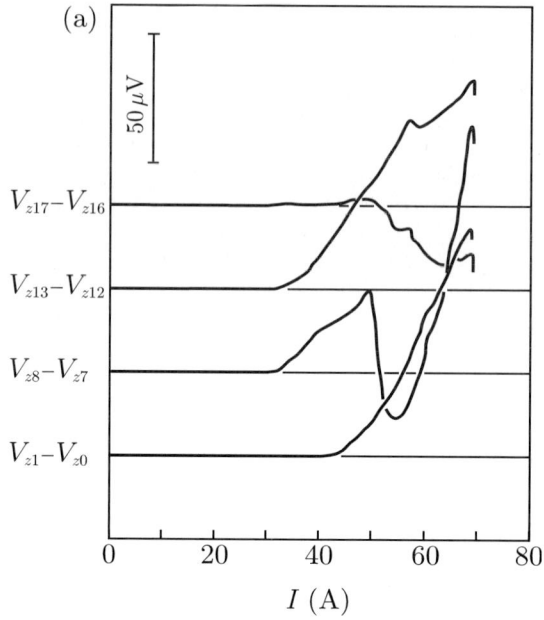
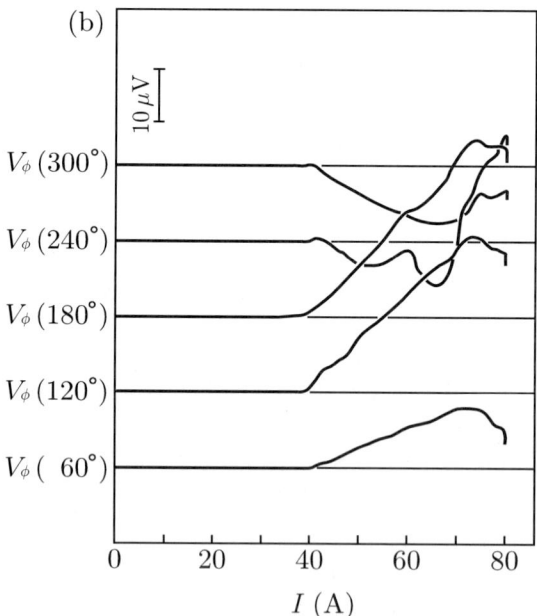

Fig. 4.28. Current-voltage curves for (**a**) longitudinal and (**b**) azimuthal directions for a cylindrical Pb-Tl specimen in a longitudinal magnetic field [34]

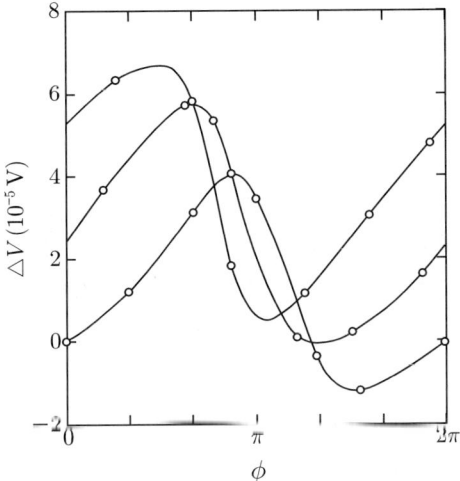

Fig. 4.29. Azimuthal variation in the surface electric potential of a cylindrical Pb-Tl specimen in a longitudinal magnetic field [5]. Each line corresponds to a different longitudinal position. This result shows a helical structure for the surface electric potential

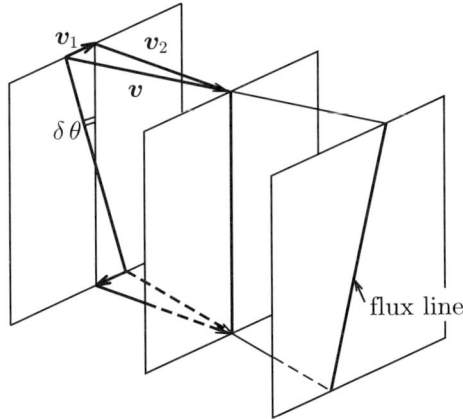

Fig. 4.30. Twisted motion of flux line. v_1 is the rotational component of motion caused by the force-free torque and v_2 is the induced translational component of motion [35]

conditions of steady state motion under given boundary conditions. There is no driving force that directly causes the translational flux motion, and hence the work done in connection with the translational motion is zero. The condition of the steady state is given by

$$\frac{\partial \theta}{\partial t} = v_2 \frac{\partial \theta}{\partial y} \tag{4.66}$$

for a slab geometry, where θ is the angle of the flux lines and the y-axis is assumed to be directed along $\boldsymbol{v}_2$. The motion of flux lines in the steady state is expected to be like the one shown by $\boldsymbol{v}$ in Fig. 4.30. If the condition is exactly symmetric, flux motion in the opposite direction is also possible, and the direction will be determined by fluctuation. Under practical experimental conditions the direction of the magnetic field will be slightly tilted from the axis of the superconductor, and the resultant small Lorentz force is expected to determine the direction of flux motion.

Figure 4.31 represents the expected flux motion inside the cylindrical superconductor [35]: (a) shows the motion of flux lines which pass through the center of the cylinder and (b) shows the direction of flux motion at each position along the length of the cylinder. This flux motion can be reproduced by twisting a uniform and translational flux flow inside the cylinder around the axis. Hereafter this shall be called the helical flux flow. A characteristic of the helical flux flow is that the term $\boldsymbol{B} \times \boldsymbol{v}$ in Eq. (4.48) brings about a negative electric field in the region N where the flux lines go out of the superconductor, and this may possibly cause a net negative electric field including the second term. In this region the Poynting vector is directed outward, and it is found that the flow of energy is directed outward. On the other hand, the Poynting vector is directed inward in region S. Such behavior is similar to the flux flow in a transverse magnetic field. As a result, the variation in the electric potential along the azimuthal direction originates from the term $\boldsymbol{B} \times \boldsymbol{v}$.

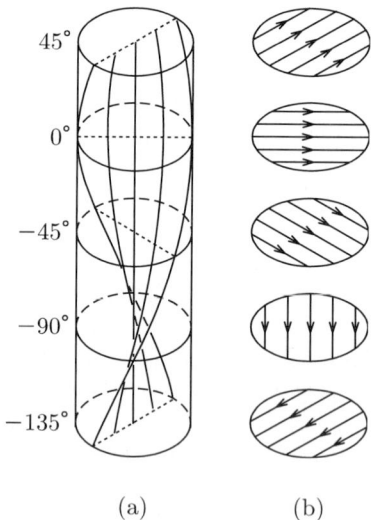

Fig. 4.31. Helical flux flow in a superconducting cylinder [35]: (**a**) motion of flux lines which pass through the center of the cylinder and (**b**) direction of flux motion at each point along the length of the cylinder. *Angles* represent the direction of flux flow

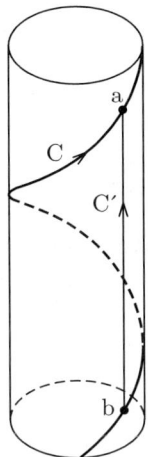

Fig. 4.32. Flux line and integral path of electric field on the surface of a superconducting cylinder

Here it will be shown that the electric field is given in the form of Eq. (4.48) in the resistive state also. It was Kogan [36] who first showed theoretically that Josephson's relation $\boldsymbol{E} = \boldsymbol{B} \times \boldsymbol{v}$ is not satisfied in a superconducting cylinder in a longitudinal magnetic field. Kogan showed that $\boldsymbol{E}$ and $\boldsymbol{B}$ are not perpendicular to each other from the cylindrical symmetry. We shall proceed with this discussion. Assume that a flux line just penetrates the superconducting cylinder as in Fig. 4.32 and treat the potential difference between two points, a and b, on the flux line. The potential difference can be obtained by curvilinearly integrating the electric field in Eq. (4.48) from b to a. In the steady condition $\partial \boldsymbol{B}/\partial t = 0$ is satisfied. Hence, it is derived from the continuity equation (2.15) that $\boldsymbol{B} \times \boldsymbol{v}$ is given by a gradient of some scalar function. Then, the curvilinear integral of $\boldsymbol{B} \times \boldsymbol{v}$ does not depend on the integral path. This means that the integral is the same whether the path is C′ or C on the flux line shown in the figure. Since $\boldsymbol{B}$ is parallel to the integral path C, this curvilinear integral is shown to be zero. Thus, we obtain

$$\int_{C'} (\boldsymbol{B} \times \boldsymbol{v}) \cdot \mathrm{d}\boldsymbol{s} = 0 \,. \tag{4.67}$$

The above result means that $\boldsymbol{B} \times \boldsymbol{v}$ does not contribute to the net potential difference in the current direction across terminals sufficiently father apart than the pitch of the helical structure. Therefore, another term which describes the observed electric field is necessary. As a result the electric field is given in the form of Eq. (4.48) [35]. This discussion shows that $\boldsymbol{B} \times \boldsymbol{v}$ does not contribute to the energy dissipation, and the important component exists in the term $-\nabla \Psi$. It is concluded that the electric field associated with the energy dissipation does not originate from the successive penetration of the azimuthal component of the magnetic flux.

Here $-\nabla\Psi$ will be calculated as a function of the velocity of the flux lines. Assume a superconducting slab and the flux motion as shown in Fig. 4.30 for simplicity. Since practical superconducting specimens have a cylindrical shape, this treatment is not exact. However, the expected flux motion does not have a general cylindrical symmetry, as shown in Fig. 4.31. Hence, the present treatment is not a bad approximation. In fact, if we look at only a part of the cylinder, the flux motion in this region is not essentially different from that in a slab. Assume that the superconducting slab is parallel to the x-z plane and occupies $0 \leq y \leq d$. The magnetic field H_e and the current I are applied along the z-axis. If the cylinder of radius R is approximated by a rectangular rod of width $2R$ and thickness d, the requirement of the same cross section leads to

$$d = \frac{\pi R}{2}. \tag{4.68}$$

In addition, it is necessary to eliminate disorder at the edge in the rectangular rod, since there is no disorder on the surface of the cylinder. Hence, only a part of the width $2R$ in a sufficiently wide slab is treated (see Fig. 4.33). Flux lines move translationally along the y-axis and rotationally in the x-z plane. If the translational motion occurs in the positive direction of the y-axis, the rotational motion occurs counterclockwise with respect to this direction. The purpose of the analysis is to estimate $-\nabla\Psi$ caused by the energy dissipation due to the flux motion. Since the energy dissipation comes from the rotational flux motion, only the energy dissipation due to the pure rotational flux motion in the superconducting slab has to be estimated.

Consider the case where the external magnetic field parallel to the superconducting slab $(-d \leq y \leq d)$ is rotated in a steady state in the x-z plane. Because of symmetry, the flux motion in the half $0 \leq y \leq d$ of the slab is treated. The magnetic flux distribution in the slab is expressed as

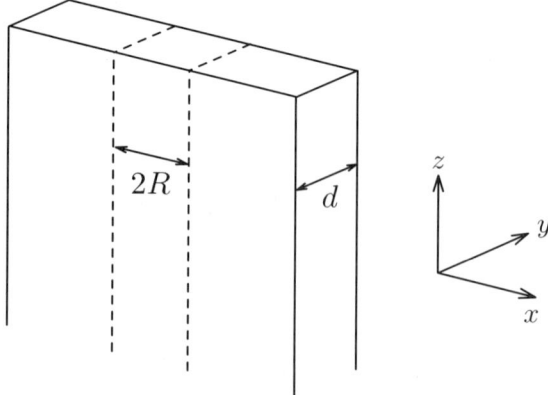

Fig. 4.33. Equivalent part of a superconducting slab for evaluation of the loss in a superconducting cylinder

$$\boldsymbol{B} = (B\sin\theta, 0, B\cos\theta) , \tag{4.69}$$

where θ is the angle of the flux lines in the x-z plane measured from the z-axis. By analogy with the flux flow in a transverse magnetic field, it can be expected that the force-free current of density $J_\|$ flows uniformly. Hence, it is derived that B is uniform and obeys $B = \mu_0[H_e^2 + (I/2\pi R)^2]^{1/2}$. The angle is given by

$$\theta = \theta_d - \frac{\mu_0 J_\|}{B}(y - d) , \tag{4.70}$$

where θ_d is the angle at the surface, $y = d$. In the case where θ_d varies constantly with time, $\boldsymbol{E} = 0$ is derived at the center ($y = 0$) from symmetry. Thus, the electric field is easily calculated:

$$E_x(y=d) = \frac{B^2\dot\theta}{\mu_0 J_\|}\left[\cos\left(\theta_d + \frac{\mu_0 J_\| d}{B}\right) - \cos\theta_d\right] ,$$

$$E_z(y=d) = \frac{B^2\dot\theta}{\mu_0 J_\|}\left[\sin\theta_d - \sin\left(\theta_d + \frac{\mu_0 J_\| d}{B}\right)\right] , \tag{4.71}$$

where $\dot\theta = d\theta/dt$ is the angular velocity of rotation of the external magnetic field. It is easily found that $E_y = 0$. As a result the input power per unit surface area of the superconducting slab is estimated as

$$P = -\frac{1}{\mu_0}(\boldsymbol{E}\times\boldsymbol{B})_y|_{y=d} = -\frac{B^3\dot\theta}{\mu_0^2 J_\|}\left[1 - \cos\left(\frac{\mu_0 J_\| d}{B}\right)\right]$$

$$\simeq -\frac{BJ_\| d^2\dot\theta}{2} , \tag{4.72}$$

when $\mu_0 J_\| d/B$ is small. (Note that $\dot\theta < 0$, since the magnetic field is rotated counterclockwise.)

Now we go back to the helical flux flow in the superconducting cylinder. In practice the rotational flux motion is accompanied by translational motion, and there exists a relationship:

$$\dot\theta = -\frac{\mu_0 J_\| v_2}{B} \tag{4.73}$$

from Eq. (4.66). Substitution of this into Eq. (4.72) leads to

$$P = \frac{\mu_0 J_\|^2 d^2 v_2}{2} . \tag{4.74}$$

The input power $2RP$ per unit length of the region of width $2R$ in the rod shown in Fig. 4.33 should be equal to the loss power $\pi R^2(-\nabla\Psi)_z J_\|$ per unit length of the cylinder. This requirement leads to [35]

$$(-\nabla\Psi)_z = \frac{\mu_0 J_\| d v_2}{2} . \tag{4.75}$$

Next the surface electric field of the superconducting cylinder will be investigated. Cylindrical coordinates (r, ϕ, z) are used, where the angle ϕ is measured from the direction of flux motion at $z = 0$. The magnetic flux density at the surface is expressed as

$$\boldsymbol{B}(R) = (B_r, B_\phi, B_z) = (0, B \sin \theta_R, B \cos \theta_R) , \quad (4.76)$$

where θ_R is the angle of the surface magnetic field from the z-axis. It seems to be acceptable to assume that the velocity of the translational flux motion v_2 is uniform. Then, from the condition $\boldsymbol{B} \cdot \boldsymbol{v} = 0$ the velocity of flux lines on the surface at $z = 0$ is given as

$$\boldsymbol{v}(R) = (v_2 \cos \phi, -v_2 \sin \phi, v_2 \tan \theta_R \sin \phi) . \quad (4.77)$$

This leads to

$$(\boldsymbol{B} \times \boldsymbol{v})_\phi |_R = Bv_2 \cos \theta_R \cos \phi , \quad (4.78\mathrm{a})$$

$$(\boldsymbol{B} \times \boldsymbol{v})_z |_R = -Bv_2 \sin \theta_R \cos \phi . \quad (4.78\mathrm{b})$$

The term $-\nabla \Psi$ does not have an azimuthal component from symmetry. Hence, the azimuthal electric field comes only from $\boldsymbol{B} \times \boldsymbol{v}$. From Eq. (4.78a) we obtain

$$V(\phi, 0) - V(\phi', 0) = -\int_{\phi'}^{\phi} Bv_2 \cos \theta_R \cos \phi R d\phi$$

$$= RBv_2 \cos \theta_R (\sin \phi' - \sin \phi) . \quad (4.79)$$

Thus, the helical symmetry leads to

$$V(\phi, z) - V(\phi', z) =$$

$$RBv_2 \cos \theta_R \left[\sin \left(\phi' - \frac{z}{R} \tan \theta_R \right) - \sin \left(\phi - \frac{z}{R} \tan \theta_R \right) \right] . \quad (4.80)$$

Next, the longitudinal electric potential difference at $\phi = \phi'$ will be calculated. The condition that $\phi - (z/R) \tan \theta_R = \mathrm{const.}$ represents equivalent positions for the flux motion. From Eq. (4.78b) the contribution from the term $\boldsymbol{B} \times \boldsymbol{v}$ to the potential difference is

$$Bv_2 \sin \theta_R \int_0^z \cos \left(\phi' - \frac{z}{R} \tan \theta_R \right) dz$$

$$= RBv_2 \cos \theta_R \left[\sin \phi' - \sin \left(\phi' - \frac{z}{R} \tan \theta_R \right) \right] . \quad (4.81)$$

On the other hand, since $\mu_0 J_\| d$ is equal to the azimuthal component of the magnetic flux density at the surface given by $B \sin \theta_R$, the contribution from the term in Eq. (4.75) leads to

$$\int_0^z (-\nabla\Psi)_z \, dz = \frac{1}{2} B v_2 z \sin\theta_R \ . \tag{4.82}$$

The longitudinal electric potential difference is given by a sum of Eqs. (4.81) and (4.82). Thus, the electric potential difference on the surface is generally given as [35]

$$\begin{aligned}\Delta V &= V(\phi, z) - V(0,0) \\ &= B v_2 \left[\frac{z}{2} \sin\theta_R - R\cos\theta_R \sin\left(\phi - \frac{z}{R}\tan\theta_R\right) \right] \ . \end{aligned} \tag{4.83}$$

The obtained electric potential difference on the surface is shown in Fig. 4.34(a). Here, Fig. 4.34(b) represents the corresponding experimental results which are the same as in Fig. 4.29 shown previously. Comparing these figures, it is found that a good agreement with experimental results is obtained including the negative electric potential difference.

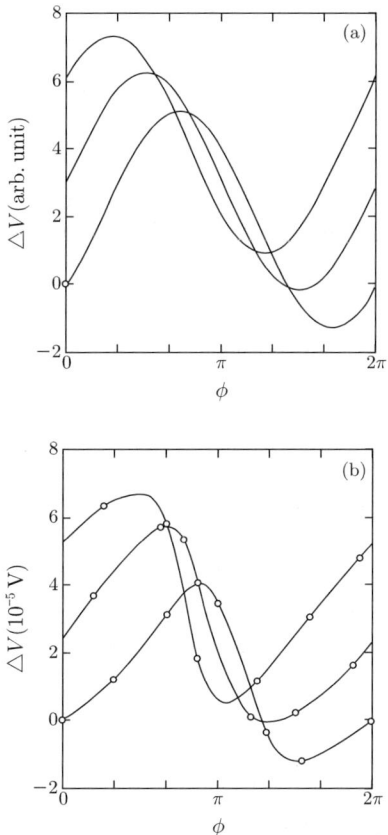

Fig. 4.34. (a) Surface electric potential given by Eq. (4.83) [35] and (b) corresponding experimental results (same as Fig. 4.29)

204   4 Longitudinal Magnetic Field Effect

As has been shown above the structure of the surface electric field of the superconducting cylinder can be explained using the helical flux flow derived from the force-free torque. In addition, the constant longitudinal magnetization and the steady electric field, which seemed to be incompatible with each other, are now found to be compatible. Peculiar phenomena were proposed to occur in region N because of the negative electric field. However, the above discussion clarifies that such a speculation is not correct. That is, $\boldsymbol{J} \cdot \boldsymbol{E}$ is positive and uniform over the entire superconductor. This can be derived from $\boldsymbol{J} \cdot (\boldsymbol{B} \times \boldsymbol{v}) = (\boldsymbol{J} \times \boldsymbol{B}) \cdot \boldsymbol{v} = 0$. Such a negative electric field results from the helical flow of the current. If we look at the electric field along the direction of the current, it is always positive.

Measurements of the radial electric field in a cylindrical specimen by Makiej et al. [37] constitute one of the experimental results which show the occurrence of the helical flux flow. This component of the electric field can be expected from the present flux flow. In other words this observation proves that the translational motion of the longitudinal component of magnetic flux occurs.

Here observed results of surface structure of the electric field are introduced for a current-carrying superconducting Pb-In slab in a longitudinal magnetic field. Arrangement of potential terminals on the surface of the slab specimen is shown in Fig. 4.35 [38]. The distance between adjacent potential terminals is 1.0 mm in the axial direction. Figure 4.36(a) shows the distribution of longitudinal electric field and L is the distance from potential terminal V8 in Fig. 4.35. Figure 4.36(b) shows the angle θ between the electric and magnetic

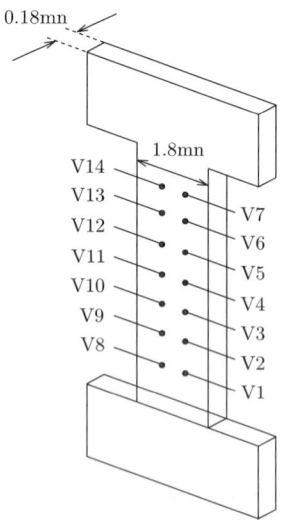

Fig. 4.35. Geometry of Pb-60at.%In specimen and arrangement of potential terminals

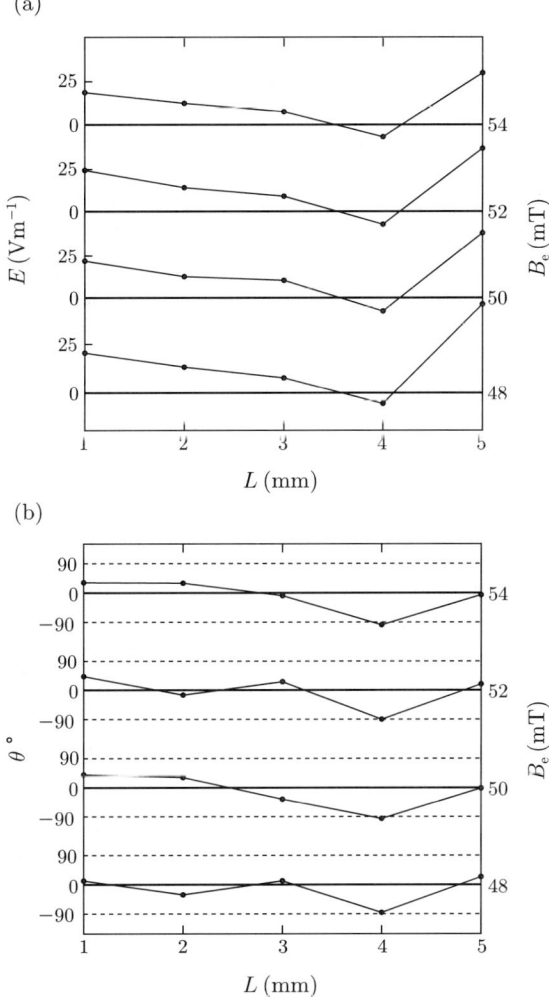

Fig. 4.36. (a) Longitudinal component of electric field and (b) angle between the electric and magnetic fields on the surface of a Pb-In slab specimen carrying a current of 31.25 A in various longitudinal magnetic fields [38]

fields on the specimen surface, and the negative θ means that the Poynting vector is directed outward the specimen.

This result shows that the Poynting vector on the surface is necessarily directed outward where the negative electric field is observed, and supports the argument in this section. In addition, it is found from Fig. 4.36(b) that the area of the region where the Poynting vector is directed outward is roughly the same as that of the inward Poynting vector within one periodic length of the structure. Hence, it is understood that a steady flow of noncompressional flux lines takes place in the superconductor.

Finally, it should be noted that what we did here was only to estimate the electric field as a function of the velocity of flux lines v_2. That is, the electric field was not directly estimated under a given condition of the current density, since we did not calculate the viscous coefficient of the viscous torque during the rotational flux motion in the resistive state. For this kind of discussion a more microscopic treatment as in Sect. 2.2 is necessary. But it is intuitively expected that the resistivity in the longitudinal magnetic field is approximately equal to the flow resistivity in the transverse magnetic field. Figure 4.37 is the longitudinal magnetic field dependence of the observed resistivity of the Pb-In slab specimen shown in Fig. 4.36 [38]. This agrees well with the predictions of the Bardeen-Stephen model given by Eq. (2.30), and is the same as for the usual flux flow resistivity in a transverse magnetic field. This strongly suggests that the electric field does not originate from the motion of the transverse magnetic flux due to the current.

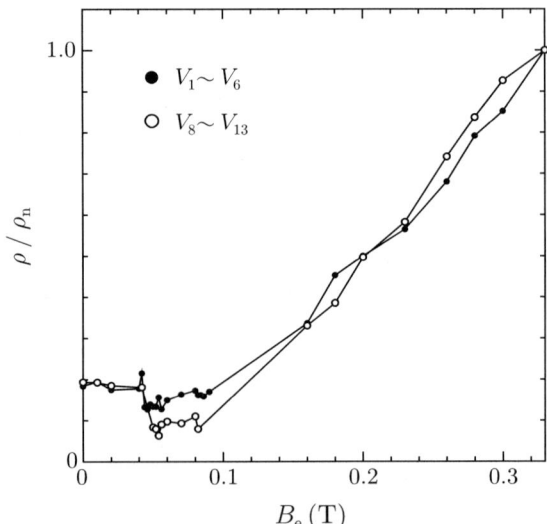

Fig. 4.37. Dependence of the resistivity averaged along the length on longitudinal magnetic field for a Pb-In slab specimen [38]

Exercises

4.1. It is assumed that a constant magnetic field H_e is applied along the z-axis to a wide pin-free superconducting slab ($0 \leq y \leq 2d$), and then, a force-free state is established by applying a current in the same direction.

Derive the magnetic field distribution and the magnetization along the z-axis during the process of establishing the force-free state. The force-free current density is given by Eq. (4.3).

4.2. Discuss the strain of the flux line structure when a force-free current flows in a superconducting cylinder ($r \leq R$).

4.3. It is assumed that a magnetic field is applied to a wide pin-free superconducting slab ($0 \leq y \leq 2d$), and then, a force-free state is established by rotating the field in the plane parallel to the slab surface. Calculate the velocity of flux lines and the induced electric field, and discuss the relationship between these results and Eq. (4.48).

4.4. Show that no solution for the velocity of flux lines exists under the conditions of Exercise 4.1, if it is restricted to the form $\bm{v} = (0, v, 0)$, assuming that there is no rotational motion.

4.5. Show that the work per unit time is given by $(-\nabla \Psi) \cdot \bm{J}$ when the electric field of Eq. (4.48) is induced by the motion of flux lines driven by the force-free torque.

4.6. Consider the case where a magnetic field parallel to a superconducting disk is rotated and the case where the disk is rotated in the opposite direction in a stationary parallel field. Discuss the equivalence of the two cases from the viewpoints of the flux-cutting model and the flux-rotation model. It is assumed that the superconductor does not contain pinning centers and that the rotation takes place at a finite angular velocity such that an eddy current can flow inside the superconductor.

References

1. D. G. Walmsley: J. Phys. F **2** (1972) 510.
2. Yu. F. Bychkov, V. G. Vereshchagin, M. T. Zuev, V. R. Karasik, G. B. Kurganov and V. A. Mal'tsev: JETP Lett. **9** (1969) 404.
3. Y. Nakayama and O. Horigami: Cryogenic Engineering (J. Cryogenic Society of Jpn.) **6** (1971) 95 [in Japanese].
4. J. R. Cave, J. E. Evetts and A. M. Campbell: J. de Phys. (Paris) **39** (1978) C6-614.
5. T. Ezaki and F. Irie: J. Phys. Soc. Jpn. **40** (1976) 382.
6. C. J. Bergeron: Appl. Phys. Lett. **3** (1963) 63.
7. B. D. Josephson: Phys. Rev. **152** (1966) 211.
8. G. W. Cullen and R. L. Novak: Appl. Phys. Lett. **4** (1964) 147.
9. A. M. Campbell and J. E. Evetts: Adv. Phys. **21** (1972) 252.
10. J. R. Clem: J. Low Temp. Phys. **38** (1980) 353.
11. E. H. Brandt: J. Low Temp. Phys. **39** (1980) 41.
12. J. R. Clem: Physica **107B** (1981) 453.
13. E. H. Brandt, J. R. Clem and D. G. Walmsley: J. Low Temp. Phys. **37** (1979) 43.
14. J. R. Clem and S. Yeh: J. Low Temp. Phys. **39** (1980) 173.
15. A. Kikitsu, Y. Hasegawa and T. Matsushita: Jpn. J. Appl. Phys. **25** (1986) 32.

16. F. Irie, T. Matsushita, S. Otabe, T. Matsuno and K. Yamafuji: Cryogenics **29** (1989) 317.
17. P. Wagenleithner: J. Low Temp. Phys. **48** (1982) 25.
18. T. Matsushita: Phys. Rev. B **38** (1988) 820.
19. G. Fillion, R. Gauthier and M. A. R. LeBlanc: Phys. Rev. Lett. **43** (1979) 86.
20. M. G. Blamire and J. E. Evetts: Phys. Rev. B **33** (1986) 5131.
21. T. Matsushita: J. Phys. Soc. Jpn. **54** (1985) 1054.
22. T. Matsushita: Phys. Lett. **86A** (1981) 123.
23. K. Yamafuji, T. Kawashima and H. Ichikawa: J. Phys. Soc. Jpn. **39** (1975) 581.
24. T. Matsushita, Y. Hasegawa and J. Miyake: J. Appl. Phys. **54** (1983) 5277.
25. V. G. Kogan: Phys. Rev. B **21** (1980) 3027.
26. T. Matsushita: Jpn. J. Appl. Phys. **26** (1987) Suppl. 26-3, p. 1503.
27. See for example, J. R. Cave and M. A. R. LeBlanc: J. Appl. Phys. **53** (1982) 1631.
28. T. Matsushita, Y. Miyamoto, A. Kikitsu and K. Yamafuji: Jpn. J. Appl. Phys. **25** (1986) L725.
29. See for example, R. Boyer, G. Fillion and M. A. R. LeBlanc: J. Appl. Phys. **51** (1980) 1692.
30. See for example, J. R. Clem and A. Perez-Gonzalev: Phys. Rev. B **30** (1984) 5041.
31. T. Matsushita, A. Kikitsu, Y. Miyamoto and E. Nishimori: *Proc. Int. Symp. Flux Pinning and Electromagnetic Properties in Superconductors*, Fukuoka, 1985, p. 200.
32. T. Matsushita, S. Ozaki, E. Nishimori and K. Yamafuji: J. Phys. Soc. Jpn. **54** (1985) 1060.
33. J. R. Cave and J. E. Evetts: Phil. Mag. B **37** (1978) 111.
34. T. Ezaki: Doctoral thesis (Kyushu University, 1976).
35. T. Matsushita and F. Irie: J. Phys. Soc. Jpn. **54** (1985) 1066.
36. V. G. Kogan: Phys. Lett. **79A** (1980) 337.
37. B. Makiej, A. Sikora, S. Golab and W. Zacharko: *Proc. Int. Discussion Meeting on Flux Pinning in Superconductors*, Sonnenberg, 1974, p. 305.
38. T. Matsushita, A. Shimogawa and M. Asano: Physica C **298** (1998) 115.

5
Measurement Methods for Critical Current Density

5.1 Four Terminal Method

The most general method for measuring the critical current density J_c, an important parameter of superconductors, is the four terminal method, in which the voltage drop V between the terminals is measured as a function of the transport current I. This is also called the resistive method. The critical current I_c is defined as the transport current at which the flow voltage clearly appears. The critical current density is given by I_c divided by the cross-sectional area S of the superconducting region: $J_c = I_c/S$. In multifilamentary superconductors, the cross-sectional area may include a metallic stabilizer and reinforcing materials.

In practice, the current-voltage curves of superconducting wires are not straight lines as in Fig. 1.13. Instead, voltage gradually rises due to various causes, which will be described later. The measurement is also subject to sensitivity limits. Hence, there is no clear point at which the flow voltage appears. To define the critical current, the following criteria are used.

(1) *Electric field criterion:* This is the simplest method. The critical current is defined by the current at which the electric field reaches a certain value (see Fig. 5.1). A value of 100 μVm^{-1} or 10 μVm^{-1} is commonly used.

(2) *Resistivity criterion:* The critical current is defined by the current at which the resistivity of the superconducting wire reaches a certain value (see Fig. 5.1). For composite superconductors with stabilizer, 10^{-13} Ωm or 10^{-14} Ωm is commonly used.

(3) *Off-set method:* The critical current is determined by the current at which a tangential line from part of the current-voltage curve crosses zero voltage (see Fig. 5.1).

A large error results from the electric field criterion and the resistivity criterion when flux creep is pronounced. Even in the case where the current-voltage curve shows an ohmic characteristic as in Eq. (3.131), a nonzero critical

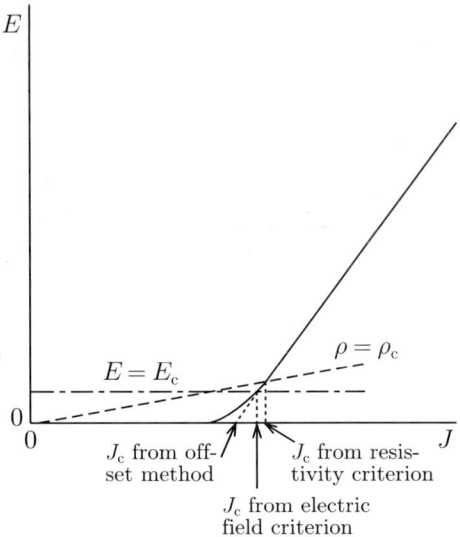

Fig. 5.1. Current-voltage curve and methods of determination of critical current density using respective criteria

current density is defined using the electric field criterion. These methods are useful only for a strongly nonlinear current-voltage curve, which rises abruptly at a nonzero current density. If the current-voltage curve is expressed as

$$V \propto I^n, \qquad (5.1)$$

the index, n, is called the n value. The n value is a supplementary parameter representing the strength of the nonlinearity. The electric field range of 1 μVm^{-1} to 100 μVm^{-1} is generally used to determine the n value. A superconducting wire with larger n is often better. Note that it is possible to reduce the induced voltage drastically by reducing the current slightly when n is high. When n is low, on the other hand, the induced voltage does not become small abruptly when the current is decreased slightly. To avoid errors when n is small, it is practical to use the offset method. Using the line tangent to the curve at the current density at the electric field criterion J_0, the critical current density determined by the offset method is

$$J'_c = \left(1 - \frac{1}{n}\right) J_0 . \qquad (5.2)$$

This gives the correct result $J'_c = 0$ for $n = 1$.

Here the meaning of the n value is discussed. The current-voltage characteristics deviate from Eq. (2.31), the relationship of which is shown in Fig. 1.13, and the voltage rises gradually near J_c. This voltage is due to both microscopic causes, such as flux creep and the nonlinearity of flux motion around the pinning potential, and macroscopic causes such as spatial nonuniformity of the critical current density and sausaging of superconducting filaments. Sausaging is a nonuniformity of the filament diameter as a result of wire drawing. Therefore, it is difficult to derive an n value directly as a physical quantity; the n value is merely a convenient parameter for practical use.

If the dispersion of J_c originates from the dispersion of T_c, its value ΔJ_c does not vary appreciably even at high temperatures and high magnetic fields. Thus, $\Delta J_c/J_c$ is larger at higher temperatures and at higher magnetic fields as shown in Fig. 5.2. In addition, the effect of flux creep becomes pronounced under these circumstances. Hence, the n value is a decreasing function of T and B.

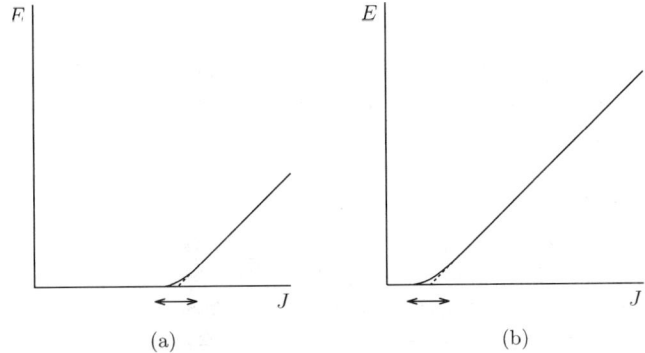

Fig. 5.2. Current-voltage curves at (**a**) low temperatures and/or low magnetic fields and (**b**) high temperatures and/or high magnetic fields. Arrows show ranges of distribution of critical current density. At high temperatures and/or high magnetic fields, the deviation is relatively large in comparison with the mean value of the critical current density, and the n value is small

Equation (5.1) insists that a true superconducting state with zero resistivity does not exist. The flux creep theory predicts that the electric field abruptly decreases exponentially in a region of ultra low electric field, far below the sensitivity of present measuring techniques. However, the electric field is not zero even in this case. This is associated with the fact that the state in which the flux lines are pinned by pinning potentials is not at equilibrium. Hence, the process of relaxation to the equilibrium state with zero current density cannot be avoided. Shortly after the discovery of high-temperature superconductors, many researchers thought that these superconductors could not be applied to the field of technology. Although the above properties are true, it is not true that these superconductors cannot be applied.

Assume that Eq. (5.1) applies approximately within some range of current. When the current is reduced by a factor p, the voltage can be reduced by a factor p^n. Hence, if the loss associated with the voltage drop can be reduced below that of an equivalent nonsuperconducting metal by reducing the current by an appropriate factor, practical application of the superconductor can be realized. For example, consider a superconducting coil that is made of 1 km of wire with a critical current of 200 A at the electric field criterion of 100 µVm^{-1}. When this coil is driven at the critical current, the voltage is 0.1 V and the loss power is 20 W. However, if the n value of the wire is 50, the loss power is reduced to 0.5 W by reducing the current to $0.93I_\text{c}$. This power may be much less than the heat transmitted into a cryostat. Thus, the coil can be applied as a superconducting device. As demonstrated by this example, if the n value is sufficiently large, even a relatively weak electric field criterion such as 100 µVm^{-1} can be used to define I_c. The n value in commercial superconducting wires exceeds 50, and $n = 21$ has been reported [1] for a Bi-2223 tape at 77.3 K in the self field.

5.2 DC Magnetization Method

The DC magnetization of a superconductor is hysteretic as mentioned in Sect. 2.5. Consider a superconducting slab with thickness $2d$. In low magnetic fields, the field dependence of the pinning force density can be approximated by Eq. (2.46). Then, the parameters α_c and γ are determined such that Eq. (2.55) fits an observed magnetization curve, and the local critical current density can be obtained from $\alpha_\text{c} \widehat{B}^{\gamma-1}$ (see Eq. (2.50)).

If the magnetization contains a diamagnetic component as mentioned in Sect. 2.6, this contribution should be eliminated. This is possible only when the diamagnetic property is known. However, even if the property is unknown, the diamagnetic effect can be approximately canceled out in the hysteresis of magnetization between increasing and decreasing fields. This is a good approximation for superconductors with a large G-L parameter κ.

The hysteresis of magnetization of a superconducting slab in a parallel external H_e is calculated from Eqs. (2.55b) and (2.55d):

$$\Delta M = \frac{2-\gamma}{3-\gamma} H_\text{p} \left\{ \left[\left(\frac{H_\text{e}}{H_\text{p}}\right)^{2-\gamma} + 1 \right]^{(3-\gamma)/(2-\gamma)} \right.$$
$$\left. + \left[\left(\frac{H_\text{e}}{H_\text{p}}\right)^{2-\gamma} - 1 \right]^{(3-\gamma)/(2-\gamma)} - 2\left(\frac{H_\text{e}}{H_\text{p}}\right)^{3-\gamma} \right\}. \quad (5.3)$$

The paramaters α_c and γ can be estimated by fitting the observed hysteresis to the above theoretical result. By contrast, the critical current density obtained by the transport method is a spatial average of the local critical current density.

The average magnetic critical current density is usually estimated from

$$J_\text{c} = \frac{\Delta M}{d}. \quad (5.4)$$

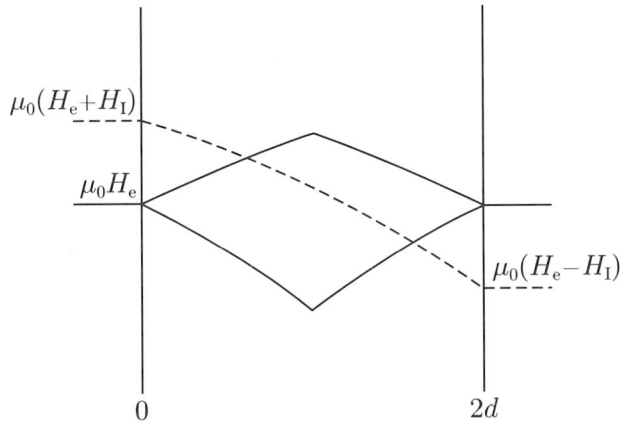

Fig. 5.3. Magnetic flux distributions in a superconducting slab during a magnetization measurement (*solid lines*) and that in the critical state during a transport measurement (*broken line*). H_I represents the self field of the current

This is correct only when the local critical current density is constant throughout the sample, i.e., when the Bean-London model holds.

The average transport critical current density $\langle J_\mathrm{c} \rangle$ at external magnetic field H_e is given by Eq. (2.61). $\langle J_\mathrm{c} \rangle$ can also be obtained from ΔM. The solid lines in Fig. 5.3 show the magnetic flux distributions in the processes of increasing and decreasing the magnetic field, and the area of the diamond-shaped region is equal to $2\mu_0 \Delta M d$. The broken line in the figure is the flux distribution when the transport current reaches the critical value in the external field H_e. When the external magnetic field is sufficiently larger than the penetration field H_p, Eq. (5.3) reduces to

$$\Delta M \simeq \frac{H_\mathrm{p}}{2-\gamma}\left(\frac{H_\mathrm{p}}{H_\mathrm{e}}\right)^{1-\gamma}\left[1+\frac{(1-\gamma)(3-2\gamma)}{12(2-\gamma)^2}\left(\frac{H_\mathrm{p}}{H_\mathrm{e}}\right)^{4-2\gamma}\right]. \quad (5.5)$$

Thus, in terms of ΔM, $\langle J_\mathrm{c} \rangle$ is given by

$$\langle J_\mathrm{c} \rangle \simeq \frac{\Delta M}{d}\left[1+\frac{(1-\gamma)(4\gamma-3)}{12}\left(\frac{\Delta M}{H_\mathrm{e}}\right)^2\right]. \quad (5.6)$$

The second term gives the correction to Eq. (5.4). This is very small when the external field is large. In the case where $\gamma = 0.5$ and $H_\mathrm{e} = 2H_\mathrm{p}$, the second term gives a correction of about 0.2%.

5.3 Campbell's Method

The shielding current density induced in the superconductor by an AC magnetic field can be estimated by measuring the penetrating flux. One method

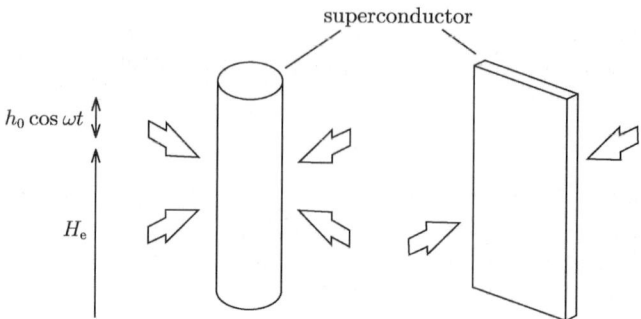

Fig. 5.4. Application of magnetic field commonly used in Campbell's method. *Arrows* show the directions of penetration of AC magnetic flux

is Campbell's method, [2] explained in this section. By using this method, not only the current density but also the relationship between the force on and the displacement of the flux lines can be derived. The analysis of the force-displacement profile is useful for investigating the reversible motion of flux lines described in Sect. 3.7 and the flux pinning properties described in Chap. 7. Other AC inductive methods will be introduced in Sect. 5.4.

Usually a DC magnetic field H_e and a small AC field $h_0 \cos \omega t$ are applied parallel to a superconducting cylinder or long slab, as shown in Fig. 5.4, to avoid the effect of demagnetization due to the specimen shape. The magnetic flux moving into and out of the specimen is measured using a pick-up coil and a reference coil. The amplitude of penetrating flux is denoted by Φ, and $\delta\Phi$ corresponds to the incremental flux change when h_0 is slightly increased by δh_0. The magnetic flux distribution is expected to be like the one shown in Fig. 5.5. The shielding current density, which is not necessarily equal to

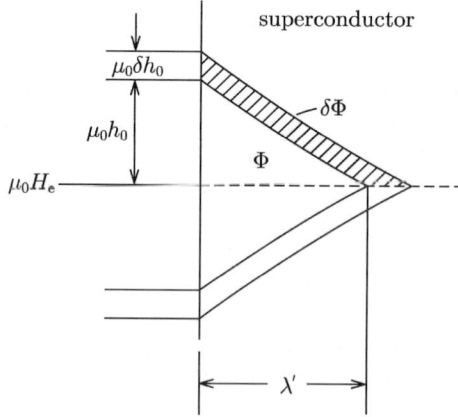

Fig. 5.5. Variation in penetration of AC magnetic flux when the AC field amplitude is slightly changed from h_0 to $h_0 + \delta h_0$. Current density is assumed to be constant

the critical value, is considered to be unchanged when the AC field amplitude increases from h_0 to $h_0 + \delta h_0$. Then, the depth of penetration of the AC field is given by

$$\lambda' = \frac{1}{2w\mu_0} \cdot \frac{\delta \Phi}{\delta h_0} \tag{5.7}$$

when the width w of the slab specimen is much larger than the thickness $2d$. In a strict sense, $2w$ in the denominator of Eq. (5.7) is replaced by the perimeter of the superconducting slab $2(w + 2d)$, when λ' is sufficiently smaller than $2d$. In the limit of small δh_0, $\delta \Phi / \delta h_0$ reduces to the derivative, $\partial \Phi / \partial h_0$. Then, Eq. (5.7) leads to

$$\lambda' = \frac{1}{2w\mu_0} \cdot \frac{\partial \Phi}{\partial h_0}. \tag{5.8}$$

In a cylindrical superconductor with radius R, a simple calculation gives

$$\lambda' = R \left[1 - \left(1 - \frac{1}{\pi R^2 \mu_0} \cdot \frac{\partial \Phi}{\partial h_0} \right)^{1/2} \right]. \tag{5.9}$$

When $\lambda' \ll R$, the right-hand side of Eq. (5.9) reduces to $(2\pi R \mu_0)^{-1} \partial \Phi / \partial h_0$. This result is understandable from the fact that the perimeter is equal to $2\pi R$. The derivative of Φ with respect to h_0 in Eqs. (5.8) and (5.9) can be obtained by expressing Φ as a polynomial of h_0.

An example of $\lambda'(h_0)$ is shown in Fig. 5.6 [3]. Except for small h_0, this λ'-h_0 characteristic can be regarded as the flux distribution in the superconductor for increasing field; the ordinate and abscissa represent the internal magnetic flux density and the depth of flux penetration, respectively. Hence, the slope of this distribution gives $\mu_0 J$:

$$J = \left(\frac{\partial \lambda'}{\partial h_0} \right)^{-1}, \tag{5.10}$$

which is equal to J_c in the critical state. In Fig. 5.6 it is found that the prediction of the Bean-London model is satisfied. Derivation of the magnetic flux distribution and the critical current density is requested in Exercise 5.1 using Eqs. (5.8) and (5.9) for the Bean-London model. Note that the penetration depth λ' is finite, $\lambda' = \lambda'_0$, deviating from the Bean-London model, when h_0 is small. This value (λ'_0) is Campbell's AC penetration depth, given by Eq. (3.92). In this region the reversible motion of flux lines mentioned in Sect. 3.7 is pronounced, and the apparent magnetic flux distribution in Fig. 5.6 is different from the real one. That is, although Eq. (5.10) predicts a large current density, this may not be correct. The real distribution is like the one shown in Fig. 3.32(a), and the current density takes a reasonable value. In this region the penetrating flux is approximately given by Eq. (3.91), and a replacement of $b(0)$ by $\mu_0 h_0$ gives

$$\Phi \simeq 2w \int_0^\infty \mu_0 h_0 \exp\left(-\frac{x}{\lambda'_0}\right) dx = 2w\mu_0 h_0 \lambda'_0 \tag{5.11}$$

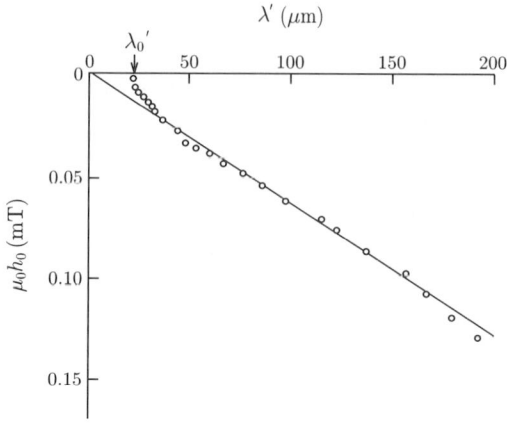

Fig. 5.6. Example of measurement of the λ' vs. h_0 characteristics using a modified Campbell's method for Nb-50at%Ta at $\mu_0 H_e = 0.336$ T [3]. The prediction of the Bean-London model holds except in the region of small h_0

for a superconducting slab sufficiently thicker than λ'_0. Substitution of this into Eq. (5.8) leads to

$$\lambda' = \lambda'_0 , \qquad (5.12)$$

which coincides with experiment.

In Campbell's method [2] the amplitude of penetrating AC flux Φ, i.e., half of the difference between the magnetic flux at $\omega t = -\pi$ and that at $\omega t = 0$ is measured. There is also a similar method, [4] in which a fundamental frequency component of the AC flux Φ' is approximately measured instead of Φ, followed by the same analysis. The resultant error due to this approximation is estimated in Exercise 5.2. In another method, [5] instantaneous values of the external AC field and penetrating AC flux, represented by $h(t)$ and $\Phi(t)$, are measured. Since the relationship between these quantities is the same as that between h_0 and Φ, $(\partial\Phi(t)/\partial t)/(\partial h(t)/\partial t)$ is equal to $\partial\Phi/\partial h_0$. Thus, Eq. (5.8) can be rewritten as

$$\lambda' = \frac{1}{2w\mu_0} \cdot \frac{\partial\Phi(t)/\partial t}{\partial h(t)/\partial t} . \qquad (5.13)$$

The denominator $(\mu_0 \partial h(t)/\partial t)$ and the numerator $(\partial\Phi(t)/\partial t)$ are the voltages measured directly by a field monitor coil and a pick-up coil, respectively. This is the wave-form analysis method. A characteristic of this method is that differentiation as in Eq. (5.8) is not necessary.

In such AC inductive methods, many more measurements and analyses are needed to determine J_c than in the four terminal method and the DC magnetization method. However, other important information can also be obtained. One of them is the relationship between the pinning force and the displacement of flux lines, which will be discussed later. Observation of an inhomogeneous current distribution is also possible, although applicable cases

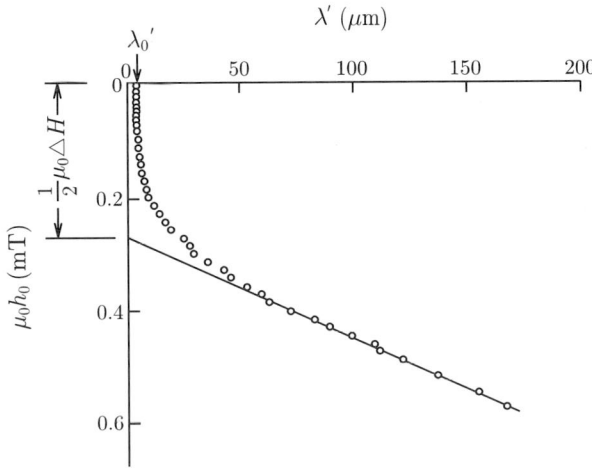

Fig. 5.7. Example of measurement of the λ' vs. h_0 characteristics using the modified Campbell's method for Nb-50at%Ta, the same specimen as that in Fig. 5.6, but at $\mu_0 H_e = 0.123$ T [6]. Extrapolation of the linear part does not go through the origin, and is thus different from Fig. 5.6. This shows a strong flux pinning near the surface

are limited. Figure 5.7 shows an example of an observed flux distribution using a modified Campbell's method for a specimen with a surface irreversibility, [6] which is discussed in Sect. 3.5. While the flux distribution is linear with a uniform critical current density in the inner region, its extrapolation does not pass through the origin, suggesting that a large magnetization is caused by a high density of shielding current flowing in the surface region. This analysis will be described later. When the shielding current flows in an inhomogeneous way, depending on a depth from the surface, as in the case of surface irreversibility, such an inhomogeneous current distribution, which cannot be obtained by the four terminal method and the DC magnetization method, can be obtained by this method. However, observable quantities are those averaged along the direction normal to the flux penetration, so any inhomogeneity along this direction cannot be obtained. Another example is the simultaneous observation of intra- and inter-grain critical current densities in sintered Y-based oxide superconductors with weakly coupled grains [7].

According to the analysis in [8], the relationship between the pinning force density F and the displacement of flux lines u is derived as follows. The equations used for this analysis are Eq. (3.88), the continuity equation for flux lines:

$$\frac{\mathrm{d}u}{\mathrm{d}x} = -\frac{b}{\mu_0 H_e} \tag{5.14}$$

and the force-balance equation between the Lorentz force F_L and the pinning force density F (see Eq. (3.89)):

218   5 Measurement Methods for Critical Current Density

$$-F = F_{\rm L} = -H_{\rm e}\frac{{\rm d}b}{{\rm d}x} + {\rm const.} \tag{5.15}$$

Note that this gives the absolute Lorentz force density, being different from Eq. (3.89), and the constant term on the right side is the value in the initial state ($b = 0$). The displacement of flux lines at the superconductor surface is initially obtained from Eq. (5.14). The displacement in the half cycle from $\omega t = -\pi$ (the initial state) to $\omega t = 0$ is positive, suggesting that flux lines move along the direction of the positive x-axis. This can be expressed in terms of the *amplitude* of observed magnetic flux Φ as

$$u(0) = -\frac{1}{\mu_0 H_{\rm e}}\int_d^0 b(x){\rm d}x = \frac{\Phi}{\mu_0 H_{\rm e} w}. \tag{5.16}$$

Assume that the initial state is the critical state with $F_{\rm L} = -\mu_0 J_{\rm c} H_{\rm e}$, as is satisfied in many experiments. Then, the pinning force density on flux lines at the surface is given by

$$-F = F_{\rm L} = -H_{\rm e}\left(\frac{\partial b}{\partial u}\cdot\frac{\partial u}{\partial x}\right)_{x=0} - \mu_0 J_{\rm c} H_{\rm e}. \tag{5.17}$$

From Eq. (5.14) and the relationship $(\partial b/\partial u)_{x=0} = [\partial b(0)/\partial\Phi]\cdot[\partial\Phi/\partial u(0)] = \mu_0 H_{\rm e}/\lambda'$, the above equation reduces to

$$-F = \frac{2\mu_0 H_{\rm e} h_0}{\lambda'} - \mu_0 J_{\rm c} H_{\rm e}. \tag{5.18}$$

This quantity is also obtained from the observed result of λ'.

Thus, $-F$ and $u(0)$ are obtained from λ' and Φ at each h_0, respectively. The force-displacement profile can be derived directly by plotting these results. Figure 5.8 shows the force-displacement profile for a Nb-Ta specimen [3], the magnetic flux distribution (the λ' vs. h_0 characteristics) of which was shown in Fig. 5.6. While the pinning force density varies linearly with the displacement of flux lines for a small displacement, it reaches a constant value asymptotically in the opposite critical state when the displacement becomes large. The $J_{\rm c}$ obtained from the saturated pinning force density is naturally equal to $J_{\rm c}$ obtained from the magnetic flux distribution.

What will be the results of the same analysis in the case of significant surface irreversibility? Figure 5.9 is the force-displacement profile [6] corresponding to the magnetic flux distribution shown in Fig. 5.7, where the pinning force density initially increases with displacement, reaching a large peak, and then deceases gradually with increasing displacement. The peak of the pinning force density originates from the strong surface pinning, and the critical current density in the surface region can be estimated from the peak value. That is, if the peak value of the pinning force density measured from the initial state is denoted by $F_{\rm m}$, the surface critical current density is given by $J_{\rm cs} \simeq F_{\rm m}/2\mu_0 H_{\rm e}$. The surface critical current density shown in Fig. 3.25 was

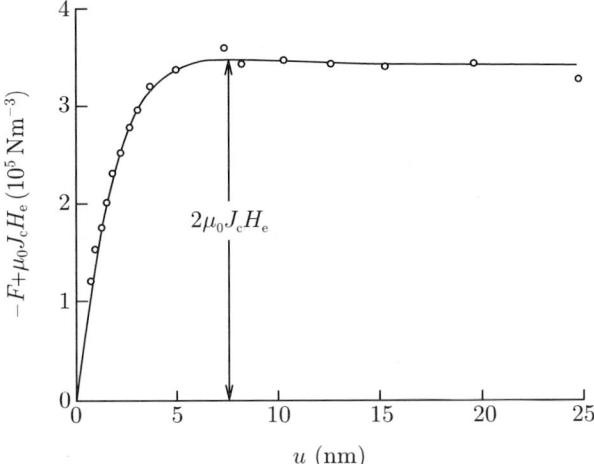

Fig. 5.8. Pinning force density vs. displacement of flux lines [3], corresponding to the λ' vs. h_0 characteristics in Fig. 5.6

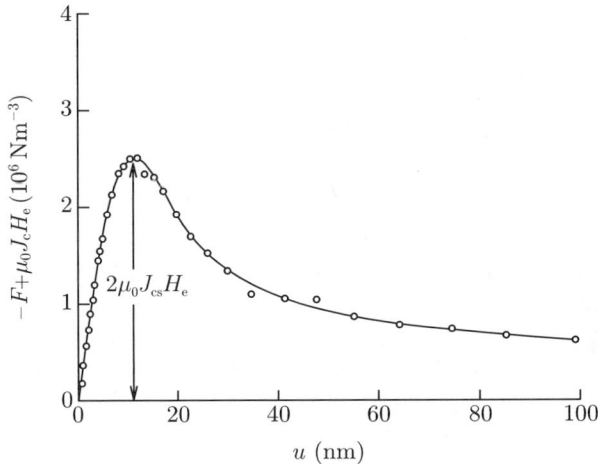

Fig. 5.9. Pinning force density vs. displacement of flux lines [6], corresponding to the λ' vs. h_0 characteristics in Fig. 5.7. A large peak of the pinning force density appears due to the strong surface pinning. J_{cs} is the critical current density in the surface region

obtained by this method. When the displacement becomes sufficiently large as in Fig. 5.9, the pinning force density asymptotically approaches the bulk value.

Within the region of small displacement where the pinning force density varies linearly with the displacement, the motion of flux lines is limited inside the pinning potentials, and the phenomenon is almost reversible, as mentioned in Sect. 3.7. This linear relationship is represented by Eq. (3.87), and the coefficient α_L, called the Labusch parameter, means the second spatial derivative of the averaged effective pinning potential. Hence, the pinning potential energy of flux lines in a unit volume is given by

$$\widehat{U}_0 = \frac{\alpha_L d_i^2}{2}. \tag{5.19}$$

Equation (3.94) holds for α_L, J_c and d_i, the interaction distance. Such information on the pinning potential can be obtained by using Campbell's methods. Thus, this method is useful for investigating electromagnetic phenomena. For example, since the pinning potential energy U_0 discussed in Sect. 3.8 is equal to $\widehat{U}_0$ multiplied by the flux bundle volume, this volume can be estimated from $\widehat{U}_0$ and U_0 obtained by an AC inductive method and by a measurement of irreversibility field, respectively. In other areas, the flux pinning mechanism is usually investigated by measuring the dependencies of the pinning force density on magnetic field and temperature (temperature scaling law), and on pinning parameters such as the elementary pinning force and the number density of defects (summation problem). Even in this case a more precise investigation is possible by measuring the dependencies of α_L or d_i on these pinning parameters (see Sects. 7.5 and 8.2).

An evaluation of the critical current density in a longitudinal magnetic field, $J_{c\parallel}$, is also possible [9] by measuring the response of a superconducting slab to a transverse AC field superimposed on the longitudinal DC field as shown in Fig. 5.10. In this case the shielding current induced by the AC field is perpendicular to the AC field, and hence, parallel to the DC field. Figure 4.26 is an example of the distribution of the transverse magnetic flux obtained by this method.

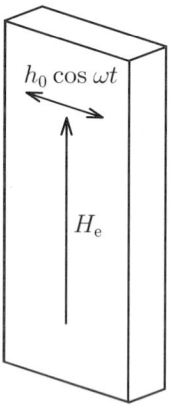

Fig. 5.10. Manner of application of magnetic field when measuring the critical current density in a longitudinal magnetic field using the modified Campbell's method. Magnetic flux moving into and out of the specimen due to the AC magnetic field is measured

An outline of the AC inductive methods such as Campbell's method was briefly given above. However, note that these methods are not always effective. For example, the method based on the irreversible critical state model does not allow correct results to be devived for superconducting specimens of a size comparable to or smaller than Campbell's AC penetration depth λ'_0 in which the reversible flux motion is pronounced. Consider the reason (Exercise 5.3). However, if the imaginary part of the complex susceptibility is measured over a wide range of AC field amplitude, the critical current density can be approximately estimated, as will be shown in the next section. On the other hand, one of the simple methods used to estimate the critical current density in such small specimens is a calculation from the hysteresis of a major DC magnetization curve.

5.4 Other AC Inductive Methods

5.4.1 Third Harmonic Analysis

The critical current density of a superconducting specimen can also be estimated by measuring the third harmonic voltage induced by an AC magnetic field [10]. For example, it is assumed that a DC field H_e and an AC field $h_0 \cos \omega t$ are applied parallel to a wide superconducting slab of thickness $2d$ ($0 \leq x \leq 2d$). If the magnetic flux density averaged within the superconducting slab is expressed as

$$\langle B \rangle = h_0 \sum_{n=0}^{\infty} \mu_n \cos(n\omega t + \theta_n) , \qquad (5.20)$$

μ_n ($n \geq 2$) represents the harmonic components of the AC permeability. These components ($n \geq 1$) are given by

$$\mu_n = (\mu_n'^2 + \mu_n''^2)^{1/2} , \qquad (5.21)$$

$$\mu_n' = \frac{1}{\pi h_0} \int_{-\pi}^{\pi} \langle B \rangle \cos n\omega t \, d\omega t , \qquad (5.22)$$

$$\mu_n'' = \frac{1}{\pi h_0} \int_{-\pi}^{\pi} \langle B \rangle \sin n\omega t \, d\omega t , \qquad (5.23)$$

where μ_n' and μ_n'' are the real and imaginary parts of the harmonic AC permeability, respectively, and there is a relationship between them:

$$\theta_n = \tan^{-1}\left(\frac{\mu_n''}{\mu_n'}\right) \qquad (5.24)$$

In the following μ_3 will be calculated assuming the Bean-London model. When $h_0 < H_p = J_c d$, the magnetic flux distribution varies as shown in

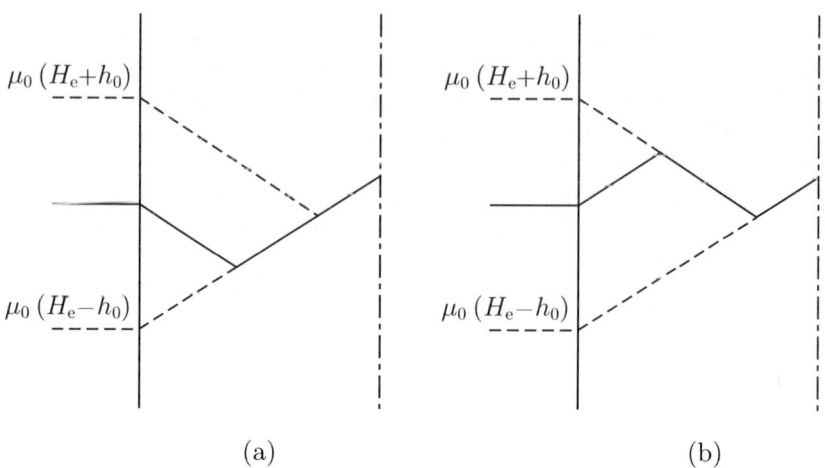

Fig. 5.11. Magnetic flux distribution in a superconducting slab of thickness $2d$ in a parallel AC magnetic field during processes of (a) increasing and (b) decreasing field. The Bean-London model is assumed

Fig. 5.11(a) and (b) with phases of $-\pi \leq \omega t < 0$ and $0 \leq \omega t < \pi$, respectively, and the spatial average of the magnetic flux density is

$$\langle B \rangle = \text{const.} + \frac{\mu_0 h_0^2}{4 J_c d}(1 + \cos \omega t)^2 \; ; \qquad -\pi \leq \omega t < 0 \, ,$$

$$= \text{const.} + \frac{\mu_0 h_0^2}{4 J_c d}[4 - (1 - \cos \omega t)^2] \; ; \qquad 0 \leq \omega t < \pi \, , \qquad (5.25)$$

where const. $= \mu_0 (H_e - h_0) + \mu_0 J_c d/2$. Substituting this into Eqs. (5.22) and (5.23), a simple calculation gives $\mu_3' = 0$ and

$$\mu_3 = -\mu_3'' = \frac{2 \mu_0 h_0}{15 \pi J_c d} \, . \qquad (5.26)$$

Hence, J_c can be estimated from the measurement of μ_3. However, note that the above result is correct only when h_0 is smaller than the penetration field $H_p = J_c d$. When h_0 is larger than H_p, the expression of μ_3 is complicated (see Exercise 5.4). In addition, correct results can be obtained only when the critical state model holds. Equation (5.26) does not hold when the effect of reversible flux motion dominates.

A method to estimate the critical current density of a superconducting thin film involves measuring an induced third harmonic voltage in a coil placed near the film surface which applies an AC magnetic field to the film [11]. In this case the coil axis is perpendicular to the film surface. However, the magnetic field is almost parallel to the film surface due to the shielding current flowing in the film. For a parallel AC magnetic field applied to the surface of a wide superconducting thin film, the third harmonic voltage induced in the coil can be simply calculated using the critical state model. In some experiments a DC magnetic field is superimposed on the superconducting thin film. The DC field only influences the value of the critical current density, but does not influence the third harmonic voltage, since a simple principle of superposition holds in electromagnetism.

It is assumed that a wide superconducting thin film occupies $0 \leq x \leq d$ and an AC magnetic field, $h_0 \cos \omega t$, due to an AC current is applied to the surface of $x = 0$. It is also assumed that the film thickness d is so small in comparison with the coil size that the thickness can be neglected in a calculation of the induced voltage in the coil.

In the case of $h_0 < J_c d$, the magnetic field at the opposite surface, $H(d)$, is zero. Hence, the current induced in a unit length of thin film along the direction of magnetic field is:

$$I'(t) = H(0) - H(d) = h_0 \cos \omega t . \tag{5.27}$$

The corresponding voltage induced in the coil is:

$$V(t) = -K \frac{dI'(t)}{dt} = K h_0 \omega \sin \omega t , \tag{5.28}$$

where K is a coefficient determined by the coil. The third harmonic voltage is estimated as

$$V_3 = (f_1^2 + f_2^2)^{1/2} , \tag{5.29}$$

where f_1 and f_2 are given by

$$f_1 = \frac{1}{2\pi} \int_0^{2\pi} V(t) \cos 3\omega t \, d\omega t , \tag{5.30}$$

$$f_2 = \frac{1}{2\pi} \int_0^{2\pi} V(t) \sin 3\omega t \, d\omega t . \tag{5.31}$$

In this case

$$V_3 = 0 \tag{5.32}$$

is easily derived.

In the case of $h_0 \geq J_c d$, we have

$$\begin{aligned} I'(t) &= J_c d - h_0 (1 - \cos \omega t); & 0 \leq \omega t < \theta_0 \\ &= -J_c d; & \theta_0 \leq \omega t < \pi , \end{aligned} \tag{5.33}$$

where θ_0 is given by

$$\theta_0 = \cos^{-1} \left(1 - \frac{2 J_c d}{h_0} \right) . \tag{5.34}$$

This leads to

$$V(t) = -Kh_0\omega \sin\omega t \;; \qquad 0 \leq \omega t < \theta_0$$
$$= 0 \;; \qquad \theta_0 \leq \omega t < \pi \;. \qquad (5.35)$$

Similar results are obtained for the latter half period, $\pi \leq \omega t < 2\pi$. After a simple calculation we have

$$f_1 = 2Kh_0\omega \int_0^{\theta_0} \sin\omega t \cos 3\omega t \, d\omega t$$
$$= 4Kh_0\omega h_{\rm p}(1-h_{\rm p})(1-8h_{\rm p}+8h_{\rm p}^2) \qquad (5.36)$$

and

$$f_2 = 2Kh_0\omega \int_0^{\theta_0} \sin\omega t \sin 3\omega t \, d\omega t$$
$$= 8Kh_0\omega \sin\theta_0 h_{\rm p}(1-h_{\rm p})(1-2h_{\rm p}) \qquad (5.37)$$

with

$$h_{\rm p} = \frac{J_{\rm c}d}{h_0} \qquad (5.38)$$

and

$$\sin\theta_0 = 2(h_{\rm p}-h_{\rm p}^2)^{1/2} \;. \qquad (5.39)$$

Thus, we obtain

$$V_3 = 4K\omega J_{\rm c}d\left(1 - \frac{J_{\rm c}d}{h_0}\right) \;. \qquad (5.40)$$

Hence, if $h_{\rm c}$ is the AC magnetic field amplitude at which the third harmonic voltage starts to appear, the critical current density of the film is estimated as

$$J_{\rm c} = \frac{h_{\rm c}}{d} \;. \qquad (5.41)$$

It should be noted, however, that this estimation is not correct, unless the film thickness is so much thicker than Campbell's AC penetration depth, given by Eq. (3.92), that the effect of reversible flux motion can be neglected. In practice, λ_0' is estimated as 0.8 μm for the case of $J_{\rm c} = 1.0 \times 10^{10}$ A/m^2 at $B = 1$ T, where Eq. (3.94) is used for $\alpha_{\rm L}$ in Eq. (3.92) and the relationship $d_{\rm i} = 2\pi a_{\rm f}$ for point-like defects is assumed. For ordinary thin films thinner than 1 μm, therefore, the measurement of the third harmonic voltage may not give a correct estimation of $J_{\rm c}$ in a DC magnetic field but may lead to an overestimation due to the effect of reversible flux motion [12], similarly to other AC measurements. The factor of overestimation is of the order of λ_0'/d, [12] and is smaller than those involved in Campbell's method (see Exercise 5.3) and AC susceptibility measurements. In the absence of a DC magnetic field, λ_0' is expected to be significantly smaller than the above estimation, and this method may be useful for the estimation of $J_{\rm c}$ even for fairly thin superconductors.

5.4.2 AC Susceptibility Measurement

The critical current density can also be estimated from a measurement of the AC susceptibility. If the magnetization of a superconducting specimen in an AC magnetic field $h_0 \cos \omega t$ is expressed as

$$M(t) = h_0 \sum_{n=0}^{\infty} (\chi'_n \cos n\omega t + \chi''_n \sin n\omega t) , \qquad (5.42)$$

χ'_n and χ''_n ($n \geq 1$) are the real and imaginary parts of the n-th AC susceptibility. In terms of $M(t)$, these parts are given by

$$\chi'_n = \frac{1}{\pi h_0} \int_{-\pi}^{\pi} M \cos n\omega t \, d\omega t , \qquad (5.43)$$

$$\chi''_n = \frac{1}{\pi h_0} \int_{-\pi}^{\pi} M \sin n\omega t \, d\omega t . \qquad (5.44)$$

These quantities are related to the AC permeabilities given by Eqs. (5.22) and (5.23) as

$$\chi'_1 = \frac{\mu'_1}{\mu_0} - 1, \qquad \chi'_n = \frac{\mu'_n}{\mu_0} \quad (n \geq 2) \qquad (5.45)$$

with

$$\chi''_n = \frac{\mu''_n}{\mu_0} \quad (n \geq 1) . \qquad (5.46)$$

These relationships are easily derived from $M = \langle B \rangle / \mu_0 - (H_e + h_0 \cos \omega t)$. Here, assume again a wide superconducting slab of thickness $2d$, in which the Bean-London model holds. Then, χ'_1 and χ''_1 can be calculated easily:

$$\chi'_1 = -1 + \frac{h_0}{2H_p} ; \qquad h_0 \leq H_p , \qquad (5.47a)$$

$$= -\frac{1}{\pi} \left(1 - \frac{h_0}{2H_p}\right) \cos^{-1}\left(1 - \frac{2H_p}{h_0}\right)$$
$$- \frac{1}{\pi} \left[1 - \frac{4H_p}{3h_0} + \frac{4}{3}\left(\frac{H_p}{h_0}\right)^2\right] \left(\frac{h_0}{H_p} - 1\right)^{1/2} ; \quad h_0 > H_p , \quad (5.47b)$$

$$\chi''_1 = \frac{2h_0}{3\pi H_p} ; \qquad h_0 \leq H_p , \qquad (5.48a)$$

$$= \frac{2H_p}{\pi h_0}\left(1 - \frac{2H_p}{3h_0}\right) ; \qquad h_0 > H_p . \qquad (5.48b)$$

The dependencies of χ'_1 and χ''_1 on the AC field amplitude are shown in Fig. 5.12(a) and (b), respectively. χ'_1 changes from -1 with increasing h_0,

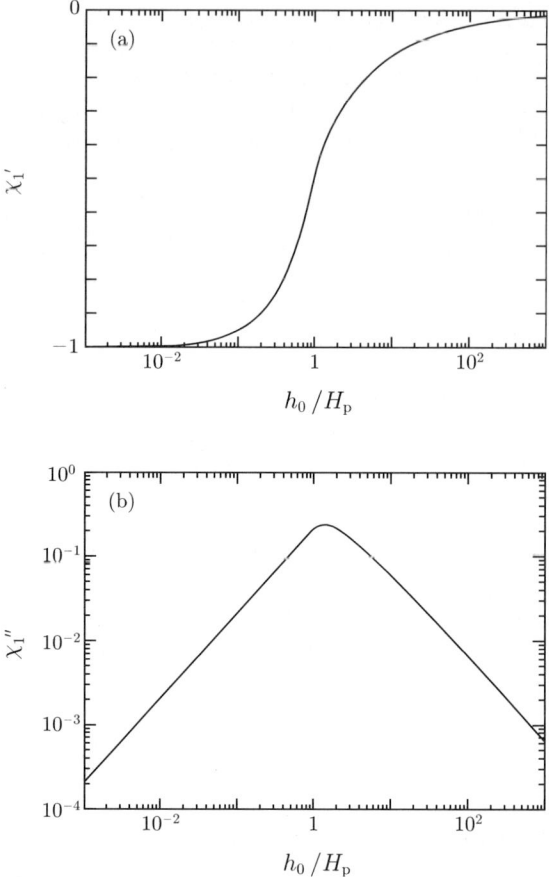

Fig. 5.12. Prediction of the Bean-London model on the AC susceptibility of a superconducting slab in a parallel AC magnetic field: (**a**) real part and (**b**) imaginary part

takes on a value $-1/2$ at $h_0 = H_p$, and then approaches 0 asymptotically. On the other hand, χ_1'' takes on a maximum value $3/4\pi$ at $h_0 = (4/3)H_p \equiv h_m$. Hence, if h_m is obtained from measurements, the critical current density can be estimated as

$$J_c = \frac{3h_m}{4d}.\qquad(5.49)$$

In most experiments, χ_1'' is measured under a variation of temperature with a constant amplitude h_0. Even in this case, the critical current density is estimated from $J_c = 3h_0/4d$ at the temperature at which χ_1'' takes on a peak value. More exactly speaking, however, the quantity obtained is nothing else besides the temperature at which J_c takes on some given value, since h_0 and d are given in such experiments. A common purpose is to know J_c under the desired condition of temperature and magnetic field. In the latter case it is required to measure χ_1'' as a function of h_0 at the given temperature and magnetic field, as shown in Eqs. (5.48a) and (5.48b).

χ_1'' is equal to μ_1'' and is directly related to the energy loss density as shown in Eq. (3.101). When the size of the superconducting specimen is comparable to or smaller than Campbell's AC penetration depth λ_0', the critical state model does not hold due to the reversible flux motion, as discussed in Sect. 3.7. In this case the Campbell model [8] is useful for analyzing the magnetic flux distribution, and the distribution can be derived by solving Eq. (3.96) numerically. Then, the magnetization M is obtained from the distribution, and χ_1' and χ_1'' are derived from Eqs. (5.43) and (5.44). The results obtained in this manner [13] are shown in Figs. 5.13 and 5.14. Figure 5.13 shows the results for $d/\lambda_0' = 10$, which corresponds to the case where the critical state model describes the magnetic behavior correctly. In fact, this result can be approximately explained by the critical state model.

By contrast, Fig. 5.14 shows the case of $d/\lambda_0' = 0.3$ where the effect of reversible flux motion is expected to be pronounced. In fact, χ_1' deviates significantly from the predicted value of -1 of the critical state model in the region of small amplitude, showing an extremely small shielding effect. The maximum value of χ_1'' is also much smaller than predicted by the critical state model with a considerable shift of the position of the maximum to a higher AC field amplitude. Thus, the value of the critical current density obtained by substitution of the observed h_m into Eq. (5.49) is considerably overestimated. In Fig. 5.14, for example, the critical current density is overestimated by a factor of about 30. Hence, it is required to judge correctly if the analysis using Eq. (5.49) is suitable for a superconductor of a relatively small size. For this purpose, it is useful to compare the maximum value of χ_1'', denoted by χ_m'', with the theoretically predicted value $3/4\pi$. That is, if χ_m'' is comparable to the predicted value, Eq. (5.49) is applicable, and if χ_m'' is much smaller than the predicted value, the use of Eq. (5.49) may lead to a serious overestimation. There have been many reports on high-temperature superconductors indicating that χ_m'' becomes smaller as the temperature is raised to the critical temperature. This is believed to originate from the reversible flux motion. Note that Campbell's AC penetration depth λ_0' takes on a larger value for a superconductor with a weaker pinning force. In high-temperature superconductors the pinning force is originally weak and becomes even weaker at a higher temperature, resulting in very large λ_0'. Hence, this reversible behavior can occur even in relatively large specimens.

To analyze the behavior under a reversible flux motion, it is required to solve numerically Eq. (3.96), which is a nonlinear differential equation. However, this is not simple. Hence, approximate formulae for AC susceptibilities are proposed here, since the dependencies of these formulae on the AC field amplitude are rather simple, as shown in Figs. 5.13 and 5.14. One of the conditions to be satisfied is that the result should approach the prediction of the critical state model, Eqs. (5.47a), (5.47b), (5.48a) and (5.48b), in the

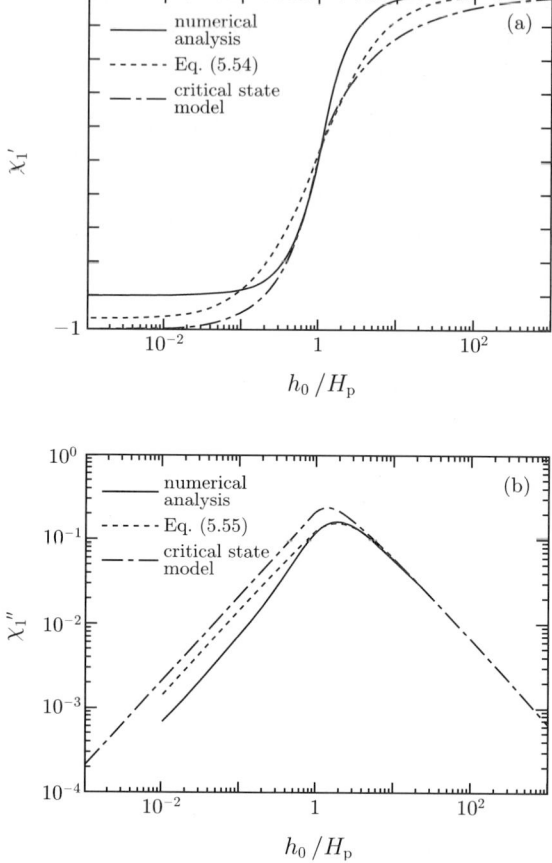

Fig. 5.13. (a) Real and (b) imaginary parts of the AC susceptibility of a superconducting slab for $d/\lambda_0' = 10$. The *solid lines* show the results of the numerical analysis of Eq. (3.96) based on the Campbell model, and the *chained lines* show the results of the Bean-London model. The *broken lines* represent the approximate formulae given by Eqs. (5.54) and (5.55). All give similar results

irreversible limit. Characteristic points for χ_1' are: $\chi_1' \to -1$ for $h_0 \ll H_p$, $\chi_1' = -1/2$ at $h_0 = H_p$ and $\chi_1' \to 0$ for $h_0 \gg H_p$. Characteristic points for χ_1'' are that it approaches Eq. (5.48a) for $h_0 \ll H_p$ and $\chi_1'' \to 2H_p/\pi h_0$ for $h_0 \gg H_p$. Candidates which satisfy the above requirements are

$$\chi_1' = -\frac{H_p}{H_p + h_0}, \qquad (5.50)$$

$$\chi_1'' = \frac{2}{\pi} \cdot \frac{H_p h_0}{3H_p^2 + h_0^2}. \qquad (5.51)$$

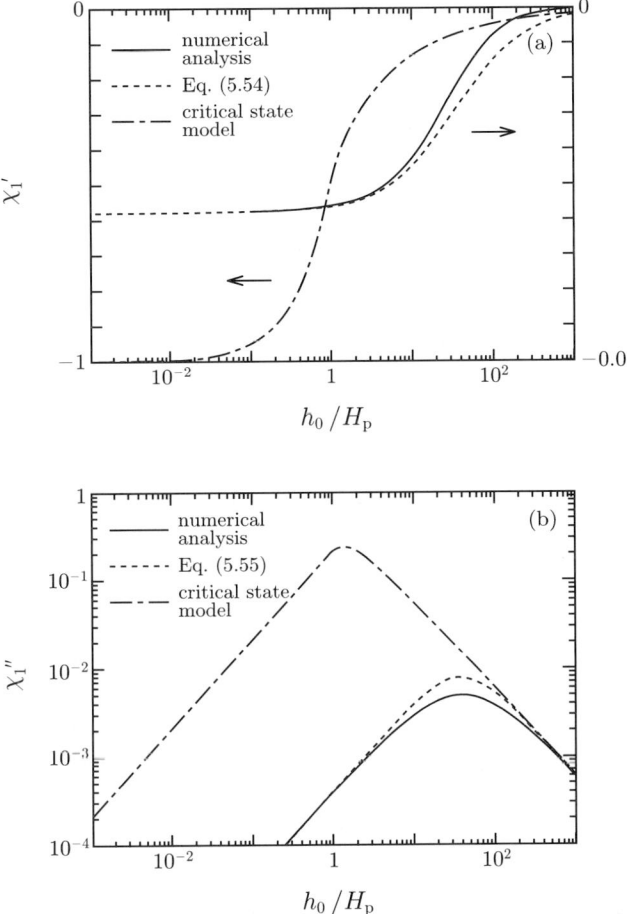

Fig. 5.14. (a) Real and (b) imaginary parts of the AC susceptibility of a superconducting slab for $d/\lambda_0' = 0.3$. The *solid, chained* and *broken lines* show the results of the numerical analysis of Eq. (3.96), the results of the critical state model and the approximate formulae, respectively. The results of the critical state model deviate greatly from the other two

$h_{\rm m} = \sqrt{3}H_{\rm p}$ and $\chi_{\rm m}'' = 1/\sqrt{3}\pi \simeq 0.184$ are obtained from Eq. (5.51), while $h_{\rm m} = (4/3)H_{\rm p}$ and $\chi_{\rm m}'' = 3/4\pi \simeq 0.239$ are obtained from Eq. (5.48b) based on the critical state model. Thus, these results are not very different from each other.

In the limit of reversible flux motion, where d is sufficiently smaller than λ_0' and h_0 is sufficiently small, a simple calculation shows that χ_1' approaches asymptotically

$$\chi_1' = -1 + \frac{\lambda_0'}{d}\tanh\left(\frac{d}{\lambda_0'}\right) \simeq -\frac{1}{3}\left(\frac{d}{\lambda_0'}\right)^2. \tag{5.52}$$

On the other hand, χ_1'' approaches

$$\chi_1'' = \frac{h_0}{6\pi H_{\rm p}}\left(\frac{d}{\lambda_0'}\right)^4. \tag{5.53}$$

In the above Eqs. (3.101), (3.109) and (5.46) were used with replacement of $d_{\rm f}$ by $2d$. When h_0 is sufficiently larger than $\widetilde{H}_{\rm p}$ given by Eq. (3.110), the behavior becomes irreversible even if d is smaller than λ_0', and χ_1' and χ_1'' approach 0 and $2H_{\rm p}/\pi h_0$, respectively.

Here approximate formulae are proposed [13]:

$$\chi_1' = -\frac{H_{\rm p}}{[1 + 3(\lambda_0'/d)^2]H_{\rm p} + h_0}, \tag{5.54}$$

$$\chi_1'' = \frac{2}{\pi}\cdot\frac{H_{\rm p}h_0}{3[1 + 2(\lambda_0'/d)^2]^2 H_{\rm p}^2 + h_0^2}. \tag{5.55}$$

These satisfy the above requirements. Figures 5.13 and 5.14 show a comparison between these formulae and the results of numerical calculation. It is found that a fairly good agreement is obtained in both limits of $d \gg \lambda_0'$ and $d \ll \lambda_0'$. The important results of these formulae are as follows: firstly, χ_1' does not reach -1 even at very low temperature, when the sample size is smaller than λ_0'. Hence, it is not correct to evaluate a superconducting volume fraction from this value of χ_1'. This is similar to the incorrect estimation of the superconducting volume fraction from DC susceptibility, as described in Sect. 3.6. Secondly, it is found from Eq. (5.55) that the AC field amplitude at which χ_1'' peaks is

$$h_{\rm m} = \sqrt{3}\left[1 + 2\left(\frac{\lambda_0'}{d}\right)^2\right]H_{\rm p} \tag{5.56}$$

and the peak value is

$$\chi_{\rm m}'' = \frac{1}{\sqrt{3}\pi[1 + 2(\lambda_0'/d)^2]}. \tag{5.57}$$

Hence, $J_{\rm c}$ cannot be estimated only from $h_{\rm m}$. This is because the value of λ_0' is unknown. However, Eqs. (5.56) and (5.57) allow us to derive

$$h_{\rm m}\chi_{\rm m}'' = \frac{H_{\rm p}}{\pi} = \frac{J_{\rm c}d}{\pi}. \tag{5.58}$$

Thus, the unknown quantity λ_0' is eliminated and $J_{\rm c}$ can be obtained from the value of the product. From this $J_{\rm c}$ value λ_0' can be determined. The value of λ_0' can also be obtained by comparing the slope of the observed minor curve of the DC magnetization with Eq. (3.111) in which $d_{\rm f}$ is replaced by $2d$.

When the size of the superconducting sample is much larger than λ_0', the critical state model holds. Even in this case Eq. (5.58) from which $J_{\rm c}$ can be determined still holds. However, the further analysis to estimate λ_0' may contain a large error.

Exercises

5.1. Assume that an AC magnetic field of amplitude h_0 is applied parallel to a superconducting slab $2d$ in thickness and w in width ($w \gg 2d$). When the Bean-London model holds in the superconducting slab, show that the penetration depth of the AC field is given by $\lambda' = h_0/J_c$, and that the h_0 vs. λ' curve represents the magnetic flux distribution for $h_0 < H_p = J_c d$, using the analysis based on Campbell's method. Show also that $\lambda' = d$ for $h_0 > H_p$.

5.2. Estimate the error of the critical current density derived from the modified Campbell's method, when the amplitude Φ of the AC magnetic flux moving into and out of the superconducting specimen is replaced by the amplitude of the component of fundamental frequency Φ'. Assume that the Bean-London model holds for the magnetic flux distribution.

5.3. Calculate the apparent value of the penetration depth λ' of an AC magnetic field for a superconducting specimen of a size smaller than Campbell's AC penetration depth λ'_0, and discuss the reason why the critical current density cannot be estimated correctly using the analysis based on Campbell's method.

5.4. Calculate μ_3 when the amplitude of the AC magnetic field h_0 is larger than the penetration field $H_p = J_c d$.

5.5. Derive Eqs. (5.47a), (5.47b), (5.48a) and (5.48b).

References

1. T. Matsushita, Y. Himeda, M. Kiuchi, J. Fujikami and K. Hayashi: IEEE Trans. Appl. Supercond. **15** (2005) 2518.
2. A. M. Campbell: J. Phys. C **2** (1969) 1492.
3. T. Matsushita, T. Honda and K. Yamafuji: Memo. Faculty of Eng., Kyushu Univ. **43** (1983) 233.
4. T. Matsushita, T. Tanaka and K. Yamafuji: J. Phys. Soc. Jpn. **46** (1979) 756.
5. R. W. Rollins, H. Küpfer and W. Gey: J. Appl. Phys. **45** (1974) 5392.
6. T. Matsushita, T. Honda and Y. Hasegawa: J. Appl. Phys. **54** (1983) 6526.
7. B. Ni, T. Munakata, T. Matsushita, M. Iwakuma, K, Funaki, M. Takeo and K. Yamafuji: Jpn. J. Appl. Phys. **27** (1988) 1658.
8. A. M. Campbell: J. Phys. C **4** (1971) 3186.
9. A. Kikitsu, Y. Hasegawa and T. Matsushita: Jpn. J. Appl. Phys. **25** (1986) 32.
10. C. P. Bean: Rev. Mod. Phys. **36** (1964) 31.
11. H. Yamasaki, Y. Mawatari and Y. Nakagawa: Appl. Phys. Lett. **82** (2003) 3275.
12. Y. Fukumoto and T. Matsushita: Supercond. Sci. Technol. **18** (2005) 861.
13. T. Matsushita, E. S. Otabe and B. Ni: Physica C **182** (1991) 95.

Exercises

5.1. Assume that an AC magnetic field of amplitude h_0 is applied parallel to a superconducting slab $2d$ in thickness and w in width ($w \gg 2d$). When the Bean-London model holds in the superconducting slab, show that the penetration depth of the AC field is given by $\lambda' = h_0/J_c$, and that the h_0 vs. λ' curve represents the magnetic flux distribution for $h_0 < H_p = J_c d$, using the analysis based on Campbell's method. Show also that $\lambda' = d$ for $h_0 > H_p$.

5.2. Estimate the error of the critical current density derived from the modified Campbell's method, when the amplitude Φ of the AC magnetic flux moving into and out of the superconducting specimen is replaced by the amplitude of the component of fundamental frequency Φ'. Assume that the Bean-London model holds for the magnetic flux distribution.

5.3. Calculate the apparent value of the penetration depth λ' of an AC magnetic field for a superconducting specimen of a size smaller than Campbell's AC penetration depth λ_0', and discuss the reason why the critical current density cannot be estimated correctly using the analysis based on Campbell's method.

5.4. Calculate μ_3 when the amplitude of the AC magnetic field h_0 is larger than the penetration field $H_p = J_c d$.

5.5. Derive Eqs. (5.47a), (5.47b), (5.48a) and (5.48b).

References

1. T. Matsushita, Y. Himeda, M. Kiuchi, J. Fujikami and K. Hayashi: IEEE Trans. Appl. Supercond. **15** (2005) 2518.
2. A. M. Campbell: J. Phys. C **2** (1969) 1492.
3. T. Matsushita, T. Honda and K. Yamafuji: Memo. Faculty of Eng., Kyushu Univ. **43** (1983) 233.
4. T. Matsushita, T. Tanaka and K. Yamafuji: J. Phys. Soc. Jpn. **46** (1979) 756.
5. R. W. Rollins, H. Küpfer and W. Gey: J. Appl. Phys. **45** (1974) 5392.
6. T. Matsushita, T. Honda and Y. Hasegawa: J. Appl. Phys. **54** (1983) 6526.
7. B. Ni, T. Munakata, T. Matsushita, M. Iwakuma, K, Funaki, M. Takeo and K. Yamafuji: Jpn. J. Appl. Phys. **27** (1988) 1658.
8. A. M. Campbell: J. Phys. C **4** (1971) 3186.
9. A. Kikitsu, Y. Hasegawa and T. Matsushita: Jpn. J. Appl. Phys. **25** (1986) 32.
10. C. P. Bean: Rev. Mod. Phys. **36** (1964) 31.
11. H. Yamasaki, Y. Mawatari and Y. Nakagawa: Appl. Phys. Lett. **82** (2003) 3275.
12. Y. Fukumoto and T. Matsushita: Supercond. Sci. Technol. **18** (2005) 861.
13. T. Matsushita, E. S. Otabe and B. Ni: Physica C **182** (1991) 95.

6
Flux Pinning Mechanisms

6.1 Elementary Pinning and the Summation Problem

Flux lines have spatial structures related to the order parameter and the magnetic flux density as shown in Fig. 1.6. When a flux line is displaced near a pinning center, it feels a variation in the energy caused by overlapping of these structures with the spatial structure of the pinning center. This is the pinning interaction, and the flux line suffers a pinning force corresponding to the gradient of the energy. The pinning force density which stops the flux line lattice from opposing the Lorentz force under a transport current is the macroscopic mean value of summed individual pinning forces. At a magnetic field sufficiently higher than H_{c1}, the flux lines repel each other and form a lattice. On the other hand, the individual pinning forces originating from randomly distributed pinning centers are directed randomly in most cases. Hence, some parts cancel each other out in the resultant pinning force density, F_p (see Fig. 6.1). That is, the pinning force density usually takes smaller values than the direct summation, $F_p = N_p f_p$, where N_p is the number density of pinning centers and f_p, called the elementary pinning force, is the maximum pinning strength of an individual pinning center. Its value depends on the strength of the elastic interaction of the flux line lattice. When the elastic interaction is strong and the flux line lattice deforms only slightly, the resultant pinning force density is small. The estimation of the pinning force density as a function of the elementary pinning force and the number density of pinning centers under given conditions of temperature and magnetic field is called the summation problem:

$$F_p = \sum_{i=1}^{N_p} f_i(B,T), \qquad (6.1)$$

where f_i is the force of the i-th pinning center and takes a value from $-f_p$ to f_p depending on the relative position of the interacting flux line with respect to the pinning center. The estimation of the pinning potential energy, U_0,

234   6 Flux Pinning Mechanisms

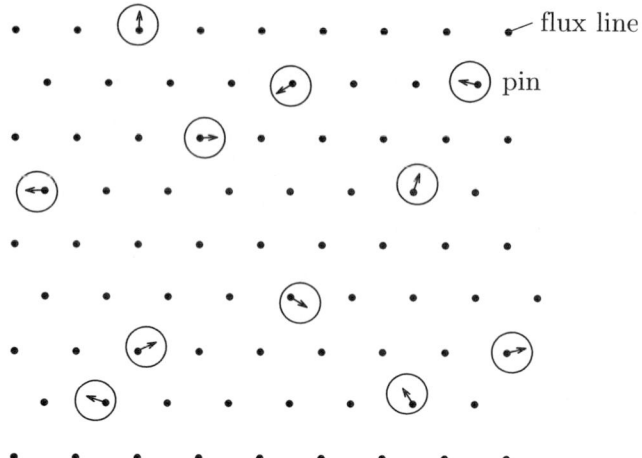

Fig. 6.1. Forces of pinning centers distributed randomly

treated in Sect. 3.8 as a function of these quantities is also a kind of summation problem.

The problem of estimating the pinning force density theoretically is usually divided into the problem of calculating the elementary pinning force and that of summing up these forces mathematically, as shown in the above. The reason why such a treatment is used is based on the experimental fact that the resultant pinning force density is almost independent of the kind and the size of pinning centers. That is, similar pinning characteristics are derived from both 0-dimensional point pinning centers and 2-dimensional grain boundaries under particular conditions of the number density of pinning centers. The characteristic features of a pinning center appear mostly in the elementary pinning force.

In this chapter the pinning mechanisms and the elementary pinning forces of various pinning centers will be discussed, and the resultant pinning force density and its properties will be described in Chap. 7.

6.2 Elementary Pinning Force

There are several definitions of the elementary pinning force. For a small pinning center which interacts with one flux line, the elementary pinning force is defined by the maximum value of its interaction force. When the energy U varies as shown in Fig. 6.2(a) during the displacement of the flux line across the pinning center, Fig. 6.2(b) shows the corresponding interaction force. The elementary pinning force is given by its maximum value:

$$f_\mathrm{p} = \left(-\frac{\partial U}{\partial x}\right)_\mathrm{max}. \tag{6.2}$$

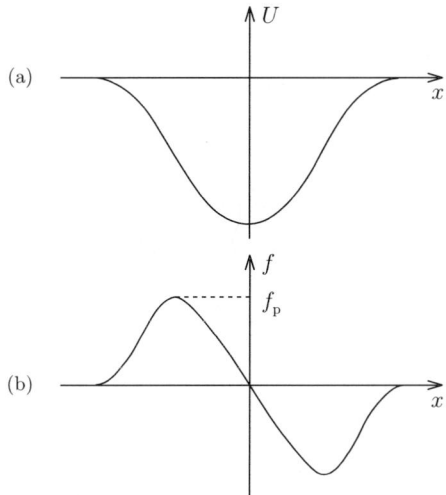

Fig. 6.2. (a) Variation in energy when a flux line passes through a pinning center and (b) the variation in the pinning force

When the pinning center is so large that it interacts with many flux lines, the definition is somewhat complicated. This is because the interaction force varies depending on the arrangement of the flux lines. Hence, for wide grain boundaries or wide interfaces between superconducting and normal regions, the elementary pinning force is sometime defined by the maximum interaction force per unit length of the flux line parallel to the boundary or the interface. In such a case the pinning force of the boundary also needs to be estimated. For such wide boundaries or long one-dimensional dislocations normal to the flux lines, the flux lines are observed [1] to fit themselves to the pinning center as shown in Fig. 6.3: a close-packed row of the flux line lattice arranges

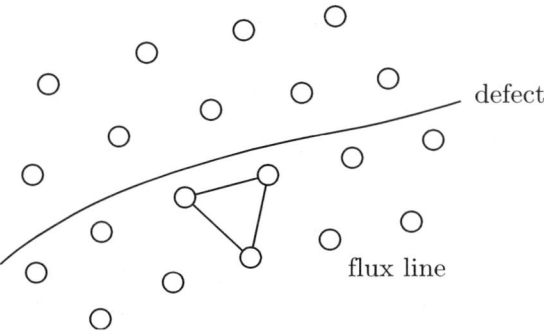

Fig. 6.3. Flux lines with a suitable arrangement for pinning

itself parallel to the pinning center. This makes the calculation of the pinning force easy. Since the deformation of the flux line lattice is small for such large pinning centers, the flux lines can arrange themselves most suitably for the pinning. This behavior can also be interpreted as the effect of the irreversible thermodynamics to minimize the energy dissipation by maximizing the pinning interaction.

For pinning centers of medium size, the flux lines are approximately treated as a rigid perfect lattice, and the elementary pinning force is estimated by Eq. (6.2) from the variation in the energy with respect to the virtual displacement of the flux line lattice [2, 3]. This is because the strain of the flux line lattice is too large for the lattice to be most suitably accommodated to a pinning center on this scale. The flux lines are considered to form a triangular lattice within this medium scale.

In the pinning center the material constants are different from those in the surrounding superconducting region. These material constants are describable phenomenologically as variations in the coefficients, α and β, in Eq. (1.21). Hence, strictly peaking, the solutions of the order parameter Ψ and the magnetic flux density b should be obtained by solving the G-L equations containing spatially varying $\alpha(\boldsymbol{r})$ and $\beta(\boldsymbol{r})$, and then, the energy U should be calculated by integrating spatially with the use of these solutions. However, this is not easy, and in the case where the variations in α and β are not large, the solutions of Ψ and b in the uniform superconducting region given by Eqs. (1.75) and (1.62) are sometimes used approximately. For example, in the case of pinning by grain boundaries, the variation in the coherence length ξ in the vicinity of the boundary is caused by the electron scattering, but the same functional form of Ψ is used for the theoretical calculation of the elementary pinning force. Generally speaking, from Eqs. (1.36), (1.37) and (1.50) we have $\alpha = -(\mu_0 e\hbar/m^*)H_{c2}$ and $\beta = 2\mu_0(e\hbar/m^*)^2\kappa^2$ and the variations in α and β can be described as the variations in H_{c2} and κ, respectively.

In the case where the material constants vary drastically, as in normal precipitates, such an approximation can no longer be used. Even in the normal precipitates, the order parameter near the interface with the superconducting region is not reduced to zero because of the proximity effect. In this case not only the order parameter in the superconducting region but that in the normal region needs to be calculated under proper boundary conditions. These boundary conditions are different dependent on whether the superconductor is "clean" or "dirty" and whether its size is larger or smaller than the electron mean free path or the coherence length. For example, the boundary conditions were defined by de Gennes [4] for the case where the size of the normal precipitates is very large and the superconductor is dirty, i.e., the electron mean free path is shorter than the coherence length, and it is known that the values of Ψ and its derivative along the direction normal to the boundary are discontinuous at the boundary on a long range scale comparable to ξ.

The treatment is similar also for the case of dielectric precipitates or voids. In this case, while Ψ is zero in the dielectric precipitate or the void, Ψ in the superconducting region should be calculated under the boundary condition of Eq. (1.33).

Thus, the estimation of the elementary pinning force of normal precipitates or dielectric precipitates is not easy in general. However, rough approximations are usually used for the cases of very large precipitates and the like. The reason for the use of rough approximations is that the exact solutions of Ψ and b when the flux line exists cannot be obtained by the method described above. Examples of these rough approximations will be discussed in Sects. 6.3–6.6.

The pinning mechanisms are classified into the condensation energy interaction, the elastic interaction, the magnetic interaction and the kinetic energy interaction according to the associated energy. In the following the mechanisms will be explained according to this classification.

6.3 Condensation Energy Interaction

In this section the case is treated where a variation in the condensation energy occurs during a displacement of flux lines resulting in the pinning interaction. Typical examples of pinning centers involving this mechanism are normal precipitates such as the α-Ti phase in Nb-Ti and grain boundaries in Nb$_3$Sn. In the following the mechanism is discussed, and the elementary pinning force is derived.

6.3.1 Normal Precipitates

We assume, for example, a wide interface between the superconducting and normal regions parallel to the flux line. It is is assumed for simplicity that the superconducting and normal regions occupy $x \geq 0$ and $x < 0$, respectively, and that the flux line is directed along the z-axis. In the case of such a large normal precipitate the tunneling or the diffusion of the superconducting electrons due to the proximity effect is limited to the vicinity of the interface, and its influence can be disregarded. Hence, there is no significant difference from the case of a dielectric precipitate. We approximate that $\Psi = 0$ for $x < 0$. If we disregard the energies of the magnetic field and current which are large but less important, the energy of the flux line can be calculated from

$$\begin{aligned} F' &= \alpha|\Psi|^2 + \frac{\beta}{2}|\Psi|^4 + |\alpha|\xi^2(\nabla|\Psi|)^2 \\ &= \mu_0 H_c^2 \left[-|\psi|^2 + \frac{1}{2}|\psi|^4 + \xi^2(\nabla|\psi|)^2 \right] \end{aligned} \quad (6.3)$$

as shown in the answer to Exercise 1.1. It is assumed that an isolated flux line is located sufficiently far from the boundary in the superconducting region. Thus, Eq. (1.75) is used for the structure of the flux line:

$$|\psi| = \tanh\left(\frac{r}{r_n}\right), \tag{6.4}$$

where r is the radius from the center of the flux line and $r_n = 1.8\xi$ for high-κ superconductors. When the flux line is absent, $|\psi| = 1$ in the whole superconducting region. Hence, the energy increase per unit length of the flux line due to its presence in the superconducting region is approximately given by

$$\Delta U = \mu_0 H_c^2 \int_0^\infty \left[\frac{1}{2} - |\psi|^2 + \frac{1}{2}|\psi|^4 + \xi^2(\nabla|\psi|)^2\right] 2\pi r dr$$

$$= \frac{2\pi}{3}\mu_0 H_c^2 \xi^2 [(k\xi)^{-2} + 2]\left(\log 2 - \frac{1}{4}\right) \simeq 1.55\pi\mu_0 H_c^2 \xi^2, \tag{6.5}$$

where the range of the integral was approximately widened to infinity for an easy calculation. Since there is no energy change in the normal region due to the presence of the flux line there, ΔU gives the energy increase when the flux line stays in the superconducting region. If the approximation is made that this energy increase takes place during a displacement of the flux line by $2r_n = 2k^{-1}$ as shown in Fig. 6.4, the elementary pinning force can be estimated as

$$f'_p \simeq \frac{\Delta U}{2r_n} = 0.430\pi\xi\mu_0 H_c^2 \tag{6.6}$$

per unit length of the flux line. This argument shows that the normal precipitate acts as an attractive pinning center and the pinning interaction occurs at the boundary.

If the common local model is applied, in which the order parameter is approximated as $|\psi| = 0$ for $r < \xi$ and $|\psi| = 1$ for $r > \xi$, the elementary pinning force is given by $f'_p = (\pi/4)\xi\mu_0 H_c^2$, which is considerably smaller than the above estimation. However, if a new local model assuming that a normal core of radius $r_n = 1.8\xi$ is introduced, the elementary pinning force is

$$f'_p = \frac{\pi}{4} r_n \mu_0 H_c^2 = 0.45\pi\xi\mu_0 H_c^2, \tag{6.7}$$

which is close to the result of Eq. (6.6).

At high fields $|\psi|$ outside the core is generally reduced from 1, and hence, the energy involved in the pinning is reduced, too, resulting in a smaller elementary pinning force. The rate of decrease is proportional to that of the main terms in the pinning energy, $|\psi|^2$ and $(\nabla|\psi|)^2$, which is approximately given by that of $\langle|\psi|^2\rangle$, namely $[2\kappa^2/(2\kappa^2-1)\beta_A](1-B/\mu_0 H_{c2}) \simeq 1 - B/\mu_0 H_{c2}$. Thus, the formula of the elementary pinning force which holds correct over a wide region of magnetic field is

$$f'_p \simeq 0.430\pi\xi\mu_0 H_c^2 \left(1 - \frac{B}{\mu_0 H_{c2}}\right). \tag{6.8}$$

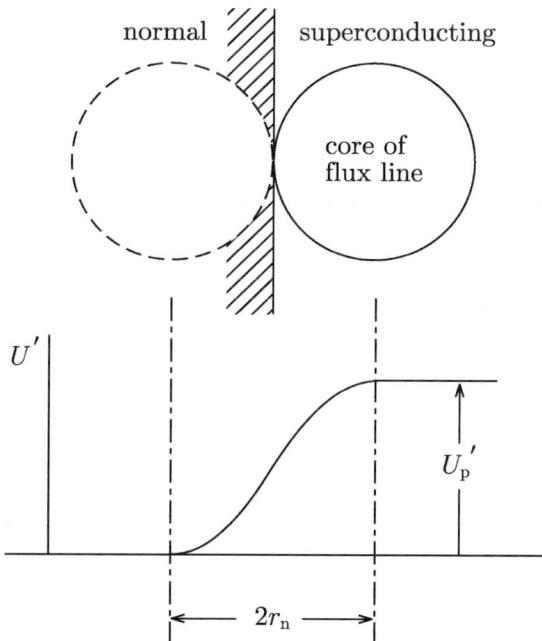

Fig. 6.4. Flux line near a large interface between superconducting and normal regions. The normal core in the normal region is imaginary

For normal precipitates sufficiently larger than the flux line spacing, $a_f = (2\phi_0/\sqrt{3}B)^{1/2}$, each normal precipitate interacts simultaneously with many flux lines. In such a case the flux lines are considered to locate themselves parallel to the interface so as to be optimum for pinning as mentioned above. Hence, for simplicity we assume a cubic normal precipitate of a size D, one surface of which is parallel to the flux lines and perpendicular to the direction of their motion due to the Lorentz force. The number of flux lines interacting with the precipitate surface is D/a_f as shown in Fig. 6.5. The elementary pinning force per flux line can be given by the value in Eq. (6.8) multiplied by the length of the flux line, D. Thus, the elementary pinning force of the precipitate is

$$f_p \simeq 0.430\pi \frac{\xi D^2 \mu_0 H_c^2}{a_f}\left(1 - \frac{B}{\mu_0 H_{c2}}\right). \tag{6.9}$$

On the other hand, the appearance of superconductivity as a consequence of the proximity effect is appreciable inside a normal precipitate smaller than the normal core of the flux line. That is, the normal precipitate behaves like a weak superconducting region. From this fact it was argued [5] that the pinning force of the condensation energy interaction is weakened. According to this proposal the elementary pinning force of a thin normal precipitate of thickness d would take a value smaller by a factor of $(d/\xi)^2$ than that of an insulating

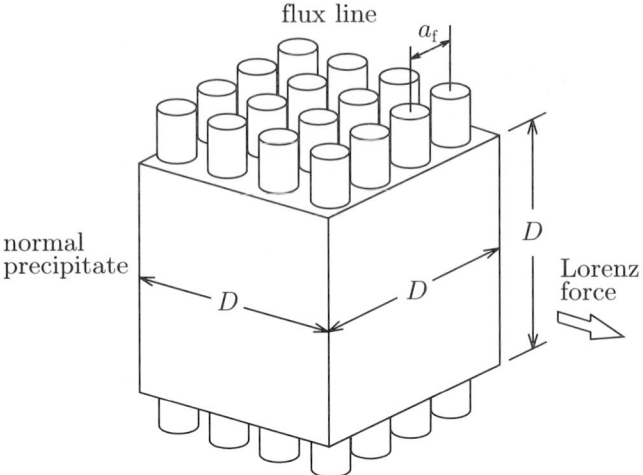

Fig. 6.5. Large normal precipitate and flux lines

precipitate of the same geometry. If this is true, the pinning force density in Nb-Ti which contains thin ribbons of normal α-Ti phase as pinning centers would not be expected to be so large. Especially in the vicinity of the critical temperature where the coherence length becomes long, it may be expected that a significant decrease of the pinning force density and a deviation from the temperature scaling law of the pinning force density, which will be introduced later, would be observed. However, such results have not been observed [6], and the pinning force density can be explained quantitatively in terms of the elementary pinning force for which the proximity effect is not taken into account [7].

How can we understand this experimental result? It can be concluded that the influence of the proximity effect on the pinning energy was incorrectly understood in [5]. For clarification of this point, the elementary pinning force will be estimated for a normal precipitate in which the proximity effect is significant using the G-L theory. For simplicity the case is treated where the superconducting and normal layers of thicknesses d_s and d_n, respectively, are assembled alternately as shown in Fig. 6.6. It is assumed that the coherence length ξ is longer than these thicknesses. An example of such a structure can be seen in α-Ti in Nb-Ti and the above assumption on the coherence length can be attained at high temperatures near the critical temperature. Equation (1.30) holds correct in the superconducting region. On the other hand, the equation which is valid in the normal region will be approximately

$$\frac{1}{2m^*}(-i\hbar\nabla + 2e\boldsymbol{A})^2\Psi + \alpha_n\Psi = 0 , \quad (6.10)$$

an equation of the Schrödinger type. In the above, α_n is a positive parameter representing the repulsive interaction between paired electrons. The

6.3 Condensation Energy Interaction

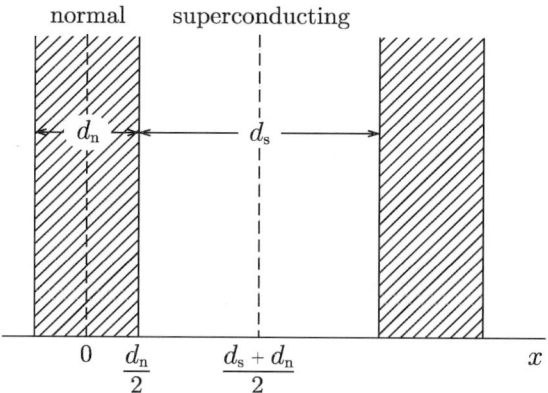

Fig. 6.6. Multi-layered structure composed of superconducting and normal layers

corresponding free energy in the normal region is given by

$$F' = \alpha_n|\Psi|^2 + |\alpha|\xi^2(\nabla|\Psi|)^2 = \mu_0 H_c^2[\theta|\psi|^2 + \xi^2(\nabla|\psi|)^2], \quad (6.11)$$

where θ is defined by

$$\theta = -\frac{\alpha_n}{\alpha} = \left(\frac{\xi}{\xi_n}\right)^{1/2} \quad (6.12)$$

with ξ_n denoting the coherence length in the normal region. We first treat the case where the flux lines do not exist. The x-axis is defined along the direction perpendicular to the layered structure and Ψ varies only along this axis. In such a one-dimensional situation the argument of the order parameter is a constant, and Ψ can be chosen as a real number. Thus, Ψ is defined as $\Psi = R|\Psi_\infty|$ with R denoting a real number. The equations for the superconducting and normal regions are

$$\frac{d^2 R}{d\eta^2} - R + R^3 = 0 \; ; \quad \frac{d_n}{2\xi} < \eta \leq \frac{d_s + d_n}{2\xi}, \quad (6.13a)$$

$$\frac{d^2 R}{d\eta^2} + \theta R = 0 \; ; \quad 0 \leq \eta < \frac{d_n}{2\xi}, \quad (6.13b)$$

where $\eta = x/\xi$.

The boundary conditions on the continuity of R at the interface between the superconducting and normal regions are necessary to solve R. In the case of the α-Ti in Nb-Ti, the electron mean free path is estimated [8] as about 10 nm and is longer than the typical thickness of the α-Ti layers. Hence, the boundary conditions of de Gennes [4] for the dirty case, according to which the values of R and its derivative are discontinuous, cannot be used. Instead of this the boundary conditions of Zaitsev [9] for the clean case which require the continuity of R and its derivative are used. The solutions of R are given

6 Flux Pinning Mechanisms

by Jacobi's elliptic function and a hyperbolic function in the superconducting and normal regions, respectively [8]. However, such a calculation is very complicated. Here the simple case is considered where both d_s and d_n are much thinner than ξ. In this case R cannot vary spatially and is uniform in both regions. Although this value of R can be obtained approximately from the exact solution, it can also be derived simply so as to minimize the G-L energy. The kinetic energy originating from the spatial variation in R can be disregarded. Then, the free energy density in the superconducting region is given by the first and second terms in Eq. (6.3), and that in the normal region is given by the first term in Eq. (6.11). Thus, we have

$$R = \left(1 - \frac{d_n}{d_s}\theta\right)^{1/2}. \tag{6.14}$$

Secondly, the case is treated where flux lines exist. The order parameter is approximately given by a superposition of the structure of the flux line of Eq. (6.4) to Eq. (6.14):

$$|\psi| = R \tanh\left(\frac{r}{r_n}\right). \tag{6.15}$$

Now the elementary pinning force of the normal layer is estimated. Usually the flux lines are not parallel to the layered structure shown in Fig. 6.6. Hence, for simplicity the flux lines are assumed to be perpendicular to the layered structure. The difference in the energy is treated between the cases where the flux line is in the state shown in Fig. 6.7(a) and that in Fig. 6.7(b). The energy is given by Eqs. (6.3) and (6.11) for the superconducting and normal regions, respectively. The kinetic energy can be disregarded, since this is the same in both regions. It is enough to treat the energy only in the region with the length d_n shown in Fig. 6.7 where the normal core meets the normal region. In case (a) a change in the energy due to the presence of the flux line in the region V_1 is calculated as

$$U_a = -2\log 2 \cdot \pi r_n^2 d_n \theta R^2 \mu_0 H_c^2.$$

In case (b) the change in the energy due to the presence of the flux line in the region V_2 is

$$U_b = \pi r_n^2 d_n R^2 \left[2\log 2 - R^2 \left(\frac{4}{3}\log 2 + \frac{1}{6}\right)\right] \mu_0 H_c^2.$$

The elementary pinning force is approximately given by the difference in the energy $U_b - U_a$ divided by $2r_n$ [10]:

6.3 Condensation Energy Interaction

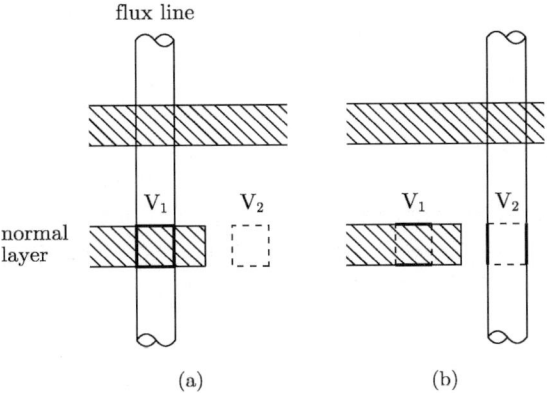

Fig. 6.7. Flux line near an edge of a normal layer

$$f_{\rm p} = \pi r_{\rm n} d_{\rm n} R^2 \left[(1+\theta)\log 2 - \frac{R^2}{12}(8\log 2 + 1)\right]\mu_0 H_{\rm c}^2$$

$$= 1.8\pi d_{\rm n}\left(1 - \frac{d_{\rm n}}{d_{\rm s}}\theta\right)\left[(1+\theta)\log 2\right.$$

$$\left. - \frac{1}{12}(8\log 2 + 1)\left(1 - \frac{d_{\rm n}}{d_{\rm s}}\theta\right)\right]\xi\mu_0 H_{\rm c}^2 . \quad (6.16)$$

The above variation in the energy takes place when the normal core passes through the edge of the normal layer, and hence, it is found that the edge interacts attractively with flux lines.

If the nonsuperconducting regions shaded in Fig. 6.6 are not the normal conducting phase but an insulating phase, 1 and 0 are substituted into R in the superconducting and nonsuperconducting regions, respectively. On the other hand, as for the contribution from the potential energy, this replacement leads to the same result as the replacement of θ by 0, since the energy of the insulating region does not depend on Ψ in the case of $\theta = 0$. In this case, the kinetic energy in the superconducting region should be considered in the interaction energy, since there is no corresponding energy in the insulating region. Hence, the elementary pinning force of the insulating layer is given by $f'_{\rm p}$ in Eq. (6.6) multiplied by the thickness $d_{\rm n}$. If this value is denoted by $f_{\rm p0}$, the elementary pinning force of the thin normal layer is

$$\frac{f_{\rm p}}{f_{\rm p0}} = 4.19\left(1 - \frac{d_{\rm n}}{d_{\rm s}}\theta\right)\left[(1+\theta)\log 2 - \frac{1}{12}(8\log 2 + 1)\left(1 - \frac{d_{\rm n}}{d_{\rm s}}\theta\right)\right] . \quad (6.17)$$

In the usual Nb-Ti wires, θ is estimated as 1.4 for normal α-Ti phase [8]. Thus, $f_{\rm p}/f_{\rm p0} = 3.83$ is derived when $d_{\rm n}/d_{\rm s} = 0.2$. This ratio is considerably larger than the ratio of 2.7 estimated using the usual local model [8]. However, it can be concluded that thin normal α-Ti layers are much stronger than insulating layers even allowing for the proximity effect. Figure 6.8 shows the dependence of $f_{\rm p}/f_{\rm p0}$ on θ when $d_{\rm n}/d_{\rm s} = 0.2$.

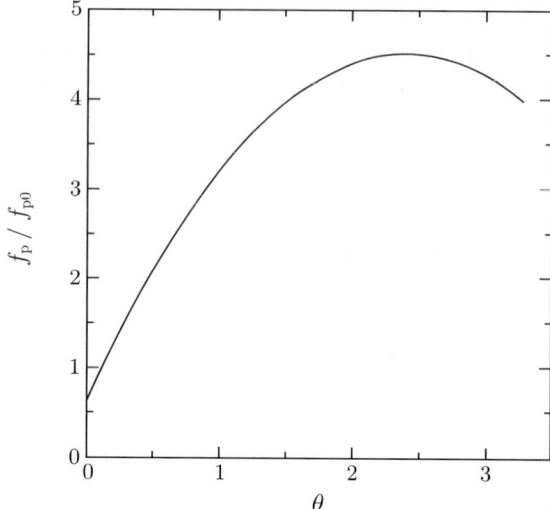

Fig. 6.8. Dependence of the elementary pinning force at the edge of a normal layer on θ for $d_\mathrm{n}/d_\mathrm{s} = 0.2$

As shown above, although the normal precipitate behaves like a superconductor due to the proximity effect, its pinning strength is not weakened. The mistake in [5] comes from a confusion of the state in which the precipitate is forced to behave like a superconductor under the environmental conditions and the intrinsic superconducting state [11]. The energy increases locally in the former case when the precipitate becomes superconducting, while the energy decreases in the latter case. It should be noted that α_n is positive, while α is negative. The reason why the precipitate becomes superconducting is that it reduces the energy globally by preventing destruction of the superconductivity in the surrounding superconducting region and minimizing the spatial variation in the order parameter around the interface. Hence, if the flux line meets the normal precipitate, the superconductivity in the precipitate is destroyed and the energy is decreased. Thus, the elementary pinning force takes a larger value.

In a strict sense the application of the G-L theory to such a phenomenon on a scale smaller than ξ may not be a good approximation. A more microscopic theoretical calculation may be necessary for a detailed estimation of the elementary pinning force. Roughly speaking, however, the proximity effect can be described even by the G-L theory in a similar way to the microscopic theory [12]. Hence, the above result is considered to be correct qualitatively.

The local model predicts [11] that $f_\mathrm{p} \geq f_\mathrm{p0}$ and $f_\mathrm{p}/f_\mathrm{p0}$ approaches 1 when θ goes to 0. On the other hand, Fig. 6.8 shows that the elementary pinning force slightly decreases due to the proximity effect in the case of small θ. This is caused by the fact that the kinetic energy is increased by the spatial variation in the induced order parameter when the normal core exists in the normal layer. When θ becomes large, the decrease in the potential energy due to presence of the normal core becomes relatively larger than the increase in the kinetic energy. Thus, the elementary pinning force increases.

6.3.2 Grain Boundary

The microstructure is different from grain boundary to grain boundary, and a general discussion on the elementary pinning force is more difficult than in the case of normal precipitates. For example, distortions only exist on an atomic scale around the grain boundaries of a pure metal superconductor. On the other hand, for grain boundaries in intermetallic superconductors formed by diffusion through boundaries, the composition deviates sometimes from stoichiometry within a certain distance from the boundary. In the former case there are only the mechanisms of electron scattering and elastic interaction. In the latter case the condensation energy interaction from a nonsuperconducting layer with a finite thickness mentioned in Subsect. 6.3.1 is also involved. In this subsection we describe mainly the electron scattering mechanism, and the interactions from nonsuperconducting layers with finite thicknesses are mentioned additionally. The elastic interaction will be treated in the next section.

If we assume an interaction of the kind treated in Subsect. 6.3.1 as the pinning mechanism of grain boundaries, the boundaries do not have a sufficient thickness as normal precipitates, and hence, the volume of the region in which the boundary overlaps the normal core of a flux line is very small. For this reason the anisotropy of the upper critical field H_{c2} among grains was first considered as a candidate for the pinning mechanism of grain boundaries [13]. That is, even if the thermodynamic critical field H_c is the same among grains, the energy of the normal core is different from grain to grain because of different coherence lengths ξ. Thus, the normal core feels a variation in the energy when it passes through the boundary, and the pinning interaction results from it. However, the pinning interaction does not arise from this mechanism for isotropic polycrystalline superconductors, and strong pinning interaction cannot be expected for superconductors with small anisotropy.

It was Zerweck [14] who first argued quantitatively that pinning by grain boundaries takes place through the electron scattering mechanism. An outline of this mechanism is as follows. A grain boundary provides an irregular variation in the potential energy for traveling electrons, which causes the scattering of electrons. Thus, the electron mean free path becomes shorter near the grain boundary, and the coherence length also becomes shorter. Hence, when the normal core of a flux line reaches the grain boundary, the diameter of the normal core with its higher energy density becomes smaller, and the energy of the normal core changes. Namely, the grain boundary interacts attractively with flux lines. In this case the properties of the superconductor influence the pinning strength. That is, the variation in the coherence length ξ caused by

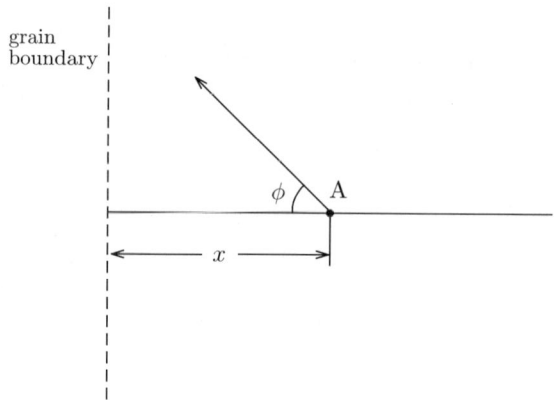

Fig. 6.9. Direction of motion of an electron from point A near the grain boundary

the scattering depends strongly on whether the superconductor is "clean" or "dirty." Naturally a larger rate of variation in ξ results in stronger pinning.

Assumed that an electron is moving through a bulk superconductor which does not contain grain boundaries. The probability that the electron can move a distance r without being scattered after the last scattering is assumed to be given by

$$p(r) = \frac{1}{l_\mathrm{b}} \exp\left(-\frac{r}{l_\mathrm{b}}\right), \tag{6.18}$$

where l_b is the bulk value of the electron mean free path. It can be easily shown that integration of r multiplied by this probability from 0 to ∞ leads to l_b. Here a sufficiently wide planar grain boundary is assumed. The expectation value of the electron mean free path at point A at a distance x from the boundary is estimated. The probability that an electron traveling from point A is scattered during the movement through r depends on the direction of the movement. The angle of the movement of the electron measured from a line perpendicular to the boundary is denoted by ϕ (see Fig. 6.9). For simplicity it is assumed that the electron is certain to be scattered at the boundary. The probability that the electron can travel a distance r without being scattered is

$$\begin{aligned} p(r,\phi) &= p(r) ; & 0 \leq r < s , \\ &= \delta(r-s)l_\mathrm{b}p(s) ; & r = s , \\ &= 0 ; & s < r \end{aligned} \tag{6.19}$$

for $0 < \phi < \pi/2$, where $s = x \sec \phi$. The factor multiplied by the delta-function in the second equation is determined from the condition of normalization and is given by the integration of $p(r)$ from s to ∞. For $\pi/2 < \phi < \pi$ the probability is simply given by

$$p(r,\phi) = p(r) . \tag{6.20}$$

Thus, the distance through which the electron can travel without being scattered from point A into the direction given by ϕ is derived as

$$l(x,\phi) = \int_0^\infty p(r,\phi) r \, dr$$
$$= l_b \left[1 - \exp\left(\frac{s}{l_b}\right)\right]; \qquad 0 \le \phi \le \frac{\pi}{2}, \qquad (6.21)$$
$$= l_b; \qquad \frac{\pi}{2} < \phi \le \pi.$$

Averaging this with respect to the whole solid angle 4π, the mean free path is obtained:

$$l(x) = \frac{1}{2}\int_0^\pi l(x,\phi) \sin\phi \, d\phi$$
$$= l_b - \frac{1}{2}\left[l_b \exp\left(-\frac{x}{l_b}\right) - x \int_{x/l_b}^\infty \exp(-z)\frac{dz}{z}\right]. \qquad (6.22)$$

The obtained variation in the electron mean free path is shown in Fig. 6.10.

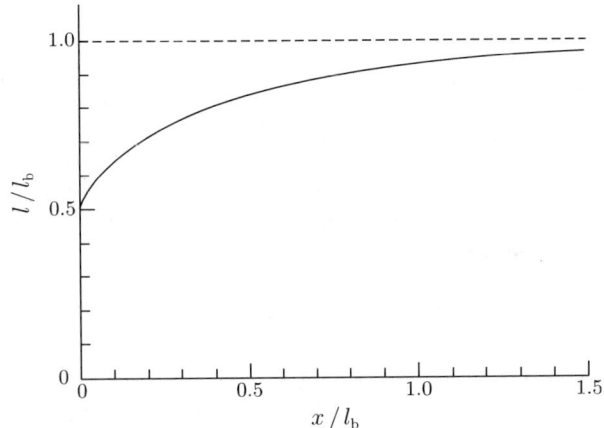

Fig. 6.10. Variation in the electron mean free path with the distance from the boundary (after Zerweck [14])

The variation in the coherence length ξ caused by the variation in the electron mean free path depends on the degree of impurity in the superconductor. Zerweck used the interpolation formula of Goodman:

$$\xi(T=0) = \frac{\xi_0}{(1 + 1.44\xi_0/l)^{1/2}} \qquad (6.23)$$

for this relation. In the above ξ_0 is the BCS coherence length independent of temperature. Figure 6.11 shows the variation in the coherence length with the

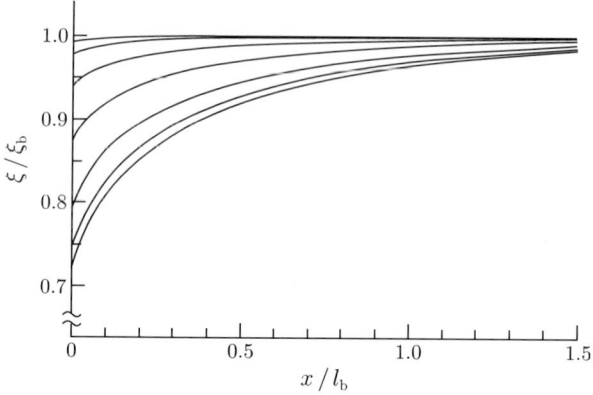

Fig. 6.11. Variation in the coherence length with the distance from the boundary (after Zerweck [14]). Respective lines are for the cases of the impurity parameter α_i of 0.01, 0.03, 0.1, 0.3, 1, 3 and 10 from the *top*

distance from the boundary, where ξ_b is the bulk coherence length and α_i in the figure is the impurity parameter of the superconductor given by

$$\alpha_i = \frac{0.882\xi_0}{l_b}. \qquad (6.24)$$

Because of the variation in the coherence length the energy varies when the flux line passes through the boundary. Zerweck [14] treated only the first and second terms of Eq. (6.3). We will follow this. It is assumed that the magnetic field is sufficiently low and that the flux line is isolated. The diameter of the normal core is $2\xi_b$ and 2ξ in the directions parallel and perpendicular to the boundary, respectively, and hence, the cross-sectional area of the normal core is $\pi\xi_b\xi$. Since the condensation energy density is $\mu_0 H_c^2/2$, the increase in the energy, i.e., the pinning energy, per unit length of the normal core is given by $U'_p = \pi\xi_b\xi\mu_0 H_c^2/2$ omitting a constant term. This leads to the elementary pinning force of the grain boundary per unit length of flux line:

$$f'_p = \frac{\pi}{2}\xi_b\mu_0 H_c^2 \left\langle \frac{d\xi}{dx} \right\rangle_m, \qquad (6.25)$$

where $\langle d\xi/dx \rangle_m$ is the maximum value of the mean variation rate of the coherence length, namely the mean variation rate between $x = 0$ and $x = \xi_b$. The obtained result is shown in Fig. 6.12, where the ordinate represents the elementary pinning force normalized by $\xi_0\mu_0 H_c^2$. If the temperature dependence of ξ is given by Eq. (1.45), the same dependence can also be added to Eq. (6.23). The factor representing this temperature dependence is the same as that of ξ_b. This means that $\langle d\xi/dx \rangle_m$ does not depend on the temperature. Hence, the temperature dependence of the elementary pinning force is given by that of $\xi_b H_c^2$.

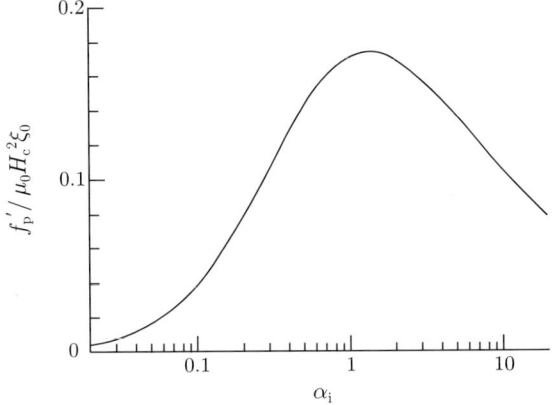

Fig. 6.12. Dependence of the elementary pinning force (per unit length of the flux line) of the grain boundary at low temperatures on the impurity parameter (after Zerweck [14])

The elementary pinning force of grain boundaries varies considerably with the impurity parameter α_i. For "clean" superconductors with small α_i or "dirty" superconductors with large α_i, the elementary pinning force is small. In these extreme cases the variation in the coherence length caused by electron scattering by boundaries is small as mentioned previously. As a result, the elementary pinning force takes a maximum around $\alpha_i \sim 1.4$ where the variation rate of the coherence length is at a maximum.

Based on the above estimation of the elementary pinning force at lower fields, Yetter et al. [15] extended the calculation to the high field region in which the flux lines form a lattice. At the same time an experimentally obtained formula was used instead of Goodman's interpolation formula given by Eq. (6.23), since this formula deviates from experiments in the "clean" limit of small α_i. Although the obtained result is qualitatively similar to that of Zerweck, the value of α_i at which the elementary pinning force takes the maximum value is shifted to about 10.

In the above treatments by Zerweck and by Yetter et al. the elementary pinning force is obtained by numerical calculation and the result is not easy to understand intuitively. For this reason Welch [16] carried out an analytical calculation and derived an approximate formula. In the beginning the spatial variation in the order parameter was approximately expressed by a simple function and the variation in the coherence length or that in the G-L parameter κ was assumed to be described by an exponential function of the distance from the boundary. Then, the variation in the energy was estimated and the elementary pinning force was calculated. The result is expressed as

$$f'_p = \frac{\pi \mu_0 H_c^2 \xi^2}{d} \cdot \left.\frac{\Delta \kappa}{\kappa}\right|_0 \cdot \frac{1}{1 + 2.0(\xi/d) + 2.32(\xi/d)^2} , \qquad (6.26)$$

where $\Delta\kappa/\kappa|_0$ is the relative variation in κ from the bulk value at the boundary and d is the characteristic length describing the spatial variation in κ. The following interpolation formula was used instead of Eq. (6.23) so as to give a correct value of the coherence length even in the "clean" limit:

$$\xi(0) = \frac{\xi_0}{(1.83 + 1.63\alpha_i)^{1/2}} . \qquad (6.27)$$

From Eqs. (1.51) and (1.52) we have $\Delta\kappa/\kappa|_0 = [\xi_b/\xi(x=0)]^2 - 1$ (the coordinates are the same as those in Fig. 6.9) assuming that H_c is unchanged. Equation (6.27) is used for ξ_b and the same equation with α_i replaced by $[l_b/l(x=0)]\alpha_i$ is used for $\xi(x=0)$. It can be easily seen that $l(x=0) = l_b/2$ from Eq. (6.22). Thus, we have

$$\left.\frac{\Delta\kappa}{\kappa}\right|_0 = \frac{1.63\alpha_i}{1.83 + 1.63\alpha_i} . \qquad (6.28)$$

Using the result of the numerical calculation that d is as large as $l/3$, the elementary pinning force is obtained from Eq. (6.26). The solid line in Fig. 6.13 shows the calculated dependence on α_i of the elementary pinning force.

Welch calculated not only for the case of the ideal boundary with zero thickness, but also the elementary pinning force for a boundary with a region of finite thickness in which the composition deviates from stoichiometry as in

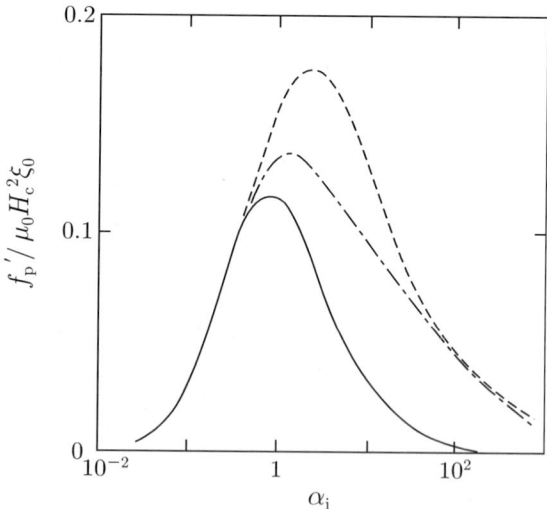

Fig. 6.13. Dependence of the elementary pinning force (per unit length of the flux line) of the grain boundary on the impurity parameter (after Welch [16]). The *solid line* is for the ideal boundary of zero thickness, the *dot dashed line* is for the boundary of thickness $0.1\xi_0$ and the *dashed line* is for the optimum boundary with the characteristic length of $d = 1.5\xi$.

6.3 Condensation Energy Interaction

compound superconductors produced by a diffusion process through boundaries. It was postulated that κ in this region was different from the bulk value by the amount given in Eq. (6.28). The result for the case where the thickness is equal to $0.1\xi_0$ is shown by the dot dashed line in Fig. 6.13. It turns out that the elementary pinning force takes a larger value for a boundary with a finite thickness. In addition Welch investigated the effect of variation in the characteristic length d. It can be proved from Eq. (6.26) that the elementary pinning force is at a maximum when d is equal to $(2.32)^{1/2}\xi \simeq 1.5\xi$. The result of f'_p in this optimum case is also shown by the dashed line in the figure.

A similar calculation was also done by Pruymboom and Kes [17]. The most important characteristic of their calculation was that the kinetic energy given by the third term in Eq. (6.3) was also included. In this treatment the term proportional to $|\Psi|^4$ was neglected and the expression of the energy was transformed using the G-L equation (see Exercise 1.2). A qualitatively similar result to that of Welch was obtained from a similar calculation. Quantitatively the obtained elementary pinning force is about twice as large as that of Welch. However, the effect of electron scattering appears originally in β through the variation in κ, and hence, the neglect of $\beta|\Psi|^4/2$ is questionable.

It is sometimes insisted that only the kinetic energy is important to the pinning energy. This is based on the following argument [18]. The order parameter Ψ is determined so as to minimize the energy given by Eq. (6.3). This equilibrium value of Ψ is represented by Ψ_e. In the vicinity of a pinning center the material constants, α and β, vary and their variations are represented by $\delta\alpha$ and $\delta\beta$. A new equilibrium value, $\Psi_\mathrm{e} + \delta\Psi_\mathrm{e}$, is correspondingly obtained. The contribution from the first and second terms in Eq. (6.3) to the energy is $\delta\alpha|\Psi_\mathrm{e}|^2 + \delta\beta|\Psi_\mathrm{e}|^4/2$. The argument is that the contribution from the variation in the equilibrium value should also be contained in the energy. According to this argument, the new term just cancels the above contribution and only the kinetic energy remains. However, this line of reasoning is clearly incorrect. Here the kinetic energy is disregarded for simplicity. From the equilibrium condition we have $|\Psi_\mathrm{e}|^2 = -\alpha/\beta$. Hence, the variation in the energy is given by

$$\delta F' = (\alpha + \delta\alpha)(|\Psi_\mathrm{e}|^2 + \delta|\Psi_\mathrm{e}|^2) + \frac{1}{2}(\beta + \delta\beta)(|\Psi_\mathrm{e}|^4 + \delta|\Psi_\mathrm{e}|^4)$$
$$- \alpha|\Psi_\mathrm{e}|^2 - \frac{1}{2}\beta|\Psi_\mathrm{e}|^4$$
$$\simeq \alpha\delta|\Psi_\mathrm{e}|^2 + \frac{1}{2}\beta\delta|\Psi_\mathrm{e}|^4 + \delta\alpha|\Psi_\mathrm{e}|^2 + \frac{1}{2}\delta\beta|\Psi_\mathrm{e}|^4$$
$$= \delta\alpha|\Psi_\mathrm{e}|^2 + \frac{1}{2}\delta\beta|\Psi_\mathrm{e}|^4$$

and is not zero. Thus, it is concluded that the new insistence on kinetic energy alone is not correct but that the original argument by Campbell and Evetts [19] is correct.

However, it is true that the contribution from the kinetic energy to the elementary pinning force is large. At the moment there is no exact theoretical calculation in which all the energy is taken into account, and further investigation is necessary.

There are experimental results which clarify the mechanism of pinning by a grain boundary. Figure 6.14 shows the dependence of critical current in a niobium bi-crystal specimen on the angle of the magnetic field [20]. A large critical current is obtained when the magnetic field is parallel to the twin boundary. This verifies the claim that the boundary acts as a pinning center. Figure 6.15 shows the magnetic field dependence of the pinning force per unit area of the twin boundary in a similar niobium bi-crystal specimen [21]. In this specimen the crystal axes are not symmetric with respect to the magnetic field, which is parallel to the boundary and perpendicular to the current. It turns out that the pinning force is different when the direction of the current or the direction of the Lorentz force is reversed, showing that the anisotropy is also involved in the pinning. That is, the mean value of the pinning force of the two directions gives the contribution from the electron scattering or the elastic interaction and half of the difference between the two pinning forces gives the contribution from the anisotropy. It can be seen from this figure that the pinning force still has a finite value even at magnetic fields above the bulk H_{c2}. This shows that the coherence length in the vicinity of the boundary is reduced due to the electron scattering by the boundary, resulting in the local enhancement of H_{c2}.

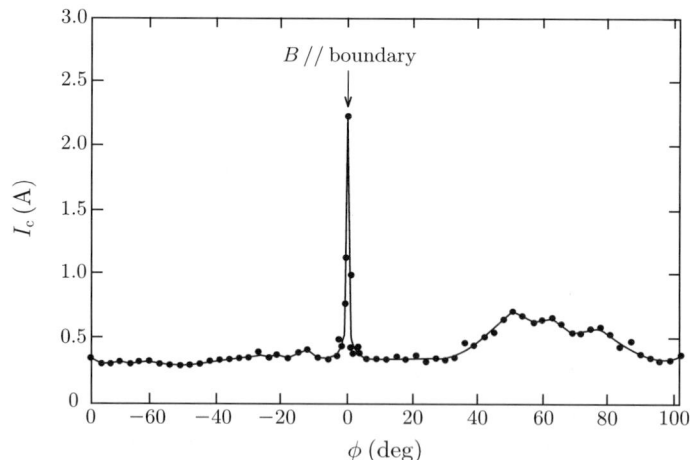

Fig. 6.14. Dependence of the critical current in a niobium bi-crystal on the angle of the magnetic field [20]. ϕ is the angle between the twin boundary and the magnetic field

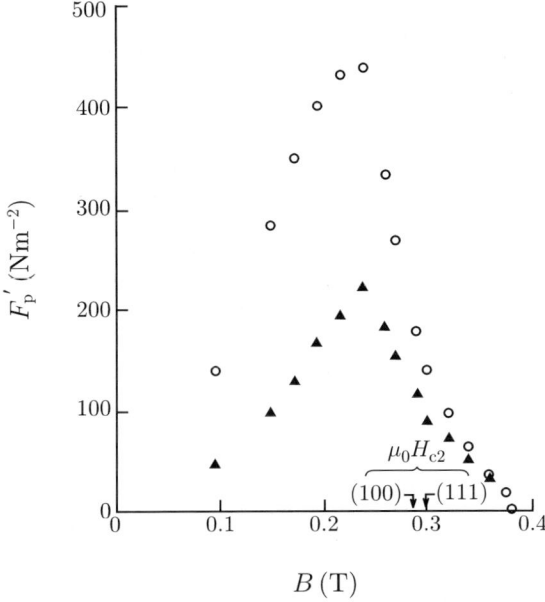

Fig. 6.15. Pinning force per unit area of the twin boundary in a niobium bi-crystal [21]. The axes of the two crystals are not symmetric with respect to the magnetic field, which is parallel to the boundary and perpendicular to the current. The different symbols represent the results for the different directions of the current

6.4 Elastic Interaction

For one-dimensional defects such as dislocations the cross-sectional area for electron scattering is small, and hence, the pinning force arising from the electron scattering mechanism is not believed to be strong. However, strains exist around such defects, and the flux pinning occurs from the interaction between the strain and the flux lines. Since the central core of a flux line is almost in the normal state, the specific volume of this region is smaller (with the relative difference on the order of 10^{-7}) than that in the surrounding superconducting region, and internal stress exists around the normal core. The interaction caused by this internal stress is called the ΔV effect. On the other hand, the elastic constant of the normal core is larger (the relative difference is of the order of 10^{-4}) than that of the surrounding superconducting region, and hence, the elastic energy due to a defect is larger when the normal core is close to it. The pinning interaction caused by the variation in this energy is called the ΔC effect or ΔE effect. The energies of interaction due to the ΔV and ΔC effects are proportional to the first and second powers of the strain of the defects, respectively. Hence, these interactions are also called the first- and second-order interactions.

Theoretical calculation of the first-order interaction was attempted for an edge dislocation parallel to the flux line [22]. Campbell and Evetts [23] proposed a simplified method of calculation for this interaction. They estimated the stress tensor of the flux line as shown below. An isolated flux line is treated and the local model is used, in which a normal core of radius ξ is assumed. The ratio of dilation in the region of the normal core to that in the surrounding region is denoted by $\epsilon_{v0}(<0)$. This is given by the dependence of the condensation energy density on the pressure in the Meissner state, $\epsilon_{v0} = (\mu_0/2)\partial H_c^2/\partial P$. It is assumed that the dilation of the normal core does not occur along the length of the flux line. Cylindrical coordinates are defined where the z-axis lies on the center of the flux line. Then, the stress tensor is expressed as

$$\tilde{\sigma}^{\mathrm{f}} = \frac{\Gamma\xi^2}{r^2}\begin{bmatrix} 1 & 0 & 0 \\ 0 & -1 & 0 \\ 0 & 0 & 0 \end{bmatrix}. \tag{6.29}$$

In the above Γ is a positive quantity given by

$$\Gamma = -\frac{\epsilon_{v0}\mu(1+\nu)}{3(1-\nu)}, \tag{6.30}$$

where μ is the shear modulus and ν is the Poisson ratio.

It is assumed that the edge dislocation is parallel to the z-axis and located at (r_0, ϕ_0) on the (r, ϕ) plane and that Burgers vector $\boldsymbol{b}_0$ is directed towards the negative x-axis. Then, according to Peach and Koehler [24], the force on the dislocation per unit length under the stress given by Eq. (6.29) is expressed as $\boldsymbol{f}'^{\mathrm{d}} = -(\tilde{\sigma}^{\mathrm{f}} \cdot \boldsymbol{b}_0) \times \boldsymbol{i}$, where $\boldsymbol{i}$ is a unit vector showing the direction of the dislocation and is along the positive z-axis. The pinning force on the flux line per unit length is equal in magnitude but directed opposite to this force. Since Burgers vector is $\boldsymbol{b}_0 = b_0(-\cos\phi_0, \sin\phi_0, 0)$ in the (x', y', z) coordinates, we have

$$\boldsymbol{f}' = \frac{\Gamma\xi^2 b_0}{r_0^2}(-\sin\phi_0, \cos\phi_0, 0) \tag{6.31}$$

(see Fig. 6.16). The magnitude of this pinning force is $\Gamma\xi^2 b_0/r_0^2$ and increases with decreasing distance between the dislocation and the flux line. From the range of applicability of the local model that was used, the lower limit of r_0 is ξ, and hence, the elementary pinning force is obtained as

$$f'_{\mathrm{p}} = \Gamma b_0 \tag{6.32}$$

per unit length of a flux line. This calculation can be applied to single element superconductors. We treat niobium at 4.2 K for example [23]. From $\epsilon_{v0} \simeq -3 \times 10^{-7}$, $\mu \simeq 3 \times 10^{10}$ Nm^{-2}, $\nu \simeq 0.3$ and $b_0 \simeq 3 \times 10^{-10}$ m, we have $f'_{\mathrm{p}} \simeq 2 \times 10^{-6}$ Nm^{-1}. This result is compared with the condensation energy interaction due to a normal precipitate. Assuming that $\mu_0 H_{\mathrm{c}} \simeq 0.16$ T and $\xi = 33$ nm ($\mu_0 H_{\mathrm{c}2} = 0.30$ T), Eq. (6.6) leads to $f'_{\mathrm{p}} \simeq 9.1 \times 10^{-4}$ Nm^{-1}. Thus,

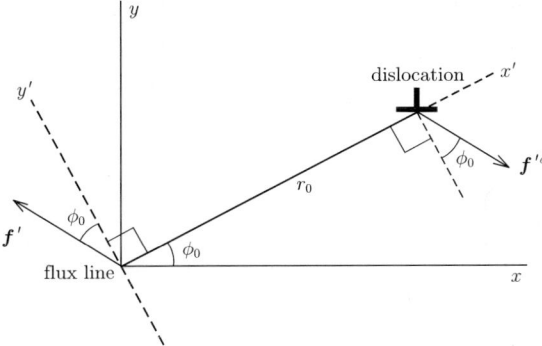

Fig. 6.16. Flux line at the origin and an edge dislocation parallel to it with the forces on the flux line and the dislocation

it is found that the pinning force due to dislocations is considerably weaker than that of normal precipitates.

The second-order interaction between a screw dislocation and a flux line was calculated by Webb [25]. Since the strain field due to a screw dislocation is a pure shear, the first-order interaction between it and the stress field due to the flux line is zero. If the shearing compliance is represented by $S_{44}(=1/\mu)$, the shear stress at a point at a distance r from the screw dislocation with the magnitude of Burgers vector b_0 is given by $\tau = b_0/2\pi r S_{44}$. The energy density of the strain is $(1/2)S_{44}\tau^2$. It is assumed that the extension of the strain field is sufficiently larger than the diameter of the normal core of the flux line and that the variation in the shearing compliance in the normal core is represented by $\delta S_{44}(>0)$. Then, the increase in the local energy density is $\delta S_{44}\tau^2/2 = (1/2)\delta S_{44}(b_0/2\pi S_{44})^2/r^2$. Here the case is treated for a screw dislocation normal to the flux line, and the distance between the screw dislocation and the flux line is represented by r_0. The local model is again used in which a normal core of radius ξ is assumed. Hence, δS_{44} is constant in the normal core and is zero outside it. The z-axis is defined along the length of the flux line. If r_0 is sufficiently larger than ξ, the normal core can be regarded as a thin line, and the increase in the total energy, i.e., the pinning energy is estimated as

$$\Delta U \simeq \frac{1}{2}\delta S_{44}\left(\frac{b_0}{2\pi S_{44}}\right)^2 \pi\xi^2 \int_{-\infty}^{\infty} \frac{dz}{r_0^2 + z^2} = \delta S_{44}\left(\frac{b_0}{2\pi S_{44}}\right)^2 \frac{\pi^2\xi^2}{2r_0}. \quad (6.33)$$

The pinning force is repulsive and its value obtained from $f = -\partial \Delta U/\partial r_0$ increases with decreasing distance r_0. Since the lower limit of r_0 is ξ, the elementary pinning force is approximately estimated as

$$f_p \simeq |f(r_0 = \xi)| = \frac{1}{8}\delta S_{44}\left(\frac{b_0}{S_{44}}\right)^2. \quad (6.34)$$

In the case of niobium at 4.2 K, substitution of $b_0 \simeq 3 \times 10^{-10}$ m, $S_{44} \simeq 3 \times 10^{-11}$ N^{-1}m^2 and $\delta S_{44} \simeq 4 \times 10^{-15}$ N^{-1}m^2 leads to [23] $f_p \simeq 5 \times 10^{-14}$ N. The elementary pinning force of a screw dislocation parallel to the flux line can also be calculated in a similar manner. A question on this will be found in Exercise 6.5.

The elastic interactions shown in the above originate from the variations in superconducting parameters. These variations are in general small enough to be treated as perturbations. Hence, a general calculation using the G-L theory can be done, and its details are described in [23]. Here the basic outline is briefly explained. As mentioned previously the variations in α and β in Eq. (6.3) can be described as the variations in H_{c2} and κ, respectively. Hence, if the normalized order parameter, $\psi = \Psi/|\Psi_\infty|$, is used, the variation in the G-L energy is written as

$$\delta F_s = -\mu_0 H_c^2 \int \left(\frac{\delta H_{c2}}{H_{c2}} |\psi|^2 - \frac{\delta \kappa^2}{2\kappa^2} |\psi|^4 \right) dV , \tag{6.35}$$

since the kinetic energy does not directly depend on the strain. The elastic energy should also be taken into account as well as the G-L energy $\delta F_s'$. The elastic energy is expressed as $\Sigma C_{ijkl}^{\mathrm{n}} \epsilon_{ij} \epsilon_{kl}$ in terms of the elastic constant tensor, $[C_{ijkl}^{\mathrm{n}}]$, in the normal state. In the above ϵ_{ij} is a component of the strain tensor and Σ represents the summation with respect to i, j, k and l. In the following the symbol of summation Σ is omitted after a conventional description. When a product of quantities with the same subscript appears, the summation with respect to this subscript should be taken. In this case δH_{c2} and $\delta \kappa^2$ arise from the strain and can be described as

$$\frac{\delta H_{c2}}{H_{c2}} = a_{ij}\epsilon_{ij} + a_{ijkl}\epsilon_{ij}\epsilon_{kl} , \tag{6.36a}$$

$$\frac{\delta \kappa^2}{\kappa^2} = b_{ij}\epsilon_{ij} + b_{ijkl}\epsilon_{ij}\epsilon_{kl} . \tag{6.36b}$$

In the above each coefficient is given by

$$a_{ij} = \frac{1}{H_{c2}} \cdot \frac{\partial H_{c2}}{\partial \epsilon_{ij}}, \quad a_{ijkl} = \frac{1}{H_{c2}} \cdot \frac{\partial^2 H_{c2}}{\partial \epsilon_{ij} \partial \epsilon_{kl}} , \tag{6.37a}$$

$$b_{ij} = \frac{1}{\kappa^2} \cdot \frac{\partial \kappa^2}{\partial \epsilon_{ij}}, \quad b_{ijkl} = \frac{1}{\kappa^2} \cdot \frac{\partial^2 \kappa^2}{\partial \epsilon_{ij} \partial \epsilon_{kl}} . \tag{6.37b}$$

The strain ϵ_{ij} can be divided into

$$\epsilon_{ij} = \epsilon_{ij}^{\mathrm{d}} + \epsilon_{ij}^{\mathrm{f}}, \tag{6.38}$$

where $\epsilon_{ij}^{\mathrm{d}}$ and $\epsilon_{ij}^{\mathrm{f}}$ are the strain due to defects and the spontaneous strain of the flux line, respectively. In general $\epsilon_{ij}^{\mathrm{f}}$ is much smaller than $\epsilon_{ij}^{\mathrm{d}}$. Omitting a constant term $C_{ijkl}^{\mathrm{n}} \epsilon_{ij}^{\mathrm{d}} \epsilon_{kl}^{\mathrm{d}}$, the sum of δF_s and the energy of strain reduces to

$$\delta F_1 = \int \epsilon_{ij}{}^{\mathrm{d}} \left[C_{ijkl}{}^{\mathrm{n}} \epsilon_{kl}{}^{\mathrm{f}} - \mu_0 H_{\mathrm{c}}^2 \left(a_{ij} |\psi|^2 - \frac{1}{2} b_{ij} |\psi|^4 \right) \right] \mathrm{d}V . \qquad (6.39)$$

This gives the energy of the first-order interaction proportional to $\epsilon_{ij}{}^{\mathrm{d}}$. On the other hand, the energy of the second-order interaction is

$$\delta F_2 = - \int \epsilon_{ij}{}^{\mathrm{d}} \epsilon_{kl}{}^{\mathrm{d}} \delta C_{ijkl} \mathrm{d}V , \qquad (6.40)$$

where δC_{ijkl} is the variation in the elastic constant in the superconducting state and is given by

$$\delta C_{ijkl} = \mu_0 H_{\mathrm{c}}^2 \left(a_{ijkl} |\psi|^2 - \frac{1}{2} b_{ijkl} |\psi|^4 \right) . \qquad (6.41)$$

Substituting the approximate expression of the flux line lattice of Eq. (1.98) into $|\psi|^2$, the elementary pinning force at high fields can be estimated.

When the superconductor is heavily worked, the defects obtained are not simple dislocations but are significantly entangled each other. Sometimes a dislocation cell structure is formed, in which a region with a low density of dislocations is surrounded by an high density region. In extremely heavily worked Nb-Ti, etc., internal low density regions and cell boundaries are similar to grains and grain boundaries in a compound superconductor. For such two-dimensional or three-dimensional pinning centers the probability for electrons to be scattered becomes high, and the condensation energy interaction due to the electron scattering mechanism seems to be more important than the elastic interaction.

We now briefly discuss the elementary pinning force of a grain boundary due to the elastic interaction. This was calculated by Kusayanagi et al. [26] Assuming a small angle twin boundary composed of a row of edge dislocations separated by the interval L, they estimated the strength of interaction between the boundary and the flux line lattice. As for the first-order interaction, the strain takes positive and negative values alternately along the boundary, resulting in almost zero elementary pinning force due to cancellation of the respective contributions. On the other hand, the second-order interaction depends on the arrangement of flux lines at the boundary. They obtained

$$f'_{\mathrm{p}} = \frac{b_0^2 \gamma_{\mathrm{e}}}{6\sqrt{3} L} \left(1 - \frac{B}{\mu_0 H_{\mathrm{c}2}} \right) \left| \log \left(\frac{2\pi r_{\mathrm{c}}}{L} \right) - 1 \right| \qquad (6.42)$$

per unit length of flux line for the arrangement of the flux line lattice shown in Fig. 6.17. In the above b_0 is the magnitude of Burgers vector, r_{c} is the cut-off radius of the dislocation, and γ_{e} is a coefficient for expressing a variation in the superconducting state as $\mu/(1-\nu) = \mu_{\mathrm{n}}/(1-\nu_{\mathrm{n}}) - \gamma_{\mathrm{e}}|\psi|^2$ and can also be expressed in terms of the elastic constants, where the subscript "n" represents a value in the normal state. For niobium at 4.2 K and $B/\mu_0 H_{\mathrm{c}2} = 0.7$, substitution of $r_{\mathrm{c}} \simeq b_0 \simeq 3 \times 10^{-10}$ m, $L \simeq 1.5 \times 10^{-9}$ m and $\gamma_{\mathrm{e}} \simeq$

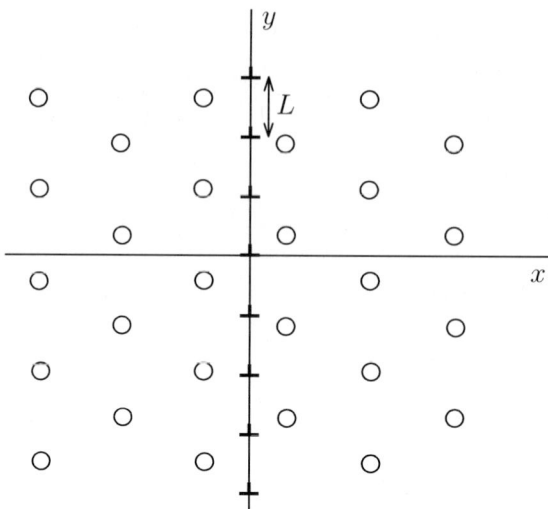

Fig. 6.17. Twin boundary and flux line lattice

5.5×10^6 Nm^{-2} leads to $f'_\mathrm{p} \simeq 5.7 \times 10^{-6}$ Nm^{-1}. This value is considerably smaller than the maximum value calculated from the electron scattering mechanism by Zerweck [14] and Yetter et al. [15], $3-5 \times 10^{-5}$ Nm^{-1}. It is also about 1/5 of the experimental result for the niobium bi-crystal shown in Fig. 6.15. Thus, electron scattering is considered to be the dominant pinning mechanism at grain boundaries in most cases.

6.5 Magnetic Interaction

Since the normal core of a flux line is surrounded by the magnetic field with a radius of the order of the penetration depth λ, the magnetic interaction with the flux line occurs for inhomogeneous regions sufficiently larger than λ. An interaction between a wide superconducting-normal interface and a parallel flux line is assumed at a sufficiently high magnetic field. The origin of the interaction is the surface barrier mentioned in Sect. 3.5. In an equilibrium state the superconducting region is slightly diamagnetic in comparison with the normal region. Its magnetization, represented by M_r, in the magnetic field $H = B/\mu_0$ in the normal region is given by Eq. (1.114) with replacement of H_e by B/μ_0. Hence, the flux line suffers a repulsive force from the interface due to the Lorentz force caused by the magnetization current flowing near the interface, when the flux line moves from the normal region to the superconducting region. This force can be written as $\phi_0|M_\mathrm{r}|\exp(-x/\lambda)/\lambda$ in terms of the flux quantum ϕ_0, where x is the distance of the position inside the superconducting region from the interface. On the other hand, an image flux line exists in the normal region opposite the flux line in the superconducting region, and an attractive force acts between them. This force is written as $K \exp(-2x/\lambda)$ with K denoting a constant. The total work done by the two forces during movement of the flux line from $x = 0$ to $x = \infty$ should be zero. From this requirement $K = 2(\phi_0|M_\mathrm{r}|)/\lambda$ is derived and the net attractive force by the interface amounts to [27]

$$f'_\mathrm{p} = \frac{\phi_0|M_\mathrm{r}|}{\lambda} = \frac{\phi_0 H_{c2}}{[(2\kappa^2 - 1)\beta_\mathrm{A} + 1]\lambda} \left(1 - \frac{B}{\mu_0 H_{c2}}\right) \quad (6.43)$$

at $x = 0$. Thus, the normal phase also works as an attractive pinning center due to the magnetic interaction.

The flux pinning strength from the magnetic interaction can be compared with that from the condensation energy interaction given by Eq. (6.8). The ratio of these elementary pinning forces is

$$\frac{f'_\mathrm{p}(\mathrm{magn})}{f'_\mathrm{p}(\mathrm{cond})} = \frac{9.30\kappa}{(2\kappa^2 - 1)\beta_\mathrm{A} + 1} \simeq \frac{4.01}{\kappa} . \quad (6.44)$$

Hence, it is concluded that the condensation energy interaction is the dominant pinning mechanism of normal precipitates for superconductors with κ larger than 4.

6.6 Kinetic Energy Interaction

The pinning performance in Nb-Ti with pinning centers of normal α-Ti is very high, and it is desired to enhance the volume fraction of α-Ti for a further improvement. However, it is empirically known that the obtained volume fraction is 15% or so at maximum in the usual fabrication process. Thus, an introduction of artificial pinning centers into Nb-Ti was examined. It was found that the critical current density reached 4.25×10^9 Am^{-2} at 4.2 K and 5 T when Nb of 27 vol.% was introduced and fabricated into thin laminar structures after the usual drawing process [28].

However, added Nb is in a superconducting state at 4.2 K, and its pinning mechanism is not a simple condensation energy interaction. In fact, if it was the condensation energy interaction, the critical current density would show a peak effect at a magnetic field in the vicinity of the upper critical field of Nb (see Sect. 7.6). The observed critical current density decreased monotonically with increasing magnetic field [28, 29] and the upper critical field was slightly lower than in conventional Nb-Ti. Such a degradation of the upper critical field can be attributed to the proximity effect between the Nb layers and the superconducting matrix. This speculation seems to be reasonable, since the thickness of the Nb layers is of the order of nanometers.

Multilayers of weakly superconducting Nb in a Nb-Ti superconducting matrix are considered, as is schematically shown in Fig. 6.6, in which the normal layers are replaced by Nb layers. The thickness of the respective layers is denoted by d_{NT} and d_{N}, and it is assumed that the Nb layers are sufficiently thin. It seldom occurs that flux lines are exactly parallel to the layered structure. Hence, the typical case is treated for simplicity where the magnetic field is normal to the layers. It is assumed that the Nb layer is in the superconducting state due to the proximity effect even in a magnetic field above its bulk upper critical field $H_{\mathrm{c2}}^{\mathrm{N}}$. This situation will continue until the magnetic field reaches the upper critical field $H_{\mathrm{c2}}^{\mathrm{av}}$ of the whole system. In addition, the case is treated for a low magnetic field region in which the flux line spacing is sufficiently large. Since the thickness of the Nb layers is thin enough, the structure of the order parameter of the normal core of an isolated flux line normal to the layers can be approximated by that in the thicker Nb-Ti layer: Eq. (6.4) with $r_{\mathrm{n}} = 1.8\xi_{\mathrm{NT}}$ where ξ_{NT} is the coherence length in the Nb-Ti layer. Since the thermodynamic critical field of Nb is approximately the same as that of Nb-Ti as is the critical temperature, the condensation energy is approximately the same between Nb-Ti and Nb. Hence, the condensation energy can be disregarded, but the kinetic energy is important when considering the pinning interaction. The energy due to the current is also disregarded, and only the third term of Eq. (6.3) is considered, similarly to the calculation in 6.3.1.

Here the theoretical analysis given in [10] is introduced. The two cases shown in Fig. 6.7(a) and (b) are considered. Here the normal layers are again replaced by Nb layers. In case (a) a normal core of a flux line crosses one more Nb layer in comparison with case (b). To discuss the difference in the energy between the two cases, it is enough to compare the kinetic energy in the regions of V_1 and V_2. In case (a) the kinetic energy comes only from region V_1 and is given by

$$U_{\mathrm{a}} = \mu_0 H_{\mathrm{c}}^2 \xi_{\mathrm{N}}^2 d_{\mathrm{N}} \int_0^\infty \left(\frac{d|\psi|}{dr}\right)^2 2\pi r dr = \frac{4\pi}{3} \mu_0 H_{\mathrm{c}}^2 \xi_{\mathrm{N}}^2 d_{\mathrm{N}} \left(\log 2 - \frac{1}{4}\right). \quad (6.45)$$

The kinetic energy in region V_2 in case (b) is similarly given by

$$U_{\mathrm{b}} = \frac{4\pi}{3} \mu_0 H_{\mathrm{c}}^2 \xi_{\mathrm{NT}}^2 d_{\mathrm{N}} \left(\log 2 - \frac{1}{4}\right). \quad (6.46)$$

Hence, the pinning energy is

$$\Delta U_{\mathrm{N}} = U_{\mathrm{a}} - U_{\mathrm{b}} = \frac{4\pi}{3} \mu_0 H_{\mathrm{c}}^2 d_{\mathrm{N}} (\xi_{\mathrm{N}}^2 - \xi_{\mathrm{NT}}^2) \left(\log 2 - \frac{1}{4}\right)$$
$$\simeq 0.591\pi \mu_0 H_{\mathrm{c}}^2 d_{\mathrm{N}} (\xi_{\mathrm{N}}^2 - \xi_{\mathrm{NT}}^2). \quad (6.47)$$

It is found that $\Delta U_{\mathrm{N}} > 0$, since $\xi_{\mathrm{N}} > \xi_{\mathrm{NT}}$. Therefore, the niobium layers act as repulsive pinning centers. The elementary pinning force of the edge of a niobium layer is

$$f_{\mathrm{pN}} \simeq \frac{\Delta U_{\mathrm{N}}}{2r_{\mathrm{n}}} = \frac{0.164\pi \mu_0 H_{\mathrm{c}}^2 d_{\mathrm{N}}}{\xi_{\mathrm{NT}}} (\xi_{\mathrm{N}}^2 - \xi_{\mathrm{NT}}^2). \quad (6.48)$$

The obtained pinning strength of the kinetic energy interaction of niobium is compared with the pinning strength of the condensation energy interaction of ordinary α-Ti in [10], and the probability of stronger pinning of niobium than α-Ti is discussed. If the upper critical field in heavily worked Nb layers is higher than 1.10 T, the pinning force of the Nb layer is expected to be stronger than that of the α-Ti layer of the same geometry.

In this case the magnetic field dependence of the elementary pinning force comes from a reduction in $\langle|\Psi|^2\rangle$ as well as the usual condensation energy interaction and is expressed as $1-(B/\mu_0 H_{c2}^{av})$. The upper critical field decreases from the value of the Nb-Ti matrix H_{c2}^{NT} due to the proximity effect as: [30]

$$H_{c2}^{av} = H_{c2}^{NT} \cdot \frac{1+d_N/d_{NT}}{1+d_N\xi_N^2/d_{NT}\xi_{NT}^2}. \qquad (6.49)$$

6.7 Improvement of Pinning Characteristics

It can be seen from Fig. 6.12 that the elementary pinning force of a grain boundary that originates from the electron scattering mechanism is at most $0.17\xi_0\mu_0 H_c^2$ per unit length of the flux line. On the other hand, Eq. (6.6) shows that the elementary pinning force of a normal precipitate that originates from the condensation energy interaction amounts to $1.35\xi\mu_0 H_c^2$. The elementary pinning force of the grain boundary takes the maximum value when the impurity parameter α_i is about 1.4, and then, $\xi(T=0)$ is approximately equal to $2\xi_0$. Hence, if the maximum elementary pinning force of the grain boundary at low temperatures is normalized to 1, the elementary pinning force of the normal precipitate is 4.0. Thus, it can be seen that the normal precipitate is stronger than the grain boundary. This is because a large condensation energy can be fully utilized as the pinning energy for normal precipitates, while the pinning energy of the grain boundary is small, since the superconducting parameters vary only slightly near the boundary due to the electron scattering.

The elementary pinning force is proportional to $H_c^2\xi$ according to both mechanisms. Hence, this quantity represents the flux pinning strength inherent in superconducting materials. If we compare these values for Nb-Ti and Nb$_3$Sn, which are practical superconducting materials, at 4.2 K, a large difference of 1 : 4.3 is obtained. This means that Nb$_3$Sn is a material with very high potential for pinning. However, the difference in the practical critical current density between these materials is small. One of the reasons for this result is the difference in the pinning centers. That is, relatively weak grain boundaries work as pinning centers in Nb$_3$Sn, while strong normal precipitates work in Nb-Ti. The difference in the number density of the effective pinning sites is also one of the reasons.

Quite high pinning efficiency in Nb-Ti is owing to the fact that the dominant pinning centers, the normal precipitates of the α-Ti phase, are not of a cubic shape but are in the shape of very thin ribbons several nanometers

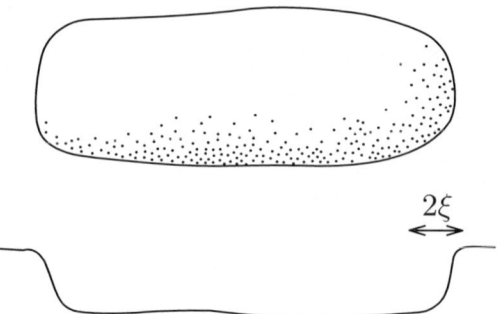

Fig. 6.18. Large normal precipitate (*upper*) and variation in energy when the flux line passes through (*lower*)

thick. For large cubic precipitates the elementary pinning force is given by Eq. (6.9). It should be noted that only the surface regions of precipitates work as pinning centers. The variation in the energy when the flux line passes through a large normal precipitate is shown in Fig. 6.18. When the flux line moves through the central part of the precipitate, the energy does not vary much, and there is no appreciable pinning interaction. Thus, it is effective for pinning to enlarge the surface area of precipitates. In fact such a structure of pinning centers is realized in practical Nb-Ti superconductors.

In practical Nb-Ti superconducting wires, α-Ti is precipitated on subband walls by heat treatment after a heavy drawing and then elongated to thin ribbons by additional drawing. For example, we assume that the dimensions of the α-Ti particles after the precipitation are $50 \times 75 \times 60$ nm^3 and that those are deformed to $4 \times 75 \times 750$ nm^3 with the longest axis along the wire by the additional drawing. Figure 6.19 shows the shapes of the precipitate and the arrangements of flux lines before and after the additional drawing. Before the drawing as in (a) the interaction per flux line is stronger, while the number of interacting flux lines is as small as three. On the other hand, after the drawing as in (b), although the interaction per flux line is smaller, the number of interacting flux lines increases drastically. As a result, the elementary pinning force in (b) is about 1.7 times as large as that in (a) [31]. The improvement of pinning characteristics in Nb-Ti is achieved in this manner by making α-Ti into thin ribbons.

The pinning efficiency in Nb$_3$Sn is not as good as in Nb-Ti because of the pinning by grain boundaries. However, this means that there is still room for drastic improvement of the property. One of the possibilities is the improvement of the elementary pinning force by addition of a third element to Nb$_3$Sn. The addition of a third element is now employed mainly for improvement in the upper critical field H_{c2}. If a suitable element which is concentrated on grain boundaries and contributes to the electron scattering is discovered, it will be effective for the improvement of the pinning characteristics.

6.7 Improvement of Pinning Characteristics

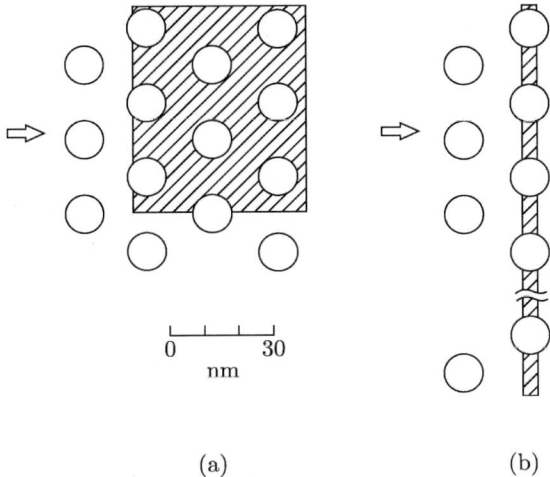

Fig. 6.19. Typical shapes of α-Ti and the arrangement of flux lines (**a**) before and (**b**) after a drawing of Nb-Ti at 4.2 K and 5 T. The *circles* represent normal cores and the *arrow* shows the direction of the Lorentz force

From the relationship between the pinning force density and the reciprocal size of grains as shown in Fig. 7.18, the elementary pinning force of the grain boundary can be estimated. According to this $f'_p = 3.1 \times 10^{-4}$ Nm^{-1} is obtained for V$_3$Ga [32] in Fig. 7.18, while we have $f'_p = 1.0 \times 10^{-4}$ Nm^{-1} for Nb$_3$Sn [33]. Such a strong pinning in V$_3$Ga reflects its excellent pinning characteristics at high fields (see Sect. 7.5). Although the reason for such a strong pinning of grain boundaries in V$_3$Ga has not yet been clarified, it attracts attention by showing the possibilities for improvement of pinning characteristics in Nb$_3$Sn.

Exercises

6.1. Calculate the elementary pinning force of a wide interface between the superconducting and normal regions parallel to a flux line assuming the local model, in which the order parameter is approximated as $|\Psi| = 0$ for $r < \xi$ and $|\Psi| = |\Psi_\infty|$ for $r > \xi$.
6.2. Derive Eq. (6.14).
6.3. Figure 6.8 shows that the elementary pinning force decreases when θ becomes too large. Consider the reason. How large is the elementary pinning force expected to be in the limit of large θ?
6.4. It is assumed that values of the upper critical field in adjacent grains are H_{c2} and $H_{c2} + \delta H_{c2}$ (δH_{c2} is sufficiently small and positive) because of the anisotropy. When the flux line moves from the grain with the higher upper critical field to the grain with the lower one at low fields, estimate

the elementary pinning force. Assume that the flux line is parallel to the interface.
6.5. Calculate the elementary pinning force of a screw dislocation parallel to a flux line based on the second-order elastic interaction.

References

1. C. P. Henning: J. Phys. F **6** (1976) 99.
2. E. J. Kramer: Phil. Mag. **33** (1976) 331.
3. N. Harada, Y. Miyamoto, T. Matsushita and K. Yamafuji: J. Phys. Soc. Jpn. **57** (1988) 3910.
4. P. G. de Gennes: Rev. Mod. Phys. **36** (1964) 225.
5. E. J. Kramer and H. C. Freyhardt: J. Appl. Phys. **51** (1980) 4903.
6. R. G. Hampshire and M. T. Taylor: J. Phys. F. **2** (1972) 89. As for the interpretation of this result, see the following. D. C. Larbalestier, D. B. Smathers, M. Daeumling, C. Meingast, W. Warnes and K. R. Marken: *Proc. Int. Symp. on Flux Pinning and Electromagnetic Properties in Superconductors*, Fukuoka, 1985, p. 58.
7. C. Meingast and D. C. Larbalestier: J. Appl. Phys. **66** (1989) 5971.
8. T. Matsushita, S. Otabe and T. Matsuno: *Adv. Cryog. Eng. Mater.* (Plenum, New York, 1990) Vol. 36, p. 263.
9. R. O. Zaitsev: Sov. Phys. JETP **23** (1966) 702.
10. T. Matsushita, M. Iwakuma, K. Funaki, K. Yamafuji, K. Matsumoto, O. Miura and Y. Tanaka: *Adv. Cryog. Eng. Mater.* (Plenum, New York, 1996) Vol. 42, p. 1103.
11. T. Matsushita: J. Appl. Phys. **54** (1983) 281.
12. T. Matsushita: J. Phys. Soc. Jpn. **51** (1982) 2755.
13. A. M. Campbell and J. E. Evetts: Adv. Phys. **21** (1972) 377. In this reference the anisotropy of H_c is treated instead of that of H_{c2}.
14. G. Zerweck: J. Low Temp. Phys. **42** (1981) 1.
15. W. E. Yetter, D. A. Thomas and E. J. Kramer: Phil. Mag. B **46** (1982) 523.
16. D. O. Welch: IEEE Trans. Magn. **MAG-21** (1985) 827.
17. A. Pruymboom and P. H. Kes: Jpn. J. Appl. Phys. **26** (1987) Supplement 26–3 1533.
18. P. H. Kes: IEEE Trans. Magn. **MAG-23** (1987) 1160.
19. A. M. Campbell and J. E. Evetts: Adv. Phys. **21** (1972) 333.
20. A. Das Gupta, C. C. Koch, D. M. Kroeger and Y. T. Chou: Phil. Mag. B **38** (1978) 367.
21. H. R. Kerchner, D. K. Christen, A. Das Gupta, S. T. Sekula, B. C. Cai and Y. T. Chou: *Proc. 17th Int. Conf. Low Temp. Phys.*, Karlsruhe, 1984, p. 463.
22. E. J. Kramer and C. L. Bauer: Phil. Mag. **15** (1967) 1189.
23. A. M. Campbell and J. E. Evetts: Adv. Phys. **21** (1972) 345.
24. M. O. Peach and J. S. Koehler: Phys. Rev. **80** (1950) 436.
25. W. W. Webb: Phys. Rev. Lett. **11** (1963) 1971.
26. E. Kusayanagi and M. Kawahara: *Extended abstract of 27th Meeting on Cryogenics and Superconductivity of the Cryogenic Society of Japan* (1981) p. 11 [in Japanese].

27. A. M. Campbell and J. E. Evetts: Adv. Phys. **21** (1972) 340.
28. K. Matsumoto, H. Takewaki, Y. Tanaka, O. Miura, K. Yamafuji, K. Funaki, M. Iwakuma and T. Matsushita: Appl. Phys. Lett. **64** (1994) 115.
29. K. Matsumoto, Y. Tanaka, K. Yamafuji, K. Funaki, M. Iwakuma and T. Matsushita: IEEE Trans. Appl. Supercond. **3** (1993) 1362.
30. T. Matsushita: *Proc. of 8th Int. Workshop on Critical Currents in Superconductors* (World Scientific, Singapore, 1996) p. 63.
31. T. Matsushita and H. Küpfer: J. Appl. Phys. **63** (1988) 5048.
32. Y. Tanaka, K. Itoh and K. Tachikawa: J. Jpn. Inst. Metals **40** (1976) 515 [in Japanese].
33. R. M. Scanlan, W. A. Fietz and E. F. Koch: J. Appl. Phys. **46** (1975) 2244.

7
Flux Pinning Characteristics

7.1 Flux Pinning Characteristics

The macroscopic pinning force density, $F_{\rm p} = J_{\rm c}B$, which works on flux lines in a unit volume is an accumulation of individual pinning interactions and depends generally on the elementary pinning force $f_{\rm p}$, the number density of pinning centers $N_{\rm p}$ and the density of flux lines, i.e., the magnetic field H (or the magnetic flux density B). It depends also on the temperature T through the elementary pinning force. The problem of analytically estimating $F_{\rm p}$ as a function of $f_{\rm p}$, $N_{\rm p}$ and B is called the summation problem. The reason why such a method is useful is that the resultant $F_{\rm p}$ does not primarily depend on the kind of pinning centers (hereafter briefly called pins), but depends mostly on $f_{\rm p}$, a parameter representing the strength of pinning. The kind of pin influences only the magnitude and the temperature dependence of $f_{\rm p}$.

When $f_{\rm p}$ or $N_{\rm p}$ increases, $F_{\rm p}$ generally increases. However, the dependences are rather complicated. This comes from the fact that the individual pinning forces originate from potentials, and hence, these forces are not directed towards a specified direction but depend on the relative position of the flux lines with respect to the pins as shown in Fig. 6.1. In the case of weak pins, for example, most of the randomly directed individual forces due to the random distribution of pins are canceled out, resulting in a very small value of $F_{\rm p}$. On the other hand, for strong pins the pinning force density has a simple dependence of

$$F_{\rm p} \propto N_{\rm p} f_{\rm p} \tag{7.1}$$

except in the high field region. This means that $F_{\rm p}$ increases with increasing $N_{\rm p}$ or $f_{\rm p}$. The dependence on the pinning parameters, $f_{\rm p}$ and $N_{\rm p}$, given by Eq. (7.1) is called a linear summation. Theories on the summation problem including the linear summation will be described in Sect. 7.3, and corresponding experimental results will be given in Sect. 7.4. Saturation and nonsaturation phenomena are particularly notable pinning characteristics where practical superconductors are concerned and will be treated in Sect. 7.5.

7 Flux Pinning Characteristics

The cancellation of the individual pinning forces of randomly distributed pins shown in Fig. 6.1 is due to interference through the elastic interaction of flux lines. That is, since flux lines strongly repel each other, each flux line cannot necessarily stay at a suitable position for pinning. However, if the pins are strong enough, the pinning forces exceed the elastic interactions among the flux lines, and each flux line can stay at a suitable position for pinning. Thus, the rate of cancellation of individual pinning forces is small, resulting in the pinning force density of Eq. (7.1). It can be seen from this argument that the elastic interaction among the flux lines is also an important factor in determining the pinning characteristics. The corresponding elastic moduli of the flux line lattice will be briefly discussed in Sect. 7.2.

Before discussing matters concerned with the summation problem, we shall briefly mention here how the resultant pinning force density varies with temperature and magnetic field. It is empirically known that these dependences are described in the form:

$$F_{\mathrm{p}}(B,T) = AH_{c2}^{m}(T)f(b) . \tag{7.2}$$

This is called the scaling law of the pinning force density (or sometimes the temperature scaling law in distinction from a similar scaling law on strain). In Eq. (7.2) A is a constant and f is a function only of the reduced magnetic field $b = B/\mu_0 H_{c2}$, and in most cases has the form of

$$f(b) = b^{\gamma}(1-b)^{\delta} . \tag{7.3}$$

One of the characteristics of the scaling law is that the temperature dependence of the pinning force density can be expressed only in terms of that of the upper critical field $H_{c2}(T)$. m, γ and δ are important parameters describing the scaling law of the pinning force density. Figure 7.1 shows the results for Pb-Bi [1] with normal precipitates, the elementary pinning force of which is given by Eq. (6.9). In this case we have $m = 2$, $\gamma = 1/2$ and $\delta = 1$. For Nb$_3$Sn showing a saturation phenomenon, $m = 2.0$–2.5, $\gamma = 1/2$ and $\delta = 2$ are known. Such a scaling law is sometimes used for evaluation of the validity of summation theories. It is also useful in practical cases for estimation of the pinning characteristics under difficult circumstances for experiments. It is useful, for example, for a measurement of critical current densities in very high magnetic fields or for a measurement of very high critical current densities at low magnetic fields. If a measurement is done at a high temperature where an equivalent magnetic field or the critical current density is sufficiently reduced, the desired characteristics can be deduced using the scaling law.

Similarly to the parabolic variation with temperature given by Eq. (1.2), the upper critical field H_{c2} varies also with the strain ϵ as

$$H_{c2}(\epsilon) = H_{c2\mathrm{m}}(1 - a\epsilon^2) \tag{7.4}$$

(see Fig. 7.2) and the pinning force density also shows a dependence [2] similar to Eq. (7.2):

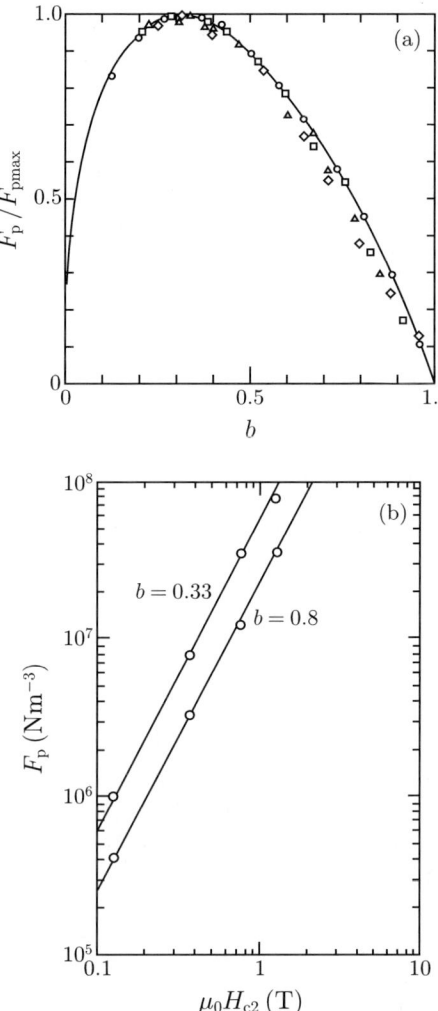

Fig. 7.1. Scaling law of pinning force density in Pb-Bi with precipitates of normal Bi phase [1]. (**a**) Relationship between normalized pinning force density and the reduced magnetic field b. The *solid line* shows Eq. (7.3) with $\gamma = 1/2$ and $\delta = 1$. (**b**) Relationship between the pinning force density at $b = 0.33$ and $b = 0.8$ and the upper critical field. The *straight lines* show Eq. (7.2) with $m = 2$

$$F_{\rm p}(B, \epsilon) = \widehat{A} H_{\rm c2}^{\widehat{m}}(\epsilon) f(b) \,. \tag{7.5}$$

This is the strain scaling law of the pinning force density. Comparing Eqs. (7.2) and (7.5), the function f of the reduced magnetic field b is the same between the two scaling laws. However, the parameters, m and $\widehat{m}$, describing dependence on temperature and strain are different. For the abovementioned Nb$_3$Sn, $\widehat{m} \simeq 1$ is obtained, while we have $m = 2.0$–2.5.

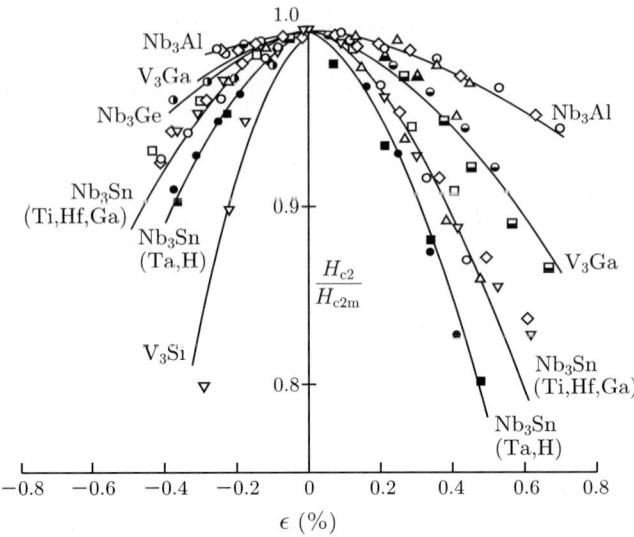

Fig. 7.2. Variation in the upper critical field vs. the strain in various superconductors [2]

The temperature dependence of the pinning force density comes from that of the elementary pinning force mentioned in Chap. 6 and that of the elastic moduli of the flux line lattice which will be mentioned in Sect. 7.2. The elementary pinning force and the elastic moduli depend only on the superconducting parameters such as thermodynamic critical field, coherence length, penetration depth, etc., and there are no other temperature-dependent factors which influence these quantities. The temperature dependence of these superconducting parameters can be approximately represented by that of H_{c2}, if the temperature dependence of the G-L parameter κ is disregarded. This brings about the result of the temperature scaling law. If the variation in the pinning force density due to the strain originates only from the variations in the superconducting parameters with the strain, it will depend only on $H_{c2}(\epsilon)$ as the summation theory predicts, and hence, the index $\widehat{m}$ should coincide with m. However, the practical situation in a superconductor under external stress is not simple. This is because there are originally local strains around pins and an additional strain is considered to be applied nonuniformly to the superconductor. Therefore a strain concentration may occur around pins, resulting in a possible variation in the elementary pinning force itself. For example, the elastic interaction given by Eqs. (6.39) and (6.40) will directly change through the variation in the strain $\epsilon_{ij}{}^d$. The differences in the strain scaling law from the temperature scaling law are speculated to originate from causes other than the variation in $H_{c2}(\epsilon)$.

7.2 Elastic Moduli of Flux Line Lattice

The displacement of flux lines is denoted by u. Then, the strain of the flux line lattice is given by

$$\epsilon_{xx} = \frac{\partial u_x}{\partial x}, \tag{7.6a}$$

$$\epsilon_{xy} = \frac{\partial u_x}{\partial y} + \frac{\partial u_y}{\partial x}. \tag{7.6b}$$

When the strain of the flux line lattice is sufficiently small, a linear relationship holds between the strain and the stress σ, and this is expressed as

$$\sigma_i = C_{ij}\epsilon_j \tag{7.7}$$

using the notation of Voigts used for crystals. In the above the subscripts, i and j, represent numbers from 1 to 6 referring to xx, yy, zz, yz, zx and xy, respectively. For example, $\sigma_6(=\sigma_{xy})$ means the component of stress along the y-axis working on the y-z plane normal to the x-axis. When a product of quantities with the same subscript appears as in the above equation (j in this case), a summation with respect to this subscript is taken. In this case the symbol representing the summation is usually omitted. The coefficients C_{ij} defined in Eq. (7.7) are the elastic moduli.

Since the flux line lattice is two-dimensional, a displacement along the length of the flux lines is meaningless. Hence, the displacement is defined to be normal to the flux lines. If the z-axis is taken to be parallel to the flux lines, $u_z = 0$ and $\epsilon_3(=\epsilon_{zz}) = 0$. Thus, from symmetry we have

$$\begin{bmatrix} \sigma_{xx} \\ \sigma_{yy} \\ \sigma_{yz} \\ \sigma_{zx} \\ \sigma_{xy} \end{bmatrix} = \begin{bmatrix} C_{11} & C_{12} & 0 & 0 & 0 \\ C_{12} & C_{11} & 0 & 0 & 0 \\ 0 & 0 & C_{44} & 0 & 0 \\ 0 & 0 & 0 & C_{44} & 0 \\ 0 & 0 & 0 & 0 & C_{66} \end{bmatrix} \begin{bmatrix} \epsilon_{xx} \\ \epsilon_{yy} \\ \epsilon_{yz} \\ \epsilon_{zx} \\ \epsilon_{xy} \end{bmatrix}. \tag{7.8}$$

In the above there is a condition of $C_{12} = C_{11} - 2C_{66}$ among the elastic moduli. Hence, the independent constants are three, i.e., C_{11}, C_{44} and C_{66}. These are the elastic moduli for the strains of uniaxial compression, bending deformation and shear, respectively (see Fig. 7.3). Labusch [3, 4] calculated these moduli. According to his calculation these are expressed as

$$C_{11} = B^2 \frac{\partial \mathcal{H}}{\partial B} + C_{66} \simeq \frac{B^2}{\mu_0}, \tag{7.9}$$

$$C_{44} = \frac{B^2}{\mu_0}, \tag{7.10}$$

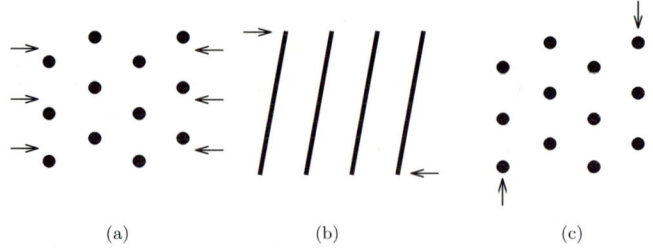

Fig. 7.3. Deformation of flux line lattice for (**a**) uniaxial compression, (**b**) longitudinal shear (bending) and (**c**) transverse shear. The corresponding shear moduli are C_{11}, C_{44} and C_{66}, respectively

$$C_{66} = \frac{\mu_0}{2} \int_0^B B^2 \frac{d^2 \mathcal{H}(B)}{dB^2} dB \; ; \quad \text{in the vicinity of } H_{c1}, \quad (7.11a)$$

$$= 0.48 \frac{\mu_0 H_c^2 \kappa^2 (2\kappa^2 - 1)}{[1 + \beta_A (2\kappa^2 - 1)]^2} (1-b)^2 \; ; \quad \text{in the vicinity of } H_{c2}. \quad (7.11b)$$

In the above $\mathcal{H}$ is the thermodynamic magnetic field and $\beta_A = 1.16$. The shear modulus C_{66} in Eq. (7.11a) is zero at $B = 0$ and increases with increasing B. At high fields it decreases with increasing B and reduces to zero at H_{c2}. The approximate formula of C_{66} in the entire field range is derived by Brandt [5] as

$$C_{66} = \mu_0 H_c^2 \frac{2\kappa^2 \beta_A^2 (2\kappa^2 - 1)}{[1 + \beta_A (2\kappa^2 - 1)]^2} \cdot \frac{b(1-b)^2}{4}$$
$$\times (1 - 0.58b + 0.29b^2) \exp\left(\frac{1-b}{3\kappa^2 b}\right), \quad (7.12a)$$

$$\simeq \frac{\mu_0 H_c^2}{4} b(1-b)^2. \quad (7.12b)$$

The lower equation is an approximation for a superconductor with an high κ value.

Contrary to the above results of Labusch, Brandt [5, 6] proposed a nonlocal theory of the elastic moduli insisting that C_{11} and C_{44} are dispersive with respect to the wave number of deformation of the flux line lattice and have very small values for a large wave number. This idea was also supported by Larkin and Ovchinnikov [7]. The moduli derived by Labusch correspond to the limit of zero wave number and are called local moduli. The foundation of the nonlocal theory is that, since the spatial variation of the local magnetic flux density in a high-κ superconductor is extremely small, as shown in Eq. (1.115), the influence of a displacement of a normal core due to a pinning interaction with the magnetic flux density is expected to be very small. In other words, the magnetic flux density and the order parameter are almost independent of each other. Hence, the increase in the magnetic energy from a deformation of the flux line lattice is very small, resulting in a very small elastic modulus. However, this idea encounters a serious problem, as will be mentioned later.

The elastic moduli are calculated from the increase in the energy when a strain is imposed on the flux line lattice. Brandt introduced a strain of wave number k and amplitude ϵ_k on the flux line lattice and obtained the elastic moduli from the superconducting energy density given by Eq. (1.112). In this equation the energy density depends only on the mean value of the local magnetic flux density. This means that any contribution from the magnetic flux density was disregarded and only contributions from terms associated with the order parameter were treated in the above calculation. That is,

$$C_{ii} = \frac{\partial F_s}{\partial \beta_A} \cdot \frac{\partial^2 \beta_A}{\partial \epsilon_k^2} , \tag{7.13}$$

where $\beta_A = \langle |\Psi|^4 \rangle / \langle |\Psi|^2 \rangle^2$. The obtained results are:

$$C_{11}(k) = B^2 \frac{\partial \mathcal{H}}{\partial B} \cdot \frac{k_h^2}{k^2 + k_h^2} \cdot \frac{k_\psi^2}{k^2 + k_\psi^2} + C_{66} , \tag{7.14}$$

$$C_{44}(k) = \frac{B^2}{\mu_0} \cdot \frac{k_h^2}{k^2 + k_h^2} + B \left(\mathcal{H} - \frac{B}{\mu_0} \right) , \tag{7.15}$$

where k_h and k_ψ are characteristic wave numbers defined respectively by

$$k_h^2 = \frac{\langle |\Psi|^2 \rangle}{\lambda^2 |\Psi_\infty|^2} = \frac{2\kappa^2(1-b)}{[1 + \beta_A(2\kappa^2 - 1)]\lambda^2} \simeq \frac{1-b}{\lambda^2} , \tag{7.16}$$

$$k_\psi^2 = \frac{2(1-b)}{\xi^2} . \tag{7.17}$$

Equations (7.14) and (7.15) coincide with the results of Labusch in the limit of $k \to 0$. Such a nonlocal property appears only for C_{11} and C_{44} with respect to the magnetic energy, as can be seen from the above explanation. The nonlocal property becomes prominent when the wave number exceeds k_h, and hence, a high-κ superconductor is more likely to satisfy this condition.

Here we shall discuss the problem in the nonlocal theory. As is well known in electromagnetism, the Lorentz force is derived from a divergence of the Maxwell stress tensor, and components of the tensor do not depend on the wave number of the deformation, even how the magnetic structure is deformed. For example, it is assumed that the magnetic flux density has only the z component B and varies only along the x-axis. In this case the Lorentz force is the magnetic pressure in the direction of the x-axis. The Maxwell stress tensor is given by

$$\tau = \mu_0^{-1} \begin{bmatrix} -B^2/2 & 0 & 0 \\ 0 & -B^2/2 & 0 \\ 0 & 0 & B^2/2 \end{bmatrix} \tag{7.18}$$

and the Lorentz force is written as

$$\boldsymbol{F}_{\mathrm{L}} = \boldsymbol{i}_x \frac{\partial}{\partial x} \tau_{xx}, \tag{7.19}$$

where $\tau_{xx} = -B^2/2\mu_0$. It is assumed that the magnetic flux density is slightly varied by δB from a mean value $\langle B \rangle$ as $B = \langle B \rangle + \delta B$. Then, we have $(\partial/\partial x)(-B^2/2\mu_0) \simeq -(\langle B \rangle/\mu_0)\partial \delta B/\partial x$. If the displacement of flux lines along the x-axis is denoted by u_x, from the continuity equation (2.15) for flux lines, the relationship between δB and u_x is written as

$$\frac{\partial u_x}{\partial x} = -\frac{\delta B}{\langle B \rangle}. \tag{7.20}$$

Hence, Eq. (7.19) leads to

$$F_{\mathrm{L}} \simeq \frac{\langle B \rangle^2}{\mu_0} \cdot \frac{\partial^2 u_x}{\partial x^2}. \tag{7.21}$$

On the other hand, this is also written as $C_{11}\partial^2 u_x/\partial x^2$ in terms of the uniaxial compression modulus C_{11}. Thus, we have

$$C_{11} = \frac{\langle B \rangle^2}{\mu_0}. \tag{7.22}$$

This result agrees with that of Labusch, Eq. (7.9), where $\mathcal{H} = B/\mu_0$ is used because of neglect of the diamagnetic effect of the high-κ superconductor.

The elastic moduli for the "magnetic flux" outside the normal cores of flux lines are generally described in terms of the components of the Maxwell stress tensor in this manner, and hence, they should be of a local nature. As can be understood from the abovementioned speculation that the magnetic flux density and the order parameter are almost independent of each other, the nonlocal theory insists that the outer "magnetic flux" and the inner normal cores form their own lattices almost independently of each other, and these lattices are deformed in different manners. That is, it is believed that the lattice of "magnetic flux" is hardly deformed, while the lattice of normal cores is easily deformed.

There is a question whether the lattices of "magnetic flux" and normal cores can really deform independently of each other. The magnetic flux density and the order parameter are originally correlated to each other in a gauge-invariant form as mentioned in Chap. 1. The quantization of magnetic flux results from this correlation. Hence, the two quantities cannot behave completely independently. In fact, the maximum point of the magnetic flux density and the zero point of the order parameter coincide with each other as shown by Eq. (1.101). This fact suggests that displacements of the two lattices are the same and the speculation that one lattice is easily deformed while the other is not is incorrect. In other words, when the lattice of normal cores is deformed, the same deformation is necessarily induced in the lattice of "magnetic flux." Therefore, the elastic moduli of the lattice of normal cores should take the local values for the lattice of "magnetic flux." For a strict proof of this conclusion the continuity equation for flux lines is necessary. The details of this proof are described in Appendix A.5, where a questionable point regarding the use of Eq. (1.112) for the energy in the estimation of the elastic moduli is also argued. A derivation of the nonlocal elastic modulus is required in Exercise 7.2 for the case where deformation of the "magnetic flux" lattice and that of the normal core lattice do not satisfy the gauge-invariant relationship. For this reason we use hereafter the local results, Eqs. (7.9) and (7.10), for the elastic moduli C_{11} and C_{44} of the flux line lattice, respectively.

7.3 Summation Problem

The summation problem has been discussed by many researchers for a long period. Although various aspects have been clarified by the theory of Larkin and Ovchinnikov [7], there still remain many problems. For a discussion on these problems some concepts are needed. For this reason the development of summation theory is explained along with the history in this section. This will be helpful for understanding the profound summation problem.

7.3.1 Statistical Theory

The summation theory for deriving the macroscopic pinning force density from individual pinning potentials was proposed first by Yamafuji and Irie, [8] who treated current-voltage characteristics in a dynamic state. However, an important point in the summation problem was clarified first by Labusch [9]. Labusch treated a static state and used a statistical method and a kind of mean field approximation. The force balance equation Labusch used is

$$\widetilde{D}_2 \bm{u}(\bm{r}) - \sum_i \nabla U(\bm{r} + \bm{u}(\bm{r}) - \bm{R}_i) + \bm{f}_\mathrm{L} = 0 , \qquad (7.23)$$

where $\bm{u}$ is a displacement of flux line lattice and $\widetilde{D}_2$ is a matrix given by

$$\widetilde{D}_2 = \begin{bmatrix} C_{11}\partial_x^2 + C_{66}\partial_y^2 + C_{44}\partial_z^2 & (C_{11} - C_{66})\partial_x\partial_y \\ (C_{11} - C_{66})\partial_x\partial_y & C_{66}\partial_x^2 + C_{11}\partial_y^2 + C_{44}\partial_z^2 \end{bmatrix} . \qquad (7.24)$$

The first term in Eq. (7.23) shows the elastic force in the flux line lattice due to its deformation, and the symbols ∂_x etc. represent the differentials $\partial/\partial x$ etc. U is the pinning energy and the second term in Eq. (7.23) represents the pinning forces of the pins at $\bm{r} = \bm{R}_i$'s. The third term is the Lorentz force. This force is originally included in the first term, and the third term represents the mean value of this force. Hence, the mean value of the first term, is zero. For example, in the case of a superconducting slab carrying a transport current in a normal magnetic field, flux lines are bent along their length and the Lorentz force is expressed as the line tension, $C_{44}\partial_z^2 u$. The first term, the Lorentz force less its mean value, gives the elastic force against a deviation from the mean curvature of the flux lines.

Labusch gradually introduced the pinning interactions and gradually deformed the flux line lattice which was not deformed in the beginning. In this situation the compliance of an observed flux line for motion under the influence of surrounding pins is a key parameter. Labusch used a continuous medium approximation for the flux line lattice and obtained the compliance from Eq. (7.23). The second term was expanded in powers of u. From the condition of equilibrium at $u = 0$ the second derivative of U is important, and its mean value is written as $\widetilde{\alpha}_\mathrm{L} u$. Here

$$\widetilde{\alpha}_\mathrm{L} = \left\langle \sum \nabla\nabla U \right\rangle \tag{7.25}$$

is called the Labusch parameter. Equation (7.25) was approximately assumed as a diagonal matrix and hereafter α_L represents the diagonal element. α_L defined in Eq. (3.87) corresponds to this diagonal element. The obtained compliance is given by

$$G'(0) \simeq \frac{1}{4}\left(\frac{B}{\pi\phi_0}\right)^{1/2}(C_{44}C_{66})^{-1/2} \tag{7.26}$$

for $(\phi_0/B)\alpha_\mathrm{L} \ll 4\pi C_{66}$, i.e., for latticelike flux lines. The pinning force density was obtained from the maximum value of the second term in Eq. (7.23) after the complete introduction of pinning interactions. Here the approximation,

$$F_\mathrm{p} = N_\mathrm{p}\left|\int \rho(X)\nabla U(X)\mathrm{d}X\right|_\mathrm{max}, \tag{7.27}$$

was used, where $\rho(X)$ was the probability of finding a pin in a small region of volume $\mathrm{d}X$ around a position X. In Eq. (7.27) the summation of individual pinning interactions distributed in a space was approximately replaced by a statistical average multiplied by the number density of pins N_p. This is considered to be a good approximation for the case of dilute pins, where the cancellation of pinning forces through the elasticity of the flux line lattice was taken into account using the parameter α_L. Labusch calculated Eq. (7.27) for a pinning potential with a particular shape and obtained the result shown in Fig. 7.4. The result shows that there exists a threshold value, $f_\mathrm{pt} \sim G'^{-1}(0)d$, for the elementary pinning force (f_p) with $2d$ denoting the size of the pins, and the pinning force density (F_p) is zero for $f_\mathrm{p} \leq f_\mathrm{pt}$. F_p takes a finite value for $f_\mathrm{p} > f_\mathrm{pt}$ and approaches

$$F_\mathrm{p} \simeq \frac{N_\mathrm{p}f_\mathrm{p}^2 G'(0)L}{2a_\mathrm{f}^2} \tag{7.28}$$

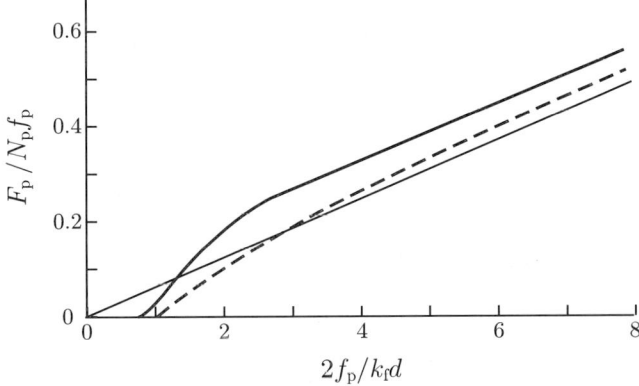

Fig. 7.4. Relationship between the pinning force density and the elementary pinning force predicted by Labusch [9] with $L = a_{\rm f}$ and $2d = a_{\rm f}/2$. The *broken line* is the result of the Lowell model, and the *thinner straight line* represents the characteristic of $F_{\rm p} = N_{\rm p} f_{\rm p}^2 / 2k_{\rm f} a_{\rm f}$. $k_{\rm f}$ corresponds to $G'(0)^{-1}$

for $f_{\rm p} \gg f_{\rm pt}$, where $a_{\rm f}$ is the flux line spacing and L is the size of a pin along the direction of the current. The resultant pinning force density is proportional to the second power of $f_{\rm p}$ in agreement with the dynamic theory of Yamafuji and Irie [8]. This proportionality is characteristic of the statistical summation as opposed to the linear summation given by Eq. (7.1). The above result of Labusch can more easily be understood using the simpler model of Lowell [10] which is described below.

Lowell treated for simplicity a one-dimensional model in which the pinning force is only varied along the direction of the Lorentz force. Here we focus on a pin at the origin and a flux line interacting with the pin. The position of the flux line is represented by x and its position when the effect of the pin is virtually switched off is represented by x_0. The flux line is displaced from x_0 to x by the pinning force, and the elastic restoring force proportional to the displacement, $(x - x_0)$, is balanced with the pinning force $f(x)$. That is, the force balance is expressed as

$$k_{\rm f}(x_0 - x) + f(x) = 0 , \qquad (7.29)$$

where $k_{\rm f}$ is the spring constant for the elastic restoring force. This equation is reduced from Eq. (2.35) in the static limit. Lowell also assumed a statistical set of such individual balances after Labusch. When a uniform flux line lattice on the superconductor is overlaid with randomly distributed pinning centers, it is believed that there is no correlation between the position of a pin and the position of the nearest uniform lattice point x_0. Hence, if we look statistically over elements of the set, the probability of finding a virtual flux line at x_0 will be uniform. On the other hand, the flux lines will in practice be drawn to pins due to the pinning interactions, resulting in a nonuniform distribution

$\rho(x)$ for the statistical set: $\rho(x)$ is expected to be high around the center of the pin. As mentioned in the last section the Lorentz force is described as the elastic force, and the statistical average of the first term of Eq. (7.29) gives the Lorentz force. That is, this term corresponds to the sum of the first and third terms of Eq. (7.23). At the same time the pinning force density is given by the maximum value of the statistical average of the second term:

$$F_\text{p} = -\frac{N_\text{p}}{a_\text{f}} \left| \int_0^{a_\text{f}} f(x(x_0)) \mathrm{d}x_0 \right|_\text{max} . \tag{7.30}$$

Here we shall investigate the statistical distribution of flux lines around the pins. After Lowell it is assumed that the pin is located at the origin in an observed element of the statistical set and the pinning force varies spatially as

$$\begin{aligned} f(x) &= \frac{2f_\text{p}}{d}(x+d) ; & -d \leq x < -\frac{d}{2} , \\ &= -\frac{2f_\text{p}}{d}x ; & -\frac{d}{2} \leq x < \frac{d}{2} , \\ &= \frac{2f_\text{p}}{d}(x-d) ; & \frac{d}{2} \leq x < d , \\ &= 0 ; & \text{otherwise} \end{aligned} \tag{7.31}$$

(see Fig. 7.5). The situation is considered in the beginning where the Lorentz force is not applied. This can be virtually achieved by reducing the pinning force to zero, resulting in a uniform flux line lattice, and then, recovering the pinning strength gradually. There are two types of resultant distribution, depending on whether the elementary pinning force f_p is smaller or larger than $k_\text{f}d/2$. Figure 7.6 is for the case of f_p smaller than $k_\text{f}d/2$ and the lower figure shows the distribution $\rho(x)$. The upper figure shows a graphical method to obtain x for given x_0. The broken line gives the elastic restoring force with the opposite sign, and the crossing point of this line with the pinning force $f(x)$ gives the position of the flux line x. In this case there is no vacant region for flux lines inside the pin. On the other hand, vacant regions exist at both edges of the pin in the case of f_p larger than $k_\text{f}d/2$ as shown in Fig. 7.7. Namely, the flux lines near the edges are pulled to the bottom region of the pinning potential due to the larger variation rate of the pinning force compared to the spring constant k_f.

The situation is next considered where the Lorentz force is applied by the transport current. Because of the Lorentz force the flux lines are displaced to the right-hand side, for example, in Figs. 7.6 and 7.7. In the case of $f_\text{p} < k_\text{f}d/2$ shown in Fig. 7.6, the flux lines come into the pin from the left-hand side and go out of the pin to the right-hand side, resulting in an unchanged distribution. When the pinning force is calculated from Eq. (7.30), we have

$$F_\text{p} = 0 . \tag{7.32}$$

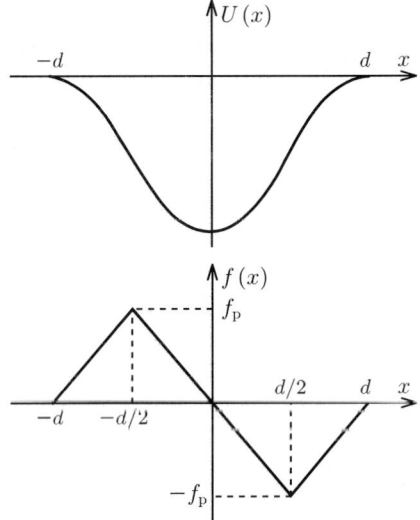

Fig. 7.5. Variation in the energy (*upper figure*) and the pinning force (*lower figure*) of a pin. The maximum value of the pinning force gives the elementary pinning force

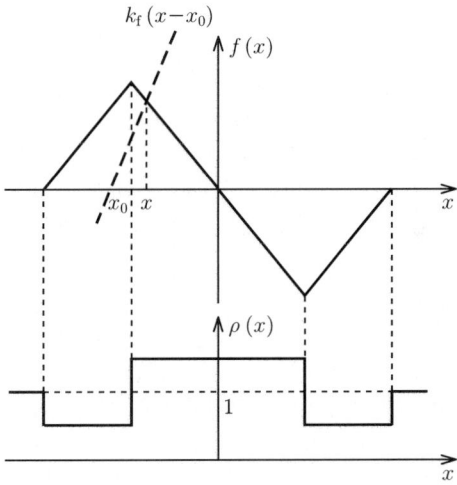

Fig. 7.6. Pinning force (*upper figure*) and distribution of flux lines around pin $\rho(x)$ (*lower figure*) for $f_\mathrm{p} < k_\mathrm{f} d/2$. ρ is normalized so as to take a value of 1 outside the pin

This can also be easily derived from the symmetric distribution of flux lines resulting in cancellation of positive and negative pinning forces. On the other hand, in the case of $f_\mathrm{p} > k_\mathrm{f} d/2$ shown in Fig. 7.7, the distribution changes with the displacement of flux lines. That is, the flux lines move to the right with movement of the unstable region to the outside of the pin, resulting an

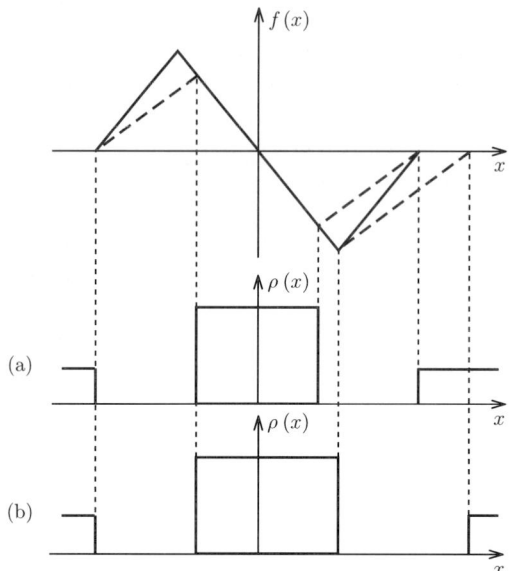

Fig. 7.7. Pinning force (*upper figure*) and distribution of flux lines around pin $\rho(x)$ (*lower figure*) for $f_p > k_f d/2$. ρ is normalized so as to take a value of 1 outside the pin. An unstable region where $\rho(x) = 0$ exists in contrast to Fig. 7.6. (a) Distribution when a transport current does not flow and (b) when the critical state is attained on application of the transport current. In the latter case the unstable region is asymmetric

asymmetrical distribution. After reaching the distribution shown in Fig. 7.7(b), the distribution no longer changes. In this situation the critical state is attained. After a simple calculation of Eq. (7.30) the pinning force density is obtained as

$$F_p = \frac{N_p}{2k_f a_f} \cdot \frac{f_p(f_p + 3f_{pt})(f_p - f_{pt})}{f_p + f_{pt}}, \qquad (7.33)$$

where

$$f_{pt} = \frac{k_f d}{2} \qquad (7.34)$$

is the minimum value of the elementary pinning force giving a nonzero pinning force density, i.e., the threshold value. Since $G'^{-1}(0)$ corresponds to k_f, the above result agrees with the condition of Labusch. In addition, Eq. (7.33) reduces to $F_p \simeq N_p f_p^2/2k_f a_f$ for $f_p \gg f_{pt}$ and this agrees with the result of Labusch, Eq. (7.28), assuming $L \sim a_f$ (see Fig. 7.4). The difference comes from the shape of individual pinning potentials. Thus, the model of Lowell can summarize the theory of Labusch and the original essence is reproduced correctly in the model.

The condition for the requirement that the pinning loss should be of the hysteresis type discussed in Sect. 2.3, i.e., $|\partial f(x)/\partial x| > k_f$, is nothing else

7.3 Summation Problem 281

than the condition for the existence of an unstable region as shown in Fig. 7.7 and agrees with the condition of nonzero pinning force density. Thus, the threshold condition for the elementary pinning force is closely related to the appearance of hysteresis loss. Namely, both the existence of a nonzero pinning force density and the hysteretic nature shown in the current-voltage characteristics in Fig. 1.13 are brought about by the abovementioned instability of flux lines. It should be noted that the randomness represented by x_0, which means the motion of x_0 with a constant velocity in a steady voltage state in the dynamic condition, is also needed for such current-voltage characteristics. On the contrary, for the approximation of a rigid body moving in a potential under a constant force due to a steady current, the unstable motion of flux lines does not occur. Calculate the corresponding current-voltage characteristics (Exercise 7.4).

In spite of a general agreement with the dynamic theory, the threshold value itself of the elementary pinning force is significantly different from experiments. In fact, it would be rather better to express that the threshold value of the elementary pinning force does not exist in practice. This was theoretically proved by Larkin and Ovchinnikov [7]. However, the instability of flux lines is necessary to derive the hysteresis loss. For discussing this point, somewhat old statistical theory and dynamic theory are introduced in this section.

In the case of dilute and isolated pins treated by Labusch and Lowell, the pinning force density increases with increasing elementary pinning force f_p in proportion to its second power. This comes from the growth of the unstable region of the right-hand side in Fig. 7.7(b). However, the growth of this unstable region is practically limited by the flux line spacing a_f. That is, even if one flux line is depinned, when the next flux line comes, the pinning interaction occurs again. As a result, the pinning force which the flux line lattice feels is not isolated as assumed in Fig. 7.5 but is periodic in its displacement with a period a_f. Thus, Campbell [11] proposed the periodic pinning force shown in Fig. 7.8. From a similar calculation the pinning force density in this case is obtained as

$$F_p = N_p \frac{f_p(f_p - f_{pt})}{f_p + f_{pt}} \; ; \qquad f_p > f_{pt} = \frac{k_f a_f}{4} \; ,$$
$$= 0 \; ; \qquad f_p < f_{pt} \; . \qquad (7.35)$$

This result shows that the pinning force density approaches the direct summation, $F_p \simeq N_p f_p$, in the limit of large f_p. Since it cannot happen that F_p exceeds the direct summation with increasing f_p in proportion to its second power, the Campbell model is appropriate. General comparisons between theories and experiments will be made in the next section.

We make comparisons here only on the threshold value of the elementary pinning force, since this is the most important point in discussing the validity of the summation theory and itself is deeply concerned with the development of the theory. Figure 7.9 shows the results on niobium specimens with various

282   7 Flux Pinning Characteristics

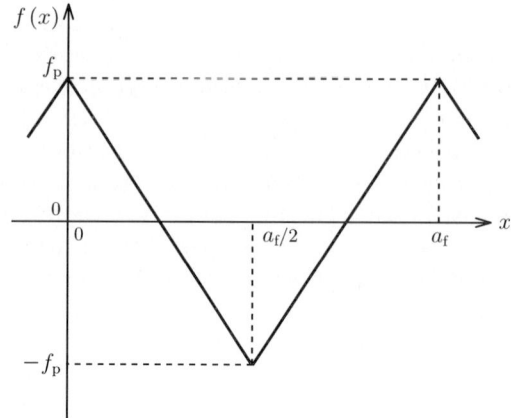

Fig. 7.8. Model of periodic pinning force by Campbell [11]

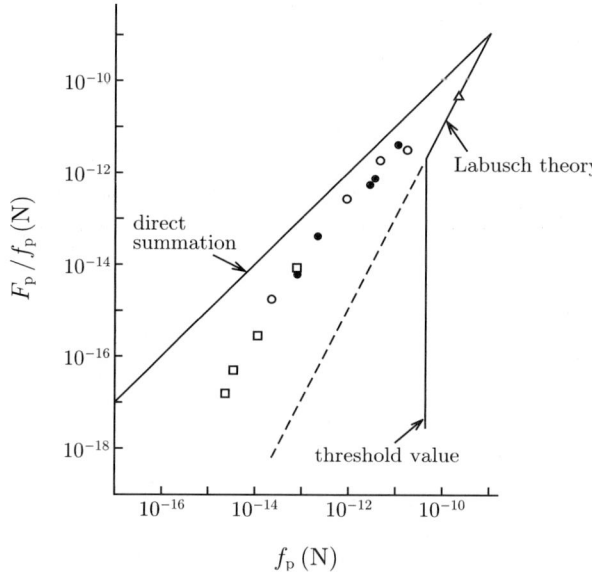

Fig. 7.9. Contribution to pinning force density from one pin F_p/N_p vs. elementary pinning force f_p for niobium at $b = 0.55$ (after Kramer [12]). The *solid line* gives the direct summation, $F_p/N_p = f_p$, and the *broken line* represents the result of Labusch theory extrapolated below the threshold value

pins (after Kramer [12]), where the abscissa is the elementary pinning force obtained theoretically for each defect and the ordinate is the pinning force density divided by the number density of the pins, i.e., the contribution per pin. The solid straight line with a slope of unity represents the direct summation, $F_p/N_p = f_p$, and the vertical solid line gives the threshold value of the Labusch theory. The broken line is an extrapolation of Labusch's result neglecting the threshold value, $F_{\rm p}/N_{\rm p} \propto f_{\rm p}^2$. It is found from the results in the figure that even defects weaker by more than four orders of magnitude than the threshold value act effectively as pins, and it seems that there is no practical threshold value. In order to overcome this disagreement between the theory and experiments, the possibility of reduction in the threshold value was discussed from the viewpoint of a decrease in the elastic moduli of the flux line lattice caused by the nonlocal property [13] or by the existence of defects [14]. Another argument was made on the possibility of easier fulfillment of the threshold condition by a collective pinning of many weak defects [15]. However, each of them could give only an insufficient correction, and the contradiction between the statistical theory and experiments was not essentially solved.

Under this circumstance the following dynamic theory, developed first by Yamafuji and Irie, [8] has attracted attention. It can be seen from Eq. (2.39) that the pinning force density is proportional to the fluctuation of the velocity of flux lines. (Note that the pinning loss power density $P_{\rm p}$ is proportional to the pinning force density $F_{\rm p}$.) Since even a very weak pin can bring about a finite fluctuation, a nonzero contribution to the pinning force can be expected, suggesting that the threshold value does not exist. The relationship between this dynamic theory and the statistical theory will be discussed in the next subsection.

7.3.2 Dynamic Theory

Yamafuji and Irie [8] treated the dynamic state of flux lines and clarified the mechanism of pinning loss using the method shown in Sect. 2.3. Since the pinning loss power density $P_{\rm p}$ is equal to $F_{\rm p}v$ with v denoting the velocity of flux lines averaged with time, the pinning force density is obtained as

$$F_{\rm p} = \frac{B^2}{\rho_{\rm f} v}[\langle \dot{x}^2 \rangle_t - v^2] , \qquad (7.36)$$

where $\rho_{\rm f}$ is the flow resistivity and $\langle \ \rangle_t$ represents an average with respect to time. Yamafuji and Irie calculated Eq. (7.36) for a triangular pinning potential and derived $F_{\rm p} = 3N_{\rm p}f_{\rm p}^2/k_{\rm f}d_{\rm p}$, where $d_{\rm p}$ is the mean interval of pins. This result is qualitatively similar to the theoretical results of Labusch and Lowell. However, the problem of the threshold value did not occur, since a triangular potential, which automatically satisfied the condition $|\partial f_{\rm p}(x)/\partial x| > k_{\rm f}$ due to infinitely large variation rate of the pinning force, was assumed and the voltage state was treated. A similar result was derived for a more general calculation by Schmid and Hauger [16].

It is believed from Eq. (7.36) that even a very weak pin which does not automatically satisfy the threshold condition can suitably affect the motion of flux lines and give rise to some fluctuation of the velocity. If this is correct, it means that the threshold value of the elementary pinning force may not exist. This is an interesting point of comparison with the statistical static theory. There, Matsushita et al. [17] calculated the motion of flux lines for the pinning model of Lowell and investigated the current-voltage characteristics. The solved equation of motion was Eq. (2.35). The form of Eq. (7.31) was assumed for the pinning force $f(x)$, and the interval between pins was assumed to be sufficiently large. The case was treated where $2d < a_\mathrm{f}$ so that one pin could not interact with several flux lines simultaneously as was done in other calculations. It is assumed that an observed flux line reaches the edge of a pin, $x = -d$, at a time, $t = 0$. Since d_p is long enough, a distortion of the flux line lattice, $(x - x_0)$, is considered to relax to a negligible small value before the flux line reaches the next pin. Hence, we can simply adopt

$$x(t) = vt - d \tag{7.37}$$

for $t < 0$. From the continuity of flux lines v should be equal to $\dot{x}_0$ as shown in Eq. (2.37). Thus, $x(t)$ can directly be solved by substituting Eqs. (7.31) and (7.37) into Eq. (2.35). The pinning force density is derived from the time average:

$$F_\mathrm{p} = -\frac{N'_\mathrm{p}}{T_\mathrm{p}} \int_0^{T_\mathrm{p}} f(x(t))\mathrm{d}t \tag{7.38}$$

using the obtained solution of $x(t)$. In the above T_p is the period of the flux line motion and is thus the time for the observed flux line to pass through the mean interval of pins:

$$T_\mathrm{p} = \frac{d_\mathrm{p}}{v}. \tag{7.39}$$

N'_p is a number of pinning events which occur in a unit volume within the period T_p and is equal to the product of the frequency for one pin to meet flux lines during this period $d_\mathrm{p}/a_\mathrm{f}$ and the number density N_p:

$$N'_\mathrm{p} = \frac{N_\mathrm{p} d_\mathrm{p}}{a_\mathrm{f}}. \tag{7.40}$$

Before starting the calculation, we shall discuss a fundamental point. The observed flux line interacts with the pin only while it runs through a pin of size $2d$, and $2d$ is assumed to be smaller than a_f as abovementioned. (This is not a necessary condition but is used only for simplicity. Other cases can also be treated in a similar manner.) Hence, it is enough to integrate from 0 to $T' = a_\mathrm{f}/v$ in Eq. (7.38), the only region in which a nonzero contribution is obtained. Thus, Eq. (7.38) leads to

$$F_\mathrm{p} = -\frac{N_\mathrm{p}}{T'} \int_0^{T'} f(x(t))\mathrm{d}t. \tag{7.41}$$

If we transform to $\mathrm{d}t = v^{-1}\mathrm{d}x_0$ using Eq. (2.37), it is proved that the time average of Eq. (7.41) is identical with the statistical average of Eq. (7.30). Namely, it can be shown that the result of the dynamic theory reduces to that of the statistical theory in the static limit without concrete calculation.

The solution of $x(t)$ is:

$$x(t) = -\frac{k_\mathrm{f} v}{\eta^* \gamma} t - q_1 + (q_1 - d)\exp(\gamma t) ; \qquad 0 \leq t < t_1 ,$$

$$= \frac{k_\mathrm{f} v}{\eta^* \gamma'} (t - t_1) + q_2 - \left(\frac{d}{2} + q_2\right)\exp[-\gamma'(t - t_1)] ; \quad t_1 \leq t < t_2 ,$$

$$= -\frac{k_\mathrm{f} v}{\eta^* \gamma}(t - t_2) - q_3 + \left(q_3 + \frac{d}{2}\right)\exp[\gamma(t - t_2)] ; \quad t_2 \leq t < t_3 ,$$

$$= vt - d + (2d - vt_3)\exp\left[-\frac{k_\mathrm{f}}{\eta^*}(t - t_3)\right] ; \qquad t_3 \leq t < \frac{d_\mathrm{p}}{v} \quad (7.42)$$

for the case $f_\mathrm{p} \neq f_\mathrm{pt}$. In the above t_1, t_2 and t_3 are the times at which the flux line reaches $x = -d/2$, $x = d/2$ and $x = d$, respectively, and can be obtained from the above solution. The constants, γ, γ', q_1, q_2 and q_3 are respectively given by

$$\gamma = \frac{k_\mathrm{f}}{\eta^*}\left(\frac{f_\mathrm{p}}{f_\mathrm{pt}} - 1\right) , \qquad (7.43)$$

$$\gamma' = \frac{k_\mathrm{f}}{\eta^*}\left(\frac{f_\mathrm{p}}{f_\mathrm{pt}} + 1\right) , \qquad (7.44)$$

$$q_1 = \frac{\eta^* v d f_\mathrm{p}}{2(f_\mathrm{p} - f_\mathrm{pt})^2} + d , \qquad (7.45)$$

$$q_2 = \frac{\eta^* v d f_\mathrm{p}}{2(f_\mathrm{p} + f_\mathrm{pt})^2} - \frac{f_\mathrm{pt}}{f_\mathrm{p} + f_\mathrm{pt}}(d - vt_1) , \qquad (7.46)$$

$$q_3 = \frac{\eta^* v d f_\mathrm{p}}{2(f_\mathrm{p} - f_\mathrm{pt})^2} - \frac{1}{f_\mathrm{p} - f_\mathrm{pt}}[d(f_\mathrm{p} + f_\mathrm{pt}) - f_\mathrm{pt} v t_2] . \qquad (7.47)$$

γ in Eq. (7.43) is equal to $1/\tau$ in Eq. (2.41).

After a simple but long calculation, Eq. (7.38) leads to

$$F_\mathrm{p} = \frac{N_\mathrm{p} f_\mathrm{p} v}{a_\mathrm{f}(f_\mathrm{p}^2 - f_\mathrm{pt}^2)}\left\{2 f_\mathrm{pt}[f_\mathrm{p}(t_1 + t_2 - 2t_3) - f_\mathrm{pt}(t_1 - 3t_2 + 2t_3)]\right.$$
$$+ \frac{v f_\mathrm{pt}}{d}[f_\mathrm{p} t_3^2 + f_\mathrm{pt}(2t_1^2 - 2t_2^2 + t_3^2)]$$
$$\left. + \eta^* v[f_\mathrm{p} t_3 + f_\mathrm{pt}(2t_1 - 2t_2 + t_3)] - 2\eta^* f_\mathrm{p} d\right\} . \qquad (7.48)$$

Hereafter t_1, t_2 and t_3 are numerically calculated and F_p is obtained. However, in the case $f_\mathrm{p} > f_\mathrm{pt}$ and sufficiently small mean velocity v, we have

$$t_1 \ll t_3 \sim t_2 \simeq \frac{d(f_\mathrm{p} + 3 f_\mathrm{pt})}{2 f_\mathrm{pt} v} . \qquad (7.49)$$

Substituting this into Eq. (7.48), the same result as Eq. (7.33) is derived. When v is small, an expansion in powers of v can be introduced. If F_p is calculated up to the first order in v, we have [17]

$$F_\mathrm{p} = F_\mathrm{ps} + \frac{N_\mathrm{p} f_\mathrm{p} \eta^* v d}{2a_\mathrm{f}(f_\mathrm{p}^2 - f_\mathrm{pt}^2) f_\mathrm{pt}} \left\{ (f_\mathrm{p} - 3f_\mathrm{pt})(f_\mathrm{p} + f_\mathrm{pt}) \right.$$
$$\left. + f_\mathrm{pt}(f_\mathrm{p} + 3f_\mathrm{pt}) \log\left[\frac{(f_\mathrm{p} - f_\mathrm{pt})^2}{f_\mathrm{p} \eta^* v} \right] \right\} , \tag{7.50}$$

where F_ps is the value given by Eq. (7.33). In the case $f_\mathrm{p} < f_\mathrm{pt}$, F_p is obtained as

$$F_\mathrm{p} = \frac{2 N_\mathrm{p} f_\mathrm{p}^2 \eta^* v d}{a_\mathrm{f}(f_\mathrm{pt}^2 - f_\mathrm{p}^2)} . \tag{7.51}$$

In the case $f_\mathrm{p} = f_\mathrm{pt}$, F_p is proportional to $v^{1/2}$ [17]. Thus, these results agree with Eq. (7.32) in the limit of zero v.

The current-voltage characteristics can be derived after Eq. (2.31) from the J_c-value equal to F_p divided by B and the E-value equal to v multiplied by B. Figure 7.10 shows these results and (a) and (b) correspond to the cases $f_\mathrm{p} > f_\mathrm{pt}$ and $f_\mathrm{p} \leq f_\mathrm{pt}$, respectively.

Even in case (b) where J_c goes to zero in the limit of zero v, if a tangential line is drawn at some voltage state on the current-voltage characteristic and this line is extrapolated to $v = 0(E = 0)$, some finite intercept remains. Namely, it seems as if a nonzero J_c value exists. However, as is seen from Eq. (7.36), $\langle \dot{x}^2 \rangle_t$ is of the order in v^2 and J_c goes to zero in the limit $v \to 0$ in the case $f_\mathrm{p} < f_\mathrm{pt}$ with no instability. $\langle \dot{x}^2 \rangle_t$ is of the order in v, and a finite J_c remains only when the instability exists. This condition is identical with that for the hysteresis loss mentioned in Sect. 2.3. In addition, when flux lines are moved from the previous critical state to the opposite direction, the pinning force density varies linearly with the displacement as shown in Fig. 3.33. Such reversible flux motion also suggests the existence of unstable regions [11]. A discussion on this point is required in Exercise 7.5.

From the above argument it is found that the dynamic theory does not resolve the large contradiction between the statistical theory and the experiments. However, the coincidence between the dynamic theory and the statistical theory itself has some meaning and is deeply concerned with the hysteresis loss, a property inherent to the flux pinning.

7.3.3 Larkin-Ovchinnikov Theory

The large contradiction with experiments on the threshold value of the elementary pinning force, which could not be resolved by the statistical theory of Labusch and the dynamic theories of Yamafuji and Irie and Matsushita et al., was first resolved by Larkin and Ovchinnikov [7].

Labusch intended to calculate the pinning force density using the statistical method and found from the statistical average on an infinite set that the pinning force density is zero for pins with an elementary pinning force below the threshold value. On the other hand, Larkin and Ovchinnikov showed that long-range order is not realized in the flux line lattice because of finite pinning correlation lengths. This means that the statistical average on an infinite

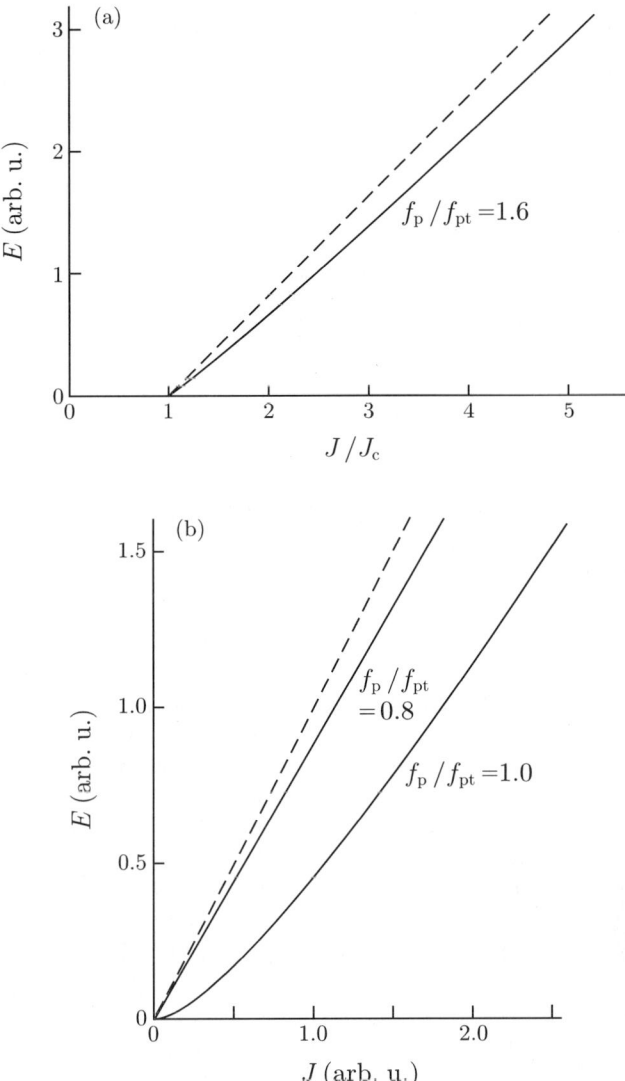

Fig. 7.10. Current-voltage characteristics calculated using the dynamic theory [17] for (a) $f_\mathrm{p} > f_\mathrm{pt}$ and (b) $f_\mathrm{p} \leq f_\mathrm{pt}$. The *broken lines* are virtual characteristics when the flow resistivity is not influenced by pinning

set under the assumption of long-range order does not give a correct result. That is, within the range of the correlation length, even if the pinning forces are randomly directed, the summation of the individual pinning forces does not cancel out completely but leads to a finite value on the order of the fluctuations. Hence, the force contributed by one pin remains finite. For example, if n pins are included in a correlated volume, the summation of these forces is of the order of $n^{1/2} f_p$, and hence, the contribution from one pin is about $n^{-1/2} f_p$. The calculation by Labusch corresponds to the case $n \to \infty$. Therefore, the pinning force density can be approximately calculated, if the pinning correlation length is obtained. They started from a force balance equation similar to Eq. (7.23) for a superconductor on a three-dimensional scale. However, the elastic force related to the uniaxial compression was neglected, since the uniaxial compression modulus C_{11} was very large and the corresponding strain was small. The nonlocal property was assumed for the bending modulus C_{44} and only the first term of Eq. (7.15) was used, neglecting the effect of diamagnetism. The problem with the assumption of nonlocal C_{44} is discussed in Sect. 7.2 and Appendix A.5. In addition, the assumption that the elastic force associated with C_{11} can be disregarded is also problematic from the viewpoint of quantitative calculation. However, we shall follow their paper, since these are not essential for discussing the problem of the threshold value of the elementary pinning force.

A superconductor of sufficiently large size is assumed. After Fourier transformation the force balance equation leads to

$$C_{66} k_\perp^2 u_k + C_{44}(k) k_z^2 u_k = (2\pi)^3 \delta(k) f_L + \sum_i f_i \exp(-i k \cdot r_i), \quad (7.52)$$

where the second term on the right-hand side is the contribution from the pins, $k_\perp$ is the wave number vector of the deformation in the plane normal to the flux line and k_z is the wave number of the deformation along the length of the flux line. After the inverse transformation the displacement u is derived. Thus, we have [18]

$$\langle |u(r) - u(0)|^2 \rangle = \frac{W(0)}{8\pi C_{66}^{3/2} C_{44}^{1/2}} \left[\left(r_\perp^2 + \frac{C_{66}}{C_{44}} z^2 \right)^{1/2} \right.$$
$$\left. + \frac{1}{4 k_h} \log \left(1 + k_0^2 r_\perp^2 + \frac{C_{66} k_0^4}{C_{44} k_h^2} z^2 \right) \right], \quad (7.53)$$

where $W(0)$ is defined by

$$W(0) = N_p \langle f^2(r) \rangle \simeq \frac{1}{2} N_p f_p^2 \quad (7.54)$$

and $r_\perp^2 = x^2 + y^2$. k_0 is an equivalent radius when the first Brillouin zone in the two-dimensional space of wave number is approximated by a circle and is given by $k_0 = (2b)^{1/2}/\xi$ with $b = B/\mu_0 H_{c2}$. C_{44} in the above means the local value $C_{44}(0)$. At low fields where the nonlocal effect of the flux line lattice is not remarkable, the second term of Eq. (7.53) can be omitted. If r_p represents the distance over which the pinning force extends (hence, r_p is a distance of the order of ξ), the transverse pinning correlation length R_c can be calculated from the condition:

$$\langle |\boldsymbol{u}(r_\perp = R_{\rm c}, z=0) - \boldsymbol{u}(r_\perp = z = 0)|^2 \rangle = r_{\rm p}^2 . \tag{7.55}$$

This condition means that, if a flux line at the origin is displaced by $r_{\rm p}$ by introducing a pin, its effect does not appear at the position separated by $R_{\rm c}$. Thus,

$$R_{\rm c} = \frac{8\pi (C_{44} C_{66}^3)^{1/2} r_{\rm p}^2}{W(0)} \tag{7.56}$$

is obtained. The pinning correlation length along the flux line $L_{\rm c}$ is obtained in a similar manner as

$$L_{\rm c} = \left(\frac{C_{44}}{C_{66}}\right)^{1/2} R_{\rm c} . \tag{7.57}$$

Within a region of the flux line lattice of sizes $R_{\rm c}$ and $L_{\rm c}$ in the transverse and longitudinal directions, respectively, the flux line lattice is proposed to have translational order, and the pinning force within this region of the volume $V_{\rm c} = R_{\rm c}^2 L_{\rm c}$ is estimated to be about $[V_{\rm c} W(0)]^{1/2}$, which is of the order of the fluctuations. Since each region of volume $V_{\rm c}$ is elastically independent of the others, the pinning force density is estimated as

$$F_{\rm p} = \frac{1}{V_{\rm c}} [V_{\rm c} W(0)]^{1/2} . \tag{7.58}$$

Substitution of Eqs. (7.56) and (7.57) leads to

$$F_{\rm p} = \frac{W^2(0)}{(8\pi)^{3/2} C_{44} C_{66}^2 r_{\rm p}^3} . \tag{7.59}$$

Thus, the threshold value of the elementary pinning force does not exist.

At higher fields important quantities vary as $C_{66} \propto (1-b)^2$ and $W(0) \propto f_{\rm p}^2 \propto (1-b)^2$ with a magnetic field, and hence, $R_{\rm c}$ decreases in proportion to $(1-b)$. On the other hand, the characteristic wave number $k_{\rm h}$ given by Eq. (7.16) decreases. Hence, if the condition $a_{\rm f} < R_{\rm c} < k_{\rm h}^{-1}$ is satisfied, the nonlocal property of the flux line lattice becomes pronounced. Then, the pinning force density varies as

$$F_{\rm p} \sim B \exp\left[-\frac{8\pi C_{44}^{1/2} C_{66}^{3/2} k_{\rm h} r_{\rm p}^2}{W(0)}\right] \tag{7.60}$$

and increases with increasing magnetic field. When the magnetic field is further increased until $R_{\rm c}$ becomes as small as $a_{\rm f}$, the flux line lattice becomes almost amorphous, and each flux line is independent of the others. From the fact that $R_{\rm c}$ is inversely proportional to $W(0)$ from Eq. (7.56), this situation can be realized even at low fields, if the pins are sufficiently strong. On the other hand, the problem of determining $L_{\rm c}$ is a one-dimensional problem, and the integral to determine $L_{\rm c}$ diverges. However, this characteristic length can be estimated from dimensional analysis as [7]

$$L_{\mathrm{c}} = \left[\frac{\pi C_{44}^2 k_{\mathrm{h}}^4 a_{\mathrm{f}}^6 r_{\mathrm{p}}^2}{W(0)}\right]^{1/3}. \tag{7.61}$$

In this case the volume of the correlated region is $V_{\mathrm{c}} = a_{\mathrm{f}}^2 L_{\mathrm{c}}$, and hence, the pinning force density is obtained from Eq. (7.58) as

$$F_{\mathrm{p}} = \left[\frac{W^2(0)}{\pi^{1/2} C_{44} k_{\mathrm{h}}^2 a_{\mathrm{f}}^6 r_{\mathrm{p}}}\right]^{1/3}. \tag{7.62}$$

Here we shall summarize the pinning force density obtained from the Larkin-Ovchinnikov theory. At low fields where the pinning force density is given by Eq. (7.59) its dependence on the pinning parameters is

$$F_{\mathrm{p}} \propto N_{\mathrm{p}}^2 f_{\mathrm{p}}^4. \tag{7.63}$$

At high fields where Eq. (7.62) holds or in the case of strong pinning, the dependence is

$$F_{\mathrm{p}} \propto N_{\mathrm{p}}^{2/3} f_{\mathrm{p}}^{4/3}. \tag{7.64}$$

These dependences are different from $F_{\mathrm{p}} \propto N_{\mathrm{p}} f_{\mathrm{p}}^2$ predicted by the statistical theory of Labusch and the dynamic theory. As for the magnetic field dependence, from $W(0) \propto (1-b)^2$, $C_{44} \propto b^2$, $C_{66} \propto b(1-b)^2$, $a_{\mathrm{f}} \propto b^{-1/2}$ and $k_{\mathrm{h}}^2 \propto (1-b)$ the pinning force density varies as

$$F_{\mathrm{p}} \propto b^{-4} \tag{7.65}$$

at low fields and as

$$F_{\mathrm{p}} \propto b^{1/3}(1-b) \tag{7.66}$$

at high fields. Figure 7.11 shows the above magnetic field dependence of the pinning force density. Except for the transient region shown by the broken line the pinning force density decreases monotonically with increasing magnetic field.

Here the case is treated where a magnetic field is applied perpendicularly to a superconducting film with thickness d thinner than the longitudinal correlation length L_{c}. The Larkin-Ovchinnikov theory is most frequently compared with experiments for such a case. In this two-dimensional case the transverse correlation length R_{c} is estimated as follows. The pinning energy density of the flux line lattice interacting with pins amounts to $r_{\mathrm{p}}[W(0)/V_{\mathrm{c}}]^{1/2} \simeq \xi[W(0)/V_{\mathrm{c}}]^{1/2}$ with $V_{\mathrm{c}} = R_{\mathrm{c}}^2 d$. On the other hand, the increase in the elastic energy density is estimated as $C_{66}(\xi/R_{\mathrm{c}})^2$, since the distortion along the length of flux line can be neglected. Hence, R_{c} is determined so as to minimize the total energy density, $\delta F = C_{66}(\xi/R_{\mathrm{c}})^2 - \xi[W(0)/d]^{1/2}/R_{\mathrm{c}}$. Thus, we have

$$R_{\mathrm{c}} = \frac{2\xi d^{1/2} C_{66}}{W^{1/2}(0)}. \tag{7.67}$$

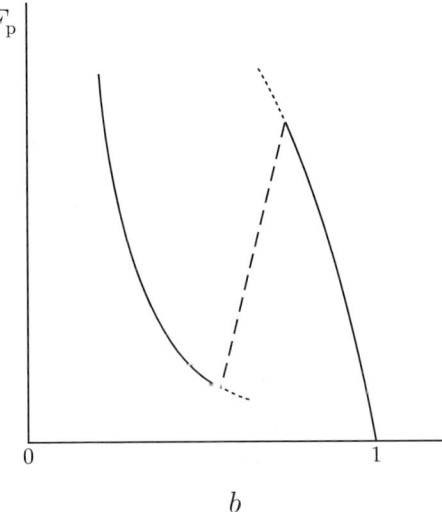

Fig. 7.11. Magnetic field dependence of the pinning force density predicted by Larkin and Ovchinnikov [7] in the case of three-dimensional superconductor with weak pinning force

Then, the pinning force density is obtained as

$$F_\mathrm{p} = \frac{W(0)}{2\xi d C_{66}}. \tag{7.68}$$

R_c decreases with magnetic field in proportion to $(1-b)$, and the pinning force density reduces to

$$F_\mathrm{p} = \frac{W^{1/2}(0)}{a_\mathrm{f} d^{1/2}} \tag{7.69}$$

at high fields where R_c becomes shorter than a_f. In the case a of low number density of pins such as $N_\mathrm{p} a_\mathrm{f}^2 d < 1$, the pinning force density is given by $F_\mathrm{p} = N_\mathrm{p} f_\mathrm{p}$.

For a superconducting thin film, L_c varies with the magnetic field and becomes longer or shorter than the film thickness. Hence, a transition between two- and three-dimensional pinning is expected to occur. Experiments associated with this kind of transition were carried out in detail by Wördenweber and Kes [19, 20]. Their experiments will be discussed in the next section, and their modification of the Larkin-Ovchinnikov theory is introduced here. Since the original Larkin-Ovchinnikov theory did not agree quantitatively with their experimental results, they proposed the following modification. First, two-dimensional pinning was assumed where the thickness of the superconducting film d was shorter than the longitudinal correlation length L_c, the value of which was estimated separately. Then, from a similar equation to Eq. (7.55) with replacement of the right-hand side by $(a_\mathrm{f}/2)^2$, the transverse correlation length was calculated as [21]

$$R_c = a_f C_{66} \left[\frac{2\pi d}{W(0)\log(w/R_c)}\right]^{1/2}, \quad (7.70)$$

where w is the width of the superconducting thin film. This result is somewhat different from Eq. (7.67) of the Larkin-Ovchinnikov theory. The volume of the region in which the short range order is maintained is $V_c = R_c^2 d$. They assumed the most nonlocal case with the largest wave number $k \sim \xi^{-1}$ for the longitudinal strain of a flux line. It was also assumed that a transition from two-dimensional pinning to three-dimensional occurs when the longitudinal correlation length L_c obtained by substituting Eq. (7.70) into Eq. (7.57) is reduced to a half of the film thickness d. This treatment seems strange, since such a transition should be determined from the more general condition that the correlation length in the three-dimensional pinning exceeds the film thickness. In the model of Wördenweber and Kes, L_c, which originally has meaning only in the three-dimensional case, is determined by this special condition in the two-dimensional case. As will be discussed in the next section, the quantitative agreement with experiments becomes better with this assumption. However, its theoretical validity is questionable.

Although the theory of Larkin and Ovchinnikov resolved the failure of the statistical theory of Labusch, their theory is in disagreement with experiments for very weak pins. That is, the obtained theoretical pinning force density is very much smaller than in experiments. The above model of Wördenweber and Kes was proposed to give a correction to the original theory. For improvement in the disagreement with experiments other approaches were also tried. These are the theory of Kerchner [22] in which the volume of the region with short-range order V_c is determined so as to maximize the pinning force density F_p, and the model of Mullock and Evetts [23] in which V_c is assumed to be decreased by defects in the flux line lattice.

7.3.4 Coherent Potential Approximation Theory

In addition to the abovementioned quantitative disagreement with experiments on the pinning force density for weak pins, the theory of Larkin and Ovchinnikov [7] also has the qualitative problem that the predicted dependence on the pinning parameters, f_p and N_p, is different from experiments as will be shown in the next section. This is probably because the pinning phenomenon is only roughly described in their theory. For example, the coherence length ξ is used for r_p in Eq. (7.55). However, judging from the fact that the value of the correlation length largely influences the pinning force density, this approximation seems to be quite rough. In addition, although a short-range order is maintained within the volume V_c, a deformation of the flux line lattice is expected to exist inside it, which will bring about a stronger pinning than $W(0)$. The existence of defects in the flux line lattice will also

7.3 Summation Problem

contribute to a quantitative improvement of the theory through a reduction in the correlation length.

On the other hand, the statistical theories such as that of Labusch [9] were proposed originally to give an exact estimation of the pinning force density. In order to resolve the quantitative problems in the Larkin-Ovchinnikov theory such a method seems to be effective. Since the statistical theory explains successfully the origins of the hysteresis loss due to pinning and the reversible motion of flux lines (see Exercise 7.5), [11] this method is expected to be useful. However, the abovementioned serious problem with the threshold value of the elementary pinning force remains in the statistical theory. Why does such a problem arise? The statistical average on the infinite set itself shown in Eq. (7.27) is originally a method for an exact calculation, and it seems to be rather strange that this method gives rise to a problem. For example, Eq. (7.33) is believed to give a correct result, if a correct value of $G'^{-1}(0)$ or k_f is given. If such a correct method is found, the abovementioned deformation of the flux line lattice inside the volume V_c can probably be treated using this method. Hence, it is considered that the evaluation of the spring constant of the flux line lattice, $G'^{-1}(0)$ or k_f, contains some problem.

Here Eq. (7.26) derived by Labusch is investigated. This value is for the case where the flux line lattice is fixed at infinity. However, the boundary condition at infinity is not satisfied in practical cases. For example, Eq. (7.26) gives a value in the limit of weak pinning, $\alpha_L \to 0$. In this limit the usual pins are not effective because of smaller elementary pinning forces than the threshold value. If this is true, since the pins are not effective, the flux flow state should be brought about even by an infinitesimal force, resulting in a divergence of the compliance. Thus, the boundary condition at infinity is not satisfied. Hence, the strong restriction at infinity is expected to be weakened drastically, and the compliance will be increased considerably. This result is understandable also from the viewpoint of the Larkin-Ovchinnikov theory. That is, since there is no long-range order in the flux line lattice as predicted in the theory, such a boundary condition at infinity is meaningless.

It is assumed that a part of the flux line lattice is displaced. The region surrounding this part is deformed and an elastic reaction occurs. However, this reaction is nothing else than the transferred reaction from the surrounding pins, since the elastic force is an internal force. That is, the force added to the flux line lattice to deform it is finally balanced with the reaction from the surrounding pins. Hence, the final result is not influenced by the elastic force so far as the deformation is elastic. It suggests that the compliance of the flux line lattice should be as large as α_L^{-1} per unit volume. Under practical conditions the compliance is considered to be slightly larger than this simple estimation because of the deformation of the flux line lattice.

Strong pins are treated here for an example. It is assumed that the number density N_p is sufficiently low. We assume a segment of flux line lattice of a proper size as in the treatment by Lowell, and the flux line lattice in the superconductor is approximated by the statistical set of such elements. Details of the theory are given in [24]. The easiest choice for the size of each segment seems to be such that each segment contains one pinning center. If the size is smaller than this, the segments are not equivalent, and if the size is larger than this, an additional summation of pinning forces is needed in each segment. The sizes of this segment in the directions of the Lorentz force, and the transverse and longitudinal directions normal to this force are represented by L_x, L_y and L_z, respectively. Hence, $L_x L_y L_z = 1/N_p$ is obtained. The phenomenon is approximated by a one-dimensional model for simplicity. One segment of the set is observed and the pinning force density is approximated by the statistical average of the force on the set. The pin is assumed to be located in the center of this segment, and the position of a flux line interacting with this pin is denoted by x. Its position when only the pinning interaction of this pin is virtually switched off and that when the pinning interactions of all surrounding pins are also virtually switched off are denoted by x_0 and Δ, respectively. Then, the force balance equation inside this segment is given in the same form as Eq. (7.29), where $k_f = G'^{-1}(0)$. Strictly speaking, since the range of the integral in the wave number space of the inverse Fourier transform is not the first Brillouin zone, i.e., from 0 to a_f^{-1}, but from about $N_p^{1/3}$ to a_f^{-1}, k_f is larger than $G'^{-1}(0)$. However, here the mean pin spacing, $d_p = N_p^{-1/3}$, is assumed to be sufficiently larger than a_f, and hence, this approximation will be allowed. When the interactions of all surrounding pins are switched on, the observed flux line is displaced from Δ to x_0. Hence, the observed segment of the flux line lattice suffers an elastic restoring force proportional to $\Delta - x_0$ from the surrounding regions. The spring constant of this restoring force is represented by K. If we confine ourselves to within this segment, the pinning force is balanced by the elastic restoring force due to the strain inside the segment, $k_f(x_0 - x)$. On the other hand, if we look at the interaction with the outside, this pinning force is balanced by the elastic restoring force, $K(\Delta - x_0)$. The former local strain corresponding to $x_0 - x$ slightly disturbs the short-range order inside the segment, and the latter strain corresponding to $\Delta - x_0$ destroys the long-range order outside the segment. In other words Labusch assumed that x_0 corresponds to a lattice point of the virtual flux line lattice with long-range order. However, Δ corresponds to this lattice point.

It should be noted here that Δ is not correlated with the distribution of pins. Therefore, the statistical average should be taken with respect to Δ. Thus, Eq. (7.29) reduces

$$k'_f(\Delta - x) + f(x) = 0 , \tag{7.71}$$

where k'_f is an effective spring constant given by [24]

$$k'^{-1}_f = k_f^{-1} + K^{-1} . \tag{7.72}$$

Thus, if the pinning force in Fig. 7.8 is assumed, the pinning force density is formally given by Eq. (7.35) with $f_{pt} = k'_f a_f / 4$.

Here K is estimated. Since K is concerned with the reaction from surrounding pins, it is expected to be proportional to the Labusch parameter α_L. This α_L is a variable to be determined consistently in the coherent potential approximation. Each region of the flux line lattice composing the statistical set is approximated to be in a mean "field" described by a common parameter. In this case this "field" means the interaction from surrounding pins through the elasticity of the flux line lattice. Referring to the Larkin-Ovchinnikov theory, we assume $L_x : L_y : L_z = C_{66}^{1/2} : C_{66}^{1/2} : C_{44}^{1/2}$. Thus, we have

$$L_x = L_y = \left(\frac{C_{66}}{C_{44}}\right)^{1/6} d_p, \quad L_z = \left(\frac{C_{44}}{C_{66}}\right)^{1/3} d_p. \tag{7.73}$$

Since d_p is sufficiently longer than a_f, and since the pinning correlation lengths, $R_c \simeq (C_{66}/\alpha_L)^{1/2}$ and $L_c \simeq (C_{44}/\alpha_L)^{1/2}$, are sufficiently smaller than L_y and L_z, respectively, for strong pinning, the elastic interaction among the segments exists only along the direction of the Lorentz force through C_{11}. These expressions of pinning correlation lengths are approximately the same as those defined by Larkin and Ovchinnikov (see Exercise 7.7). In practice L_x is considered not to exceed $(C_{11}/\alpha_L)^{1/2} \simeq L_c$, and hence, the elastic interaction exists along this direction. In the Larkin-Ovchinnikov theory this interaction was disregarded. This point will be taken up in Subsect. 7.4.3. The displacement of flux lines along the x-axis varies as $\exp(-x/\lambda_0')$ with $\lambda_0' = (C_{11}/\alpha_L)^{1/2}$ denoting Campbell's AC penetration depth. Thus, K is estimated as

$$K \simeq \alpha_L L_y L_z \lambda_0' \exp\left(-\frac{L_x}{\lambda_0'}\right). \tag{7.74}$$

On the other hand, α_L itself is also a quantity to be estimated by summation. As can be expected, α_L should be zero, if the pinning interactions are ineffective because the elementary pinning force is below the threshold value. In this case we have $k_f' = 0$ from Eq. (7.72) and the contradiction mentioned above is resolved. α_L is concerned with the pinning force density as given by Eq. (3.94). As for the interaction distance d_i, we shall use a generally known relationship:

$$d_i = \frac{a_f}{\zeta}, \tag{7.75}$$

where ζ is a constant dependent on the kind of pinning centers and $\zeta \simeq 2\pi$ is derived for pointlike defects [25]. Thus, F_p and α_L can be solved self-consistently from Eqs. (3.94), (7.35) and (7.73)–(7.75). If we write as $f_{pt} = tf_p$, after a simple calculation we have

$$\beta \cdot \frac{4f_p}{k_f a_f} \cdot \frac{1-t}{1+t} = \frac{t^2}{[(k_f a_f/4f_p) - t]^2}, \tag{7.76}$$

where β is given by

$$\beta = \frac{1}{16}\left(\frac{2}{\sqrt{3}\pi}\right)^{1/2}\left(\frac{C_{44}}{C_{66}}\right)^{5/6} \frac{\zeta d_p}{a_f} \exp\left(-\frac{2L_x}{\lambda_0'}\right). \tag{7.77}$$

In the above $C_{11} \simeq C_{44}$ was used and $G'^{-1}(0)$ given by Eq. (7.26) was substituted into $k_{\rm f}$.

In the case where the pins are not so strong that $k_{\rm f} a_{\rm f}/4f_{\rm p} > 1$, the solution of t can be obtained graphically as shown in Fig. 7.12(a). Since β is sufficiently larger than 1, t has a value close to 1 and we have

$$t \simeq 1 - \frac{8f_{\rm p}}{\beta k_{\rm f} a_{\rm f}} . \tag{7.78}$$

The pinning force density is

$$F_{\rm p} = \frac{4N_{\rm p} f_{\rm p}^2}{\beta k_{\rm f} a_{\rm f}} . \tag{7.79}$$

It can be seen from this result that a threshold value of the elementary pinning force does not exist even in the case of fairly weak pins. At the same time the obtained pinning force density agrees formally with the results of the statistical theory, neglecting the threshold value, and with the dynamic theory. However, this case is not consistent with the initial assumption of strong pins, and an application of this result to very weak pins has a quantitative problem.

Secondly the case of strong pins is treated. It is assumed that $4f_{\rm p}/k_{\rm f} a_{\rm f}$ is sufficiently larger than 1. It can be seen from Fig. 7.12(b) that the solution of t also exists in this case. Since the solution is given as

$$t \simeq \frac{k_{\rm f} a_{\rm f}}{4f_{\rm p}} , \tag{7.80}$$

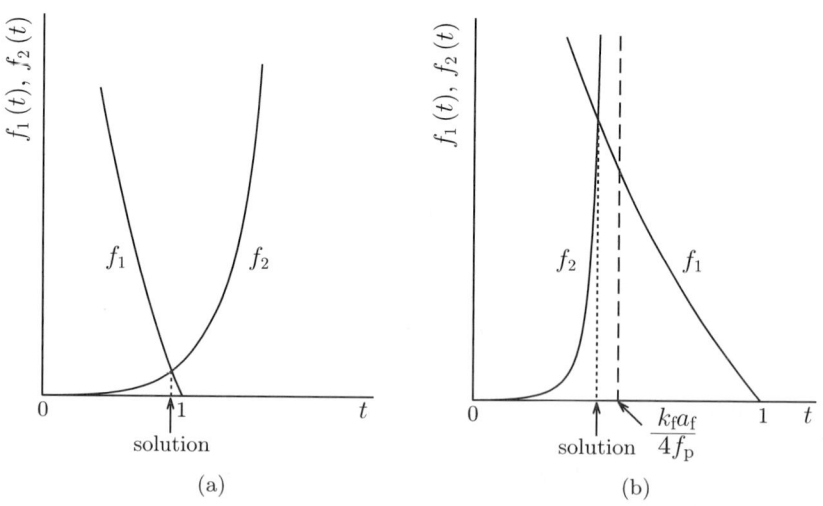

Fig. 7.12. Graphic solution of Eq. (7.76) for **(a)** relatively weak pins and **(b)** strong pins. f_1 and f_2 represent the *left* and *right sides* of Eq. (7.76), respectively. A solution is found to exist in each case

the pinning force density obeys the linear summation [24]:

$$F_\mathrm{p} = \eta_\mathrm{e} N_\mathrm{p} f_\mathrm{p} . \qquad (7.81)$$

In the above η_e is a pinning efficiency given by

$$\eta_\mathrm{e} = \frac{1 - (k_\mathrm{f} a_\mathrm{f}/4 f_\mathrm{p})}{1 + (k_\mathrm{f} a_\mathrm{f}/4 f_\mathrm{p})} . \qquad (7.82)$$

Thus, when the pins are very strong while their number density is not too high, the direct summation $F_\mathrm{p} = N_\mathrm{p} f_\mathrm{p}$ holds approximately within a certain accuracy.

The above result of the coherent potential approximation theory is summarized. Since t is not zero, the threshold value of the elementary pinning force is not zero. However, it depends on the elementary pinning force and never exceeds the elementary pinning force. Thus, it can be said that the threshold value of the elementary pinning force does not exist. This explains the experimental result in Fig. 7.9. In addition, from the relationship between the elementary pinning force and its threshold value, the pinning loss is shown to be of the hysteresis type caused by the unstable location of flux lines and agrees with generally known characteristics.

Here we shall treat pinning by large normal precipitates as an example of the direct summation. The mean diameter of the precipitates is denoted by D. The elementary pinning force is given by Eq. (6.9). This gives

$$F_\mathrm{p} = 0.430\pi \frac{N_\mathrm{p} \xi D^2 \mu_0 H_\mathrm{c}^2}{a_\mathrm{f}} \left(1 - \frac{B}{\mu_0 H_\mathrm{c2}}\right) . \qquad (7.83\mathrm{a})$$

If the elementary pinning force derived from the local model for the normal core with diameter 2ξ is used (see Exercise 6.1), the pinning force density is given by

$$F_\mathrm{p} = \frac{\pi N_\mathrm{p} \xi D^2 \mu_0 H_\mathrm{c}^2}{4 a_\mathrm{f}} \left(1 - \frac{B}{\mu_0 H_\mathrm{c2}}\right) . \qquad (7.83\mathrm{b})$$

It can be shown from these results that the temperature scaling law holds for the pinning force density with the pinning parameters of $m = 2$, $\gamma = 1/2$ and $\delta = 1$. This result agrees with the experimental results shown in Fig. 7.1(a) and (b).

7.4 Comparison with Experiments

In this section the results of the Larkin-Ovchinnikov theory [7] and the coherent potential approximation theory [24], which resolve the threshold problem, are compared with experiments. However, experimental results suitable for comparison are rather difficult to find. Hence, the comparison cannot be made systematically but is restricted to only certain aspects. In the following the comparison is divided into qualitative aspects, such as dependences on the number density of pins and the elementary pinning force, and quantitative aspects, and then the results are discussed. Finally the problems that arise from the theories are discussed.

7.4.1 Qualitative Comparison

The results are first compared on the dependence on the number density of pins N_p. For three-dimensional pinning, F_p is predicted to be proportional to N_p^2 and $N_\mathrm{p}^{2/3}$ for weak collective pinning and strong pinning, respectively, in the Larkin-Ovchinnikov theory. On the other hand, F_p is predicted to be proportional to N_p both for the weak and strong pinning when N_p is not too high in the coherent potential approximation theory. An example of experimental result supporting the former prediction is a report on pinning by irradiated defects from fast neutrons. Figure 7.13 shows the result for Nb-Ta [26]. It can be seen that, although a constant part remains due to residual pinning, $F_\mathrm{p}^{1/2}$ increases linearly with the irradiation dose. Figure 7.14 shows

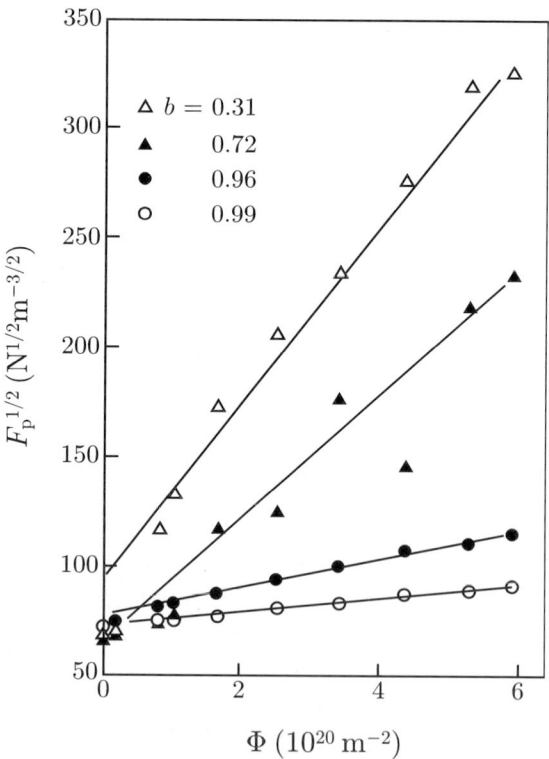

Fig. 7.13. Variation in pinning force density in irradiated Nb-20at%Ta at 5.0 K as a function of the fast neutron dose [26]

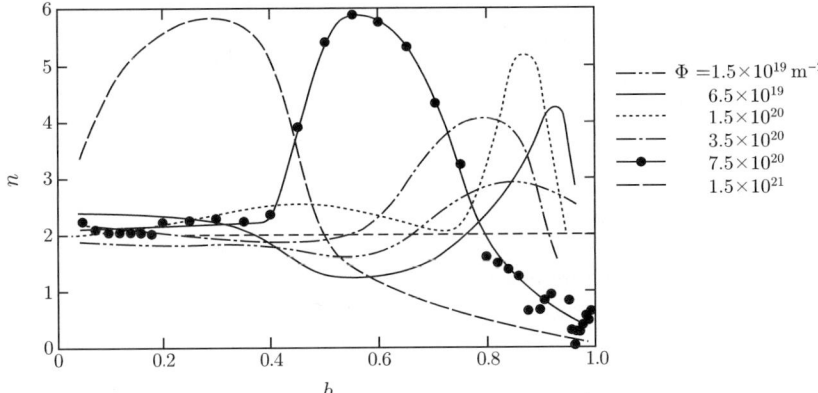

Fig. 7.14. Parameter showing the dose dependence of the pinning force density in V$_3$Si irradiated by fast neutrons [27]

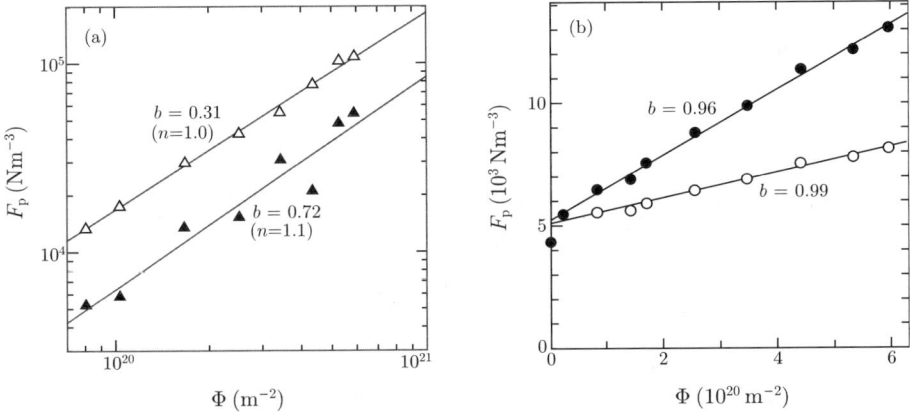

Fig. 7.15. Replots of Fig. 7.13. Pinning force density vs. dose of neutrons for **(a)** $b = 0.31$, $b = 0.72$ and **(b)** $b = 0.96$, $b = 0.99$

the results of the irradiation dose on V$_3$Si [27], where the dependence is expressed as $F_\mathrm{p} \propto N_\mathrm{p}^n$. According to this result $n \simeq 2$ is obtained at low fields when N_p is not so large.

On the other hand, there are questionable points in interpreting these results. For example, Fig. 7.15(a) and (b) are replottings of the results in Fig. 7.13. In (a) at $b = 0.31$ and $b = 0.72$, the pinning force density appears to increase linearly with the dose of neutrons. Also in (b) at $b = 0.96$ and $b = 0.99$, the linear relationship seems to hold between the pinning force density and the dose of irradiation. The dependence of the pinning force density on the dose of neutrons in the region of $n \simeq 2$ in Fig. 7.14 (its data will be shown later in Fig. 7.20(a)) is replotted in Fig. 7.16, from which n is evaluated to be close to 1 [28]. Thus, the assertion in [26] and [27] cannot be completely

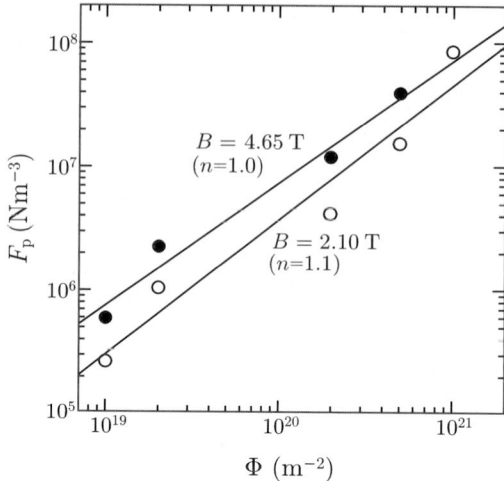

Fig. 7.16. Pinning force density vs. dose of neutrons in the region of $n \simeq 2$ in Fig. 7.14 ($\Phi < 1 \times 10^{21}$ m^{-2} at $B = 2.10$ T and $\Phi < 5 \times 10^{20}$ m^{-2} at $B = 4.65$ T)

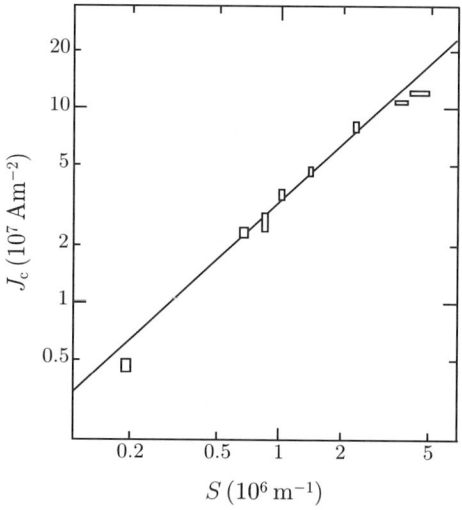

Fig. 7.17. Critical current density at $B = 1$ T vs. effective surface area of normal Bi precipitates in a unit volume, $S = N_{\mathrm{p}} D^2$, in Pb-Bi specimens [29]

accepted, and hence, it cannot be concluded that these are just the dependence on number density of pins which agrees with the Larkin-Ovchinnikov theory.

Contrary to the above results, experimental results which support the predictions of the coherent potential approximation theory, $F_{\mathrm{p}} \propto N_{\mathrm{p}}$, are found in the category of fairly strong pinning. Figure 7.17 shows the case of Pb-Bi with precipitates of normal Bi phase [29]. The critical current density is proportional to the effective surface area of precipitates of the Bi

7.4 Comparison with Experiments

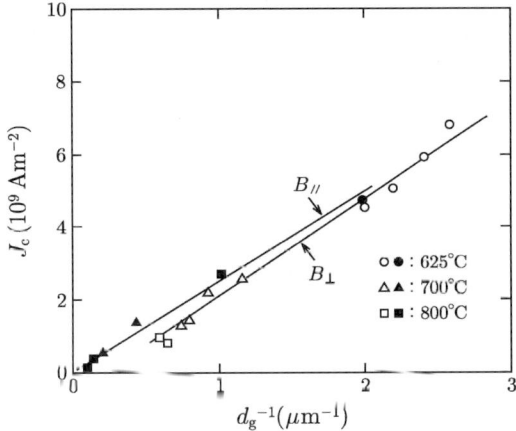

Fig. 7.18. Critical current density at $B = 6.5$ T vs. reciprocal grain size d_g^{-1} in V$_3$Ga [31]

phase in a unit volume, $S = N_\mathrm{p} D^2$, over a wide range, showing agreement with Eq. (7.83). The pinning by nonsuperconducting 211 phase particles in Y-123 high-temperature superconductor [30] also agrees with this prediction. Figure 7.18 shows the relationship [31] between the critical current density in V$_3$Ga and the inverse of the grain size d_g, and a proportionality is obtained. Here, the pinning by grain boundaries is similar to that by the surface of normal precipitates, and if we note that d_g and d_g^{-3} correspond to D and N_p, respectively, it is derived that F_p is proportional to d_g^{-1}. A similar result was also obtained for Nb$_3$Sn [32].

The dependence of the pinning force density only on the elementary pinning force f_p is difficult to investigate, and there is no suitable experiment. This is because, when the size of pins is changed to change f_p, N_p also changes. Hence, it is necessary to investigate the two dependences on f_p and N_p simultaneously. If we summarize the results of the two theories again, F_p is proportional to $N_\mathrm{p}^2 f_\mathrm{p}^4$ and $N_\mathrm{p}^{2/3} f_\mathrm{p}^{4/3}$ for weak pinning at low fields and for strong pinning, respectively, in the Larkin-Ovchinnikov theory. In the coherent potential approximation theory, on the other hand, F_p is proportional to $N_\mathrm{p} f_\mathrm{p}^2$ and $N_\mathrm{p} f_\mathrm{p}$ for relatively weak pinning and for strong pinning, respectively. The results on niobium [12] shown in Fig. 7.9 are suitable for a systematic comparison. This result is not a direct comparison for the above functions and is not perfect because f_p and N_p are not independent of each other as mentioned above. However, all the data meet on a single master curve, and hence, this seems to show the functions predicted by the coherent potential approximation theory. In Fig. 7.19 the experimental result for Nb-Ta with normal precipitates of Nb$_2$N phase is compared with the two theories [33]. It turns out that agreement is not obtained with experiment either qualitatively or quantitatively for the Larkin-Ovchinnikov theory. On the other hand, two groups of experimental results with considerably different pinning parameters

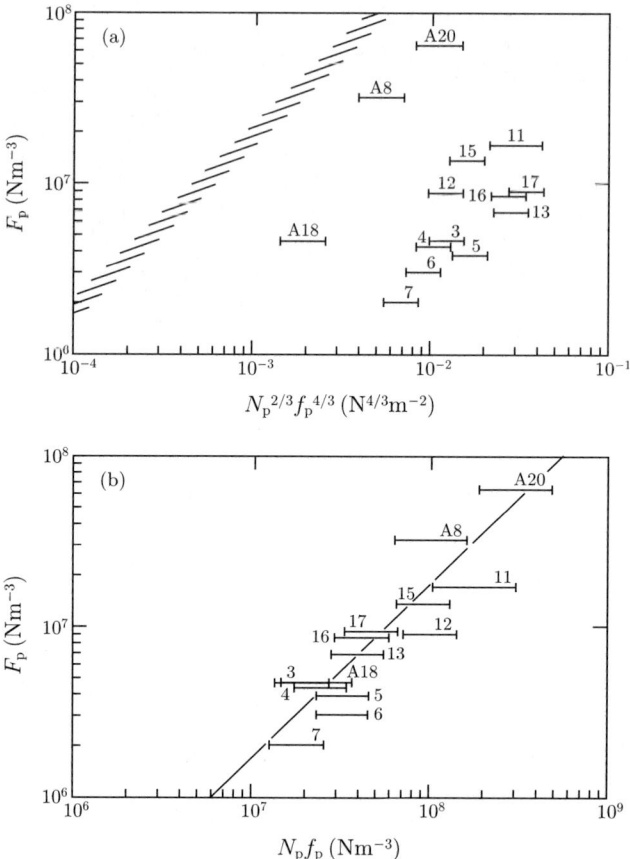

Fig. 7.19. Comparison of pinning force density with predicted functions at $b = 0.65$ for Nb-Ta with normal Nb_2N precipitates [33] for (**a**) the Larkin-Ovchinnikov theory (*shaded area* shows the theoretical prediction) and (**b**) the coherent potential approximation theory with $\eta_e = 0.17$. Data labeled A8, A18, A20 are experimental results by Antesberger and Ullmaier [34]

[33, 34] can be simultaneously explained by the coherent potential approximation theory. In Fig. 7.19(b) the straight line represents Eq. (7.81) with $\eta_e = 0.17$ so as to get a good fit with experiments. The theoretical estimation of η_e will be given in Subsect. 7.4.2.

Although the number of experimental results suitable for comparison with the theories is not sufficient, it seems from the above results that the coherent potential approximation theory is in better agreement with experiments, especially for strong pinning, than the Larkin-Ovchinnikov theory. As an example of other experimental results which support the coherent potential approximation theory, we find the temperature scaling law of the pinning force density in Pb-Bi with precipitates of normal Bi phase shown in Fig. 7.1.

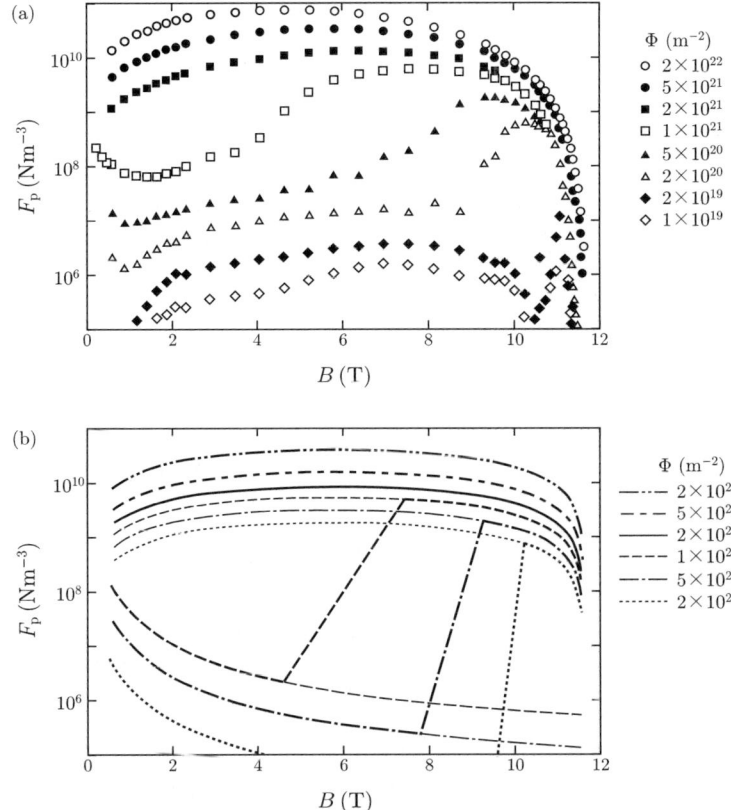

Fig. 7.20. (a) Variation in the pinning force density in V_3Si by variation in the irradiation dose of fast neutrons [27] and (b) the corresponding prediction of the Larkin-Ovchinnikov theory. All measurements were done at temperatures at which $\mu_0 H_{c2}$ was equal to 11.6 T

On the other hand, the abovementioned experimental result [27] on neutron irradiated V_3Si was compared with the Larkin-Ovchinnikov theory. Figure 7.20(a) and (b) show the experimental result and the corresponding theoretical result, respectively. The elementary pinning force of defects produced by neutron irradiation was estimated so as to get a good agreement in the case of an high number density of defects. Although the theoretical result is considerably smaller than the experimental result for a low number density of pins as will be mentioned later, the theory generally explains the trend of experimental results.

Experiments compared with the Larkin-Ovchinnikov theory in most detail are those on thin films by Wördenweber and Kes [19, 20]. For example, Fig. 7.21 shows the dependence of the pinning force density on the film thickness of Nb_3Ge [35]. This agrees with the prediction of the Larkin-Ovchinnikov

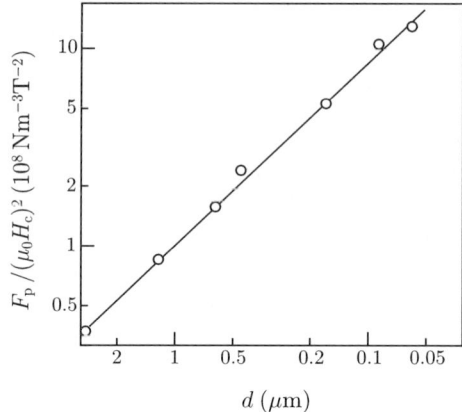

Fig. 7.21. Dependence of pinning force density on film thickness in Nb_3Ge thin films at $b = 0.4$ and at $T/T_c = 0.7$ [35]. The *solid line* shows the prediction of Eq. (7.68)

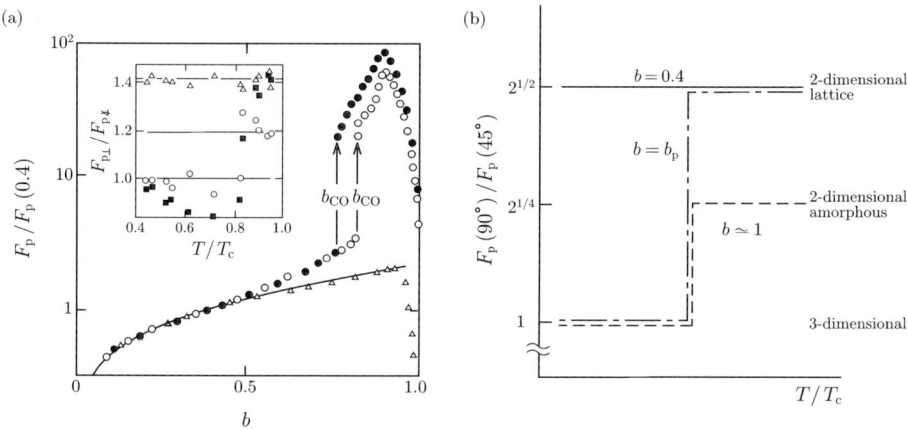

Fig. 7.22. (a) Magnetic field dependence of pinning force density in Nb_3Ge thin film 7.9 μm thick [20] in a normal magnetic field at $T/T_c = 0.44$ (*circles*), in a magnetic field tilted by 45° from the direction of current at the same temperature (*solid circles*) and in a normal magnetic field at $T/T_c = 0.95$ (*triangles*). The *solid line* is the prediction of two-dimensional collective pinning theory. The inset shows the ratio of two pinning forces for different directions of the magnetic field at $b = 0.4$ (*triangles*), the peak field (*solid squares*) and $b \simeq 1$ (*circles*). (b) Theoretical prediction corresponding to the inset in (a)

theory, Eq. (7.68). Figure 7.22(a) gives the magnetic field dependence of the pinning force density in a Nb_3Ge thin film 7.9 μm thick in magnetic fields normal to the surface and tilted by 45° from the direction of the current. The inset shows the ratio of the pinning force density when the magnetic field

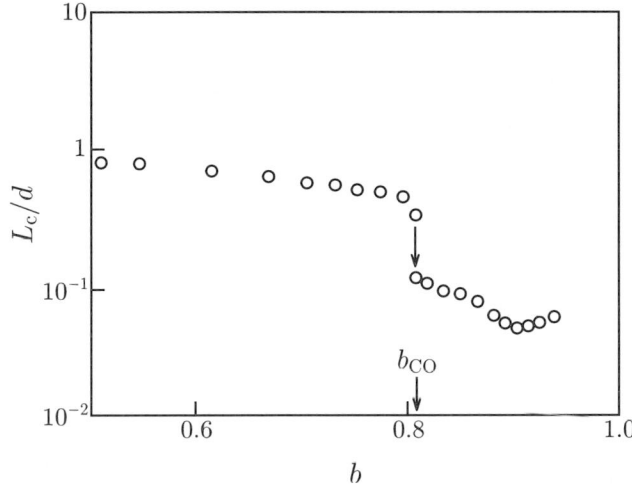

Fig. 7.23. Magnetic field dependence of L_c/d in Nb_3Ge thin film estimated at $T/T_c = 0.44$ in a normal magnetic field [20]

is normal to the current and that when the field is tilted by 45° from the current. Figure 7.22(b) is the corresponding theoretical prediction. According to the theory, F_p is independent of the film thickness d for three-dimensional pinning, proportional to d^{-1} as in Eq. (7.68) for the two-dimensional pinning of latticelike flux lines, and proportional to $d^{-1/2}$ as in Eq. (7.69) for the two-dimensional pinning of amorphous flux lines. Since the film thickness increases equivalently by a factor of $2^{1/2}$ when the magnetic field is tilted by 45° from the film, the results in Fig. 7.22(b) are predicted. Thus, a good agreement is obtained between the theory and the experiment.

It is expected in the theory that a transition from the two-dimensional pinning to the three-dimensional occurs when the longitudinal correlation length L_c, obtained from Eqs. (7.57) and (7.70), decreases with increasing magnetic field to a half of the film thickness d. A significant enhancement of the pinning force density was observed at the field $b = b_{CO}$, at which the transition is speculated to occur. The value of L_c estimated using $W(0)$ obtained from the experimental result of F_p at $b = 0.4$ is shown in Fig. 7.23 [20]. In this estimation $V_c = R_c^2 d$ and Eq. (7.57) were used for $b < b_{CO}$, and $V_c = R_c^2 L_c$ and Eq. (7.57) were used for $b > b_{CO}$. According to this result $L_c \simeq d/2$ was obtained at $b = b_{CO}$, and hence, it was asserted that the transition between the two-dimensional pinning and the three-dimensional had occurred. In this view the reason why L_c decreased discontinuously at the transition point $b = b_{CO}$ with increasing magnetic field was because the two-dimensional flux line lattice became unstable and went into an entangled "spaghetti" state with penetration of screw dislocations nucleated near the film surface. However, if such a discontinuous transition of the first order really occurs, hysteresis may be expected to exist between increasing and decreasing magnetic fields. However, such a hysteresis has not yet been observed.

As for the result shown in Fig. 7.22(a), it is considered that two-dimensional pinning is attained over the entire magnetic field region at high temperatures near the critical temperature. In fact, a pronounced peak effect accompanied by the transition to three-dimensional pinning was not observed. Since the nonlocal property of the flux line lattice and the dependence of $W(0) \propto H_c^2 H_{c2}$ were assumed, the longitudinal correlation length was predicted to be proportional to H_{c2}^{-1}.

7.4.2 Quantitative Comparison

A quantitative comparison of theories with experiments has already been made on some results in the last subsection. If the results are summarized briefly, the Larkin-Ovchinnikov theory gives a very small pinning force density for weak pinning, as can be seen from the comparison in Fig. 7.20. This is related to the prediction that F_p is proportional to $N_p^2 f_p^4$, and hence, the above trend is emphasized especially for weak pinning. On the other hand, the theory gives a larger pinning force density than is seen in experiments for strong pinning as in the case of normal precipitates shown in Fig. 7.19(a). The theory predicts a stronger f_p-dependence ($F_p \propto N_p^{2/3} f_p^{4/3}$) than the experimental result ($F_p \propto N_p f_p$) in this case, too.

Such a trend can clearly be seen from the comparison with the experimental result on niobium [36] shown in Fig. 7.24 in which the results of the Larkin-Ovchinnikov theory for the cases of $N_p = 1 \times 10^{20}$ m^{-3} and $N_p = 1 \times 10^{22}$ m^{-3} are compared. Namely, the theoretical results are very much smaller in the region of small f_p and larger in the region of large f_p than the experimental results. Thus, it is concluded that the Larkin-Ovchinnikov theory does not agree quantitatively with the experiments.

On the other hand, some papers reported that the Larkin-Ovchinnikov theory agreed with experiments. One of them is the pinning of three-dimensional amorphous flux lines at high fields in niobium and vanadium by voids, which were nucleated by neutron irradiation at high temperatures [37]. In this paper the theoretical result of Thuneberg [38] on the electron scattering mechanism was used for estimating the elementary pinning force of pointlike defects and the Larkin-Ovchinnikov theory with the nonlocal elastic moduli of the flux line lattice was used for the summation. Hence, there is a possibility that an overestimation by the nonlocal theory and an underestimation by the Larkin-Ovchinnikov theory are by chance canceled. It was reported that a quantitative agreement was obtained also for a series of experiments on thin films [19, 20]. However, nonlocal elastic moduli were assumed in the analysis. From this point and the ambiguity about practical defects acting as pinning centers it cannot be concluded if this theory agrees quantitatively with experiments.

Here the coherent potential approximation theory is compared quantitatively with experiments. For example, the result on specimen 13 in

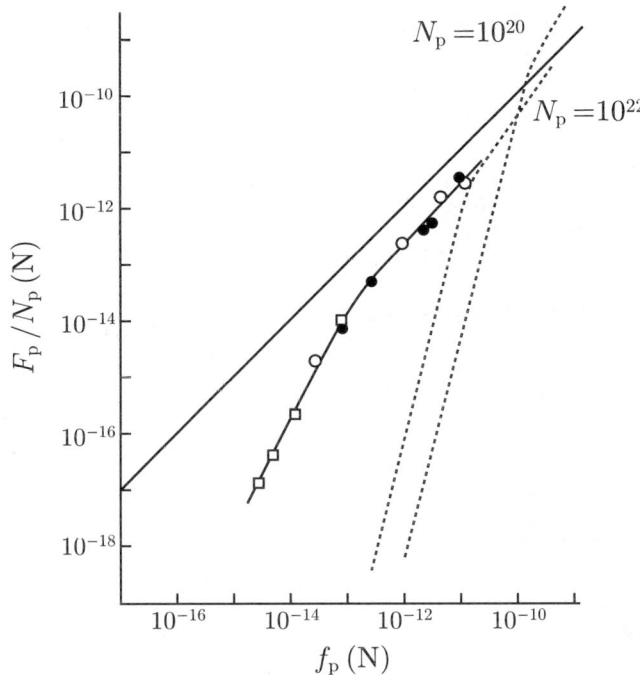

Fig. 7.24. Comparison [36] of experimental results at $b = 0.6$ on the same niobium specimens shown in Fig. 7.9 with the results of the Larkin-Ovchinnikov theory for $N_{\rm p} = 1 \times 10^{20}$ m^{-3} and $N_{\rm p} = 1 \times 10^{22}$ m^{-3}

Fig. 7.19(b) is compared with theory. $\mu_0 H_{\rm c2}$ of this specimen is 0.449 T and $\mu_0 H_{\rm c}$ is estimated as 0.084 T at 4.2 K. The mean number density of pins is $N_{\rm p} \simeq 0.35 \times 10^{18}$ m^{-3} and the elementary pinning force is estimated as $f_{\rm p} \simeq 1.2 \times 10^{-10}$ N at $b = 0.65$ ($B = 0.292$ T, $a_{\rm f} = 9.05 \times 10^{-8}$ m) [33]. From Eqs. (7.10) and (7.12b) $C_{44} = 6.79 \times 10^4$ Nm^{-2} and $C_{66} = 1.12 \times 10^2$ Nm^{-2} are derived. We have $k_{\rm f} = 1/G'(0) = 3.29 \times 10^{-3}$ Nm^{-1} from Eq. (7.26) and $k_{\rm f} a_{\rm f}/4 f_{\rm p} = 0.62$. Thus, $\eta_{\rm e} = 0.23$ is obtained from Eq. (7.82). This is close to the experimental result, $\eta_{\rm e} = 0.17$, shown by the straight line in Fig. 7.19(b). It can be seen that the experimental result for strong pinning can be explained almost quantitatively by the coherent potential approximation theory. Since the present theory is not applicable to weak pinning, comparisons cannot be made for experimental results in this category. A new theoretical approach is necessary for this purpose.

7.4.3 Problems in Summation Theories

The pinning force density is globally obtained through the estimated elastic correlation length of the flux line lattice in the theoretical method of collective pinning of Larkin and Ovchinnikov, while it is calculated from the statistical average by taking account of interactions among pins using the mean field method in the coherent potential approximation theory. It is expected that the two methods would ultimately derive similar results by improving the accuracy of approximations. However, some approximations are still not good, and apparent problems remain at the moment. Here we shall mainly describe the problems in the Larkin-Ovchinnikov theory and its modification by Wördenweber and Kes [19, 20], since these theories are considered to contain apparent problems.

One of them is the assumption of nonlocal elastic moduli of the flux line lattice. This is discussed in detail in Appendix A.5, and we do not mention it further here.

Comparing Eqs. (7.23) and (7.52), it turns out that the uniaxial compression, a part of the Lorentz force, is not taken into account in the Larkin-Ovchinnikov theory. This is clearly incorrect. According to the original paper, this was neglected since the compression strain was small along the direction of the Lorentz force because of the large C_{11} in comparison with C_{66}. However, this asserts that a strong compression force exists, contrary to the original assumption. In fact it can be shown that a compression force of the same magnitude exists from the partition of the strain energy. This force is known as a hydrostatic pressure even in a liquid in which the shear modulus is zero. It is also shown that an elastic interaction with a longer correlation length, $(C_{11}/C_{66})^{1/2}R_c \simeq L_c$, exists in practice along the direction of the Lorentz force, and a theoretical calculation has already been done under this condition [39]. Under this compression force clusters of flux lines with short-range order interact with each other. The pinning force is originally the variation rate of the pinning energy against the displacement in the direction of the Lorentz force, and hence, the compression force is necessarily associated with this displacement. Each cluster cannot always stay at a suitable position for pinning because of this interaction, and its pinning force is distributed in the range of $-f$ to f as shown in Subsect. 7.3.4, with f denoting the maximum force on each cluster. On the other hand, it was assumed in the Larkin-Ovchinnikov theory that each cluster was independent of the others and that its pinning force takes the value of f. However, there is no theoretical proof.

If the correction on the nonlocal elastic moduli is done and if the elastic interaction due to the compression is taken into account in the collective pinning theory, it will surely bring about a smaller pinning force density. In the category of weak pinning the theoretical pinning force density is already very small in the low field region where the local elastic moduli are used. Hence, the difference between theory and experiments is further enhanced by the introduction of the compression interaction. However, defects and strains are considered to exist even inside a region with a short-range order in the flux line lattice. Thus, the residual pinning force after the cancellation will be considerably larger than that which only results from fluctuations. This may compensate for the abovementioned reduction in the pinning force density due to the corrections. In fact, the softening of C_{66} is remarkable due to defects

7.4 Comparison with Experiments

in the flux line lattice and the influence of thermal activation of flux lines, and the resulting enhancement of the pinning efficiency is significant (see Appendix A.8). However, it should be noted that C_{66} cannot be theoretically obtained by usual methods, as will be discussed in Sect. 7.7. For this reason it is evident that any theoretical prediction may fail to explain experimental results, unless the reduction in C_{66} from C_{66}^0 for the perfect flux line lattice is not taken into account. The most typical case may be found in the prediction of the collective pinning theory for weak pinning forces. Hence, for the purpose of more exact estimation of the pinning force density the coherent potential approximation theory, in which the size of the treated region can be reduced below the dimension of short-range order, seems to be more suitable. On the other hand, a thermodynamic method in which the dimension of short-range order is assumed to be determined so that the pinning force is maximized was proposed by Kerchner [22]. This has attracted attention since it may break through the limit of the Larkin-Ovchinnikov-type theory. However, the softening of C_{66} is not considered in the Kerchner model and this influence should be taken into account in the maximization of the pinning force density.

Wördenweber and Kes developed their collective pinning theory and argued for a transition between two- and three-dimensional pinning in thin films. The fundamental problem with this theoretical treatment has already been mentioned, and we do not repeat it here. Instead of this, we shall discuss why such a transition cannot correctly be described by the original Larkin-Ovchinnikov theory. If we summarize the pinning in two-dimensional latticelike, three-dimensional latticelike and three-dimensional amorphous flux lines in a thin film that is predicted by the Larkin-Ovchinnikov theory, the results are schematically shown in Fig. 7.25. According to these results the transition between two-dimensional latticelike pinning and three-dimensional latticelike pinning does not occur (and even if it dose occur, it is quite near H_{c2}). Hence, the pronounced peak effects observed in experiment cannot be explained as the transition. This is due to the fact that the effectiveness of pinning is too different between the two- and three-dimensional cases even in the latticelike state. We shall here discuss this from a purely theoretical viewpoint. In the Larkin-Ovchinnikov theory the elastic interactions taken into account are the bending associated with C_{44} and the shear associated with C_{66} for the three-dimensional case, while only the shear is considered in the two-dimensional case. Hence, the strength of elastic interaction is too different resulting in a large difference in the pinning force density between the two cases. It is believed that the compression elastic interaction associated with C_{11} should be taken into account for both cases. Then, the strength of the elastic interaction would not be so different between the two- and three-dimensional cases, and the pinning force densities in each case would become closer than before. The dimensional transition of pinning might be successfully explained by this correction. In this case we will face the problem of too small a pinning force density. This is in fact a common problem in collective pinning theory. One of the methods to solve this has already been mentioned above.

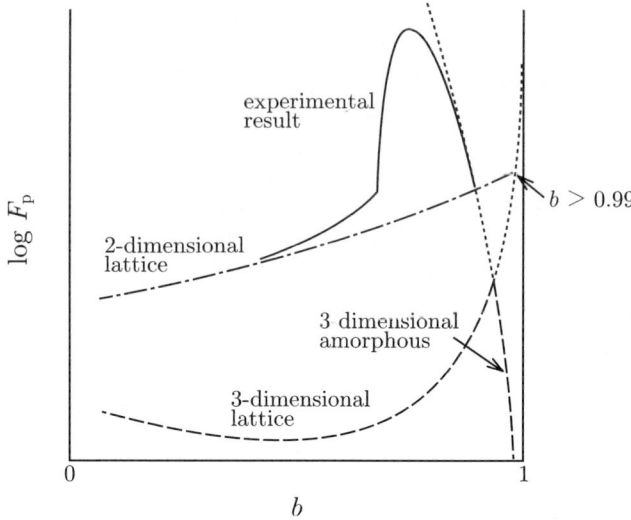

Fig. 7.25. Pinning force densities in superconducting thin film for two-dimensional latticelike pinning, three-dimensional latticelike pinning and three-dimensional amorphous pinning predicted by Larkin and Ovchinnikov

7.5 Saturation Phenomenon

7.5.1 Saturation and Nonsaturation

The pinning force density $F_{\rm p}$ generally depends closely on the microstructure of pins, and $F_{\rm p}$ increases when the elementary pinning force $f_{\rm p}$ is strengthened or the number density of pins $N_{\rm p}$ is increased. However, it is sometimes observed that, when $f_{\rm p}$ and/or $N_{\rm p}$ are increased, $F_{\rm p}$ does not change much at high fields, while $F_{\rm p}$ increases at low fields. This phenomenon at high fields is called saturation. Figure 7.26(a) shows the case of Nb$_3$Sn [40]. Although $F_{\rm p}$ varies at low fields due to a variation in the effective $N_{\rm p}$ resulting from a variation in grain size, it is almost unchanged at high fields. It is also known for Nb-Ti that $F_{\rm p}$ does not increase at high fields in spite of increasing $f_{\rm p}$ during the process of precipitation of α-Ti phase by heat treatment [41] (see Fig. 7.26(b)). As can be seen from the fact that the saturation is observed in commercial superconductors, this phenomenon occurs in the case of fairly strong pinning. Under saturation the normalized magnetic field at which $F_{\rm p}$ is maximized becomes lower when $f_{\rm p}$ and/or $N_{\rm p}$ are increased. One of the characteristics of the saturation is that the parameter δ in Eq. (7.3) representing the magnetic field dependence of $F_{\rm p}$ at high fields is about 2.

On the other hand, it is also normal to observe cases where the pinning force density varies with variation in the microstructure of the pins without showing saturation at high fields, although the value itself is of the same order of magnitude as that in the saturation. Figure 7.27 shows the characteristics

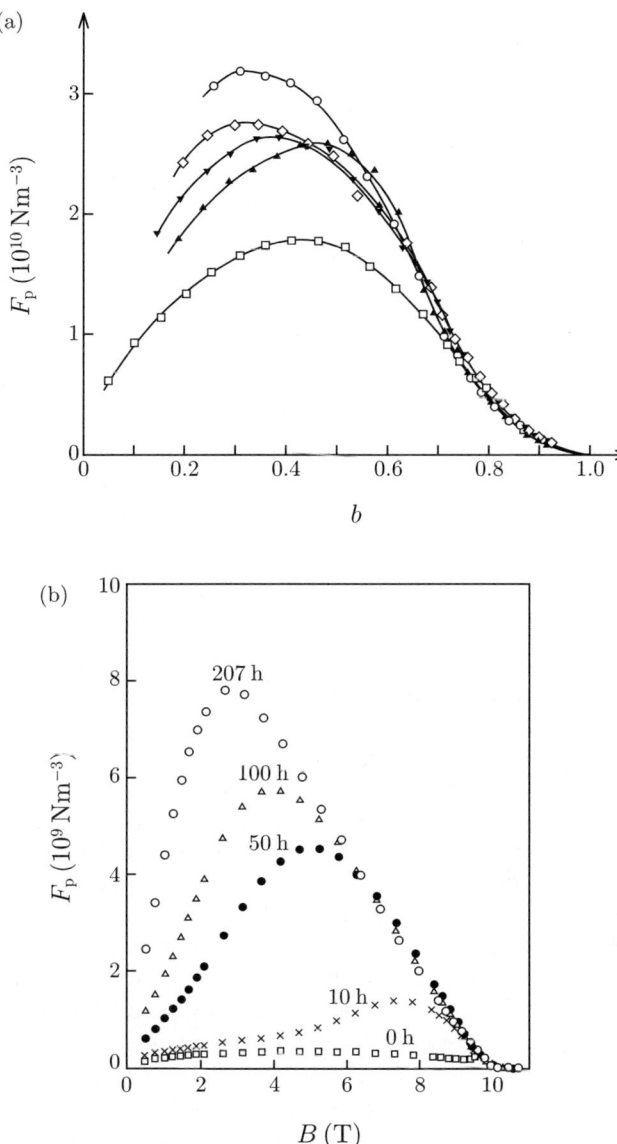

Fig. 7.26. Examples of the pinning force density showing saturation: (**a**) Nb$_3$Sn [40] and (**b**) Nb-Ti during precipitation of α-Ti by heat treatment [41]

of Nb-Ti wires [42] for which F_p has different values over the entire region of magnetic field due to differences in the amount and/or structure of precipitates of α-Ti phase working as pinning centers. These are the results for Nb-Ti wires that were heavily drawn after precipitation of α-Ti phase. The normalnized magnetic field at which F_p is maximized is almost the same and

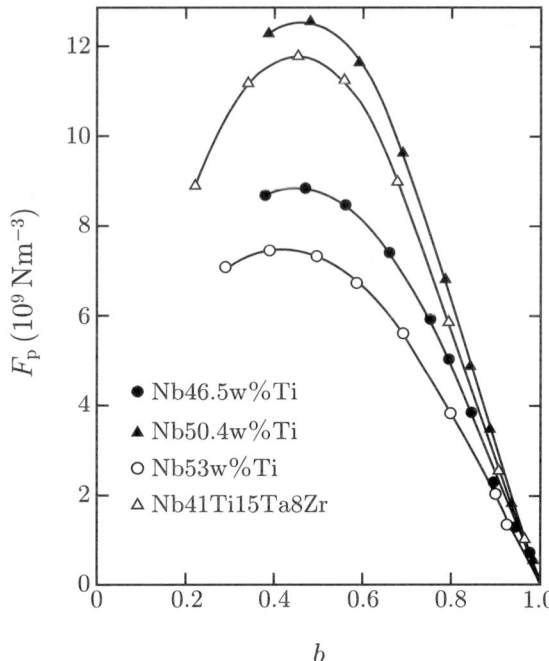

Fig. 7.27. Nonsaturation of the pinning force density in Nb-Ti wires [42]

the function $f(b)$ expressing the dependence of F_p on the normalized field is also almost the same. This phenomenon is called nonsaturation as opposed to the saturation discussed above. One of the characteristics of nonsaturation is that the magnetic field dependence of F_p at high fields is characterized by $\delta \simeq 1$ in Eq. (7.3).

Most high field superconductors such as the commercial ones show either saturation or nonsaturation. For example, Nb_3Sn belongs to the saturation type and Nb-Ti and V_3Ga to the nonsaturation type. From the viewpoint of applications of superconductors at high fields the nonsaturation type with smaller degradation of critical current at high fields is preferable. Hence, it is desirable to improve the pinning characteristics of Nb_3Sn, which has superior superconducting properties to Nb-Ti and is more economical than V_3Ga, from the saturation type to the nonsaturation type.

There is no pure theory which can explain the saturation, but some models have been proposed. In this section these models are introduced and experimental results of Campbell's method that are useful for verification of the validity of the models are discussed.

7.5.2 The Kramer Model

The first pinning model for saturation and nonsaturation was proposed by Kramer [40, 43]. According to his model, when the number density of pins is sufficiently high, strains of the flux line lattice due to individual pins heavily overlap, and each flux line seems to be pinned as if by a line pin parallel to itself. This situation is called line pinning. Kramer expressed the pinning force per unit length of flux line as $\hat{f}_p = w f_p$. He assumed that the equivalent number of pins w fluctuates widely from one flux line to another. Then, weakly pinned flux lines (with small w) are largely displaced by the Lorentz force, resulting in shearing deformations around strongly pinned flux lines (with large w). Since the magnitude of this displacement u is proportional to $\hat{f}_p/C_{66}$, it is found from $\hat{f}_p \propto (1-b)$ and $C_{66} \propto (1-b)^2$ that u increases in proportion to $(1-b)^{-1}$ with increasing field. The probability for the flux line to encounter a stronger line pin increases in proportion to the displacement. Thus, a group of strongly pinned flux lines is considered to be formed around a flux line that was originally strongly pinned. The process by which the flux line lattice is deformed with increasing field so that it fits a structure of distributed pins is called synchronization. The pinning force of this group is proportional to u and is expressed as

$$\tilde{f}_p = \frac{\alpha_K \hat{f}_p^{\,2}}{C_{66}}, \tag{7.84}$$

where α_K is a constant.

Groups of flux lines pinned with the forces given by Eq. (7.84) are distributed in space. Kramer used the results of the dynamic pinning theory of Yamafuji and Irie [8] for evaluating the pinning force density. This is equivalent to taking the statistical average of groups as a result. Hence, the pinning force density is proportional to $\tilde{f}_p^2/C_{66}$ and is given by

$$F_p = \frac{\beta_K \rho_p \hat{f}_p^{\,4}}{C_{66}^3 a_f}, \tag{7.85}$$

where β_K is a constant and ρ_p is the area density of strong line pins. As to the magnetic field dependence, Eq. (7.85) leads to

$$F_p = K_p b^{1/2} (1-b)^{-2} \tag{7.86}$$

except for the low field region. Thus, the pinning force density increases with increasing magnetic field caused by the synchronization. In the above, K_p is a constant dependent on the temperature and its temperature dependence is given by $K_p \propto H_{c2}^{5/2}$.

According to Kramer the pinning characteristics at higher fields are classified into two types depending on the strength of pinning. When the pinning force of a strongly pinned flux line is weaker than the maximum shear force of the flux line lattice, the synchronization is completely attained without

314   7 Flux Pinning Characteristics

plastic shear deformation of the flux line lattice. In this case the pinning force density is given by the linear summation in Eq. (7.1). Thus, the pinning force density depends on the structure of pins, and its magnetic field dependence is expressed as $(1-b)$. These features satisfy the nonsaturation conditions (see Fig. 7.28(a)).

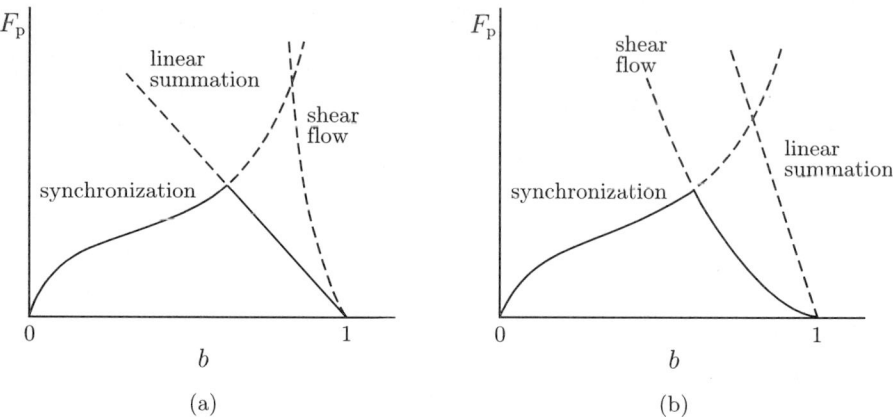

Fig. 7.28. Variations in pinning mechanisms and pinning characteristics predicted by Kramer for the cases of (a) nonsaturation and (b) saturation

On the other hand, when the pinning force is stronger than the maximum shear force, the flux line lattice starts a shearing flow, resulting in a voltage state before the synchronization is completed. The pinning force density in this case is calculated as

$$F_{\rm p} = \frac{C_{66}}{12\pi^2(1-a_{\rm f}\rho_{\rm p}^{1/2})^2 a_{\rm f}} .\tag{7.87}$$

The characteristics of this pinning force density are that it does not depend much on the area density of line pins $\rho_{\rm p}$ over its wide range from a very low value to the high limit $1/4a_{\rm f}^2$ (if the mean spacing of line pins becomes smaller than $2a_{\rm f}$, the shear flow does not occur), and that its magnetic field dependence is given by

$$F_{\rm p} = K_{\rm s} b^{1/2}(1-b)^2 .\tag{7.88}$$

That is, the magnetic field dependence comes mostly from that of the shear modulus. Thus, Eq. (7.87) satisfies the characteristics of the saturation conditions. Variations in the pinning mechanism and the pinning characteristics in this category are shown in Fig. 7.28(b). In Eq. (7.88) $K_{\rm s}$ is a constant dependent on temperature and is proportional to $H_{\rm c2}^{5/2}$ as well as $K_{\rm p}$. Hence, it is derived from Eqs. (7.86) and (7.88) that the pinning force density can be described in the form of the usual temperature scaling law in which the

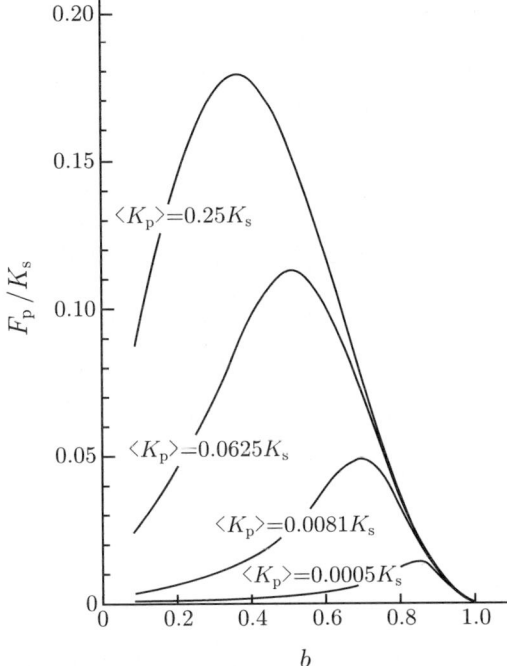

Fig. 7.29. Variation in the pinning force density due to variation in the pinning strength predicted by Kramer [43]. The distribution of K_p representing the pinning strength is assumed to obey the Poisson distribution

dependences on temperature and normalized magnetic field can be separated. Figure 7.29 shows the predicted variation in the pinning force density with strengthening pinning force [43]. This resembles qualitatively the experimental results in Fig. 7.26.

Equation (7.88) is called Kramer's formula and sometimes used for estimation of the upper critical field H_{c2}, etc. That is, this equation reduces to

$$F_p^{1/2} B^{-1/4} = J_c^{1/2} B^{1/4} = K_s^{1/2} (\mu_0 H_{c2})^{-1/4} \left(1 - \frac{B}{\mu_0 H_{c2}}\right), \qquad (7.89)$$

and hence, by extrapolating the linear part of the experimental data on $J_c^{1/2} B^{1/4}$ vs. B, $\mu_0 H_{c2}$ can be obtained from a value of B at which $J_c^{1/2} B^{1/4}$ reaches zero.

According to the Kramer model the pinning characteristic of saturation gives the upper limit of the pinning force density, and it is predicted that a further improvement of the critical current density at high fields is impossible.

7.5.3 Model of Evetts et al.

Because the pinning force density of Nb_3Sn does not agree quantitatively with the saturation characteristic of the Kramer model, Evetts and Plummer [44] and Dew-Hughes [45] proposed their model in which an occurrence of the shearing flow of flux lines and the resultant pinning characteristic are determined by the morphology of the pins. According to this model, since the pins in a Nb-Ti wire are ribbonlike α-Ti precipitates and dislocation cell walls elongated along the length of the wire, flux lines should overcome the pins as shown in Fig. 7.30(a) to reach a voltage state. Thus, it leads to nonsaturation. On the other hand, Nb_3Sn has an isotropic grain structure, and the occurrence of a shearing flow of flux lines along grain boundaries is possible as shown by the thick line in Fig. 7.30(b). Thus, the pinning force density is expected to be close to saturation, although the pinning force density varies depending on the pinning morphology.

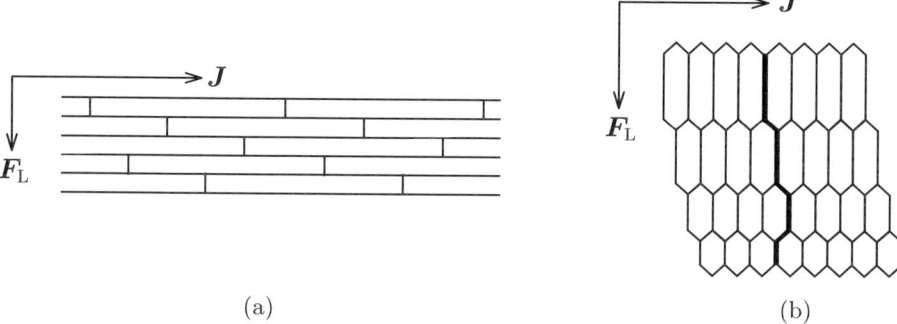

Fig. 7.30. Explanation of the model of Dew-Hughes [45]. Morphology of pins in (a) Nb-Ti and (b) Nb_3Sn. Arrows show the directions of the current and the Lorentz force

7.5.4 Comparison Between Models and Experiments

As abovementioned, it is assumed that the shearing flow of flux lines occurs depending on the pinning strength in the Kramer model or on the pinning morphology in the model of Evetts, Plummer and Dew-Hughes, and that the resultant pinning characteristic, saturation or nonsaturation, is determined by the occurrence of shearing flow. It is expected in the Kramer model that the pinning force density in saturation is larger than that in nonsaturation. However, many experimental results are contradictory to this assertion. For example, saturation is observed in a process of precipitation of α-Ti by heat treatment in Nb-Ti as shown in Fig. 7.26(b). When this material is cold worked, the pinning force density increases, accompanied by a transition from saturation

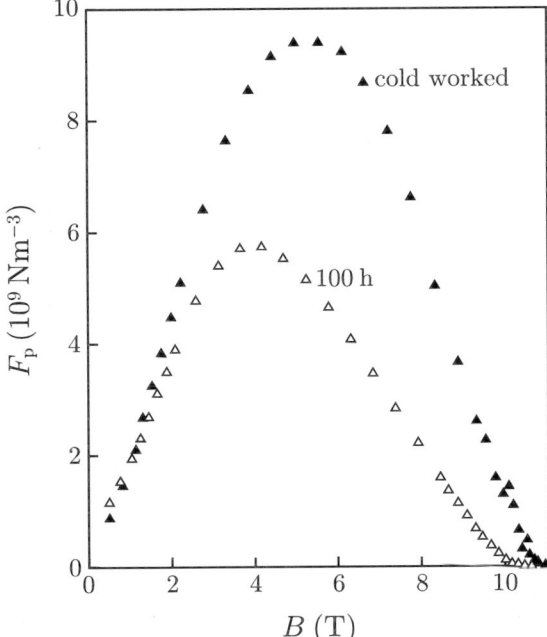

Fig. 7.31. Variation of the pinning force density by cold working Nb-Ti [41], which shows saturation in Fig. 7.26(b)

to nonsaturation [41] as shown in Fig. 7.31. This is opposite to the prediction of the Kramer model. On the other hand, although the pins in V_3Ga are grain boundaries just as in Nb_3Sn, the fact that its pinning characteristic is close to that of nonsaturation does not agree with the prediction of the model of Evetts et al.

Examples of pinning characteristics showing saturation and nonsaturation are classified with respect to the kind of pins in Table 7.1 [46]. It turns out that saturation and nonsaturation are general phenomena independent of superconductor, the G-L parameter (κ), and kind and morphology of pins. Therefore, it can be said that the pinning characteristic cannot only be determined by the morphology of pins as assumed by Evetts et al. The results in Table 7.1 suggest generally that pins resulting in nonsaturation are stronger than those resulting in saturation. For example, a void of a size of about 10 nm has a stronger elementary pinning force than a dislocation loop in niobium. Even for similar grain boundaries it is also known [47] that the elementary pinning force in V_3Ga is stronger than that in Nb_3Sn. In addition, it is shown by a calculation [41] that the elementary pinning force of α-Ti precipitates in Nb-Ti is stronger for a long thin ribbon shape after a heavy drawing than for an isotropic geometry before the drawing. (See Sect. 6.7 for the above two cases.) Thus, all of these results are contradictory to the predictions of the

Table 7.1. Pins and pinning characteristics in various superconductors [46]

	Saturation	Nonsaturation
point pin (0-dimensional)	Nb(dislocation loop) V_3Si(cascade defect)	Nb(void)
planar pin (2-dimensional)	Nb-Ta(subband wall) Nb_3Sn(grain boundary)	Pb-Bi(S-N interface) V_3Ga(grain boundary)
normal precipitate (2~3-dimensional)	Nb-Ti(platelike)	Nb-Ti(ribbonlike) Nb-Ta(platelike)

Kramer model. The reason why saturation is observed in Nb-Ti even for similar platelike normal precipitates, while nonsaturation is observed in Nb-Ta, is that the elementary pinning force in Nb-Ti is relatively weaker due to a shorter coherence length ξ. The difference in the magnetic field strength is also one of the reasons.

While nonsaturation is a normal phenomenon in which the linear summation holds for the pinning force density, saturation is a peculiar phenomenon. So, we shall introduce the result of Campbell's method which can provide us with useful information on the flux line lattice for deducing the mechanism of the peculiar saturation phenomenon. One merit of this method is that it clarifies not only the pinning force density but also the Labusch parameter α_L and the interaction distance d_i mentioned in Sect. 5.3. These quantities provide us useful information on the flux line lattice. Figure 7.32(a) and (b) show the results on α_L and d_i, respectively, in the Nb-Ti specimen in Fig. 7.26(b) which shows saturation [41]. The figure indicates that α_L increases with the pinning strength and is approximately proportional to $(1-b)$ at high fields. If the shearing flow model of Kramer held correct, α_L would be independent of the pinning strength and would be proportional to C_{66}, and hence, to $(1-b)^2$ (see Exercise 7.6). In addition, while its theoretical value is at most 1.6×10^2 Nm^{-2} at $b = 0.7$, the observed value amounts to 1.0×10^3 Nm^{-2} and is about seven times as large as the theoretical limit. This means that the elastic restoring force against the motion of flux lines cannot be the shear force of the flux line lattice. At the same time the interaction distance d_i shown in Fig. 7.32(b) is not proportional to $a_f \propto B^{-1/2}$ but decreases linearly as $(1-b)$ at high fields. This also shows that the phenomenon is not so simple as assumed in the Kramer model. In addition, d_i decreases with strengthening pinning, while α_L increases. Similar results were also obtained for Nb-Ta [48]. These results show that, while the elasticity of the flux line lattice is enhanced with strengthening pinning, the elastic limit of strain is lowered. That is, the flux line lattice becomes hard but brittle at the same time, keeping the proof stress unchanged as shown in Fig. 7.33.

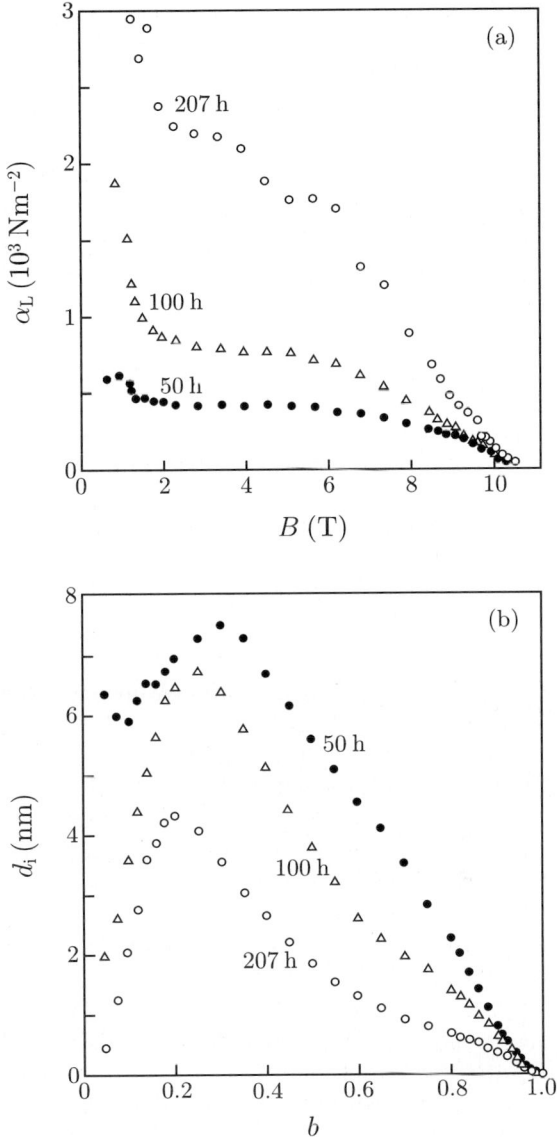

Fig. 7.32. (a) Labusch parameter α_L and (b) the interaction distance d_i in Nb-Ti [41] showing saturation in Fig. 7.26(b)

In the case of nonsaturation, on the other hand, while α_L behaves similarly to the case of saturation, the decrease of d_i at high fields is less significant. As a result, the pinning force density increases at high fields with strengthening pinning. That is, the characteristic brittleness of the flux line lattice in saturation disappears.

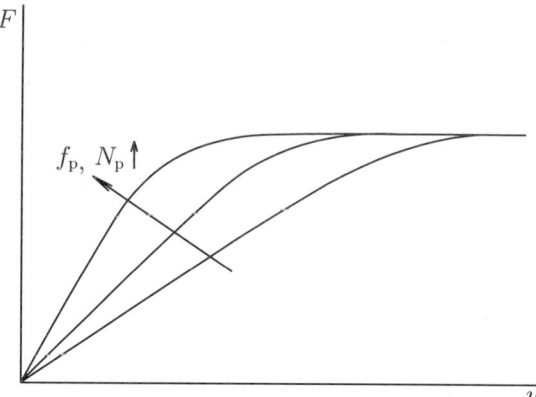

Fig. 7.33. Variation in pinning force density vs. displacement characteristic by increment of f_p or N_p in the case of saturation

7.5.5 Avalanching Flow Model

Based on the above experimental results using Campbell's method the avalanching flow model [41, 48, 49] was proposed to explain saturation and nonsaturation. The details of α_L and d_i derived from this model are described in Appendix A.6. A brief summary of it is given here. When a Lorentz force with a certain strength is applied on the flux line lattice, local plastic deformations are proposed to occur around lattice defects. In this model it is assumed that if the pins are not sufficiently strong, the whole flux line lattice becomes unstable under local plastic deformations and an avalanching flux flow occurs (see Fig. 7.34). In this case, when the pinning becomes stronger, the number density of lattice defects increases, and the flux line lattice becomes more unstable, resulting in an easier occurrence of avalanching flux flow. Thus, the increase of α_L and the decrease of d_i are canceled out giving rise to saturation. On the other hand, when the pinning strength exceeds some critical level, local plastic deformations cannot develop to global instability due to the strong shielding effect of surrounding pins. Hence, the flux flow occurs only when a Lorentz force corresponding to the strength of the pins is applied, and nonsaturation results.

According to this model saturation and nonsaturation occur for various defects and the key factor which determines the characteristic is the pinning strength. That is, nonsaturation occurs for sufficiently strong pins, and the prediction of the Kramer model that a pinning characteristic superior to the saturation cannot be realized is incorrect. It is empirically known that nonsaturation is difficult to attain in superconductors with high κ values. This is because the elementary pinning force is weaker due to the shorter coherence length ξ as mentioned above. The above argument only applies from the viewpoint of pinning strength, i.e., the elementary pinning force f_p. It should

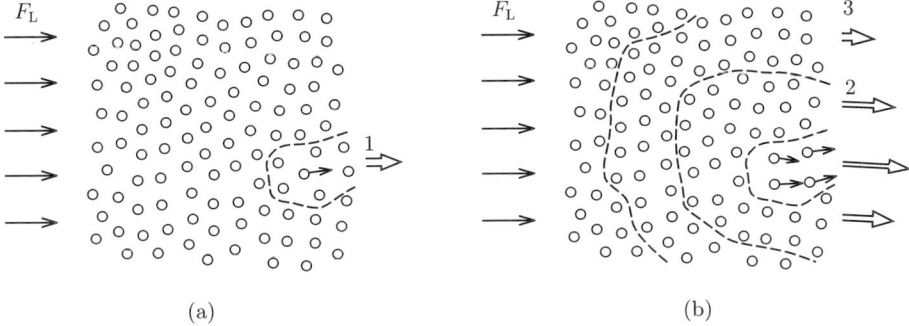

Fig. 7.34. (a) Local plastic deformation of flux line lattice around a defect and (b) development to avalanching flux flow [49]

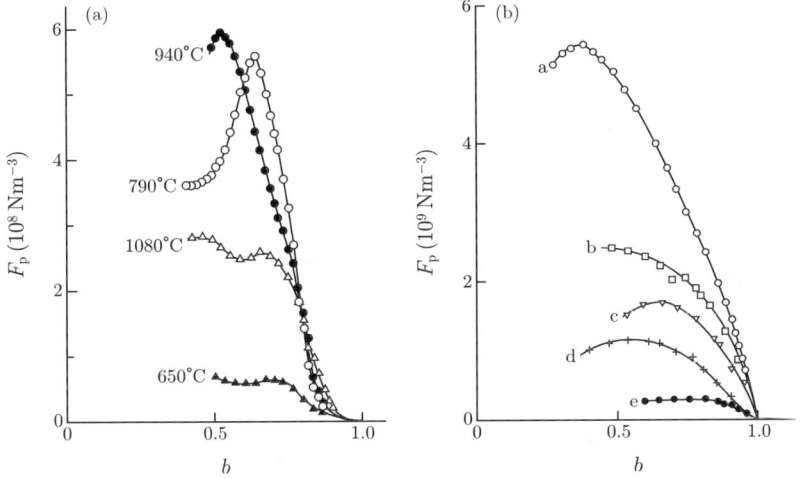

Fig. 7.35. Pinning force density in niobium with large voids for the cases of (a) saturation with low number density of pins [50] and (b) nonsaturation with high number density of pins [51]

be noted that the number density of pins N_p is also important in determination of the pinning characteristic. Figure 7.35 shows the results for pinning by strong voids in niobium [50, 51]. It can be seen that the characteristic is of saturation type for small N_p, while nonsaturation is attained for large N_p.

Figure 7.36 summarizes the expected variation in F_p over wide ranges of pinning parameters, f_p and N_p [41]. The summation theory is valid except in the region of saturation. Namely, the linear summation and the statistical summation described by the coherent potential approximation theory hold in the regions beyond and below the saturation, respectively. In the figure a variation from point a to point a' is seen in an example of Nb-Ti (see Fig. 7.26(b)) showing saturation, and that from point a' to a'' is seen in an

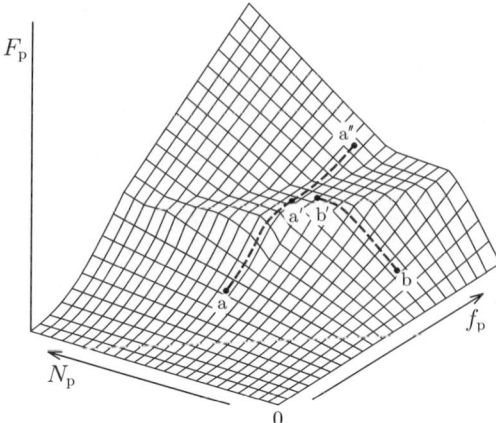

Fig. 7.36. Expected dependences of the pinning force density F_p on pinning parameters, f_p and N_p [41]. In the region where F_p is constant saturation occurs

example of Nb-Ti (see Fig. 7.31) showing the transition from saturation to nonsaturation. An example from point b to point b' is found for Nb_3Sn or neutron irradiated V_3Si [52]. The variation in niobium shown in Fig. 7.35 will be clear.

7.6 Peak Effect and Related Phenomena

The peak effect is a peculiar pinning phenomenon. That is, the critical current density has a peak at a certain magnetic field (peak field), while it usually decreases monotonically with increasing magnetic field. The peak effect is observed not only in metallic superconductors but also in high-temperature superconductors. The peak field is close to the upper critical field for superconductors with a small G-L parameter κ, and is shifted to lower field according to increasing κ. Especially in high-temperature superconductors with very large κ the peak effect occurs at fairly low fields (see Fig. 7.37) [53–55]. The peak effect in Bi-2212 superconductor with an extremely large anisotropy caused by a layered crystalline structure will be described in Sect. 8.2.

The following mechanisms have been proposed for explanation of the peak effect: (1) matching, (2) elementary pinning by a weakly superconducting region, (3) reduction in elastic moduli of the flux line lattice due to the nonlocal nature and (4) synchronization of the flux line lattice, etc.

The matching mechanism is that the critical current density has a peak at a magnetic field at which the flux line spacing is equal to the pin spacing. Figure 7.38 shows the critical current density of a Pb-Bi thin film [56] prepared by vapor deposition with a periodically varied concentration of bismuth. The peak effect is observed not only at the matching field at which the flux

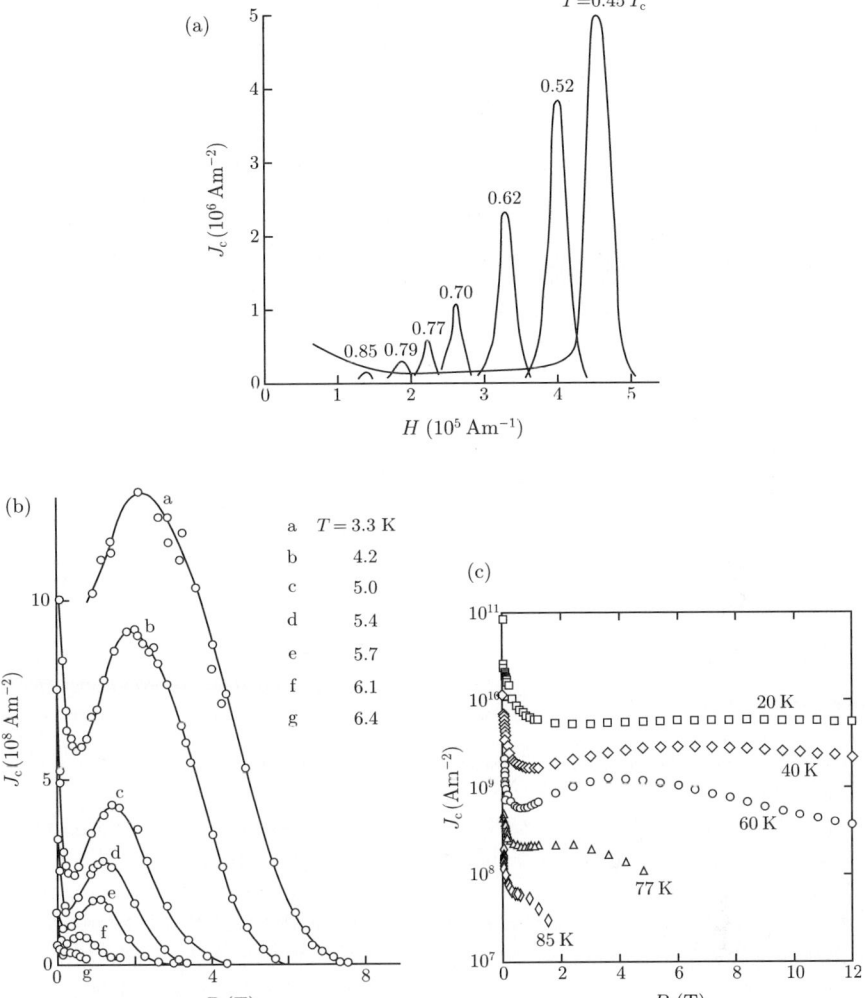

Fig. 7.37. Peak effects in various superconductors: (a) Nb-50at%Ta [53] with low κ, (b) Ti-22at%Nb [54] with high κ and (c) Y-Ba-Cu-O high-temperature superconductor in a magnetic field parallel to the c-axis [55]

line spacing is equal to the period of variation of the bismuth concentration, but also at harmonic fields. However, such a matching mechanism is usually observed only for a specimen with a highly periodic arrangement of relatively weak pins. The peak fields are rather low in most cases. Another characteristic of this peak effect is that the peak field is independent of temperature.

In most cases, however, the peak field varies with temperature as shown in Fig. 7.37(a), and the pinning force density obeys the scaling law within a certain range of temperature (except in the vicinity of the critical temperature,

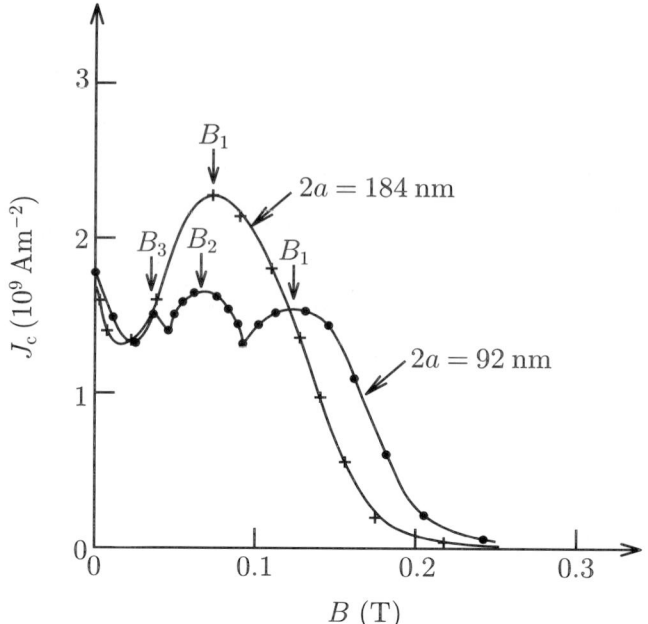

Fig. 7.38. Critical current density of Pb-Bi thin film with a pinning structure introduced by periodic variation in Bi concentration [56]. $2a$ is the period of concentration variations, and the flux line spacing coincides with this period at the peak field B_1. Harmonic peaks are also observed at B_2 and B_3

for example). In addition, the peak field is likely to move to lower fields with strengthening pinning as shown in Fig. 7.20(a). This means that the trend is just opposite to the matching mechanism. That is, since the flux line spacing at matching becomes shorter when the number density of pins increases, this mechanism predicts a shift of the peak to higher field.

If the pins are of a more weakly superconducting phase than the matrix superconductor, the peak effect is expected to be observed at the magnetic field where the pins change from a superconducting state to a normal state. Figure 7.39 shows the magnetizations of two phases when pinning regions have the same thermodynamic critical field but a slightly smaller κ value than the matrix. The elementary pinning force originates from the magnetic interaction, which is proportional to the difference in the magnetization, is zero at point "a" and has a maximum at point 'b', the upper critical field of the pinning region [57]. Thus, the peak effect arises from a variation in the elementary pinning force itself with magnetic field, i.e., the so-called field-induced pinning mechanism. Figure 7.40 shows the results [58] of magnetization of $Pb_{57}In_{22}Sn_{21}$ at 2.0 K and 4.2 K. It has been proposed that the weaker pinning at 2.0 K than at 4.2 K in the low field region originates from pinning by the superconducting tin phase and that the peak effect at higher

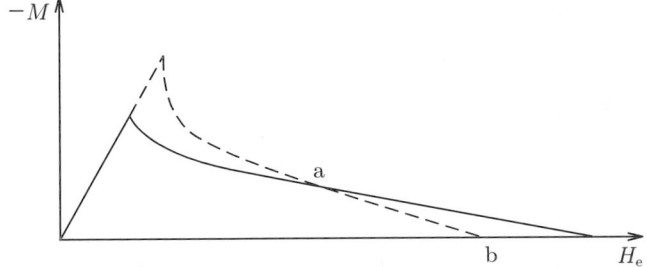

Fig. 7.39. Magnetizations of superconducting matrix (*solid line*) and inclusions (*broken line*) with the same thermodynamic critical field but a slightly smaller κ value

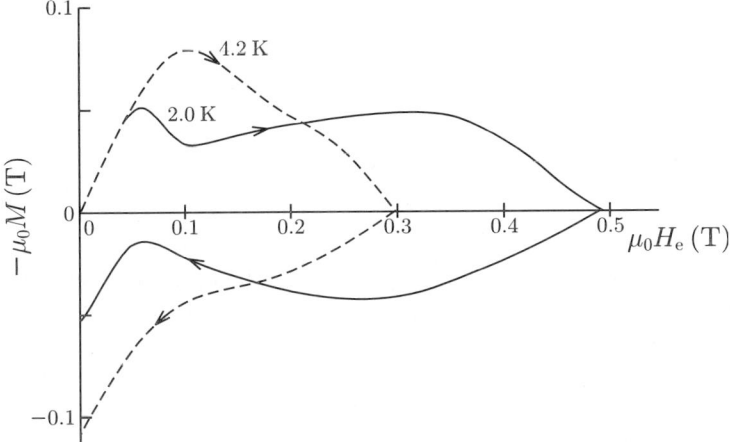

Fig. 7.40. Magnetization curves of $Pb_{57}In_{22}Sn_{21}$ at 2.0 K and 4.2 K [58]

field is caused by the transition of the tin phase to the normal state. The peak effect of $YBa_2Cu_3O_7$ shown in Fig. 7.37(c) is speculated to be caused by a similar mechanism related to of oxygen deficient regions ($YBa_2Cu_3O_{6.5}$) [59]. However, there is a problem in this speculation as will be mentioned later.

It is predicted in the Larkin-Ovchinnikov theory [7] that the flux bundle size determined by the pinning correlation lengths decreases rapidly with increasing magnetic field due to a reduction in the elastic moduli, C_{11} and C_{44}, originating from the nonlocal property, and resulting in the enhancement of the critical current density. However, this reduction in the nonlocal elastic moduli, i.e., the key factor of this mechanism, contains a theoretical problem (see Sect. 7.2 and Appendix A.5). In addition, the appearance of a sharp peak of critical current density observed in low-κ superconductors cannot be explained by this mechanism, since the decrease in the elastic moduli is not appreciable in such superconductors. In high-κ superconductors such as high-temperature superconductors, the peak effect is observed at relatively low fields. There is no theoretical foundation for a decrease in the elastic moduli at such low fields.

Finally the mechanism of synchronization is discussed here. As mentioned in the last section, this is a mechanism such that the flux lines rearrange themselves to fit the structure of pins due to a reduction in the shear modulus C_{66} with increasing field, resulting in a more strongly pinned state. This mechanism was first proposed by Pippard [60] and then developed by Kramer [43]. Although many problems are contained in this model concerned with the saturation discussed in the last section, the concept of synchronization is developed to a model of the order-disorder transition of flux lines [61]. Namely, a proliferation of dislocations into the flux line lattice occurs at the transition, resulting in a more strongly pinned state of the flux line lattice. Since the transition field at which the peak effect starts has a strong dependence on the pinning strength, as shown in Fig. 7.20(a), it is considered that the pinning energy is involved in the transition [62].

Changes in the peak effect in Y-123 single crystals with varying amounts of oxygen deficiency and in V_3Si single crystals with varying doses of neutron irradiation are compared over a wide range in Fig. 7.41 [63]. These results for similar pointlike defects are quite similar to each other. Thus, it can be said that the peak effect is of a general nature independent of the kind of pinning centers. This supports a speculation that the peak effect is brought about by the order-disorder transition of flux lines. It can also be concluded that the result shown in this figure cannot be explained by a simple field-induced pinning mechanism.

Here we shall introduce a new experimental result on Sm-123 powder specimens [64]. Figure 7.42 shows that the peak effect of the magnetization current disappears for specimens of an average size smaller than the longitudinal pinning correlation length L, which is approximately equal to Campbell's AC penetration depth λ'_0, since $C_{44} \simeq C_{11}$. If this peak effect comes from the elementary pinning force due to the field-induced pinning mechanism, the peak effect will not be influenced by the specimen size. When the specimen size is below the pinning correlation length, it is considered that a collective pinning of lower dimension occurs, resulting in a higher pinning efficiency than in three-dimensional pinning in the bulk case (see Eq. (7.69) which shows that J_c is inversely proportional to the square root of the specimen diameter). Hence, even when the magnetic field reaches the transition field, the transition itself disappears, since the flux lines are already in the disordered state with the higher pinning efficiency, resulting in a disappearance of the peak effect.

Many experimental results suggest that a plastic deformation of the flux line lattice occurs in the disordered state around the peak field. One of the proofs is pronounced flux flow noise [65] observed at a magnetic field just below the peak field at which the critical current density increases with increasing field as shown in Fig. 7.43. This result shows a variation in the flux line structure accompanied by plastic deformation with increasing critical current density and is consistent with the assumptions of the synchronization

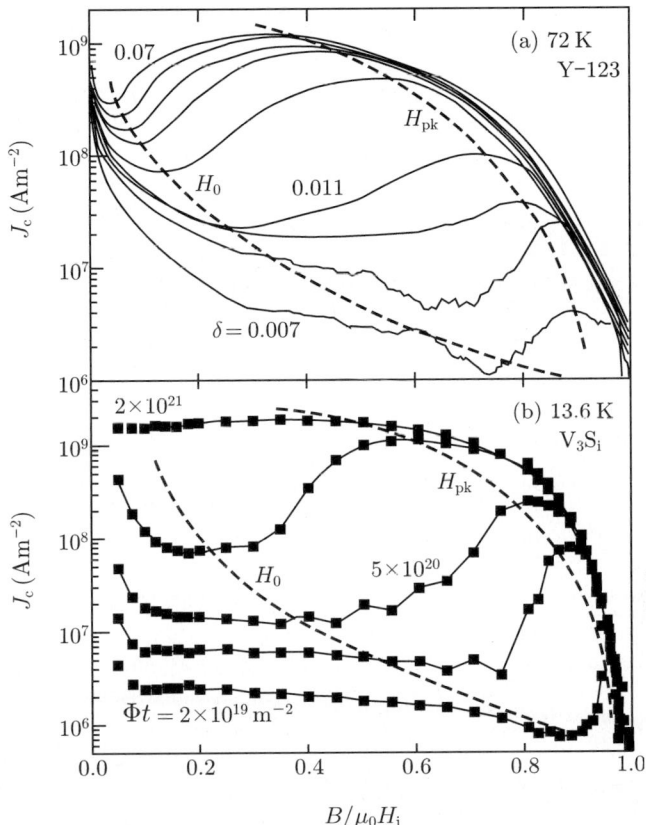

Fig. 7.41. (a) Variation in critical current density in Y-123 single crystal at 72 K under variation in the amount of oxygen deficiency and (b) variation in the critical current density in V_3Si single crystal at 13.6 K under variation in the dose of neutron irradiation [63]

mechanism. This variation is not the simple elastic deformation assumed by Kramer.

A history effect is also observed in this region of magnetic field. That is, the critical current density is not uniquely determined by the final condition of the magnetic field and temperature but depends also on the path through which the final condition is reached, as shown in Fig. 7.44 [66]. Figure 7.44(b) shows the results on a niobium single crystal rod with a triangular cross-section after a tensile deformation of 1%. When the sample is cooled in a magnetic field, the critical current density is largest and the current-voltage characteristic is concave upward with a strong nonlinearity. When the magnetic field is reduced from a sufficiently high value at a fixed temperature, the critical current density is smaller than that obtained in the field-cooled process. The critical current density is lowest, and the current-voltage

328   7 Flux Pinning Characteristics

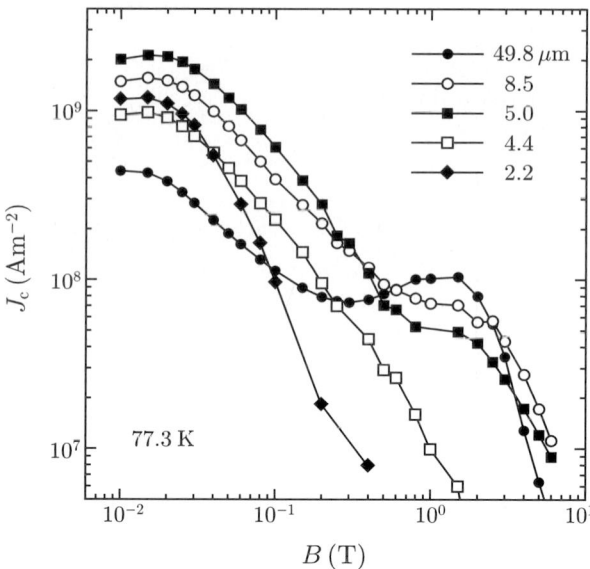

Fig. 7.42. Critical current density of superconducting Sm-123 powder specimens with different particle sizes at 77.3 K [64]. When the pinning correlation length exceeds the particle size around the peak field, the peak effect disappears

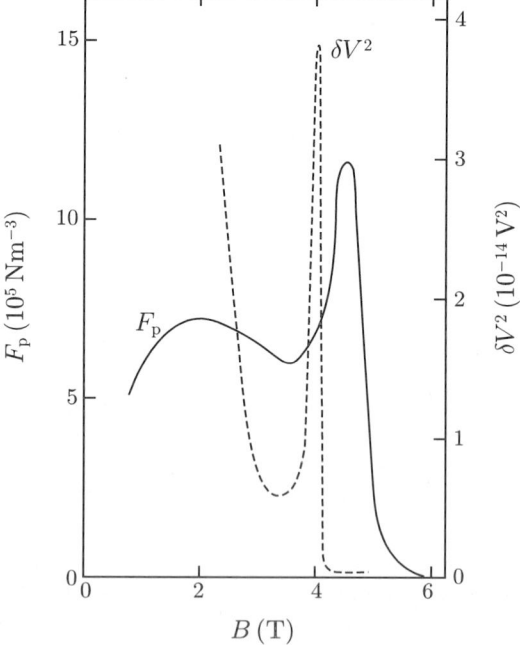

Fig. 7.43. Large flux flow noise observed in Pb-In [65] at a magnetic field just below the peak field

7.6 Peak Effect and Related Phenomena

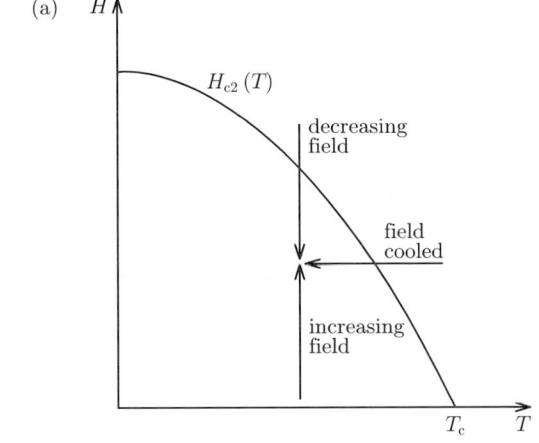

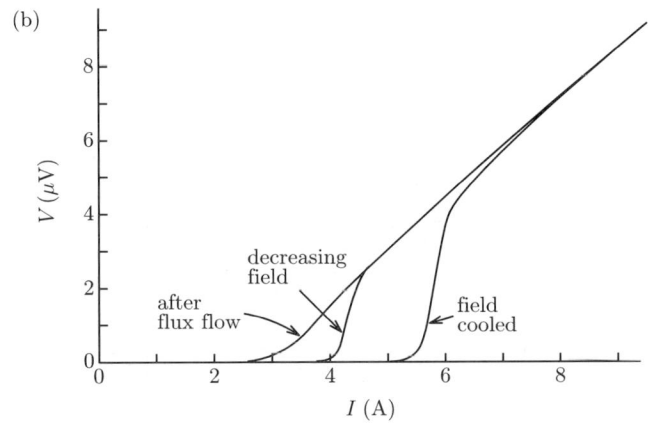

Fig. 7.44. (a) Various paths to the condition of measurement and (b) corresponding current-voltage curves for a niobium specimen [66]

characteristic is almost linear when the magnetic field is increased at a constant temperature. If a state of sufficiently high voltage is attained after any other state, the current-voltage characteristic is the same as that in the increasing field process. This history effect makes it clear that the state of the flux line lattice varies considerably depending on the process leading to the final condition of measurement. Especially in the field-cooled process, the current-voltage characteristic shown in Fig. 7.44(b) deviates from linearity, and hence, the flux line lattice is not believed to be in the perfectly random state which is describable by the statistical average. Therefore, the synchronization caused by the order-disorder transition seems to really involve a change in the state of flux lines to fit a distribution of pins.

330   7 Flux Pinning Characteristics

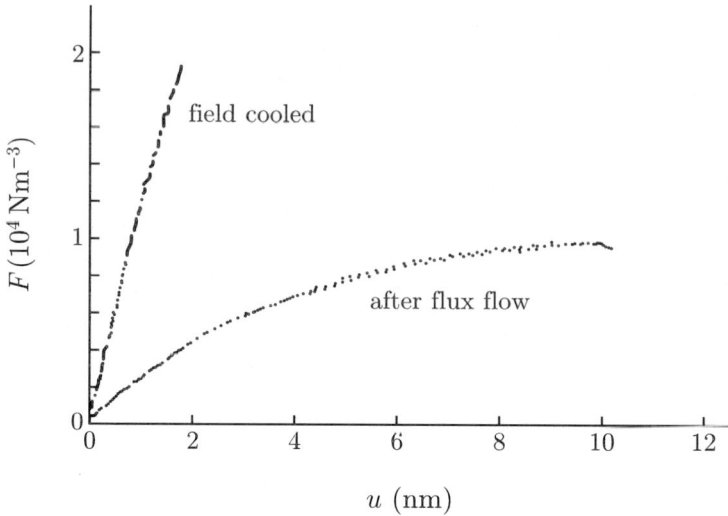

Fig. 7.45. Pinning force vs. displacement characteristics in the field-cooled process and after flux flow for Nb-Ta showing an history effect [67]. In the field-cooled process the critical current density is larger, and the variation in the pinning force density with displacement is larger

Figure 7.45 shows the pinning force vs. displacement of flux lines in Nb-Ta showing an history effect [67]. It was found that the flux line lattice is hardly deformed in the field-cooled process, while it is easily deformed after flux flow. Figure 7.46 shows that the peak effect of the critical current density can be introduced by introducing defects into the flux line lattice within a range of magnetic field in which the state of the flux line lattice can easily be varied [35]. That is, after a measurement at $T/T_c = t = 0.7$ in a certain magnetic field the temperature was reduced to $t = 0.6$ and the measurement was done again. Then, the temperature was brought back to $t = 0.7$, the magnetic field was varied, and then the measurement was repeated. This result shows that the flux line lattice forms an advantageous structure for pinning at the peak field and at $t = 0.7$. Such a structure can be maintained and the peak effect occurs again even when the temperature is reduced to $t = 0.6$. This result also supports the order-disorder transition of flux lines with respect to the peak effect. Similar history effects are observed in the magnetic field range between the minimum and the maximum of critical current density also for RE-123 superconductors which show the peak effect [68], as in Fig. 7.37(c).

Figure 7.47(a) shows summarized results of history effects on various superconductors investigated by Küpfer and Gey [69]. The ordinate is the ratio of flux line spacing a_f to the mean pin spacing d_p, and the abscissa is the ratio of the calculated displacement of a flux line caused by one pin u to a_f. The partially filled symbols in the figure show the rate of observation of the history effect. According to this result the history effect is not observed in

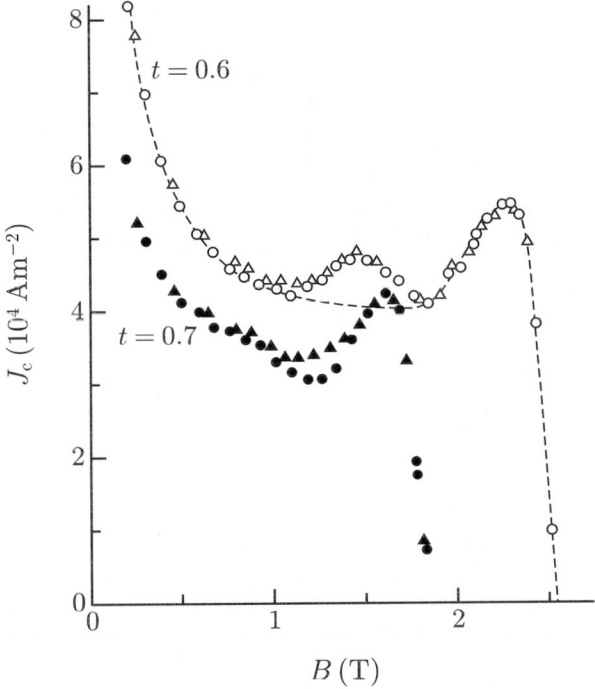

Fig. 7.46. History effect in Nb$_3$Ge thin film [35]. After measurement at a normalized temperature $T/T_c = t = 0.7$ in a certain magnetic field, the temperature is reduced to $t = 0.6$ and the measurement is repeated. Then, the temperature is brought back to $t = 0.7$, the magnetic field is varied and the measurement is done again. This whole procedure is repeated. *Circles* and *triangles* show results at increasing and decreasing field processes, respectively. The *broken line* shows the result when only the magnetic field is changed under isothermal conditions

an amorphous state of flux lines where f_p is very large or in a lattice state where both f_p and N_p are small. The history effect is observed only in an intermediate state (see Fig. 7.47(b)). This speculation on the state of the flux lines is consistent with the above experimental results on flux flow noise, etc.

7.7 Pinning Potential Energy

In this section a method is treated to theoretically estimate the pinning potential energy U_0, which is important in determination of the relaxation rate of the critical current density J_c and the irreversibility line, i.e., the boundary between the reversible region with zero J_c and the irreversible region with nonzero J_c. As mentioned in Sect. 5.3 the averaged pinning potential energy of flux lines per unit volume is expressed as $\widehat{U}_0 = \alpha_L d_i^2/2$ in Eq. (5.19) in terms of the Labusch parameter α_L and the interaction distance d_i. Since each flux

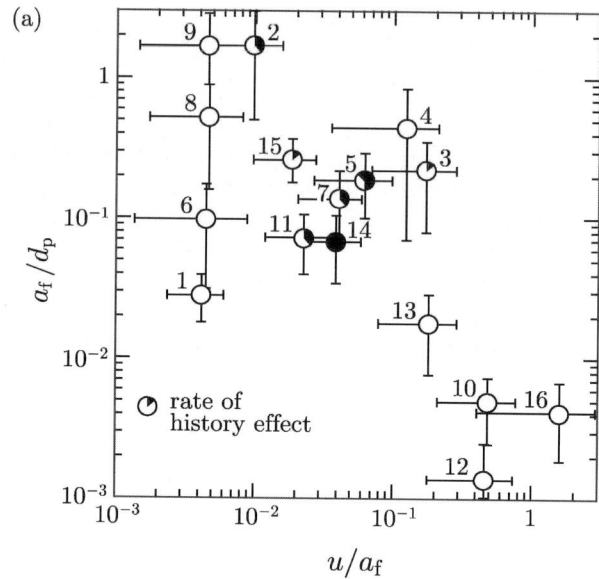

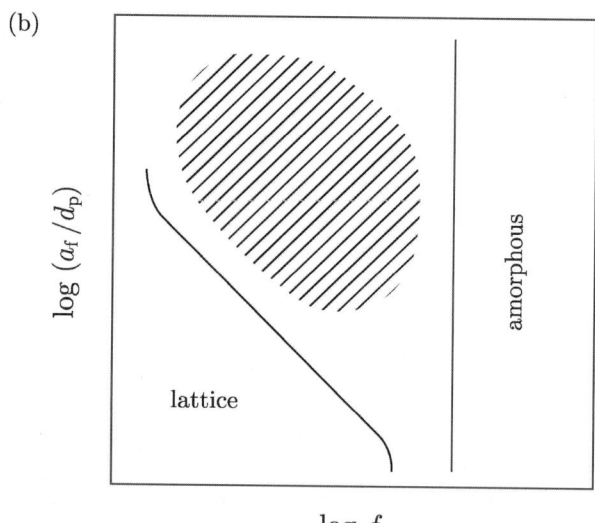

Fig. 7.47. (a) Dependence on the history effect of the critical current density on the number density of pins N_p and the elementary pinning force f_p [69]. The ordinate represents the flux line spacing a_f divided by the mean pin spacing d_p and is proportional to $N_p^{1/3}$. The abscissa is the displacement of a flux line u divided by a_f and is proportional to f_p. A filled fraction of symbol shows the relative difference in critical current densities between increasing and decreasing field processes. (b) Summary of the results in (a) from the viewpoint of the state of the flux line lattice. The history effect is observed only in an intermediate state between amorphous and lattice states

bundle is expected to be independent of the others, $\widehat{U}_0$ is given by the pinning potential energy of each flux bundle divided by its mean volume. It should be noted here that observed values of α_L and d_i, which are related to the critical current density through Eq. (3.94), are influenced by the flux creep, as is the critical current density deteriorated from the value (J_c0) obtained by the summation theory. However, the pinning potential energy necessary for analyzing creep phenomena would be a virtual one without the influence of flux creep. Hence, we should use virtual values of α_L and d_i in the creep-free case when we evaluate $\widehat{U}_0$. This will be discussed later.

The pinning potential energy of a flux bundle is given by the product of $\widehat{U}_0$ in Eq. (5.19) and its volume V. Then, a method of estimating the volume is described. Since the flux bundle is a cluster of flux lines which move coherently, it is considered to correspond to the region in which short-range translational order is maintained. Hence, a simple argument leads to the result that the sizes of flux bundles will be given by the virtual pinning correlation lengths of the flux line lattice in the creep-free case. The longitudinal pinning correlation length is derived for an example. It is assumed that flux lines are directed along the z-axis and that the flux line lattice is deformed by pins within an elastic range. When the flux lines are slightly displaced by u in the direction normal to the z-axis, the force balance is written as

$$C_{44}\frac{\partial^2 u}{\partial z^2} = \alpha_\mathrm{L} u, \tag{7.90}$$

where the left- and right-hand sides are the elastic restoring force density and the pinning force density, respectively. This is easily solved and we have

$$u(z) = u(0)\exp\left(-\frac{|z|}{L_0}\right), \tag{7.91}$$

where

$$L_0 = \left(\frac{C_{44}}{\alpha_\mathrm{L}}\right)^{1/2} \simeq \left(\frac{Ba_\mathrm{f}}{\mu_0 \zeta J_\mathrm{c0}}\right)^{1/2} \tag{7.92}$$

is the pinning correlation length in the longitudinal direction. In the above, Eq. (7.75) and Eq. (3.94) in which J_c is replaced by J_c0 were used. The transverse pinning correlation length can be obtained in a similar manner as

$$R_0 = \left(\frac{C_{66}}{\alpha_\mathrm{L}}\right)^{1/2} \simeq \left(\frac{C_{66}a_\mathrm{f}}{\zeta J_\mathrm{c0} B}\right)^{1/2}. \tag{7.93}$$

Thus, the volume of the flux bundle is estimated as

$$V = L_0 R_0^2 \tag{7.94}$$

for a bulk superconductor with a size larger than L_0 and R_0. The above lengths, L_0 and R_0, are equivalent to L_c and R_c given by Eqs. (7.57) and

(7.56) in the virtual creep-free case, respectively. On the other hand, the observed pinning correlation lengths, L and R, are given by Eqs. (7.92) and (7.93) with substitution of J_c into J_{c0}.

A longer pinning correlation length, $L_0' = (C_{11}/\alpha_L)^{1/2}$, exists along the direction of the Lorentz force due to the magnetic pressure. The reason why R_0 is chosen instead of L_0' as the flux bundle size in this direction is that the motion of flux lines in flux creep is not a global and continuous motion of the whole flux line lattice as in flux flow but the local and discontinuous motion of a cluster of flux lines. That is, the motion of flux lines will contain a plastic shear. Such a motion is proposed to occur mostly within a restricted region around lattice defects and the size of this region is expected to be much smaller than L_0'. This is a different point from the collective creep theory argument by Feigel'man et al. [39] If the size along this direction is given by L_0' as assumed by them, U_0 is too large to explain quantitatively using practical J_c values.

In a case where flux creep is significant, as in high-temperature superconductors at high temperatures, the condition of the flux line lattice is largely disordered, and the shear modulus C_{66} takes on a much smaller value than that in Eq. (7.12) for a perfect flux line lattice. The practical value of C_{66} is zero when the flux line lattice is melted. Since the condition of the flux line lattice is determined as a result of flux creep, the values of C_{66} and R_0 cannot be known beforehand. Even if the condition of the flux line lattice is known, it is very hard to determine these values. Hence, it is necessary to determine R_0 using a different method. One of the candidates is the method of irreversible thermodynamics which is sometimes used in this book (see Sect. 4.6 and Appendix A.3). That is, R_0 is considered to be determined such that the critical current density might take on a maximum value to minimize energy dissipation under flux creep. If we represent R_0 as

$$R_0 = g a_f, \quad (7.95)$$

g^2 gives the number of flux lines in a flux bundle. According to the result of the derivation based on the above principle shown in Appendix A.8, we have [70]

$$g^2 = g_e^2 \left[\frac{5k_B T}{2U_e} \log \left(\frac{B a_f v_0}{E_c} \right) \right]^{4/3}, \quad (7.96)$$

where g_e^2 and U_e are the values of g^2 and U_0 for a perfect flux line lattice. The flux creep property of high-temperature superconductors depends strongly on the dimensionality of the superconductor through the flux bundle size. This will be discussed in detail in Sects. 8.3 and 8.5.

Here we shall mention matters to be noted in the above theoretical analysis. One of them is that the physical quantities associated with flux pinning, i.e., α_L and d_i, are necessary to derive U_0. These quantities must be virtual ones in the creep-free case as mentioned above. Another point is that, although these quantities can be theoretically obtained if the pins are known, the material parameters, α and β in Eq. (1.21), of pins are unclear, and theoretical calculations are impossible in most high-temperature superconductors. Among them the interaction distance does not depend much on the kind of pins and is known in most cases to be expressed as in Eq. (7.75). In fact a relationship of this form is confirmed by experiments on various metallic superconductors in the region where the influence of flux creep is not significant. Using this relationship α_L can be expressed in terms of the virtual critical current density J_{c0} in the creep-free case. As a result, the pinning potential energy U_0 can be described in terms of J_{c0} alone. Even in this case, if the dominant pins are unknown as in high-temperature superconductors, J_{c0} cannot be calculated. However, if the critical current density is observed at sufficiently low temperatures where the influence of flux creep is not large, it may be approximately used for J_{c0}. That is, the virtual critical current density at high temperatures can be estimated using the scaling law of critical current density at sufficiently low temperatures. This is a practical method by which ambiguity of pinning centers can be compensated to some extent by experimental results. On the other hand, in the case of significant flux creep, J_{c0} should be determined so as to explain $J_c(B,T)$ over a wide range of temperature. This will be discussed in Sect. 8.3.

Thus, the pinning potential energy of the flux bundle is given by [64, 71]

$$U_0 = \frac{g^2}{2(\sqrt{3}/2)^{7/4}\zeta^{3/2}} \left(\frac{\phi_0^7 J_{c0}^2}{\mu_0^2 B}\right)^{1/4} = \frac{0.835 k_B g^2 J_{c0}^{1/2}}{\zeta^{3/2} B^{1/4}} \tag{7.97}$$

for a bulk superconductor, where $(1/2)(2/\sqrt{3})^{7/4}(\phi_0^7/\mu_0^2)^{1/4} \simeq 0.835 k_B$ was used for a numerical equation. In such a bulk case the pinning potential energy is independent of the size of the superconductor. This is the case where the size of the superconductor along the direction of magnetic field is larger than L_0 given by Eq. (7.92). On the other hand, for a superconducting thin film of thickness d smaller than L_0 in a perpendicular magnetic field, the flux bundle volume is

$$V = dR_0^2 \tag{7.98}$$

and the pinning potential energy is given by

$$U_0 = \frac{4.23 k_B g^2 J_{c0} d}{\zeta B^{1/2}} \tag{7.99}$$

(Exercise 7.8). In the above the following numerical equation was used: $(1/2)(2/\sqrt{3})^{3/2}\phi_0^{3/2} \simeq 4.23 k_B$. This result holds also for superconducting powder of a size smaller than L_0.

It should be noted that the pinning potential energy is different from the apparent one obtained from magnetic relaxation measurements as mentioned in Sect. 3.8. It is irreversibility field what is directly related to U_0. The irreversibility field obtained from U_0 will be discussed for high-temperature superconductors in Sect. 8.5.

336   7 Flux Pinning Characteristics

The above treatment to estimate U_0 is formally the same as in collective creep theory [39]. However, there are many points of difference. One of them is the difference in the flux bundle size along the direction of the Lorentz force as abovementioned. Another point is that the summation for J_c is treated separately here, since the collective pinning theory cannot explain the experimental results on J_c. This treatment may seem to be contradictory. However, such an assumption is incorrect. Although the two theories are the same with regards to the correlation lengths under the given J_c, their standpoint is different only for the influence of the correlation lengths on the summation of J_c. The resultant large difference from the theory of Feigel'man et al. is that the maximum value of the activation energy U is given by U_0 and does not diverge even at zero current density, unless J_{c0} does not reduce to zero.

Exercises

7.1. It is assumed that, when a strain ϵ is added to a superconductor, a stress concentration occurs around pins and the pinning force density is increased by a factor of $(1 + c\epsilon^2)$ due to the pinning mechanism of elastic interaction except for the contribution from the variation of H_{c2}. In the above c is a small constant value. Discuss the difference between the temperature scaling law and the strain scaling law of pinning force density in this case.

7.2. It is assumed that a periodic displacement u^* of a single mode of wave number k is added to a flux line lattice of $|\Psi|^2$ given by Eq. (1.98) along the x-axis normal to the close-packed row. In the above the wave number k is sufficiently smaller than ξ^{-1}. If the local magnetic flux density is given by Eq. (1.101), show that the quantization of magnetic flux is not fulfilled under this variation. At the same time calculate the uniaxial compression modulus $C_{11}(k)$.

7.3. Derive Eqs. (7.50) and (7.51).

7.4. The force balance equation describing the motion of a rigid body in a periodic potential of period $2\pi/k_p$ under a constant driving force f is written as

$$f + f_p \cos k_p x - \eta^* \dot{x} = 0 \, ,$$

where x is the position of the rigid body and the third term is the viscous force. Solve this and calculate the current-voltage characteristic. Compare the obtained result with the result shown in Fig. 7.10(a).

7.5. Derive the variation in the pinning force density against the displacement of flux lines using the statistical theory for the pinning force given by Eq. (7.31) and discuss the reversible motion of flux lines.

7.6. Calculate the Labusch parameter α_L when the Kramer model holds correct. Assume the arrangement of line pins shown in Fig. 7.48. (*Hint*: Calculate the shearing displacement of flux lines under the driving force in

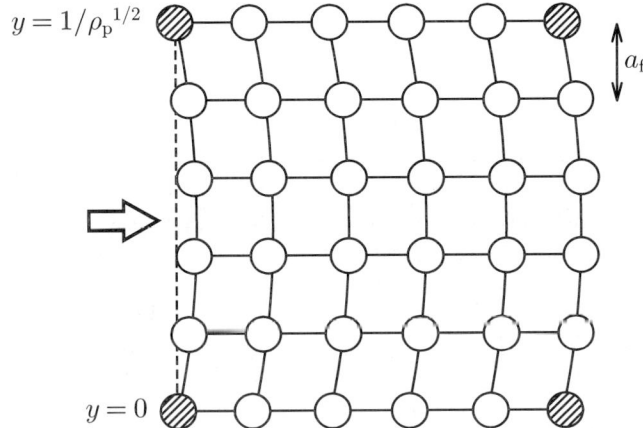

Fig. 7.48. Deformation of flux line lattice pinned by line pins distributed like a superlattice under the driving force. Shadowed flux lines are strongly pinned by line pins

the direction shown by the arrow. Derive α_L from a relationship between the average displacement and the driving force.)

7.7. Show that the pinning correlation lengths, L_c and R_c, given by Eqs. (7.57) and (7.56) are approximately the same as the characteristic lengths, L_0 and R_0, defined by Eqs. (7.92) and (7.93), respectively.

7.8. Derive Eq. (7.99) for a superconducting thin film of thickness d thinner than L_0 in a perpendicular magnetic field.

References

1. A. M. Campbell and J. E. Evetts: Adv. Phys. **21** (1972) 372.
2. J. W. Ekin: *Adv. Cryog. Eng. Mater.* (Plenum, New York, 1984) Vol. 30, p. 823.
3. R. Labusch: Phys. Status Solidi: **19** (1967) 715.
4. R. Labusch: Phys. Status Solidi: **32** (1969) 439.
5. E. H. Brandt: Phys. Rev. B **34** (1986) 6514.
6. E. H. Brandt: J. Low Tem. Phys. **26** (1977) 709.
7. A. I. Larkin and Yu. N. Ovchinnikov: J. Low Temp. Phys. **34** (1979) 409.
8. K. Yamafuji and F. Irie: Phys. Lett. **25A** (1967) 387.
9. R. Labusch: Crystal Lattice Defects: **1** (1969) 1.
10. J. Lowell: J. Phys. F **2** (1972) 547.
11. A. M. Campbell: Phil. Mag. B **37** (1978) 149.
12. E. J. Kramer: J. Nucl. Mater. **72** (1978) 5.
13. R. Schmucker and E. H. Brandt: Phys. Status Solidi (b) **79** (1977) 479.
14. E. J. Kramer: J. Appl. Phys. **49** (1978) 742.
15. R. Labusch: J. Nucl. Mater. **72** (1978) 28.
16. A. Schmid and W. Hauger: J. Low Temp. Phys. **11** (1973) 667.

17. T. Matsushita, E. Kusayanagi and K. Yamafuji: J. Phys. Soc. Jpn. **46** (1979) 1101.
18. E. H. Brandt: *Proc. Int. Symp. on Flux Pinning and Electromagnetic Properties in Superconductors*, Fukuoka, 1985, p. 42.
19. R. Wördenweber and P. H. Kes: Physica **135B** (1985) 136.
20. R. Wördenweber and P. H. Kes: Phys. Rev. B **34** (1986) 494.
21. P. H. Kes and C. C. Tsuei: Phys. Rev. B **28** (1983) 5126.
22. H. R. Kerchner: J. Low Temp. Phys. **50** (1983) 337.
23. S. J. Mullock and J. E. Evetts: J. Appl. Phys. **57** (1985) 2588.
24. T. Matsushita: Physica C **243** (1995) 312.
25. A. M. Campbell, H. Küpfer and R. Meier-Hirmer: *Proc. Int. Symp. on Flux Pinning and Electromagnetic Properties in Superconductors*, Fukuoka, 1985, p. 54.
26. H. R. Kerchner, D. K. Christen, C. E. Klabunde, S. T. Sekula and R. R. Coltman, Jr.: Phys. Rev. B **27** (1983) 5467.
27. R. Meier-Hirmer, H. Küpfer and H. Scheurer: Phys. Rev. B **31** (1985) 183.
28. In private communication, H. Küpfer clarified that there was a mistake in plotting in Fig. 7.14.
29. A. M. Campbell, J. E. Evetts and D. Dew-Hughes: Phil. Mag. **18** (1968) 313.
30. H. Fujimoto, M. Murakami, N. Nakamura, S. Gotoh, A. Kondoh, N. Koshizuka and S. Tanaka: *Adv. Supercond. IV* (Springer, Tokyo, 1992) p. 339.
31. Y. Tanaka, K. Itoh and K. Tachikawa: J. Jpn. Inst. Metals **40** (1976) 515 [in Japanese].
32. R. M. Scanlan, W. A. Fietz and E. F. Koch: J. Appl. Phys. **46** (1975) 2244.
33. T. Matsushita, N. Harada and K. Yamafuji: Cryogenics **29** (1989) 328.
34. G. Antesberger and H. Ullmaier: Phil. Mag. **29** (1974) 1101.
35. R. Wördenweber, P. H. Kes and C. C. Tsuei: Phys. Rev. B **33** (1986) 3172.
36. E. J. Kramer and H. C. Freyhardt: J. Appl. Phys. **51** (1980) 4930.
37. G. P. van der Meij and P. H. Kes: Phys. Rev. B **29** (1984) 6233.
38. E. V. Thuneberg: J. Low Temp. Phys. **57** (1984) 415.
39. M. V. Feigel'man, V. B. Geshkenbein, A. I. Larkin and V. M. Vinokur: Phys. Rev. Lett. **63** (1989) 2303.
40. E. J. Kramer: J. Electron. Mater. **4** (1975) 839 (Data from: R. E. Enstrom, J. R. Appert: J. Appl. Phys. **43** (1972) 1915).
41. T. Matsushita and H. Küpfer: J. Appl. Phys. **63** (1988) 5048.
42. D. C. Larbalestier: *Superconductor, Materials Science – Metallurgy, Fabrication, and Applications* (Plenum, New York, 1981) p. 133.
43. E. J. Kramer: J. Appl. Phys. **44** (1973) 1360.
44. J. E. Evetts and C. J. G. Plummer: *Proc. Int. Symp. on Flux Pinning and Electromagnetic Properties in Superconductors*, Fukuoka, 1985, p. 146.
45. D. Dew-Hughes: IEEE Trans. Magn. **MAG-23** (1987) 1172.
46. T. Matsushita and J. W. Ekin: *Composite Superconductors*, ed. K. Osamura (Marcel Dekker, New York, 1993) p. 79.
47. T. Matsushita, A. Kikitsu, H. Sakata, K. Yamafuji and M. Nagata: Jpn. J. Appl. Phys. **25** (1986) L792.
48. T. Matsushita, M. Itoh, A. Kikitsu and Y. Miyamoto: Phys. Rev. B **33** (1986) 3134.
49. N. Harada, Y. Miyamoto, T. Matsushita and K. Yamafuji: J. Phys. Soc. Jpn. **57** (1988) 3910.

50. C. C. Koch, H. C. Freyhardt and J. O. Scarbrough: IEEE Trans. Magn. **MAG-13** (1977) 828.
51. H. C. Freyhardt: J. Low Temp. Phys. **32** (1978) 101.
52. H. Küpfer, R. Meier-Hirmer and W. Schauer: *Adv. Cryog. Eng. Mater.* (Plenum, New York, 1988) Vol. 34, p. 725.
53. K. E. Osborne: Phil. Mag. **23** (1971) 1113.
54. Yu. F. Bychkov, V. G. Vereshchagin, V. R. Karasik, G. B. Kurganov: Sov. Phys. JETP **29** (1969) p. 276.
55. H. Küpfer, I. Apfelstedt, R. Flükiger, C. Keller, R. Meier-Hirmer, B. Runtsch, A. Turowski, U. Wiech and T. Wolf: Cryogenics **29** (1989) 268.
56. H. Raffy, J. C. Renard and E. Guyon: Solid State Commun. **11** (1972) 1679.
57. A. M. Campbell and J. E. Evetts: Adv. Phys. **21** (1972) 378.
58. J. D. Livingston: Appl. Phys. Lett. **8** (1966) 319.
59. M. Daeumling, J. M. Seuntjens and D. C. Larbalestier: Nature **346** (1990) 332.
60. A. B. Pippard: Phil. Mag. **19** (1969) 217.
61. D. Ertas and R. D. Nelson: Physica C **272** (1996) 79.
62. E. H. Brandt and G. P. Mikitik: Supercond. Sci. Technol. **14** (2001) 651.
63. H. Küpfer, Th. Wolf, C. Lessing, A. A. Zhukov, X. Lancon, R. Meier-Hirmer, W. Schauer and H. Wühl: Phys. Rev. B **58** (1998) 2886.
64. T. Matsushita, E. S. Otabe, H. Wada, Y. Takamaha and H. Yamauchi: Physica C **397** (2003) 38.
65. F. Habbal and W. C. H. Joiner: J. de Phys. (Paris) **39** (1978) C6–641.
66. M. Steingart, A. G. Putz and E. J. Kramer: J. Appl. Phys. **44** (1973) 5580.
67. J. E. Evetts and S. J. Mullock: *Proc. Int. Symp. on Flux Pinning and Electromagnetic Properties in Superconductors*, Fukuoka, 1985, p. 94.
68. A. A. Zhukov, S. Kokkaliaris, P. A. J. de Groot, M. J. Higgins, S. Bhattacharya, R. Gagnon and L. Taillefer: Phys. Rev. B **61** (2000) R886.
69. H. Küpfer and W. Gey: Phil. Mag. **36** (1977) 859.
70. T. Matsushita: Physica C **217** (1993) 461.
71. N. Ihara and T. Matsushita: Physica C **257** (1996) 223.

8
High-Temperature Superconductors

8.1 Anisotropy of Superconductors

One of the properties of the crystal structure of oxide superconductors with high critical temperature is an alternating multi-layer of CuO_2 planes which generate superconductivity and charge-reservoir blocks which are almost insulating. This structure causes a large anisotropy in normal conducting and superconducting properties. Lawrence and Doniach [1] proposed a phenomenological theory for an alternating multi-layer of thin superconductors and insulators before the discovery of high-temperature superconductivity. This model is considered to be approximately applicable for high-temperature superconductors. In this case each superconducting layer is numbered and the two-dimensional order parameter defined in the n-th layer is represented as $\Psi_n(x, y)$ with the z-axis defined normal to the layers, i.e., along the c-axis. Assuming a case where a magnetic field is not applied, the vector potential can be neglected. Then, the kinetic energy density given by Eq. (1.20) is generalized to the anisotropic case as

$$\frac{\hbar^2}{2m_{ab}^*}\left(\left|\frac{\partial \Psi_n}{\partial x}\right|^2 + \left|\frac{\partial \Psi_n}{\partial y}\right|^2\right) + \frac{\hbar^2}{2m_c^* s^2}|\Psi_n - \Psi_{n-1}|^2 \ , \tag{8.1}$$

where m_{ab}^* and m_c^* are the effective masses of superconducting electrons moving in the a-b plane and along the c-axis, respectively, and s is the distance between superconducting layers. The spatial variation along the c-axis is approximated by the difference between the adjacent layers. Inserting the vector potential, the free energy density is given by

$$F = \sum_n \int_{V_n} \left[\alpha|\Psi_n|^2 + \frac{1}{2}\beta|\Psi_n|^4 + \frac{1}{2m_{ab}^*}|(-i\hbar\nabla' + 2e\boldsymbol{A}')\Psi_n|^2 \right.$$
$$\left. + \frac{\hbar^2}{2m_c^* s^2}\left|\Psi_n - \Psi_{n-1}\exp\left(\frac{2ieA_z s}{\hbar}\right)\right|^2\right] dV \ , \tag{8.2}$$

where ∇' and $\boldsymbol{A}'$ are two-dimensional vectors in the a-b plane and A_z is the z-component of the vector potential. The integral is taken in the n-th superconducting layer. Minimizing this energy density with respect to Ψ_n^*, the Lawrence-Doniach equation is obtained:

$$\alpha\Psi_n + \beta|\Psi_n|^2\Psi_n + \frac{1}{2m_{ab}^*}(-i\hbar\nabla' + 2e\boldsymbol{A}')^2\Psi_n$$
$$-\frac{\hbar^2}{2m_c^* s^2}\left[\Psi_{n+1}\exp\left(-\frac{2ieA_z s}{\hbar}\right) - 2\Psi_n + \Psi_{n-1}\exp\left(\frac{2ieA_z s}{\hbar}\right)\right] = 0. \quad (8.3)$$

If the spatial variation along the z-axis is smooth, the discontinuous function Ψ_n can approximately be replaced by a smooth function Ψ. The difference is replaced by a derivative and the Lawrence-Doniach equation of (8.3) leads to the anisotropic Ginzburg-Landau equation:

$$\alpha\Psi + \beta|\Psi|^2\Psi + \frac{1}{2}(-i\hbar\nabla + 2e\boldsymbol{A}) \cdot \left[\frac{1}{m^*}\right] \cdot (-i\hbar\nabla + 2e\boldsymbol{A})\Psi = 0, \quad (8.4)$$

where $[1/m^*]$ is a tensor defined by

$$\left[\frac{1}{m^*}\right] = \begin{bmatrix} 1/m_{ab}^* & 0 & 0 \\ 0 & 1/m_{ab}^* & 0 \\ 0 & 0 & 1/m_c^* \end{bmatrix}. \quad (8.5)$$

Anisotropic superconducting parameters are obtained from the anisotropic effective mass. From a treatment similar to that in Chap. 1 the coherence length is derived as

$$\xi_i = \frac{\hbar}{(2m_i^*|\alpha|)^{1/2}}, \quad (8.6)$$

where the subscript i denotes either ab or c, representing quantities in the a-b plane or those along the c-axis. Thus, the upper critical field is given by

$$H_{c2}^{ab} = \frac{\phi_0}{2\pi\mu_0\xi_{ab}\xi_c}, \quad H_{c2}^c = \frac{\phi_0}{2\pi\mu_0\xi_{ab}^2}. \quad (8.7)$$

From Eq. (1.43) the penetration depth is

$$\lambda_i = \frac{\hbar}{2\sqrt{2}e\mu_0 H_c \xi_i}. \quad (8.8)$$

In the above the thermodynamic critical field H_c is isotropic.

Here the anisotropy parameter is defined by

$$\gamma_a = \left(\frac{m_c^*}{m_{ab}^*}\right)^{1/2}. \quad (8.9)$$

Then, anisotropies of various superconducting parameters are expressed as

$$\frac{\xi_{ab}}{\xi_c} = \frac{\lambda_c}{\lambda_{ab}} = \frac{H_{c2}^{ab}}{H_{c2}^c} = \gamma_{\mathrm{a}} . \tag{8.10}$$

As for the lower critical field, neglecting the anisotropy of $\log \kappa$, the relationships of $H_{c1}^{ab} \propto \xi_{ab}\xi_c$ and $H_{c1}^c \propto \xi_{ab}^2$ lead to

$$\frac{H_{c1}^c}{H_{c1}^{ab}} \simeq \gamma_{\mathrm{a}} . \tag{8.11}$$

Parameters of typical high-temperature superconductors are listed in Table 8.1.

Table 8.1. Superconducting parameters of high-temperature superconductors

Superconductor	T_c (K)	$\mu_0 H_{c2}^{ab}(0)$ (T)	$\mu_0 H_{c2}^c(0)$ (T)	$\xi_{ab}(0)$ (nm)	$\xi_c(0)$ (nm)	κ_{ab}	κ_c
Y-123	93	670	102	1.80	0.27	67	355
Bi-2212	91	>530	19	4.16	<0.15	–	–
Hg-1212	128	454	113	1.71	0.42	114	466
Hg-1223	138	389	88	1.93	0.44	76	339

In the case of field direction in terms of angle θ from the c-axis, the upper critical field obeys the relationship:

$$\begin{aligned}H_{c2}(\theta) &= H_{c2}^c (\cos^2\theta + \gamma_{\mathrm{a}}^{-2}\sin^2\theta)^{-1/2} \\ &= H_{c2}^c \left[\cos^2\theta + \left(\frac{H_{c2}^c}{H_{c2}^{ab}}\right)^2 \sin^2\theta\right]^{-1/2} .\end{aligned} \tag{8.12}$$

The anisotropic transverse cross-sectional structure of an isolated flux line is shown in Fig. 8.1, where (a) and (b) are the cases of magnetic field along the a- and c-axes, respectively. Both the normal core and the surrounding magnetic flux are of ellipsoidal structure with a shorter radius in the direction of the c-axis in the case of (a), while both are isotropic in the case of (b).

In general, the electric conductivity of block layers in the normal state is low and their thickness s is large in superconductors with large anisotropy. In particular, ξ_c is shorter than s in the most anisotropic superconductor (Bi, Pb)$_2$Sr$_2$CaCu$_2$O$_8$(Bi-2212), and the three-dimensional anisotropic Ginzburg-Landau model cannot be applied to this case. When a magnetic field is applied parallel to the a-b plane of such a superconductor, the normal core of a flux line usually stays in the block layer as shown in Fig. 8.2(a). This is called a Josephson vortex. In this case the transverse cross-section of the flux line is anisotropic similarly to the case in Fig. 8.1(a). The size of the normal core is s along the c-axis and is about $\gamma_{\mathrm{a}} a_c$ in the a-b plane, where a_c is the lattice

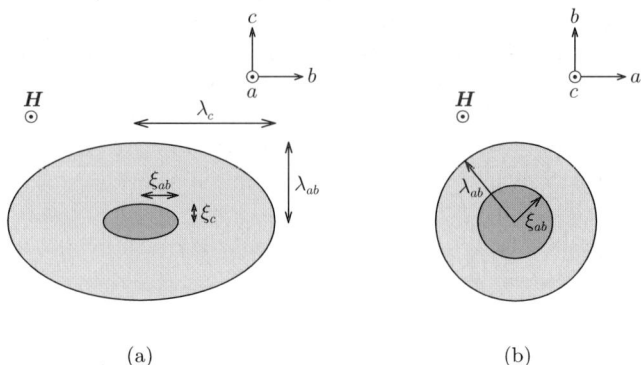

Fig. 8.1. Transverse cross-sectional structure of isolated flux line in anisotropic three-dimensional superconductor: (a) flux line along the a-axis and (b) along the c-axis

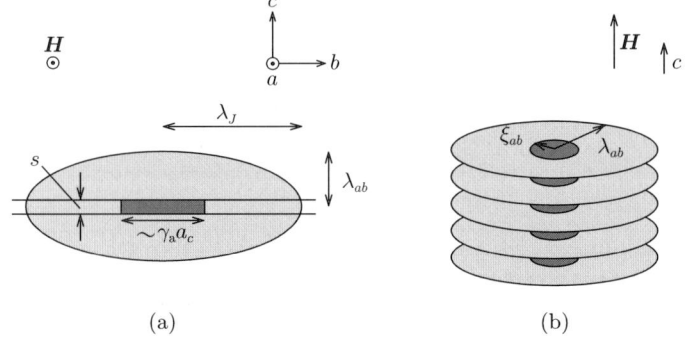

Fig. 8.2. Structure of isolated flux lines in two-dimensional superconductor: (a) Josephson vortex along to the a-axis and (b) a pancake vortex along the c-axis

parameter along the c-axis. The size of the magnetic flux along the c-axis is λ_{ab} and that in the a-b plane is the Josephson penetration depth given by

$$\lambda_\mathrm{J} = \left(\frac{\phi_0}{2\pi\mu_0 j_c s}\right)^{1/2} , \tag{8.13}$$

where the effective thickness of the junction D in Eq. (A.65) in Appendix A.7 has been replaced by the distance between the superconducting layers s.

When the magnetic field is applied along the c-axis, the flux lines are two-dimensional structures formed in the CuO_2 planes. Their structure is the same as in Fig. 8.1(b). These are weakly coupled with each other along the c-axis by the Josephson coupling energy given by the second term of Eq. (8.1). This flux line is called a pancake vortex (see Fig. 8.2(b)).

According to the mean-field theory of Ginzburg and Landau, the upper critical field is defined as the transition point between the superconducting mixed state and the normal state. However, a clear phase transition does not seem to occur on the phase boundary predicted by this mean field theory because of the very strong influence of fluctuations caused by high temperature and two-dimensionality. It should be noted, however, that this fact does not make the upper critical field derived from the mean field theory meaningless. If we look at some physical quantity in a sufficiently low field region where the influence of fluctuation is not serious, the theory describes its magnetic field dependence. For example, the magnetization is predicted to vary linearly with magnetic field, and if this variation is extrapolated to high fields, the magnetization seems to become zero at the "upper critical field." That is, the magnetic field dependence of the magnetization at low fields can be correctly expressed in terms of the "upper critical field."

8.2 Phase Diagram of Flux Lines

There has been much detailed discussion, and many things have been clarified on the state of flux lines in the mixed state of type-II superconductors, especially of high-temperature superconductors. This state is classified in principle by magnetic field and temperature as shown in Fig. 1.2(b). In high-temperature superconductors, however, the state of the flux lines is strongly influenced by the pinning interactions from a high density of included defects. Hence, the pinning interaction is not simply treated as a perturbation but as one of the external variables such as magnetic field and temperature. In addition, there are influences of the dimensionality of the superconductor, the size of the specimen, the kind of pinning centers, etc. Hence, the situation is very complicated. Only the most important aspects are treated in this section for simplicity.

For this reason, we consider the pinning energy U_p in addition to two other energies, i.e, the thermal energy $U_T = k_B T$ and the elastic energy U_E necessary for an argument on the state of flux lines. Hence, we can simply expect three phase transitions determined mainly by two of three energies [2]. The third energy will give some perturbation to the transition except in the region near the critical point where the three energies are comparable to each other in magnitude.

The elastic energy is the energy of the flux line system itself, and is generally associated with the elastic moduli described in Sect. 7.2. The Josephson coupling energy between superconducting layers given by Eq. (8.1) is sometimes treated separately. However, this is a part of the kinetic energy due to the spatial variation in the order parameter. Thus it is included in the elastic energy here.

8.2.1 Melting Transition

The most fundamental transition is the melting transition determined by U_T and U_E, and it is known that this transition is first order. In this case it is considered that the melting occurs when the mean displacement of flux lines from their lattice points due to thermal activation reaches approximately 15% of the lattice parameter, similarly to Lindemann's criterion for melting of solids. This transition occurs only in a superconductor in which U_p is sufficiently smaller than U_E.

8.2.2 Vortex Glass-Liquid Transition

The transition determined by U_T and U_p is the vortex glass-liquid transition (hereafter called the G-L transition) and is known to be second order [3]. It was initially considered to be independent of the flux pinning property [4]: glass and liquid states of flux lines existed first, and the state then determined the flux pinning property. According to the G-L transition model, the coherence length and the relaxation time in the vortex glass state are expected to diverge as $\xi_\mathrm{g} \sim |T - T_\mathrm{g}|^{-\nu}$ and $\tau \sim \xi_\mathrm{g}^z$, respectively, in the vicinity of the transition temperature T_g. In the above ν and z are the static and dynamic critical indices, respectively. Then, if the resistivity E/J and the current density J are replotted in the relationship $(E/J)/|T-T_\mathrm{g}|^{\nu(z+2-D)}$ vs. $J/|T-T_\mathrm{g}|^{\nu(D-1)}$, taking account of the dimension D of the state of the flux lines, a scaling of the E-J curves is obtained. Figure 8.3 shows the E-J curves and the result of scaling for an Y-123 thin film in a magnetic field of 4 T along the c-axis [5]. It can be seen that all curves meet on the two master curves representing the glass and liquid states of the flux lines. It is considered in the model that ν takes a value of 1–2 and z takes a value of 4–5. It has been reported that [6] if the relationship with $D = 3$ is used for an analysis of two-dimensional flux lines in Bi-based superconductors, ν is smaller than 1 and z is larger than 10. However, the use of the appropriate parameter, $D = 2$, leads to the critical indices in the above range [6]. This indicates that the G-L transition model correctly describes the transition phenomena of flux lines including the dimension.

However, it seems appropriate to consider that the cause and the result are opposite, judging from the fact that the transition is determined by the thermal energy and the pinning energy. Namely, we postulate that the flux lines are in the glass state when they are effectively pinned and in the liquid state when they are not. In fact, the transition itself is strongly influenced by the flux pinning: the transition point T_g is determined by the flux pinning strength. In addition, the static critical index ν determined by the scaling of E-J curves corresponds to half of the pinning parameter δ', which describes the temperature (or magnetic field) dependence of the critical current density as $J_\mathrm{c} \propto (T_\mathrm{g} - T)^{\delta'}$ (see Fig. 8.4) [7]. The dynamic critical index z decreases with broadening distribution width of the flux pinning strength [8] and is determined by one of the Weibull parameters describing the nonuniformity [9]. Namely, the z value is larger for a superconductor with a sharper distribution of the flux pinning strength. In fact, $z = 11.8$ and $\nu = 0.7$ are obtained even for three-dimensional flux lines in a homogeneous Y-123 thin film [9]. On the

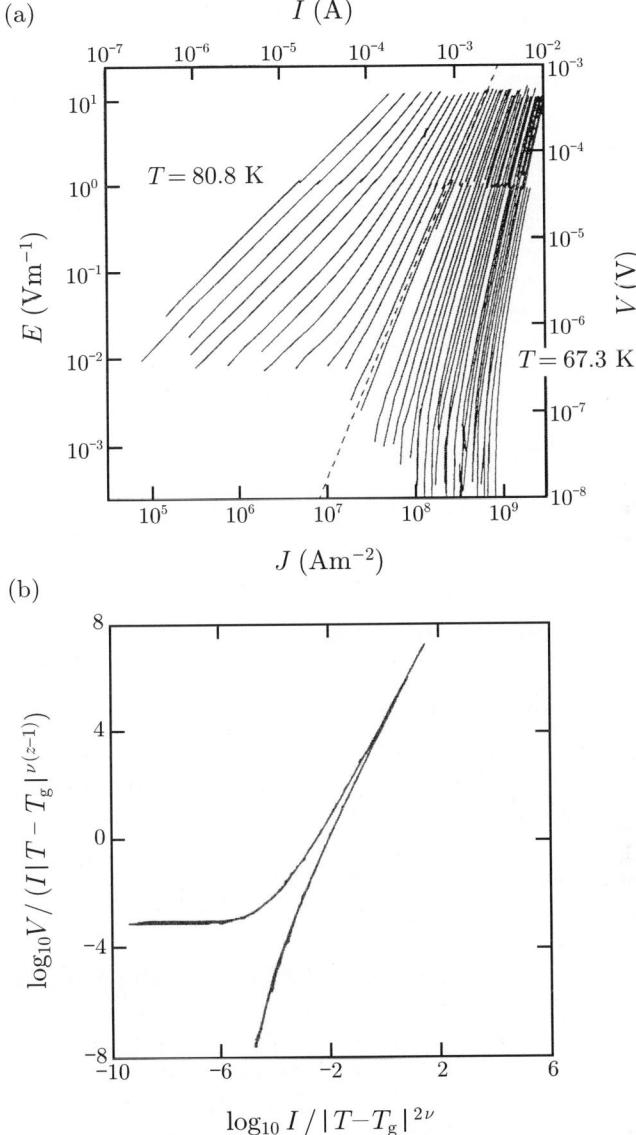

Fig. 8.3. (a) E-J curves and (b) the scaling ($\nu = 1.7$, $z = 4.8$) in an Y-123 thin film in a magnetic field of $B = 4$ T along the c-axis [5]

other hand, z takes the value of 3–4 and ν takes the value of 2–3 even for the two-dimensional case as in Bi-based superconductors, when the specimens are inhomogeneous [10]. Therefore, it cannot be said that the dimensionality of the flux lines influences the G-L transition. One consequence is a trend that the z value in Bi-based superconductors is larger. The reason is considered to

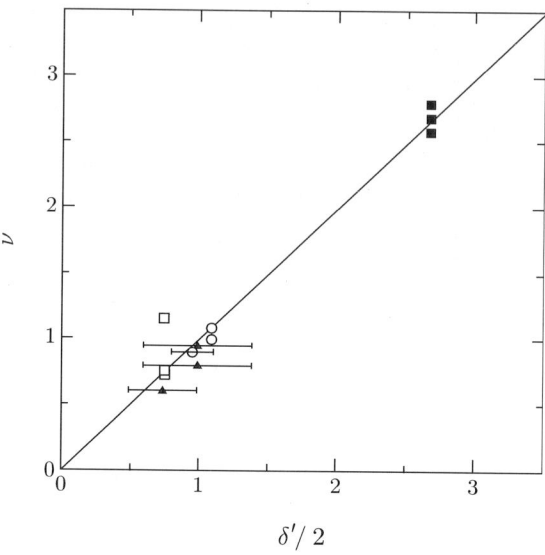

Fig. 8.4. Relationship between ν and $\delta'/2$ for Bi-2223 tapes (*open circles*), Bi-2212 tapes (*solid triangles*) and dip-coated tapes (*solid squares*) and Y-123 thin films (*open squares*) [7]. The *straight line* shows $\nu = \delta'/2$

be that the transition temperature is low and the variation rate of the pinning force with temperature is very large there.

Originally, the static critical index ν was associated with a divergence of the correlation length, which determines the G-L transition, as $(T_g - T)^{-\nu}$ in the vicinity of the transition temperature T_g. The pinning correlation length, which leads to Eq. (7.92) in the creep-free case, is approximately the same as Campbell's AC penetration depth of Eq. (3.92) and diverges as $(T_g - T)^{-\delta'/2}$ at the transition temperature. Hence, the result of Fig. 8.4 indicating the relationship $\nu = \delta'/2$ makes it clear that the correlation which governs the transition is the flux pinning. This is the reason why ν depends on the flux pinning phenomena. The divergence of the correlation length when the magnetic field changes is also expressed as $(H_g - H)^{-\nu}$ with the same static critical index ν. Here, H_g is the transition field, and $H_g(T)$ is the inverse function of $T_g(H)$. Hence, the scaling of E-J curves with varying magnetic field is similar to that with varying temperature.

Hence, the scaling of E-J curves can be explained by the mechanism of flux creep and flow. An example of the scaling is shown in Fig. 8.5, and the theoretical results on the transition line and the critical indices are compared with experiments in Fig. 8.6 [11].

The irreversibility line, $T_i(H)$ or $H_i(T)$, discussed from the viewpoint of practical applications, is the same characteristic as the G-L transition line, but is determined by a different criterion, i.e., the temperature or magnetic field at which the critical current density obtained using a proper electric

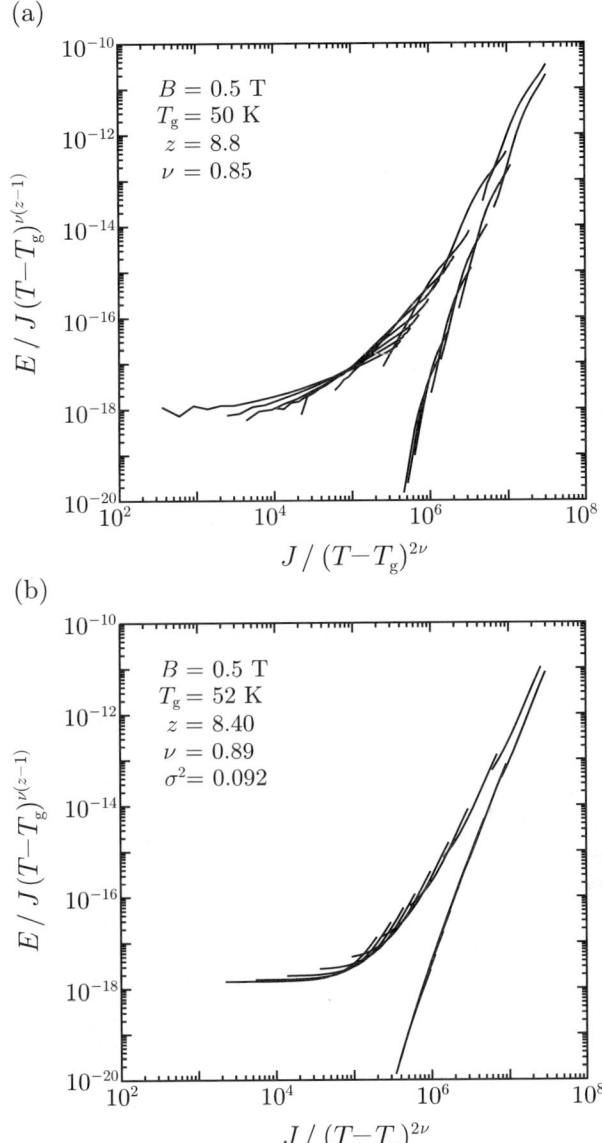

Fig. 8.5. Scaling of E-J curves of a Bi-2223 tape in a magnetic field of $B = 0.5$ T along the c-axis: [11] (**a**) experimental result and (**b**) theoretical result of the flux creep-flow model. Small deviation from exact scaling seems to be caused by imperfect compensation of the resistance of silver

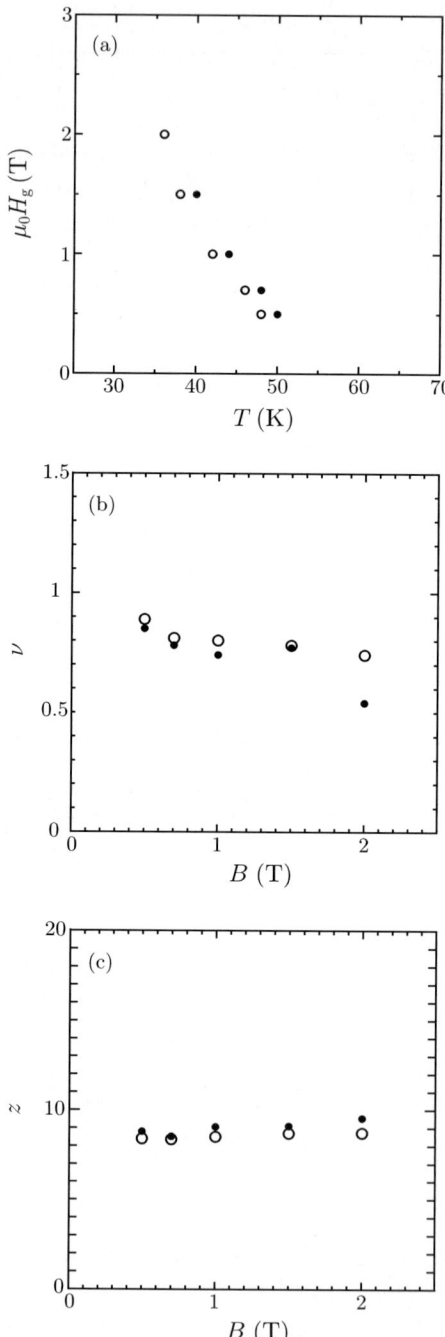

Fig. 8.6. (a) Transition line, (b) static critical index and (c) dynamic critical index of the Bi-2223 tape shown in Fig. 8.5 [11]. *Solid* and *open* symbols are experimental results and theoretical results of the flux creep-flow model, respectively

field criterion is reduced to some threshold value. Hence, there is a strong correlation between the G-L transition line and the irreversibility line. The irreversibility field can only be explained by the mechanism of flux creep as will be described in Sect. 8.5. The G-L transition magnetic field is significantly influenced by the level of electric field at which it is determined, similarly to the irreversibility magnetic field [12, 13]. Figure 8.7 represents the E-J curves of a Bi-2223 tape at 70 K over a wide range of electric field [14], and shows that, even if the log E-log J curve is concave upward, suggesting a glass state at some electric field level, it becomes concave downward suggesting a liquid state at a much lower level of electric field. This behavior is related to the electric field dependence of the irreversibility field which will be discussed in Sect. 8.5 (see Fig. 8.43). The scaling parameters at 70 K are: $\mu_0 H_g = 310$ mT, $\nu = 0.68$ and $z = 9.5$ in the high electric field region and $\mu_0 H_g = 56$ mT, $\nu = 0.80$ and $z = 14.5$ in the low electric field region [15]. Thus, not only the transition point but also the critical indices change with electric field. A similar change is also observed in an Y-123 thin film [13], and $T_g = 88.4$ K, $\nu = 1.4$ and $z = 8.2$ are obtained in a high electric field region, while $T_g = 56.6$ K, $\nu = 0.6$ and $z = 22.5$ are obtained in a low electric field region at $B = 0.52$ T. Such a feature is considered to be caused by the mechanism of TAFF and a nonuniform distribution of flux pinning strength.

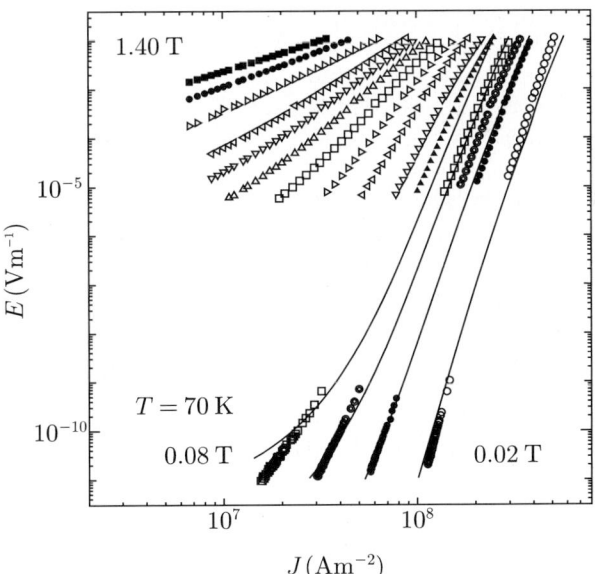

Fig. 8.7. E-J curves of a Bi-2223 tape at 70 K [14]. Results at high and low electric fields are obtained using the four terminal method and the relaxation of DC magnetization, respectively. *Solid lines* are theoretical results of the flux creep-flow model

Since such a transition changes depending on the level of electric field or the time of observation, the phase diagram of flux lines is different for different kinds of measurement. For example, the mean velocity of flux lines in a magnetic flux density of $B = 1$ T is 0.1 mms^{-1} at an electric field of 1×10^{-4} Vm^{-1} as in the resistive measurement, and is 0.1 nms^{-1} or only 3 mm in a year at an electric field of 1×10^{-10} Vm^{-1} as in the magnetization measurement. In the latter case, it is hard to understand over a short period if the flux lines are moving or not. The same thing will occur in other transitions.

At this transition, it has been theoretically explained that the degree of disorder of flux lines due to the pinning is reduced significantly as the temperature is increased from the transition point T_g, approaching a perfect flux line lattice, if the effect of melting can be disregarded [3, 7]. This prediction agrees with an observation of flux lines in the liquid state using a Lorentz microscope [16]. In practice, the behavior of flux lines in the vicinity of the transition point is significantly influenced by the spatial distribution of the pinning strength, and the E-J characteristics and their scaling at an electric field level above 10^{-4} Vm^{-1} observed by the usual resistive method can well be explained by a percolation model [17], in which the thermally activated flux motion is described by an effective flux flow. The percolation model will be described in Subsect. 8.4.4.

8.2.3 Order-Disorder Transition

The transition determined by U_E and U_p is an order-disorder transition. One of the features of this transition is the peak effect of critical current density in a magnetic field parallel to the c-axis. In three-dimensional superconductors such as Y-123, the transition field at which the peak effect starts decreases as the pinning becomes stronger as shown in Fig. 7.41. This is a common behavior in metallic and high-temperature superconductors. Brandt and Mikitik [18] calculated the transition line using a similar Lindemann criterion to that for the melting transition. As for the flux pinning, it is assumed that each flux bundle composed of multiple flux lines is pinned at high temperatures, while the single vortex pinning mechanism is adopted at low temperatures. The corresponding deformation of flux lines is shear, and the flux lines are expected to be in a lattice-like Bragg-glass state and an amorphous-like vortex-glass state at low and high fields, respectively. It was predicted that the transition is first order. Although the transition width is very wide as is seen from Fig. 7.41, this is considered to be caused by an inhomogeneous pinning strength and the distribution of magnetic field in the specimens. Observation of the history effect accompanied by the peak effect [19] shows that this is a first-order transition.

On the other hand, the two-dimensional Bi-2212 shows a sharp peak effect at low magnetic field along the c-axis as in Fig. 8.8 [20]. The peak field is almost independent of the temperature. Usually the magnetization measurement is done for a specimen of a finite size, and hence, the magnetic field

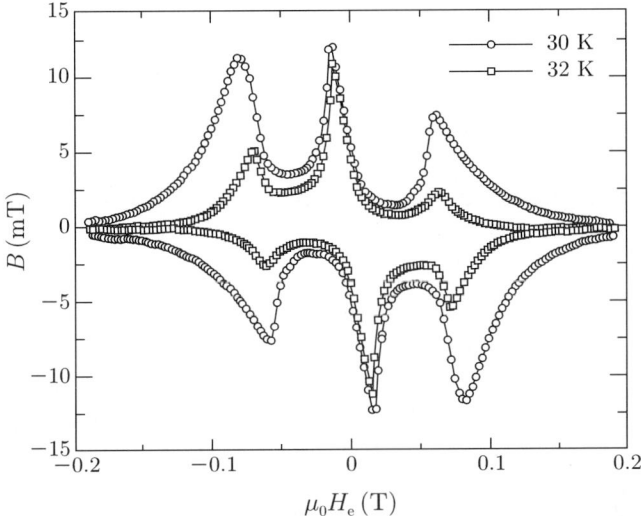

Fig. 8.8. Local magnetic flux density in a Bi-2212 single crystal measured using a small Hall probe [20]

is not uniform throughout the specimen, resulting in some transition width. Figure 8.9 shows the observed critical current density of an overdoped Bi-2212 single crystal using Campbell's method with a small AC magnetic field in the a-b plane superimposed on a DC magnetic field initially applied along the c-axis [21]. It is found that the critical current density changes discontinuously with magnetic field. In addition, there are two stable states with different values of the critical current density under the same conditions of temperature and magnetic field. This proves that this transition is also first order. One of the two states is a state where flux lines are slightly deformed and weakly pinned as depicted in Fig. 8.10(a). This state continues from a lower magnetic field. The other state is a state where flux lines are significantly deformed and strongly pinned, and is attained at a higher magnetic field. It is proposed that a double potential well is formed by a combination of the elastic energy and the pinning energy. This will bring about the results in Fig. 8.9. At a sufficiently low temperature, the critical current density increases dramatically with a disappearance of the peak effect. This seems to result from the formation of a single potential well as shown in Fig. 8.10(b) owing to a rapid enhancement of the pinning energy which comes from the enlarged condensation energy at low temperatures (see Fig. 8.55).

The reason why such a measurement is possible is that the shielding effect against the DC magnetic field along the c-axis is eliminated by the superposition of a small AC magnetic field normal to the DC field as shown in Fig. 3.14, resulting in the penetration of the uniform DC magnetic field. This transition is also a kind of order-disorder transition and takes place accompanied by a

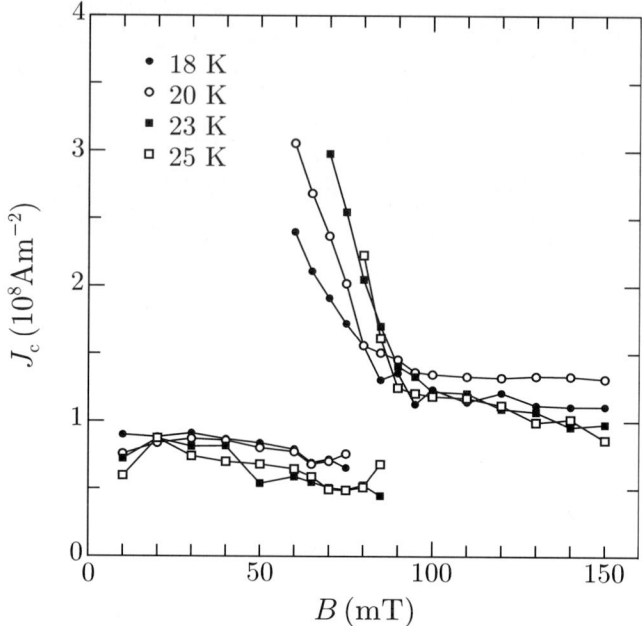

Fig. 8.9. Critical current density in a Bi-2212 single crystal measured by using Campbell's method [21]

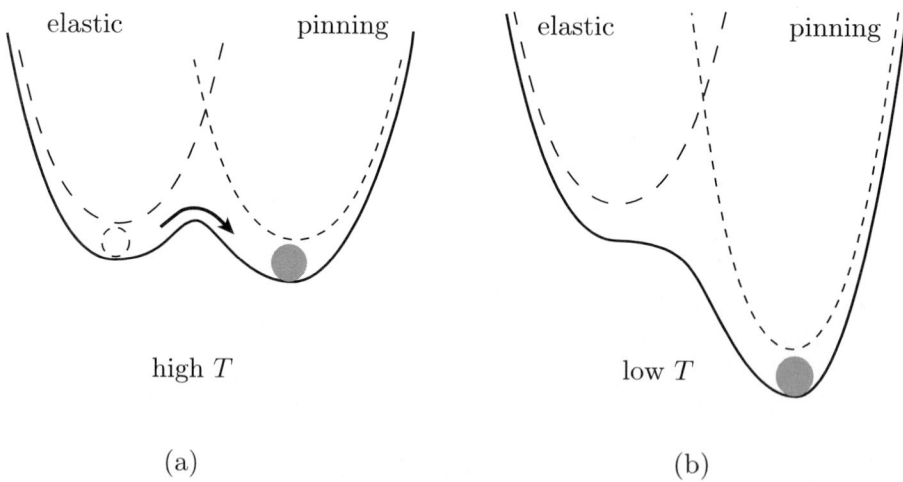

Fig. 8.10. Free energy composed of elastic energy and pinning energy: (**a**) a double potential well with two stable states in the region of dimensional crossover of flux lines and (**b**) a single potential well due to strong pinning at low temperatures

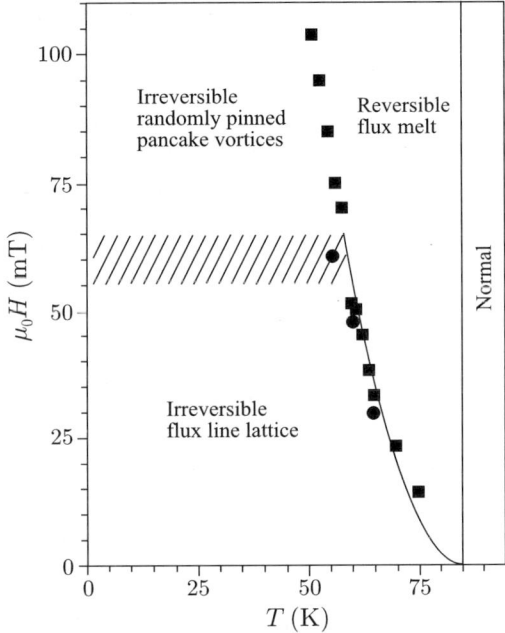

Fig. 8.11. Phase diagram of flux lines in a Bi-2212 single crystal. At a magnetic field higher than the *shaded region*, the neutron diffraction pattern of flux lines disappears, suggesting that flux lines change to two-dimensional pancake vortices [22]

crossover of flux lines from the three-dimensional state to the two-dimensional with an increase in the magnetic field. The dimensional crossover was deduced from the experimental result [22] that the neutron diffraction pattern disappears in the low temperature region for a magnetic field parallel to the c-axis above the hatched region in Fig. 8.11. The corresponding deformation of flux lines is along the length of the flux lines, i.e., a bending. The interaction distance d_i shown in Fig. 8.12(b) changes drastically at the transition field, while the Labusch parameter α_L shown in Fig. 8.12(a) does not appreciably change [21]. This shows that flux lines become flexible owing to the crossover, resulting in an enhanced threshold for depinning, although the shape of the pinning potential does not change appreciably. This supports the order-disorder transition explanation for the origin of the peak effect. The theoretical predictions of the usual collective pinning of point-like defects, $\alpha_L \propto B^{3/2}$ and $d_i \propto B^{-1/2}$, are shown for comparison by the dashed lines in these figures.

The characteristic peak effect is not observed for heavily underdoped Bi-2212 superconductors with a very large anisotropy parameter. The crossover field is very low for these superconductors. Hence, the disappearance of the peak effect is considered to be caused by the fact that flux lines are pinned with a high efficiency independently of each other at such low fields, and a

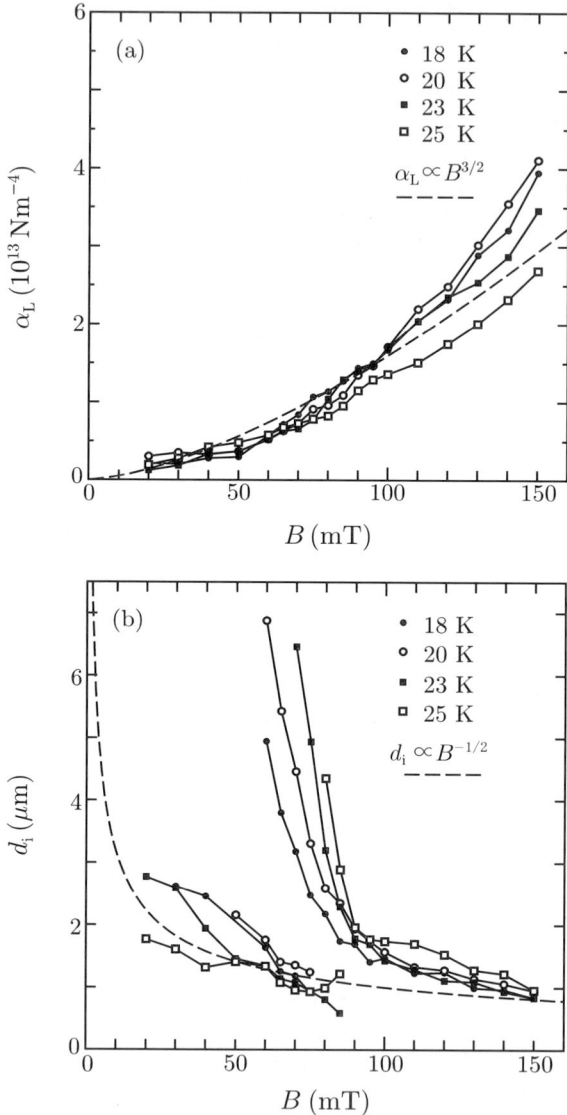

Fig. 8.12. (a) Labusch parameter and (b) interaction distance in the same Bi-2212 single crystal as in Fig. 8.9 [21]. The *dashed lines* in these figures represent theoretical predictions of collective pinning of point-like defects, $\alpha_L \propto B^{3/2}$ and $d_i \propto B^{-1/2}$, respectively

further enhancement of the pinning efficiency is not attained even when the crossover occurs.

8.2.4 Phase Diagram of Flux Lines in Each Superconductor

The phase diagram of flux lines is slightly different between three-dimensional Y-123 and two-dimensional Bi-2212 as mentioned above. Figure 8.13 shows the phase diagram for a twin-free Y-123 single crystal with a weak pinning force [23]. The solid line with solid circles which is drawn from the critical temperature to a higher field with decreasing temperature is the melting transition line $H_{\rm m}(T)$. This line disappears at the critical point around 73 K. The dotted line with solid triangles which extends further is the G-L transition line $H_{\rm g}(T)$. The solid line with solid squares at lower temperature and at lower field is the order-disorder transition line $H_{\rm dis}(T)$, at which the critical current starts to increase with increasing magnetic field. This line extends to a high field at elevated temperature and terminates at the critical point. The dimensional crossover of flux lines is expected [23] to occur at around 80 T. The dotted line with open triangles $H_{\rm pk}(T)$ represents the peak field of the critical current density and is shown for comparison. The dotted line with open circles represents the irreversibility line $H_{\rm i}(T)$, which extends to a high field with decreasing temperature, reaches the critical point and meets the G-L transition line at low temperatures. In the original paper [23] the irreversibility line is not considered as a phase transition. However, the pinning correlation length diverges on this line as abovementioned, and this shows

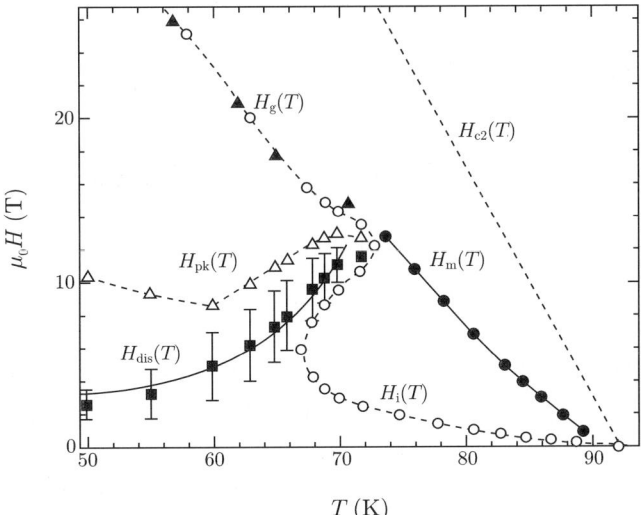

Fig. 8.13. Phase diagram of flux lines in twin-free Y-123 single crystal [23]. $H_{\rm pk}$ is the magnetic field at which the critical current density has a peak

that the irreversibility line is a second-order transition. In fact, it agrees with the G-L transition at low temperatures. Hence, it seems to be reasonable to treat $H_\mathrm{g}(T)$ and $H_\mathrm{i}(T)$ as a single transition curve, although there exists a difference due to the different definitions.

The state of the flux lines is the Bragg-glass state in the regime below $H_\mathrm{dis}(T)$ and $H_\mathrm{i}(T)$ in the H-T plane, the vortex glass state in the regime between $H_\mathrm{dis}(T)$ and $H_\mathrm{g}(T)$, and the vortex liquid state in the regime above $H_\mathrm{g}(T)$ and $H_\mathrm{m}(T)$. It is considered in [23] that flux lines are in the Bragg-glass state in the regime between $H_\mathrm{i}(T)$ and $H_\mathrm{m}(T)$. However, it seems to be reasonable to regard this as a crystalline phase (the Abrikosov vortex state), [24] since the pinning correlation length diverges.

If the flux pinning becomes strong, the melting transition line $H_\mathrm{m}(T)$ does not change appreciably, but the irreversibility line $H_\mathrm{i}(T)$ moves to a higher temperature and merges with the melting transition line. As a result the critical point (H_cp), the intersection of $H_\mathrm{m}(T)$ and $H_\mathrm{i}(T)$, moves in the direction of higher temperature and lower field, as shown in Fig. 8.14(b) [24]. The order-disorder transition line $H_\mathrm{dis}(T)$ follows the movement of the critical point. If the flux pinning strength is further increased, the critical point reaches the critical temperature, the melting line disappears and the G-L transition line $H_\mathrm{g}(T)$ extends to the critical temperature, as shown in Fig. 8.14(c). The order-disorder transition line $H_\mathrm{dis}(T)$ also reaches the critical temperature.

A typical phase diagram of flux lines in an optimally doped single crystal of most two-dimensional Bi-2212 superconductor is shown in Fig. 8.15 [25]. This is essentially the same as that of Y-123 single crystals with relatively weak pinning forces except that the order-disorder transition comes from the

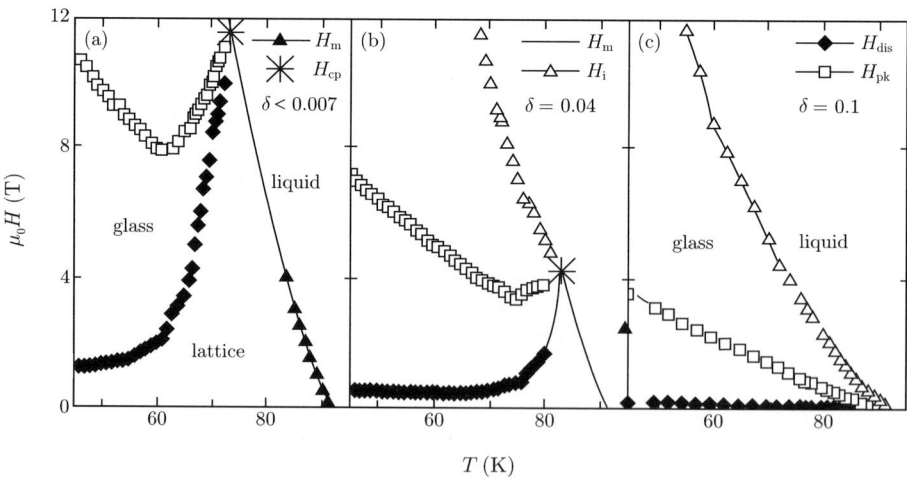

Fig. 8.14. Change in phase diagram of flux lines in Y-123 single crystal with twin boundaries by enhancement of flux pinning strength by increasing oxygen deficiency δ [24]

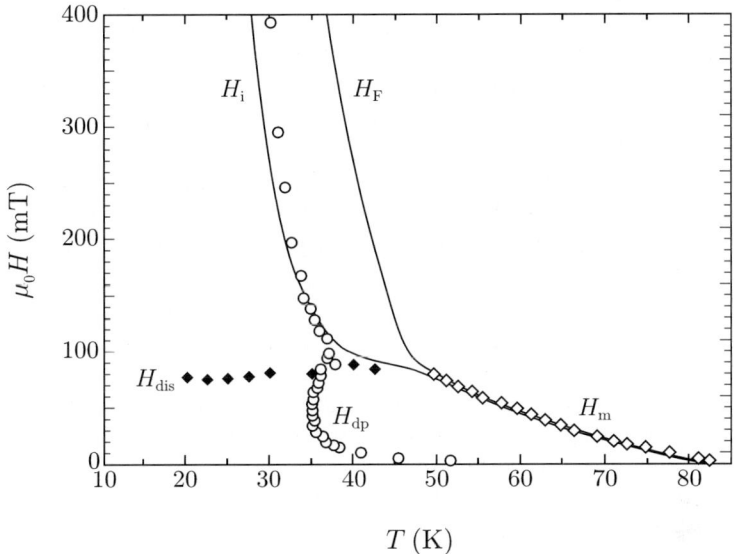

Fig. 8.15. Phase diagram of flux lines in optimally doped Bi-2212 single crystal [25]

dimensional crossover of flux lines and the transition field is almost independent of temperature. The order-disorder transition field at low temperatures is about 80 mT and this low value is also a characteristic of Bi-2212. Its value becomes extremely low for specimens with very large anisotropy parameters in underdoped states. In fact, it is given by [26] $\mu_0 H_{\mathrm{dis}} = \phi_0/(\gamma_{\mathrm{a}} s)^2$. H_{i} in Fig. 8.15 was defined as the field at which observed M-T curves of increasing and decreasing temperature merged, while H_{F} was defined as the field at which observed M-H curves of increasing and decreasing field merged. Hence, some flux pinning effect remains in the region between H_{i} and H_{F}. This difference might come from the difference in the electric field strength during the measurements. That is, in the M-T measurement the electric field strength is proportional to $\mathrm{d}M/\mathrm{d}t = (\mathrm{d}M/\mathrm{d}H)(\mathrm{d}H/\mathrm{d}t)$ which is typically three orders of magnitude smaller than $\mathrm{d}H/\mathrm{d}t$ in the M-H measurement (see Subsect. 8.5.3). However, the final conclusion has not yet been obtained. H_{dp} is the boundary between the Bragg glass state with strong pinning forces and the Abrikosov vortex state with weak pinning forces. This line coincides with the irreversibility line above the critical point.

8.2.5 Size Effect

For three-dimensional superconductors like Sm-123, the peak effect disappears when the specimen size is smaller than the pinning correlation length, as described in Sect. 7.6 (see Fig. 7.42). For two-dimensional Bi-2212 the peak effect also disappears when the film thickness becomes small, as will be discussed in Subsect. 8.4.2. In this case, however, the critical thickness is not the pinning correlation length, which is on the order of 10 μm, but smaller than 1 μm [27]. Such a size effect and how it differs between two- and three-dimensional superconductors will be discussed in more detail again in Subsect. 8.4.2. These results suggest that the order-disorder transition itself disappears when the specimen becomes small, as already mentioned in Sect. 7.6.

The G-L transition originates from thermal activation of flux lines which determines the irreversibility line as mentioned above. Hence, the transition line and the behavior of flux lines in the vicinity of this line can be expressed in terms of the E-J characteristic in the theoretical model of flux creep and flow. The important parameter for this description is the pinning potential energy U_0 discussed in Sect. 7.7. The U_0 depends on the size of the superconductor as shown in Eqs. (7.97) and (7.99). Thus, the G-L transition is also affected by the specimen size, and the transition field or the transition temperature decreases for a superconductor smaller than the critical size, as can be easily supposed. The corresponding critical size is the pinning correlation length, which is independent of the dimensionality of the superconductor and is different from the critical size for the order-disorder transition in two-dimensional superconductors. This difference will be discussed in Subsect. 8.4.2.

It is considered that the melting transition is also influenced by the size of the superconductor. However, the details have not yet been clarified.

The phase diagram of flux lines should be represented with respect to three-dimensional axes of the temperature, the magnetic field and the flux pinning strength, as can be seen from the above argument. Other factors which influence the diagram are the dimensionality and the size of the superconductor and the electric field strength. The type of pinning center is also one of the factors. More detailed states in the liquid phase are also now being examined.

8.2.6 Other Theoretical Predictions

Here the features which can easily be predicted on the basis of the phase diagram of flux lines are described.

The first item is the argument that the irreversibility line which is extended from the critical point to higher temperatures and lower magnetic fields is the G-L transition of the second order. Experiments to prove this prediction such as a scaling of E-J curves are required.

This prediction means that the melting line and the G-L transition line, which start from the critical temperature, meet at the critical point, in the case where the pinning is not so strong. Since the G-L transition line is extended from the critical point to the higher field region, the melting line is also expected to be extended in a similar manner, but below the G-L transition line [3]. In this region the transition is a change of flux lines from the

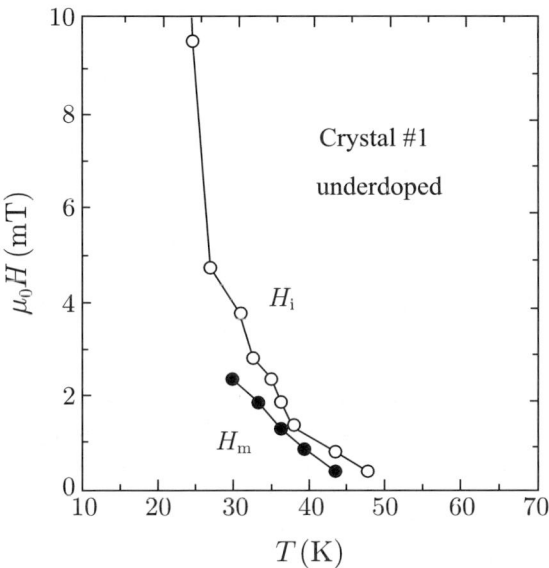

Fig. 8.16. Irreversibility and melting transition lines of Bi-2212 single crystal in an underdoped state [28]

crystalline state to the amorphous state, and hence, observations were considered to be difficult [2]. However, the melting transition was observed at a magnetic field lower than the irreversibility field for an underdoped Bi-2212 specimen with an extremely large anisotropy parameter as shown in Fig. 8.16 [28]. However, the possibility exists of strong surface pinning which might cause a much higher irreversibility field than the bulk one as was pointed out. Hence, a current distribution in a similarly underdoped Bi-2212 single crystal was measured, and it was found that the current is not localized in the surface region but flows fairly uniformly throughout the whole specimen [29]. Therefore, the claim in [28] is expected to be valid. This result indicates the possibility that the melting line can be observed in the temperature region lower than the critical point as mentional above. Further experiments are necessary for confirmation of this prediction.

8.3 Weak Links of Grain Boundaries

In sintered Y-123 superconductors fabricated just after the discovery of high-temperature superconductors, it was found that the critical current density measured by the resistive four terminal method was much lower than the value estimated from the magnetization hysteresis. Sintered specimens were of polycrystalline structure, and grain boundaries severely restricted the transport current, while a closed current of fairly high density flowed inside grains of single crystalline structure.

In later years, Y-123 bulk superconductors which were composed of big single grains were successfully fabricated by the melt process, and Y-123 coated conductors with a highly oriented polycrystalline structure were grown on textured substrates. The weak link property in these superconductors was much improved.

Typical misorientation of crystal axes at the grain boundary is shown in Fig. 8.17. The main reasons for much stronger restriction by grain boundaries in high-temperature superconductors than in metallic superconductors are the short coherence length and the low carrier density which restricts the number of superconducting electrons tunneling through grain boundaries. Another reason is the d-wave symmetry of the superconductivity: the tunneling probability of superconducting electron pairs across a grain boundary reduces with increasing misorientation angle between grains. A higher tunneling barrier due to the high sensitivity of the superconductivity to various kinds of disorder such as local stress and impurity is also one of the reasons. A deterioration of superconductivity occurs at most grain boundaries for this reason. For example, a [001] tilt boundary with a small misorientation angle can be

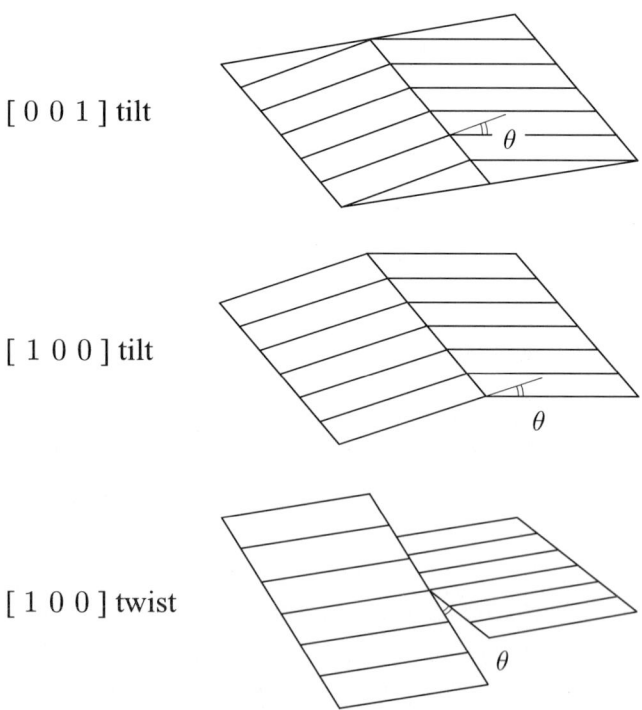

Fig. 8.17. Typical misorientation of crystal axes at grain boundary

regarded as a row of edge dislocations and there is speculation [30] that the superconductivity will disappear at the core of the dislocation due to strong stress, local deviation from chemical stoichiometry and variation in the hole concentration. In addition, it is also considered [31] that the electronic band structure is bent and the boundary becomes insulating due to local variation of the work function and/or the existence of impurity atoms.

If the grain boundary misorientation angle is θ, the critical current density generally obeys:

$$J_c \propto \exp\left(-\frac{\theta}{\theta_0}\right) \qquad (8.14)$$

as shown in Fig. 8.18 [32]. For [001] tilt and [100] tilt boundaries, θ_0 amounts to 4–5°. On the other hand, its value is slightly lower than this for [100] twist boundaries. Thus, the critical current density is significantly reduced with increasing misorientation angle. Hence, the crystal axis orientation, especially that of the c-axis, is important. The abovementioned Y-123 bulk and coated conductors are examples of improvement of the critical current property by alignment of crystal axes.

On the other hand, it was recently reported that the weak link property was improved by Ca doping. Figure 8.19(a) shows a dependence of the critical current density on the Ca concentration at 4.2 K for $Y_{1-x}Ca_xBa_2Cu_3O_{7-\delta}$

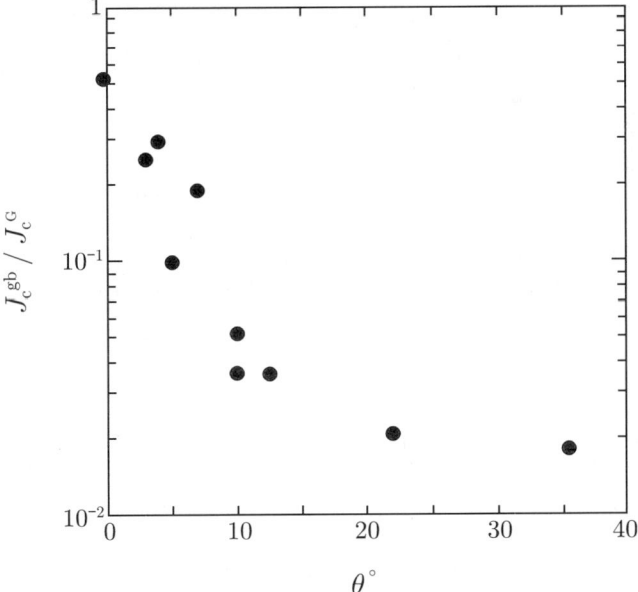

Fig. 8.18. Relationship of critical current density and misorientation angle of [001] tilt grain boundary in Y-123 thin film at 4.2–5.0 K in the absence of magnetic field [32]

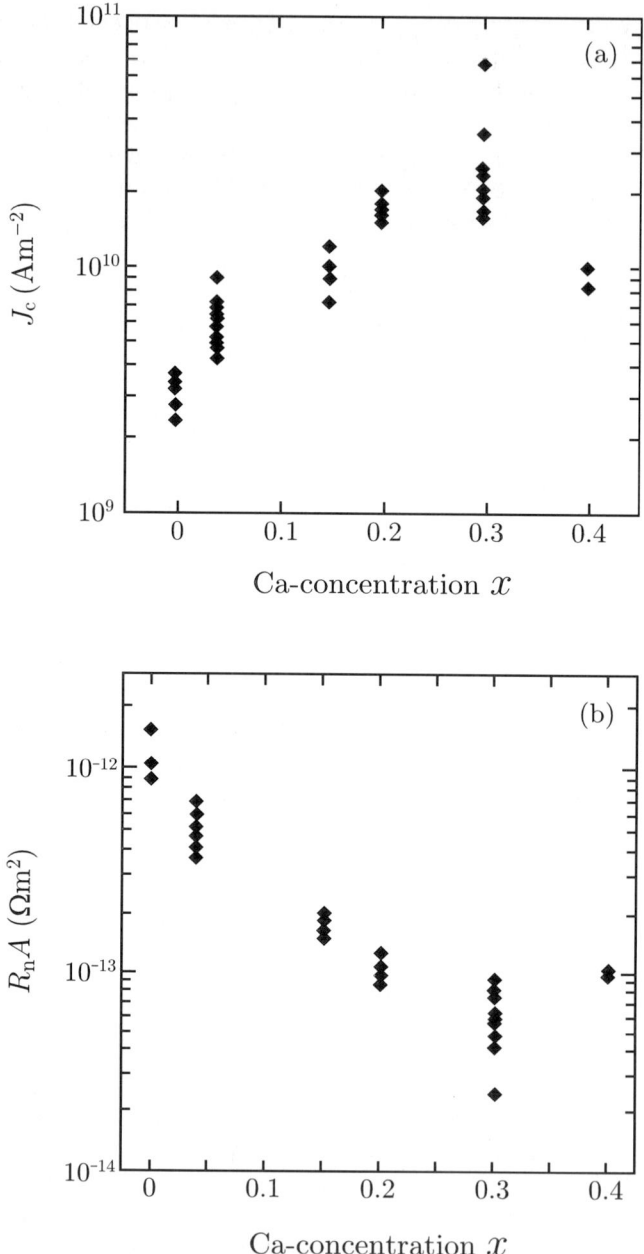

Fig. 8.19. Dependence of (**a**) critical current density at 4.2 K and (**b**) normal state resistivity of 24° [001] tilt grain boundaries in $Y_{1-x}Ca_xBa_2Cu_3O_{7-\delta}$ thin films on Ca concentration [31]

thin films deposited on STO substrates with 24° [001] tilt grain boundaries [31]. This result shows that the critical current density at a low temperature is significantly improved by doping with a high concentration of Ca. This is considered to result from an overdoped state of the CuO_2 planes due to the substitution of Y by Ca and from a reduction of the built-in potential at the boundary owing to the smaller valence of Ca^{2+} than Y^{+3}. Figure 8.19(b) shows the normal state resistivity of the same boundaries, and the resistivity decreases significantly with increasing Ca concentration. This suggests that the transport property of the boundary is improved by the Ca doping.

However, the problem with this method is that the critical temperature is lowered by substitution of Y by Ca. The improvement of critical current density is not appreciable at high temperatures such as 77 K, although it is significant at low temperatures. Hence, it is necessary to optimize the doping conditions. For example, it is expected to be effective for the purpose to deposit a thin layer of (Y, Ca)-123 on the Y-123 film followed by a subsequent heat treatment to make Ca diffuse into grain boundaries [33]. Since the diffusion constant of Ca in grain boundaries is larger by a factor of 10^3 than that in grains, this method allows a diffusion of Ca only in grain boundaries. Hence, it is expected to improve the weak link property of grain boundaries without degrading the critical temperature of the whole superconductor.

8.4 Electromagnetic Properties

The electromagnetic properties in high-temperature superconductors are essentially the same as those in metallic superconductors: most of them are described by the critical state model, while in superconductors smaller than Campbell's AC penetration depth they obey the Campbell model and deviate from the critical state model due to the reversible motion of flux lines. However, some electromagnetic properties are inherent to high-temperature superconductors, and these will be introduced in this section.

8.4.1 Anisotropy

High-temperature superconductors are composed of superconducting layers and almost insulating block layers, and this results in a large anisotropy of the critical current density. Namely, the value of the critical current density flowing along the c-axis J_c^c is much lower than that in the a-b plane J_c^{ab}. This anisotropy is larger for a more two-dimensional superconductor.

In addition, even the critical current density flowing in the same direction takes different values depending on the direction of the magnetic field as shown in Fig. 8.20 [34]. Generally speaking, the critical current density is smaller in a magnetic field parallel to the c-axis than in a field in the a-b plane. This difference comes from the anisotropy of superconducting electron mass described in Sect. 8.1 and the solid line in the figure shows the theoretical prediction with

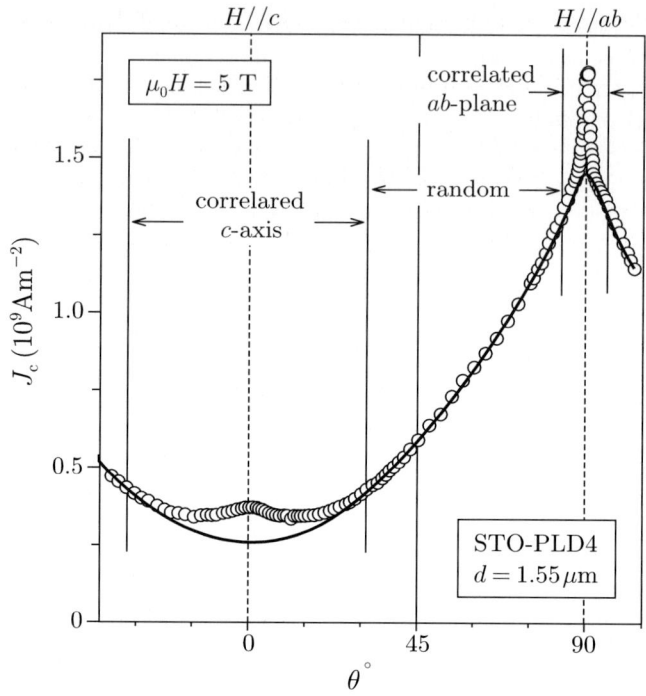

Fig. 8.20. Magnetic field anisotropy of critical current density in an Y-123 PLD thin film deposited on MgO single crystal substrate (75.5 K, 5 T) [34]. The *solid line* shows the theoretical prediction with the anisotropy parameter $\gamma_a = 5$ for the electron mass

the effective magnetic field; $\widetilde{H} = H\epsilon(\theta)$, $\epsilon(\theta) = (\cos^2\theta + \gamma_a^{-2}\sin^2\theta)^{1/2}$ with $\gamma_a = 5$ for Y-123 supercunductor. The cause of the deviation from the theory at around $\theta = 0°$ and $90°$ will be described in Subsect. 8.6.1. At high fields, the anisotropy becomes more significant due to the effect of the irreversibility field.

The electromagnetic phenomena are also anisotropic and complicated because of such anisotropies of the critical current density. For example, the magnetization current in a magnetic field along the a-axis flows in the direction of the b- and c-axes of the superconductor. Hence, the extended Bean model which takes account of the anisotropic critical current density is necessary to describe the magnetization. When an AC magnetic field is applied to a superconducting Bi-system tape with a large anisotropy, the AC loss is mostly determined only by the c-axis component of the AC magnetic field. Hence, when a coil is wound with such a superconducting tape, the direction of the magnetic field is different between the central part and the edge of the coil, resulting in large differences in the critical current density and the loss density in the different parts.

8.4.2 Differences in the Size Effect due to the Dimensionality

As mentioned in Sect. 8.2, the critical size of the superconductor for the irreversibility field, i.e., the G-L transition field, is the virtual pinning correlation length L_0 given by Eq. (7.92) for both Bi-2212 and Y-123, independently of the dimensionality of the superconductor. The details will be explained in Sect. 8.5. On the other hand, the critical size for the order-disorder transition associated with the peak effect is much smaller than the pinning correlation length in two-dimensional Bi-2212, while it is given by the pinning correlation length in three-dimensional Sm-123 as shown in Sect. 7.6. Figure 8.21 shows the critical current density of Bi-2212 single crystal specimens with different thicknesses [27]. It is found that the peak effect is not observed for films thinner than 0.5 μm, while the peak effect occurs for specimens

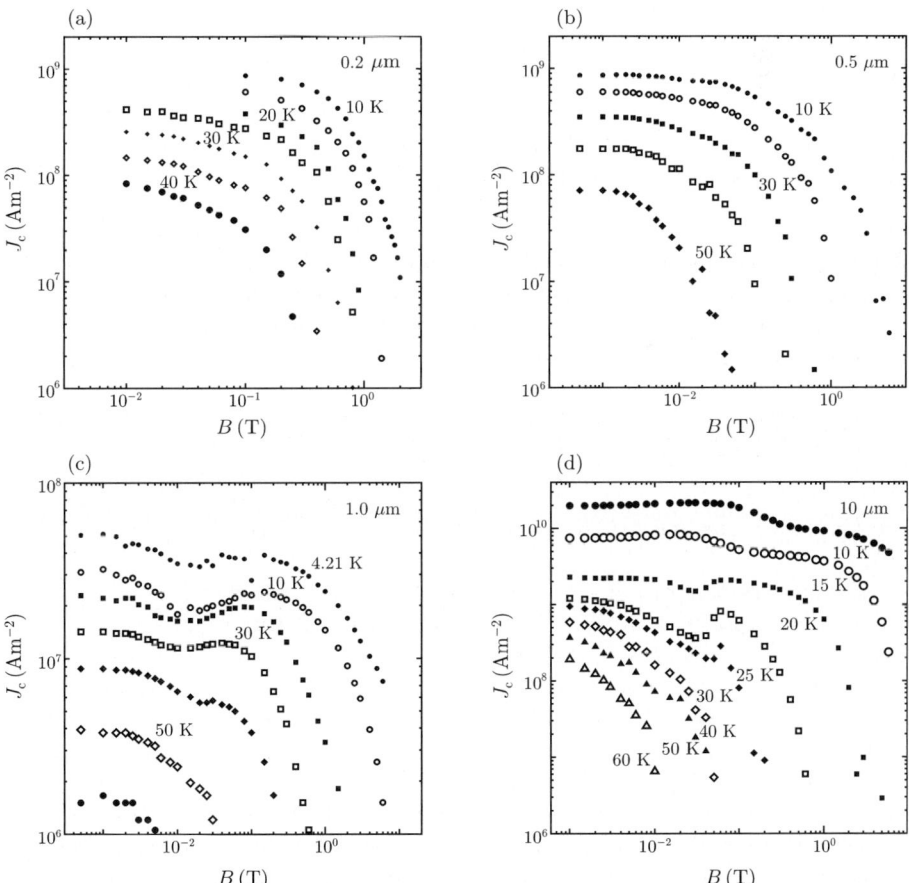

Fig. 8.21. Critical current density of Bi-2212 single crystal specimens with different thicknesses [27]

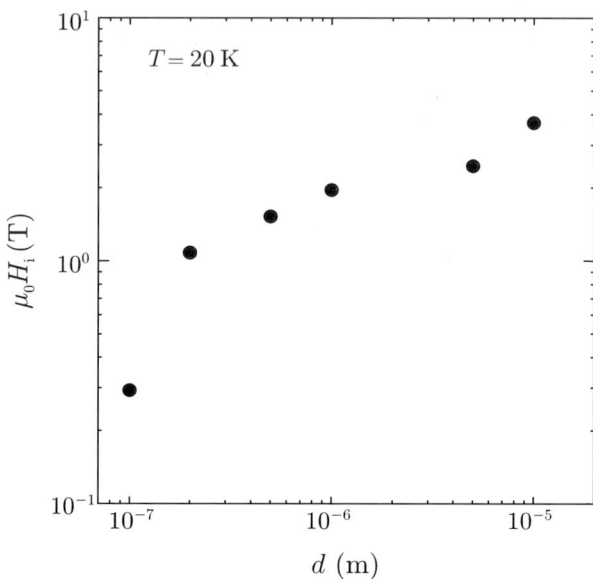

Fig. 8.22. Variation in irreversibility field with the thickness for the same Bi-2212 specimens in Fig. 8.21 at 20 K [27]

thicker than 1 μm. This suggests that, even when the dimensional crossover of flux lines takes place, its characteristic length is not shorter than 0.5 μm.

Figure 8.22 shows the thickness dependence of the irreversibility field of the same specimens at 20 K [27]. The irreversibility field increases monotonically with the thickness and seems to be saturated at the critical size L_0, which is expected to exceed 10 μm. Hence, the critical sizes are different between the G-L transition and the order-disorder transition in Bi-2212. Here is another example which indicates a long coupling of flux lines along the c-axis in a Bi-2212 superconductor, deviating from the prediction of the pancake vortex model. Figure 8.23(a) shows the pinning correlation length obtained by Campbell's method for a Bi-2212 single crystal in a magnetic field parallel to the c-axis [35]. The correlation length is longer than 10 μm even when flux lines are in the two-dimensional state and increases with increasing magnetic field and/or temperature. Figure 8.23(b) shows the theoretical prediction of the pinning correlation length of

$$L = \left(\frac{Ba_{\mathrm{f}}}{2\pi\mu_0 J_{\mathrm{c}}}\right)^{1/2}, \quad (8.15)$$

in which the observed J_c values are substituted and the relationship $\zeta = 2\pi$ for pointlike defects is used for the interaction distance in Eq. (7.75). It can be seen that the two results agree well with each other. Similar agreement is obtained also for three-dimensional Y-123 superconductors [36, 37]. Therefore, Eq. (8.15) holds generally, independently of the dimensionality of the

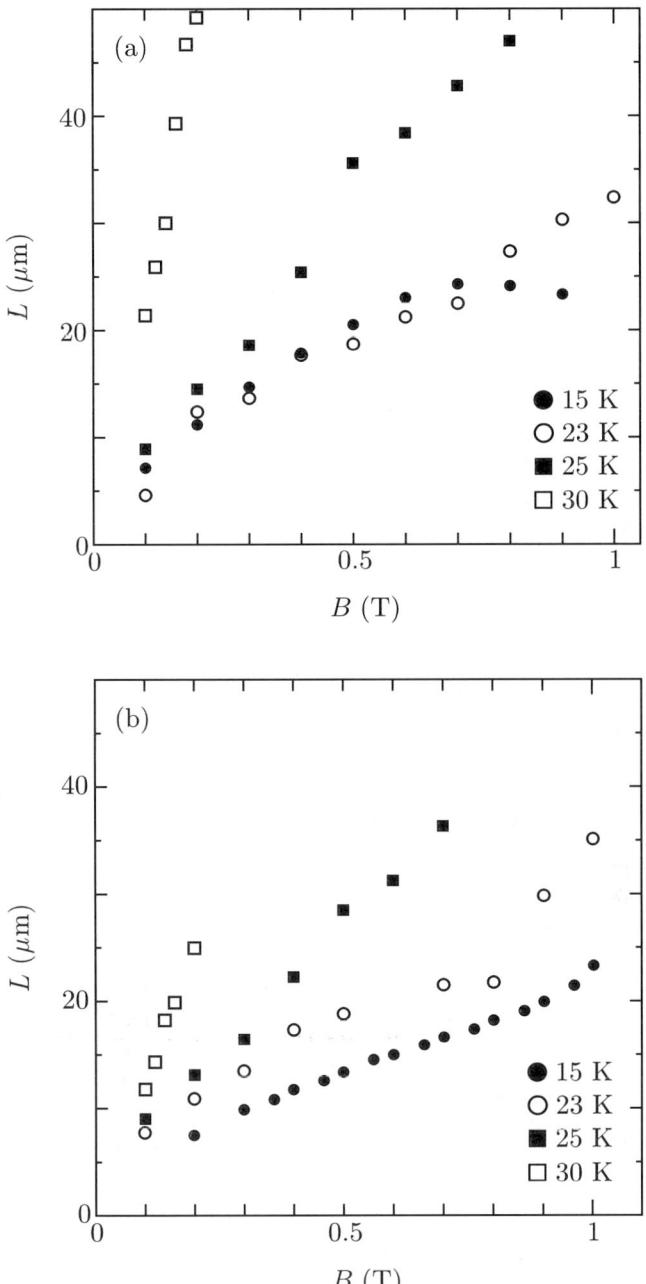

Fig. 8.23. Pinning correlation length in a Bi-2212 single crystal in a magnetic field along the c-axis [35]: (**a**) experimental results obtained by using Campbell's method and (**b**) theoretical estimates using Eq. (8.15) with observed J_c value

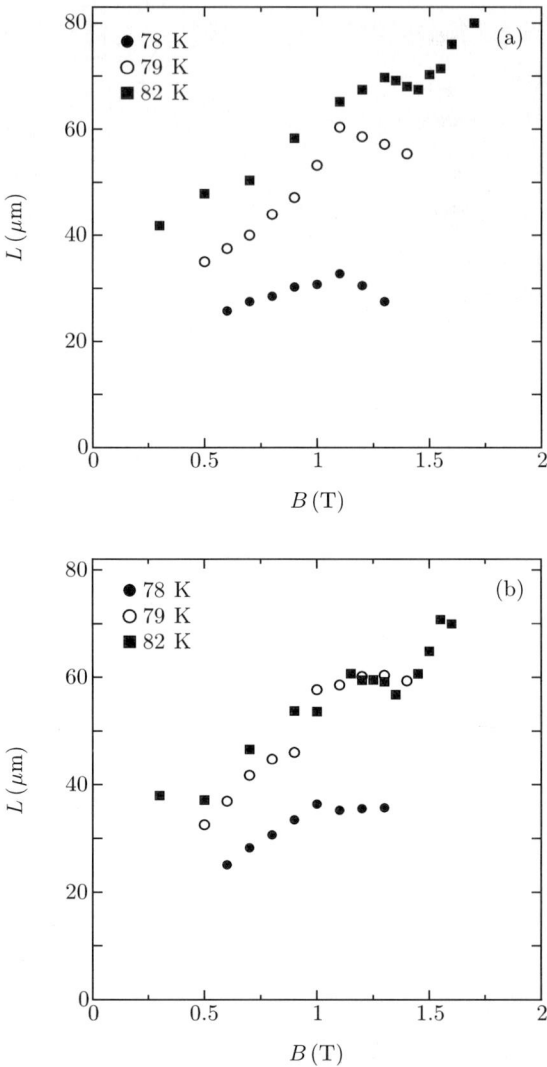

Fig. 8.24. Pinning correlation length in an Y-123 single crystal in a magnetic field along the c-axis [38]: (**a**) experimental results obtained by using Campbell's method and (**b**) theoretical estimates

superconductor. Figures 8.24(a) and (b) are an example for an Y-123 single crystal showing the peak effect [38], and L takes maximum and minimum values at magnetic fields at which the critical current density takes minimum and maximum values, respectively.

The critical size of the dimensional crossover of flux lines can also be approximately estimated from the rate of discontinuous variation in the critical current density at the peak effect. Namely, since the critical current density increased approximately by a factor of 4 with the variation from the three-dimensional vortex state to the two-dimensional one, the virtual critical current density J_{c0} is expected to increase approximately by the same factor. Thus, the mechanism of the collective pinning indicates that the correlation length of the normal core inside the vortex decreased from L_0 to $L_0/4^2 = L_0/16$. The observed L was about 35 μm [21]. Since L_0 is expected to be shorter than L, it seems to be reasonable that $L_0/16$ is close to the critical thickness for the peak effect observed for thin films.

The critical size of the normal core of a flux line in most two-dimensional Bi-2212 superconductors was estimated similarly from independent experiments on the peak effect. Hence, the estimated critical size seems to be valid. This value shows that the pancake normal cores in CuO_2 planes do not behave independently of each other along the c-axis but are coupled over a fairly long distance even in two-dimensional superconductors. In the three-dimensional superconductors such as Y-123, on the other hand, the critical size of the irreversibility field coincides with that of the peak effect, suggesting that the outer magnetic flux and the inner normal core behave as one body. Figure 8.25 schematically depicts the difference in the deformation between the normal core of a flux line and the surrounding magnetic structure due to the difference in the state of flux lines.

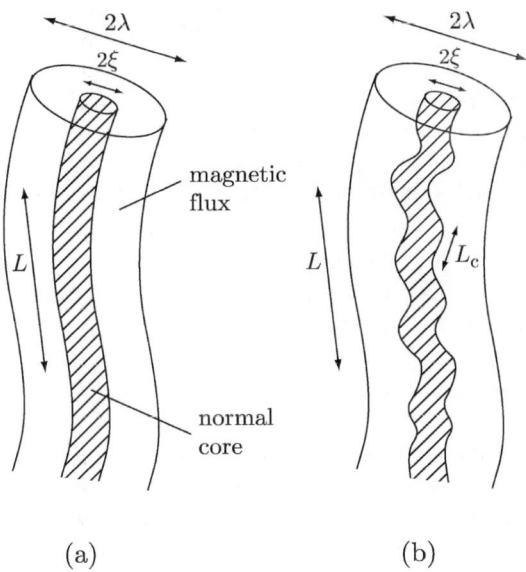

Fig. 8.25. Schematic illustration of deformed flux line in Bi-2212 superconductor: (**a**) three-dimensional state at low magnetic fields and (**b**) two-dimensional state at high magnetic fields

According to Vinokur et al. [26] the interlayer coupling length is given by

$$r_{3D} \simeq (\gamma_a s a_f)^{1/2} . \tag{8.16}$$

Substitution of typical values for Bi-2212, i.e., $\gamma_a = 100$, $s = 1.5$ nm and $B \simeq 0.1$ T ($a_f \simeq 150$ nm) for the peak field leads to $r_{3D} \simeq 0.15$ μm. This value is slightly shorter than the critical size for the peak effect, which is on the order of 0.5 μm, but is of the same order of magnitude. Hence, the characteristic length of the normal core in a two-dimensional vortex might be given by the interlayer coupling length.

However, the interlayer coupling length of Eq. (8.16) takes on very small values for three-dimensional superconductors with small anisotropy parameters, and any phenomena associated with such critical sizes have not yet been observed.

The characteristic lengths of Bi-2212 superconductors in the two-dimensional vortex state are now summarized. In each instance of thermal agitation the normal cores of the flux lines are interacting with pinning centers with a strength corresponding to the virtual critical current density J_{c0}, which is not influenced by the flux creep, and are deformed along their length characterized by a distance L_c less than 1 μm. The magnetic flux in the outer region of the normal core is deformed along the length by a distance L_0. Hence, a segment of the flux lines of this length moves by a distance of about a_f when hopping by the thermal agitation. Thus, the longitudinal flux bundle size is given by L_0. When the flux motion is averaged over a time period during an observation in experiments, the critical current density is decreased to J_c due to the flux creep and the pinning correlation length is increased to L.

The longitudinal flux bundle size in a three-dimensional superconductor is also given by L_0. However, if the size of a specimen is less than L, the pinning efficiency takes a higher value due to a weaker interference among pinning interactions in the stage of determination of J_c, resulting in a disappearance of the peak effect.

8.4.3 Flux Creep

High-temperature superconductors are frequently used at high temperatures, and the effect of flux creep is strong. The strength of the effect changes from superconductor to superconductor depending on the dimensionality. In two-dimensional superconductors the irreversibility field, the characteristic most strongly influenced by the flux creep, takes on significantly small values. Figure 8.26 shows the relationship between the irreversibility field at fixed normalized temperatures and the anisotropy parameter [39]. Extremely small irreversibility fields in two-dimensional superconductors are caused by their small condensation energy density. As schematically shown in Fig. 8.27, the superconductivity in the block layer is very weak and this layer is thick in a two-dimensional superconductor, resulting in a small effective condensation

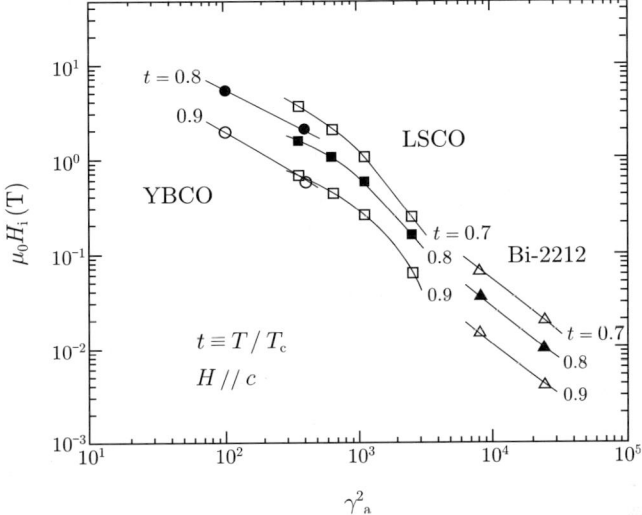

Fig. 8.26. Relationship between the irreversibility field along the c-axis at the same reduced temperatures and the anisotropy parameter [39]

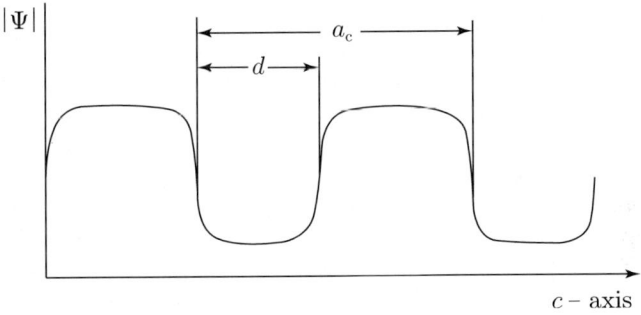

Fig. 8.27. Schematic variation in the superconducting order parameter along the c-axis

energy density when the magnetic field is applied parallel to the c-axis. This leads to very weak pinning.

Another effect of the small condensation energy density is a small transverse flux bundle size. As shown in Eq. (7.93) the transverse flux bundle size is determined by the shear modulus of the flux lines C_{66}, which is related to the shear energy of the flux line lattice, i.e., the energy increase due to the deformation of the structure of the order parameter. In a two-dimensional superconductor C_{66} is very small due to the small mean value of the order parameter, resulting in a small transverse flux bundle size. Figure 8.28 depicts how the flux bundle changes with the dimensionality of the superconductor under an assumption of the same flux pinning strength, i.e., the same longitudinal size [36].

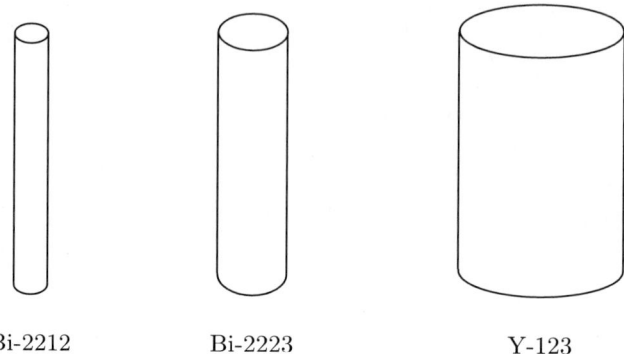

Fig. 8.28. Schematic illustration of different geometries of flux bundles due to differences in the dimensionality of the superconductor when the pinning strength is the same [36]

Thus, the small condensation energy density causes the weak pinning and the small flux bundle size, and both of them then bring about the small pinning potential energy. Hence, the influence of flux creep is severe in a two-dimensional superconductor. Such an influence of the dimensionality of the superconductor appears in J_{c0} and g^2 in U_0 in Eqs. (7.97) and (7.99). Practical examples of the influence of the dimensionality on the irreversibility field will be described in Sect. 8.5.

8.4.4 E-J Curve

One of the characteristic electromagnetic phenomena in high-temperature superconductors is a gradual rise in the electric field at the critical point in the E-J curve represented by the quite small value of n in Eq. (5.1) in comparison with metallic superconductors. This seems to be caused by a wide statistical distribution of the flux pinning strength as well as the strong effect of flux creep. High-temperature superconductors have complicated crystal structures composed of many kinds of elements, and hence, a high number density of crystalline defects, especially oxygen deficiencies, is expected to exist. These defects will act as weak pinning centers. Defects of bigger size such as dislocations or stacking faults of fairly high number density will also be contained in the superconductor, acting as stronger pinning centers. In addition, very strong pinning centers such as normal precipitates will also be included. As a result it is expected that the pinning strength is very widely distributed.

In addition, there are regions where the superconducting current finds it hard to flow due to weak links at grain boundaries. This also causes inhomogeneity in the superconductor. As a result, the distribution of the local critical current density where the weak links are included is extremely wide, and the observed critical current density changes greatly with the electric field strength. Figure 8.29 shows the pinning force densities of a Bi-2223 tape

8.4 Electromagnetic Properties 375

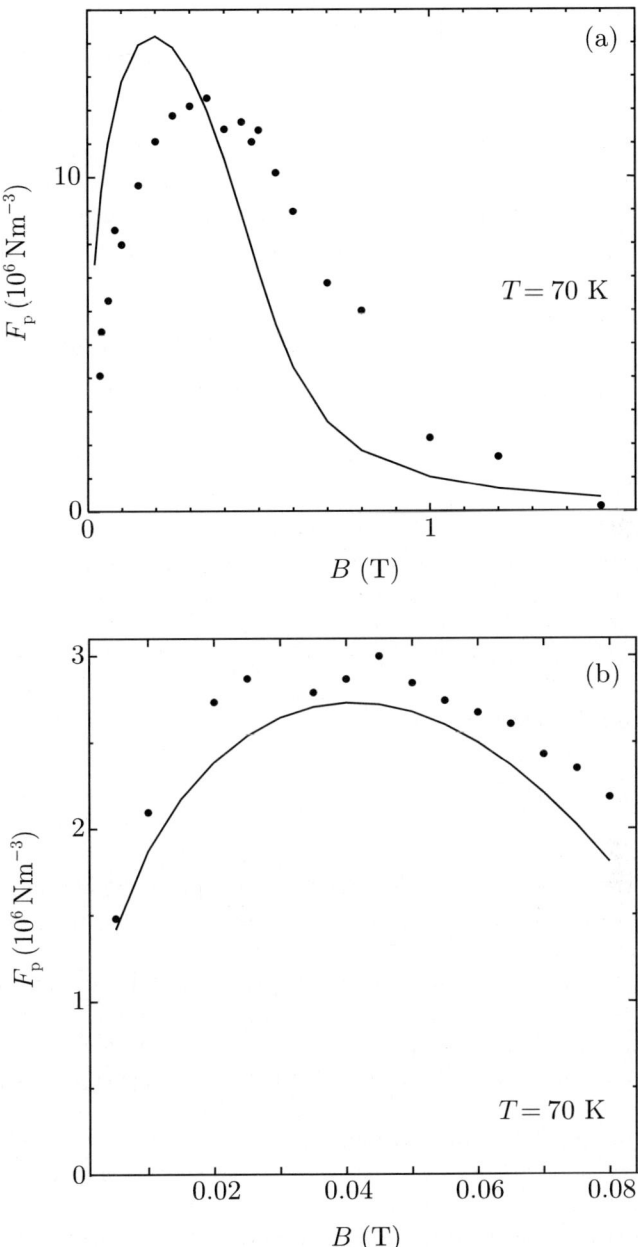

Fig. 8.29. Flux pinning strength of a Bi-2223 tape [40] in a magnetic field along the c-axis obtained by (**a**) the four terminal method at 1.0×10^{-4} Vm^{-1} and by (**b**) the DC magnetization method at 1.0×10^{-10} Vm^{-1}. The *solid lines* are the theoretical results of the flux creep-flow model

observed by the four terminal method at 1.0×10^{-4} Vm^{-1} and by the DC magnetization method at 1.0×10^{-10} Vm^{-1} [40]. It is found that these results are completely different.

In inhomogeneous superconductors the motion of flux lines is not uniform as in an ideal superconductor with a uniform critical current density but is complicated. Such a motion of flux lines can be correctly described by the percolation model [17]. In this model the activated flux motion due to flux creep is expressed as an effective flux flow in a shallow pinning potential, and it is assumed that the probability that flux lines pinned with nonuniform strength will be depinned by a Lorentz force is expressed by a Weibull function. Thus, the distribution function of the critical current density J_c is given by

$$P(J_c) = \frac{m_0}{J_0}\left(\frac{J_c - J_{cm}}{J_0}\right)^{m_0-1} \exp\left[-\left(\frac{J_c - J_{cm}}{J_0}\right)^{m_0}\right] ; \; J_c \geq J_{cm} ,$$
$$= 0 ; \hspace{4cm} J_c < J_{cm} , \quad (8.17)$$

where J_{cm} is the minimum value of J_c and J_0 shows the distribution width. m_0 is a parameter that determines the structure of the distribution and is related to the dynamic critical index z in the scaling of the E-J curve as $m_0 = (z-1)/2$. Thus, the E-J curve is expressed as

$$E(J) = \frac{\rho_f J}{m_0 + 1}\left(\frac{J}{J_0}\right)^{m_0}\left(1 - \frac{J_{cm}}{J}\right)^{m_0+1} ; \hspace{1cm} B \leq \mu_0 H_g ,$$
$$= \frac{\rho_f |J_{cm}|}{m_0 + 1}\left(\frac{|J_{cm}|}{J_0}\right)^{m_0}\left[\left(1 + \frac{J}{|J_{cm}|}\right)^{m_0+1} - 1\right] ; \; B > \mu_0 H_g . \quad (8.18)$$

In the above H_g is the G-L transition field, and it should be noted that $J_{cm} < 0$ for $B > \mu_0 H_g$. In Fig. 8.30 the theoretical results with adjusted parameters J_{cm}, J_0, m_0 and H_g are compared with experiments for an Y-123 thin film [9]. It is found that the percolation model describes the experiments well. The E-J curves given by the percolation model are approximately explained by the mechanism of flux creep and flow [14], and this agreement proves the validity of the percolation model.

When the results obtained at various temperatures are summarized for J_{cm} and $J_k = J_{cm} + J_0$, which corresponds to a representative value of J_c, a relationship similar to the scaling law of pinning force density is obtained:

$$J_{cm} B = A H_g(T)^{\zeta'}\left[\frac{B}{\mu_0 H_g(T)}\right]^\gamma \left[1 - \frac{B}{\mu_0 H_g(T)}\right]^\delta . \quad (8.19)$$

An example is shown in Fig. 8.31. A similar relationship is obtained also for J_k.

Since the percolation model describes simply the E-J characteristics in a suitable range of electric field and over a wide range of temperature and magnetic field, it is very useful for the application of superconductors.

8.4 Electromagnetic Properties   377

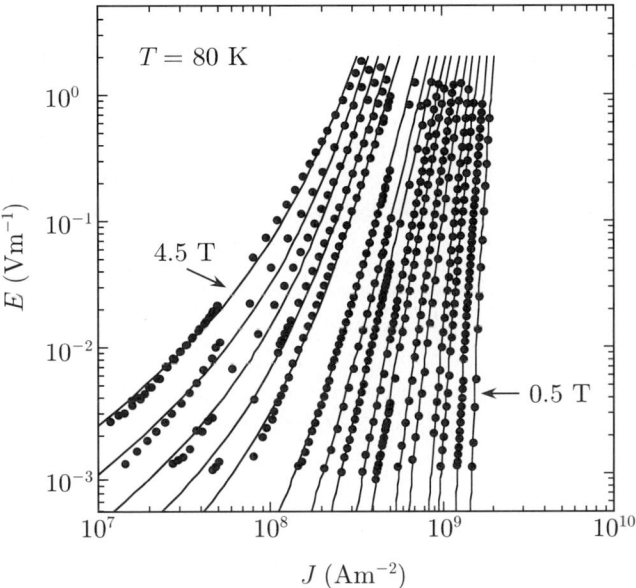

Fig. 8.30. Comparison of E-J curves between experiments on an Y-123 single crystal thin film and the percolation model (*solid lines*) [9]

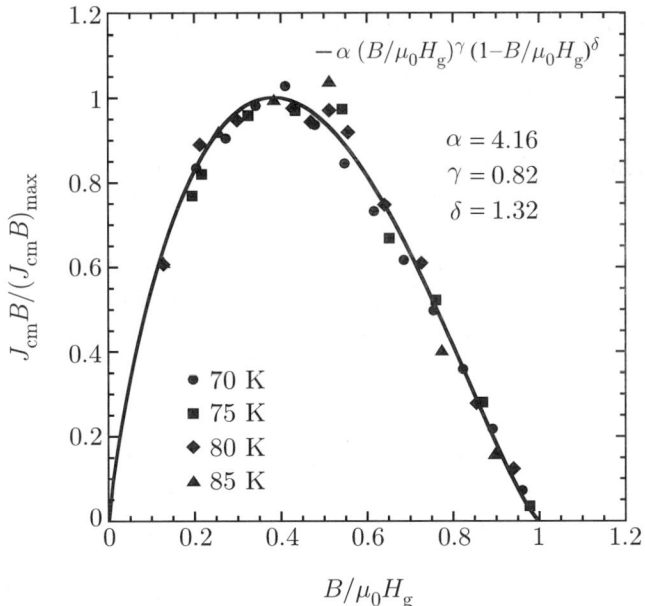

Fig. 8.31. Scaling law of the minimum pinning force density $J_{\mathrm{cm}}B$ of an Y-123 single crystal thin film in Fig. 8.30 [9]

8.4.5 Josephson Plasma

Two-dimensional high-temperature superconductors such as Bi-2212 are composed of superconducting CuO_2 planes and insulating block layers, and form stacked intrinsic Josephson junctions. Hence, it is expected that the Josephson current flowing along the c-axis across the block layers is coupled with electromagnetic fields, resulting in a unique excitation. This excited wave is called the Josephson plasma wave. The characteristic feature of the Josephson plasma wave is that the resonance frequency is not very high, since the electric charges cannot move quickly across the block layers. Hence, the resonance frequency is sufficiently lower than the superconducting gap frequency that the excitation of quasiparticles does not occur. As a result the Josephson plasma can be stably excited without damping. There are two Josephson plasma waves, i.e., the longitudinal wave which propagates along the c-axis and the transverse wave which propagates in the a-b plane.

The longitudinal Josephson plasma wave is considered at first. If the phases of the order parameter in the n-th and $(n+1)$-th superconducting layers are represented by ϕ_n and ϕ_{n+1}, respectively, the difference in gauge-invariant phase between these layers is:

$$\theta_{n+1,n} = \phi_{n+1} - \phi_n - \frac{2\pi}{\phi_0} \int_n^{n+1} A_z \mathrm{d}z , \qquad (8.20)$$

where the z-axis is parallel to the c-axis. Hence, the Josephson current density along the c-axis is given by

$$J_\mathrm{s} = j_\mathrm{c} \sin \theta_{n+1,n} . \qquad (8.21)$$

The wave number of the longitudinal plasma wave is denoted by k. When its wave length $2\pi/k$ is sufficiently longer than the distance between the superconducting layers s, the phase difference can be approximated by a continuous function which varies along the $c(z)$-axis: it will be given as

$$\theta(z) = \theta_0 \exp \mathrm{i}(kz - \omega t) . \qquad (8.22)$$

When these equations are coupled with Maxwell equations, we have: [41]

$$\frac{\partial^2 \theta}{\partial t^2} = v_\mathrm{B}^2 \frac{\partial^2 \theta}{\partial z^2} - \omega_\mathrm{p} - 2\theta , \qquad (8.23)$$

where v_B is a propagation velocity of the phase difference along the z-axis and ω_p is the Josephson plasma frequency given by

$$\omega_\mathrm{p} = \frac{\bar{c}}{\lambda_\mathrm{J}} = \frac{c}{\epsilon_\mathrm{s}^{1/2} \lambda_\mathrm{J}} . \qquad (8.24)$$

In the above $c = 1/(\epsilon_0 \mu_0)^{1/2}$ and $\bar{c} = c/\epsilon_\mathrm{s}^{1/2}$ are the light velocities in the vacuum and in the block layers, respectively, with ϵ_s denoting the relative dielectric constant of the block layers, and λ_J is given by Eq. (8.13). Substituting

Eq. (8.22) into Eq. (8.23) derives the dispersion relation of the longitudinal Josephson plasma frequency:

$$\omega = (\omega_p^2 + v_B^2 k^2)^{1/2} . \tag{8.25}$$

In the case of the transverse Josephson plasma wave, the dispersion relation of the plasma frequency is given by

$$\omega = \left(\omega_p^2 + \frac{c^2 k^2}{\epsilon_s}\right)^{1/2} . \tag{8.26}$$

Because $c^2/\epsilon_s \gg v_B^2$, the dispersion of the transverse wave is much stronger. The dispersion relations of the two plasma waves are compared in Fig. 8.32.

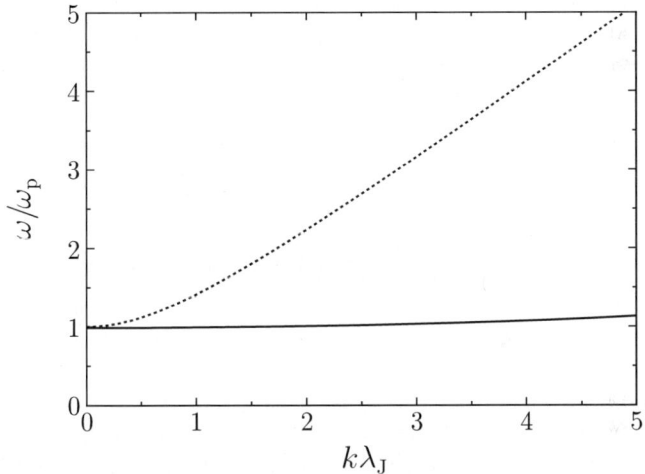

Fig. 8.32. Dispersion relations of Josephson plasma frequency [41]. The *solid* and *dotted lines* represent longitudinal and transverse plasma waves

A proof of the excitation of Josephson plasma waves is obtained from measurements of absorption of electromagnetic microwaves, etc. Figure 8.33 shows an observed result [42] of the surface resistance of a Bi-2212 single crystal in an AC electric field of 45 GHz in the direction of the *c*-axis during a sweep of DC magnetic field in the same direction. Sharp peaks show that the longitudinal Josephson plasma oscillation takes place.

With the aid of measurements of Josephson plasma waves the behavior or the state of flux lines mainly in the liquid state has been investigated in detail.

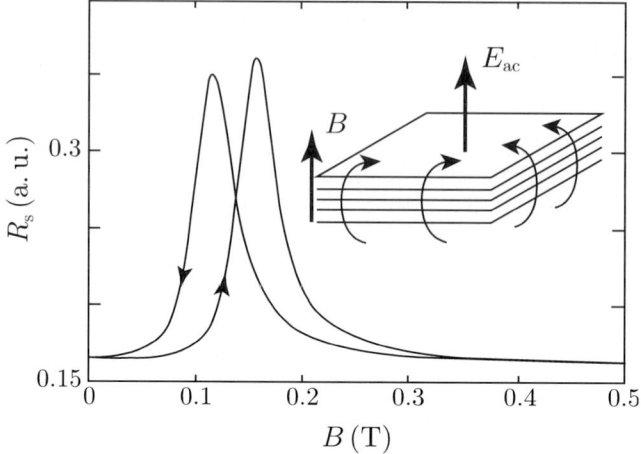

Fig. 8.33. Observed result of surface resistance of a Bi-2212 single crystal when an AC electric field of 45 GHz and a DC magnetic field are applied along the c-axis [42]

8.5 Irreversibility Field

The irreversibility field, a characteristic field at which the critical current density reduces to zero, is one of the important parameters of the superconductor for its application. This is strongly associated with the G-L transition field as will be discussed.

The irreversibility field is determined by the mechanism of flux creep, since most of the induced electric field does not originate from the flux flow but from the flux creep at the electric field strength of 1×10^{-4} Vm^{-1} at which the critical current is usually determined [14]. Hence, the irreversibility field can be derived from the E-J characteristics caused by the flux creep. Here the statistical distribution of the local critical current density is neglected for simplicity. This allows us to show analytically various aspects of the irreversibility field. Later a theoretical treatment in which the distribution of the critical current density is taken into account will be introduced and the results will be compared with experiments. A discussion will also be given on the association with the G-L transition.

8.5.1 Analytic Solution of Irreversibility Field

According to the flux creep model, Eq. (3.129) gives the condition of the irreversibility field, if the flux motion in the opposite direction to a Lorentz force is neglected. This is allowed except in the vicinity of the TAFF region. The analytic expression of the irreversibility field is obtained by substituting

Eq. (7.97) or (7.99) into this equation. It is assumed that the scaling law of the virtual critical current density in the creep-free case J_{c0} is expressed as*

$$J_{c0}(B,T) = A \left[1 - \left(\frac{T}{T_c}\right)^2\right]^{m'} B^{\gamma-1} \left(1 - \frac{B}{\mu_0 H_{c2}}\right)^{\delta}. \quad (8.27)$$

Usually the irreversibility field is sufficiently lower than the upper critical field, and hence, the factor $(1 - B/\mu_0 H_{c2})^{\delta}$ in the above equation can be regarded as 1. Then, the irreversibility field is obtained:

$$(\mu_0 H_i)^{(3-2\gamma)/2} = \left(\frac{K}{T}\right)^2 \left[1 - \left(\frac{T}{T_c}\right)^2\right]^{m'} \equiv (\mu_0 H_{\text{imax}})^{(3-2\gamma)/2}; \quad d \geq L_0,$$

$$= \frac{K'}{T} \left[1 - \left(\frac{T}{T_c}\right)^2\right]^{m'}; \quad d < L_0. \quad (8.28)$$

In the above H_{imax} is the irreversibility field of a bulk superconductor, and the constants K and K' are given by

$$K = \frac{0.835 g^2 A^{1/2}}{\zeta^{3/2} \log(B a_{\text{f}} \nu_0 / E_c)}, \quad (8.29)$$

$$K' = \frac{4.23 g^2 A d}{\zeta \log(B a_{\text{f}} \nu_0 / E_c)}. \quad (8.30)$$

Various aspects of the irreversibility field, i.e., its dependencies on various factors, will be described later. This enables us to qualitatively understand these aspects. In case of an exact comparison with experiments, a theoretical analysis involving the statistical distribution of the pinning force, which is the distribution of A in most cases, is necessary as will be shown in the next subsection.

8.5.2 Effect of Distribution of Pinning Strength

As described above, it is necessary to take account of the wide distribution of the effective virtual critical current density J_{c0} for a description of the practical pinning property over a wide range of electric field strength. This is directly associated with the wide distribution of J_c in the percolation model. Here we introduced a theoretical treatment of the flux creep-flow model [43]

---

* For high-temperature superconductors the temperature dependence of J_{c0} is sometimes expresses as $[1 - (T/T_c)]^{m'}$ instead of Eq. (8.27). This seems to be more suitable for Bi-2212 superconductors in which the temperature dependence of thermodynamic critical field is linear (see Fig. 8.52). These temperature dependencies are not different appreciably in the vicinity of the critical temperature, although they are remarkably different at low temperatures.

in which the distribution of J_{c0} is considered. It is assumed for simplicity that only A, representing the magnitude of J_{c0}, is distributed.

There are many other pinning parameters, i.e., m', γ and δ in addition to A. Hence, a simple distribution function for A is desirable. However, it does not obey the gaussian distribution. Here we assume the distribution function as:

$$f(A) = G \exp\left[-\frac{(\log A - \log A_{\rm m})^2}{2\sigma^2}\right], \qquad (8.31)$$

where $A_{\rm m}$ is the most probable value of A, σ is a parameter representing the distribution width and G is a normalization constant.

When the virtual critical current density J_{c0} changes locally, flux flow takes place where the flowing current density J is larger than J_{c0}, and flux creep occurs where J is smaller than J_{c0}. Here the normalized current density is defined:

$$j = \frac{J}{J_{c0}}. \qquad (8.32)$$

then, the electric field due to the flux flow is given by

$$\begin{aligned} E_{\rm ff} &= 0; & j \leq 1, \\ &= \rho_{\rm f} J_{c0}(j-1): & j > 1. \end{aligned} \qquad (8.33)$$

On the other hand, the electric field due to the flux creep is given by Eq. (3.115). Here, taking account of its contribution in the flux flow state, we approximate it as

$$\begin{aligned} E_{\rm fc} &= Ba_{\rm f}\nu_0 \exp\left[-\frac{U(j)}{k_{\rm B}T}\right]\left[1 - \exp\left(-\frac{\pi U_0 j}{k_{\rm B}T}\right)\right]; & j \leq 1, \\ &= Ba_{\rm f}\nu_0 \left[1 - \exp\left(-\frac{\pi U_0 j}{k_{\rm B}T}\right)\right]; & j > 1. \end{aligned} \qquad (8.34)$$

Equation (3.125) is used for the activation energy U. The local electric field from the two contributions is approximated by

$$E' = (E_{\rm ff}^2 + E_{\rm fc}^2)^{1/2}. \qquad (8.35)$$

This gives $E_{\rm fc}$ for $j \leq 1$ and is approximately equal to $E_{\rm ff}$ for $j \gg 1$. Then, the electric field when the current density J is applied is calculated from

$$E(J) = \int_0^\infty E' f(A) \mathrm{d}A. \qquad (8.36)$$

The E-J curves in Figs. 8.5(b) and 8.7 are obtained in this manner. Figures 8.29 and 8.53(b) show the pinning force density corresponding to the critical current density which is determined from the theoretical critical current density using the electric field criterion $E_{\rm c}$, as in experiments. The irreversibility field is also determined as the magnetic field at which the obtained critical current density reduces to some threshold value, as in experiments.

8.5.3 Comparison with Flux Creep-Flow Model

For an application of a high-temperature superconductor, design of the pinning properties under circumstances in which the superconductor is to be used is necessary. Hence, it is desirable that the practical pinning properties including the irreversibility field, the upper limit of the magnetic field for application, can be described theoretically. In this subsection, some experimental results on the irreversibility field for various superconductors will be compared with each theoretical prediction of the flux creep-flow model.

(a) *Temperature Dependence*

Initially the temperature dependence of the irreversibility field, i.e., the irreversibility line is discussed. Equation (8.28) leads to

$$H_\mathrm{i}(T) \propto \left[1 - \left(\frac{T}{T_\mathrm{c}}\right)^2\right]^n \tag{8.37}$$

with

$$n = \frac{2m'}{3 - 2\gamma} \tag{8.38}$$

in the vicinity of the critical temperature independently of the size of the superconductor. In Fig. 8.34 the observed parameter n is compared with the

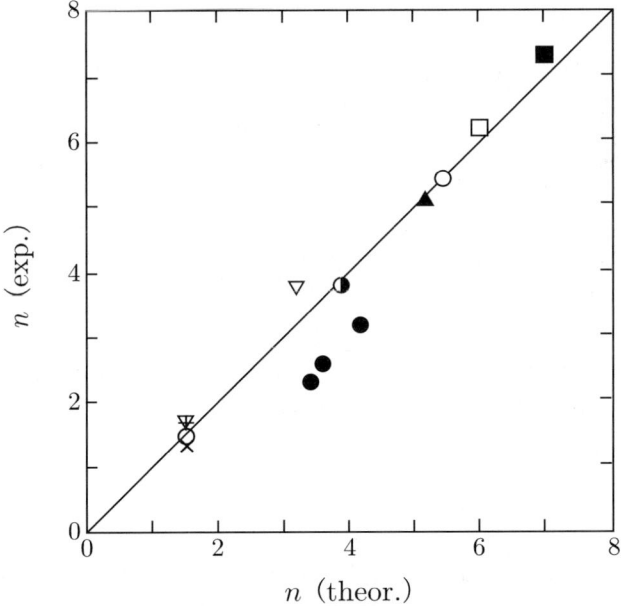

Fig. 8.34. Parameter n representing the temperature dependence of the irreversibility field [44]. Ordinate and abscissa are experimental values and theoretical values obtained from Eq. (8.38), respectively

corresponding value determined from m' and γ, which are obtained at low temperatures [44]. It is found that a good agreement is obtained over a wide range of the parameter. At low temperatures the temperature dependence becomes stronger due to the factor of T^{-2} or T^{-1} on the right hand side of Eq. (8.28).

(b) *Dependence on Flux Pinning Strength*

The dependence of the irreversibility field on the flux pinning strength is mainly expressed as that on the parameter A. As mentioned in Sect. 8.4, the influence of the dimensionality of the superconductor is also included, and the pinning strength is different for different superconductors even for the same defect. In this sense, the influences of the flux pinning strength and the dimensionality cannot be strictly distinguished. However, we focus only on the dependence on A, the parameter representing the pinning strength. In this case it should be noted that g^2 in Eq. (7.96) is also influenced by A. For simplicity the superconductor is assumed to be sufficiently large. From the relationship of $\alpha_\mathrm{L} \propto A$ we have $g_\mathrm{e}^2 \propto A^{-1}$. Substitution of this and $U_\mathrm{e} \propto g_\mathrm{e}^2 A^{1/2} \propto A^{-1/2}$ into Eq. (7.97) leads to

$$g^2 \propto A^{-1/3} . \tag{8.39}$$

Hence, in the case of $\gamma = 1/2$ observed for many superconductors we have

$$H_\mathrm{imax} \propto g^4 A \propto A^{1/3} . \tag{8.40}$$

Figure 8.35 shows the relationship between the irreversibility field perpendicular to the c-axis at $T/T_\mathrm{c} = 0.75$ and A for various high-temperature superconductors [45]. The straight lines represent the relationship of Eq. (8.40). Good agreement with experiments is obtained for Bi-2212 and Bi-2223.

It should be noted that Eq. (8.40) holds, if the irreversibility field is much lower than the upper critical field. If not, the irreversibility field is restricted by the upper critical field. In this case the irreversibility field should be calculated without neglecting the factor $(1 - B/\mu_0 H_\mathrm{c2})^\delta$ in Eq. (8.27) (see Exercise 8.5).

(c) *Influence of Dimensionality of Superconductor*

The dimensionality of the superconductor influences the irreversibility field not only through g^2 but also through A as described in Subsect. 8.4.3. Here we confine ourselves to the influence of the dimensionality only through g^2, so as to be consistent with the former subsection. This parameter is responsible for the difference between the straight lines representing the characteristic of each superconductor in Fig. 8.35. In a magnetic field normal to the c-axis, the typical value of g^2 is about 2, 4 and 6 for Bi-2212 bulk, Bi-2223 tape and Y-123 bulk, respectively. Examples of calculation of g^2 is given in Appendix A.8. In the case of $\gamma = 1/2$, we have $H_\mathrm{imax} \propto g^4$, which approximately explains the experimental results in the figure. Experimental and corresponding theoretical results on the irreversibility field normal to the c-axis for various

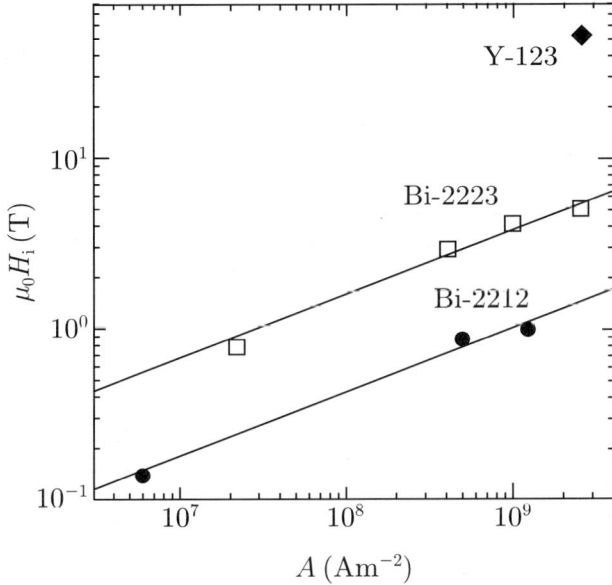

Fig. 8.35. Pinning strength dependence of the irreversibility field normal to the c-axis at $T/T_c = 0.75$ for various superconductors [45]. A corresponds to J_{c0} at $T = 0$ K and $B = 1$ T

superconductors are shown in Fig. 8.36, and the parameters used in the calculation are listed in Table 8.2 [45]. These results are for sufficiently large superconductors, and the agreement implies that the irreversibility line is determined by the dimensionality of the superconductor, the pinning strength and the pinning parameters such as m' and γ.

In a magnetic field along the c-axis, the irreversibility field is qualitatively the same, although it is much lower. The irreversibility lines of typical Y-123 bulk, [46] Bi-2223 tape [47] and Bi-2212 tape [48] are shown in Fig. 8.37. In each superconductor the observed result is explained by the theoretical results of the flux creep-flow model shown by the solid lines. The parameters used in the calculation are given in Table 8.3. The value of A_m is occasionally the same among three superconductors. On the other hand, the value of g^2 is about 1 in Bi-2212, ranged between 1 and 2 in Bi-2223 and above 3 in Y-123, which are similar to other measurements. This suggests that the influence of flux creep is less significant for a more three-dimensional superconductor. In most cases of Bi-2212 superconductors theoretical estimations using Eq. (7.98) are less than 1, and the practical minimum value, $g^2 = 1$, is used. For sintered powder specimens with similar particle sizes, it is known [49] that the irreversibility field of Yb-123 is higher than that of Y-123 at the same reduced temperature. This suggests that the dimensionality of Yb-123 is higher than that of Y-123, and this is likely to be caused by a more uniform carrier distribution along the c-axis in the block layer in Yb-123.

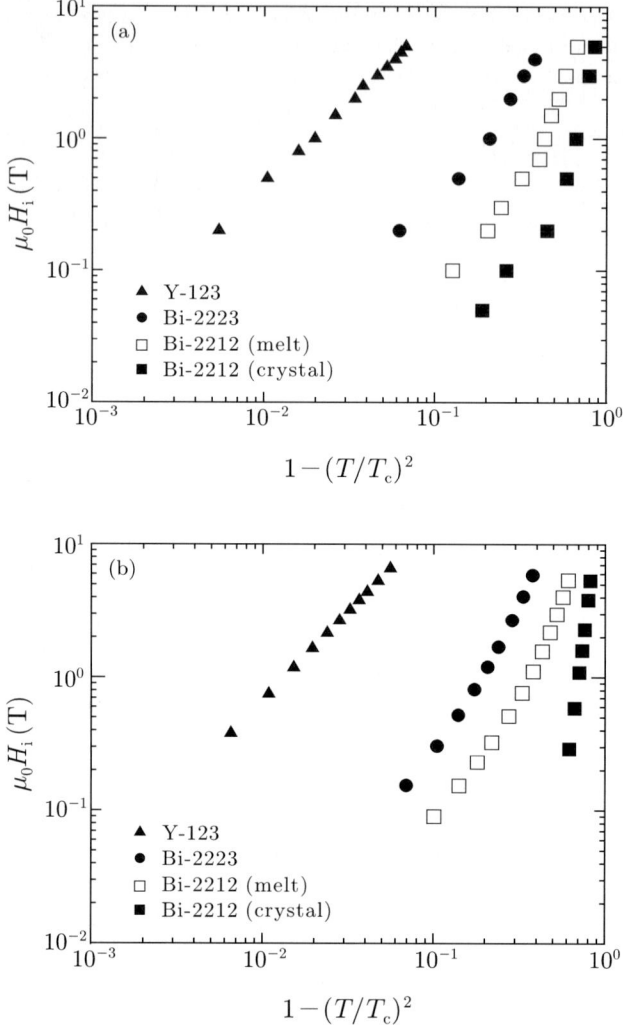

Fig. 8.36. (a) Experimental and (b) theoretical results of the irreversibility line in a magnetic field normal to the c-axis for various superconductors [45]

Here we show an example of the anisotropy of the irreversibility field. Figure 8.38 illustrates the results obtained from J_c estimated using Campbell's method for a Bi-2223 tape [50]. Although this anisotropy is larger than that obtained using the four terminal method, it is smaller than expected from the anisotropy of the electron mass. This is caused by a misalignment of grains in the superconductor: even if a magnetic field is parallel to the tape surface, it is not parallel to the a-b plane of each grain, resulting in a suppressed irreversibility field.

Table 8.2. Parameters of various superconductors for theoretical calculation of irreversibility field normal to the c-axis shown in Fig. 8.36(b) [45]. Two kinds of pinning centers are assumed for (Pb,Bi)-2223 tape and Bi-2212 melt specimen, and the distribution of A is not considered

	Y-123 (melt)	(Pb, Bi)-2223 (tape)	Bi-2212 (melt)	Bi-2212 (crystal)
T_c (K)	92.0	107.7	92.8	77.8
$\mu_0 H_{c2}^{ab}(0)$ (T)	670	1000	690	690
$\rho_n(T_c)$ (Ωm)	2.0×10^{-6}	1.0×10^{-4}	1.0×10^{-4}	1.0×10^{-4}
A	2.58×10^9	$2.54 \times 10^9 /$ 6.57×10^8	$7.52 \times 10^8 /$ 3.33×10^8	1.71×10^6
m'	1.5	3.0/1.5	2.25/1.5	3.10
γ	0.5	0.63/0.50	0.79/0.50	0.98
δ	2.0	2.0	2.0	2.0
g_e^2	44.4	6.7/1.55	2.2	620
g^2	6.0	4.0/1.5	2.0	14*
ζ	4	$2\pi/4$	2π	2π

*$g^2 = 14$ is assumed so as to get a good fit, although $g^2 = 4.0$ is theoretically predicted.

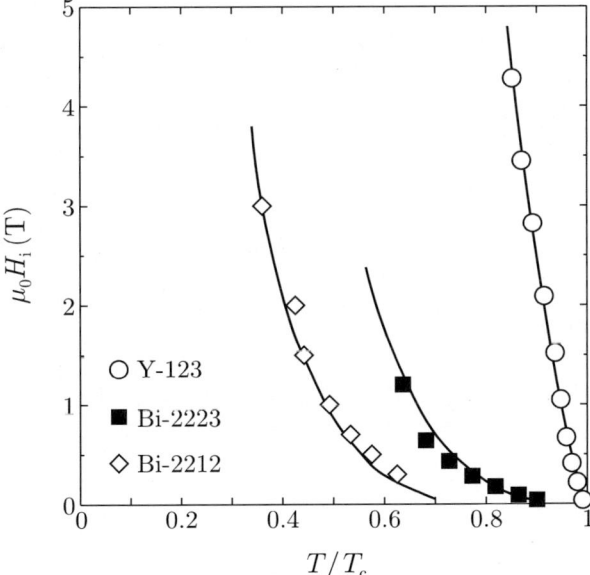

Fig. 8.37. Irreversibility lines in a magnetic field parallel to the c-axis for various superconductors [46–48]. The *solid lines* are theoretical results of the flux creep-flow model

8 High-Temperature Superconductors

Table 8.3. Parameters of various superconductors for theoretical calculation of irreversibility field parallel to the c-axis shown in Fig. 8.37 [46–48]

	T_c (K)	$\mu_0 H_{c2}^c(0)$ (T)	$\rho_n(T_c)$ ($\mu\Omega$m)	A_m	σ^2	m'	γ	δ	g^2
Bi-2212	90.0	34.5	100	1.00×10^9	0.08	3.9	0.90	2.0	1.00
Bi-2223	110.0	50.0	100	1.00×10^9	0.10	2.6	0.70	2.0	1.40
Y-123	90.8	80.5	1.5	1.00×10^9	0.04	1.5	0.50	2.0	4.32

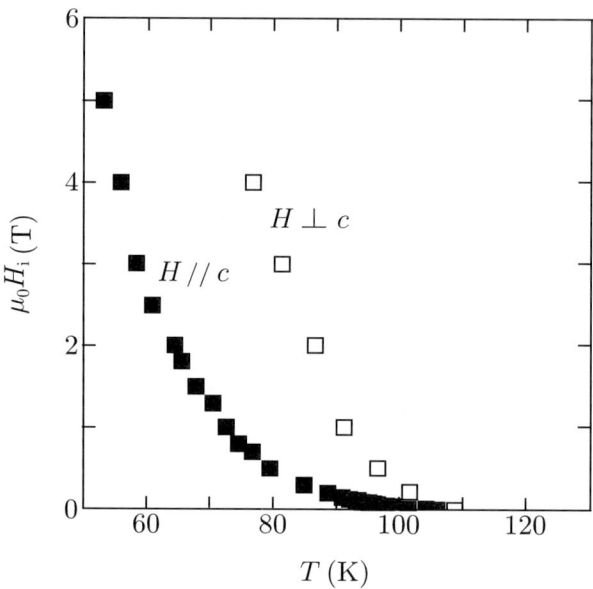

Fig. 8.38. Anisotropy of irreversibility field of a Bi-2223 tape [50]

(d) Size Dependence

The dependence of the size of the superconductor can be seen in Eq. (8.28). This is expected to be more simply expressed as:

$$\left(\frac{H_i}{H_{i\max}}\right)^{(3-2\gamma)/2} = 1; \qquad d \geq L_0,$$

$$= \frac{d}{L_0}; \qquad d < L_0. \qquad (8.41)$$

The derivation of Eq. (8.41) is asked in Exercise 8.4.

In Fig. 8.39 the irreversibility field along the c-axis is compared for a typical Bi-2223 tape and a Bi-2223 thin film with a high J_c value [45]. It is found that the irreversibility field of the thin film, which has a J_c value about five times larger at 4.2 K and 1.0 T, is much lower than that of the tape. The theory explains this difference, although the distribution of J_{c0} is not considered.

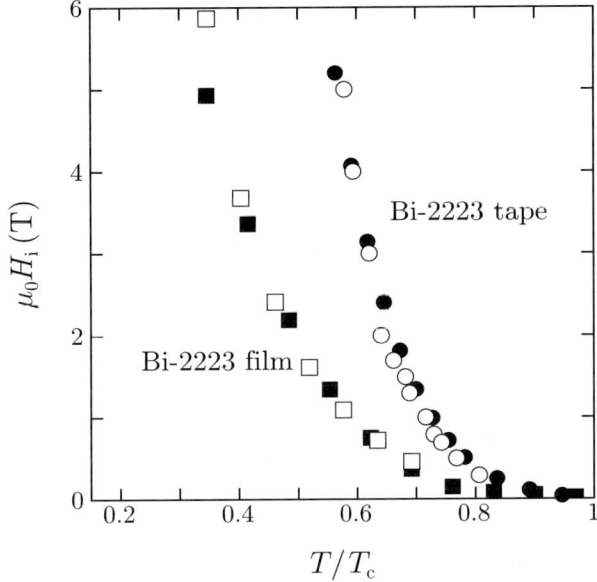

Fig. 8.39. Comparison of irreversibility line in a magnetic field parallel to the c-axis for a typical Bi-2223 tape and a high J_c Bi-2223 thin film [45]. *Open* and *solid symbols* show experimental and theoretical results, respectively

Figure 8.40 displays the irreversibility field of Sm-123 superconducting powders with different average particle sizes at 77.3 K [51]. The irreversibility field increases with increasing particle size and tends to be saturated when the particle size exceeds the pinning correlation length L_0, as predicted by the theory. In fact, Fig. 8.41 is a replot of the result in the form of Eq. (8.41): it shows that the analytic result of Eq. (8.41) holds also for the case of distributed pinning strength. In the figure each theoretical line shows a virtual variation in the ideal case where only the particle size changes with the same pinning parameters. In the practical case, even if the pinning centers and the number density are unchanged, J_{c0} changes with the particle size when the size is below L_0.

The phenomenon that the irreversibility field reduces with the size of a small superconductor is observed also for metallic superconductors with very fine filaments [52].

(e) *Electric Field Dependence*

As was shown in Subsect. 8.4.4, the critical current density depends appreciably on the electric field criterion E_c. Hence, the irreversibility field also depends on the electric field criterion. This can be explained from the electric field criterion in K and K' given by Eqs. (8.29) and (8.30), respectively. Namely, when E_c becomes small as in a DC magnetization measurement, K and K' become small, resulting in a reduction of the irreversibility field.

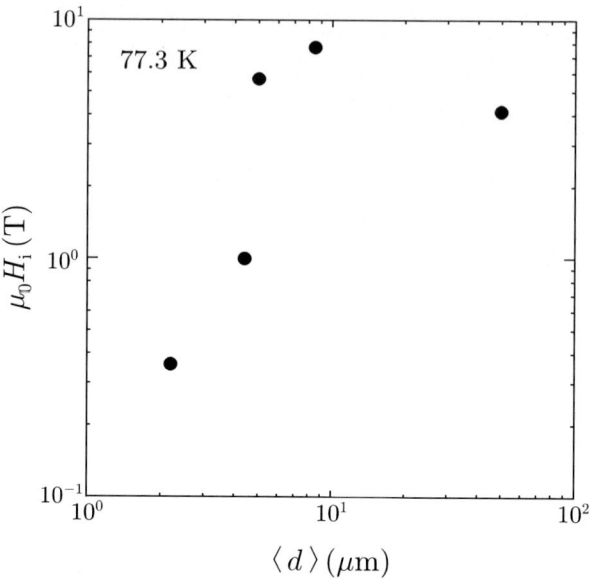

Fig. 8.40. Irreversibility field at 77.3 K vs. particle size of superconducting Sm-123 powders [51]

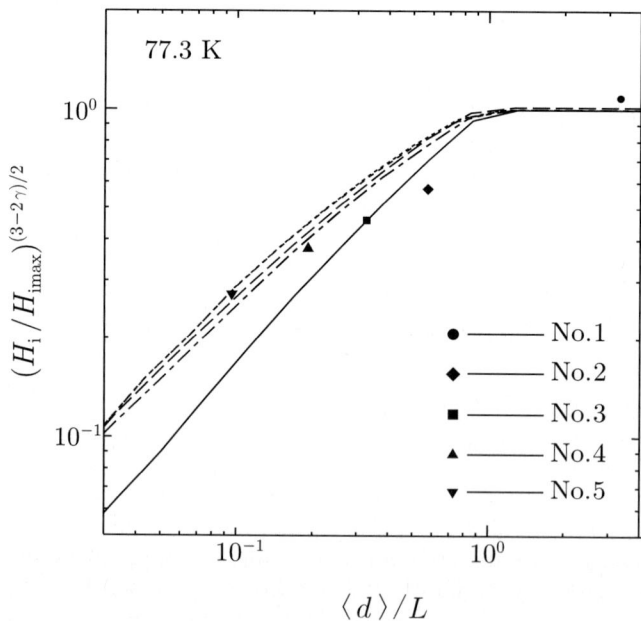

Fig. 8.41. Replot of Fig. 8.40 in the form of Eq. (8.41) [51]. Each *line* is a theoretical estimate for each sample

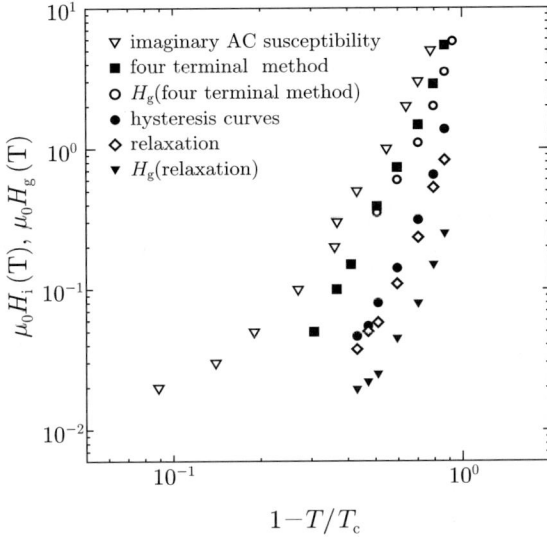

Fig. 8.42. Irreversibility lines of a Bi-2223 tape measured by various methods [12]. The G-L transition field obtained in different ranges of electric field is also shown for comparison

Figure 8.42 shows the irreversibility lines of the same Bi-2223 tape measured by various methods [12]: the differences come from the different values of the electric field strength at the measurement. Figure 8.43 is a replot of the irreversibility field at 70 K as a function of the electric field. The results of the G-L transition field obtained for different ranges of the electric field are also shown for comparison. The G-L transition field obtained from the scaling of the E-J curves at low electric fields measured by a relaxation of the magnetization was mentioned in Subsect. 8.2.2. It can be seen from this result that the irreversibility field changes dramatically with the electric field at the measurement. The solid line in the figure is a theoretical prediction of the flux creep-flow model, and it is found that this model correctly describes the phenomenon.

8.5.4 Relation with G-L Transition

It was initially believed that there were intrinsically glass and liquid states for flux lines and that the flux lines could be pinned effectively and ineffectively when they were in the glass and liquid states, respectively. However, the causality is opposite, and the flux pinning determines the state of the flux lines as described in detail in Subsect. 8.2.2. In fact, the G-L transition field or temperature can be determined and the scaling of the E-J curves can be explained as shown in Fig. 8.5 by the flux creep-flow model, when the pinning parameters are given. The G-L transition theory might in principle be useful

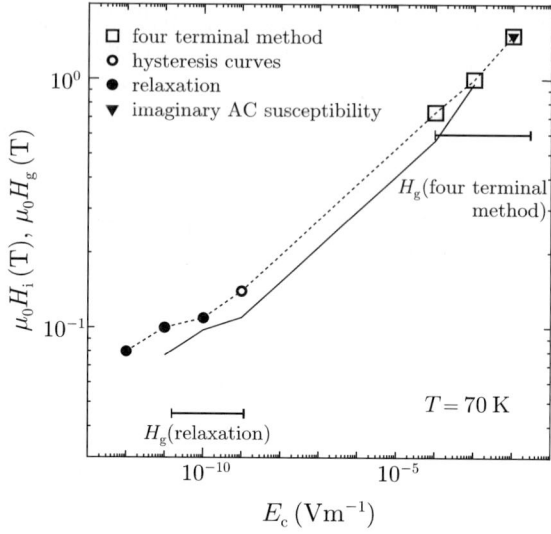

Fig. 8.43. Relationship between the irreversibility field at 70 K and the electric field strength [12]. The *solid line* is a theoretical prediction of the flux creep-flow model

for a description of the behavior of flux lines when the transition takes place. However, it should be noted that it cannot determine the transition point. In addition, the effect of flux pinning is not taken into account for the two critical indices in the original theory. By the way, it has been clarified from the viewpoint of flux pinning [3] that a derivative of the disorder of flux lines with respect to temperature or magnetic field is discontinuous at the transition point as a typical example of the difference between the two states.

In other words, the two states are distinguished by the effectiveness of the flux pinning: a finite critical current density is observed in the glass state, while flux lines are driven to the resistive state by application of an infinitesimal current in the liquid state. This is the definition of the irreversibility temperature or magnetic field. Hence, it is concluded that the G-L transition field is identical with the irreversibility field. In this sense the G-L transition can be called the thermal depinning transition.

However, the practical results are different between the G-L transition field and the irreversibility field due to the difference in the method of determination. To describe the state of flux lines in a given regime of electric field the G-L transition is more rigorous. However, many measurements and a detailed analysis are needed for the determination of the G-L transition field. Also, from the viewpoint of theoretical analysis using the flux creep-flow model, the scaling of theoretical E-J curves is necessary, similarly to the analysis of experimental results for the G-L transition field, while the irreversibility field can be directly and simply determined as shown in Subsect. 8.5.1. For this reason the irreversibility field, which can be easily determined with a suitable current criterion, is mostly used in the field of applications.

The irreversibility line and the G-L transition line of a Bi-2223 tape are compared in Fig. 8.44 [53]. It is found that the irreversibility line exists at higher temperature or magnetic field in the B-T plane. The theoretical results of the two lines determined with the same pinning parameters are also shown in the figure for comparison. Agreement is fairly good, and it can be seen that the transport phenomena in this range can be correctly described by the theoretical model. A similar agreement between the theory and experiment is obtained for a Bi-2212 tape [48].

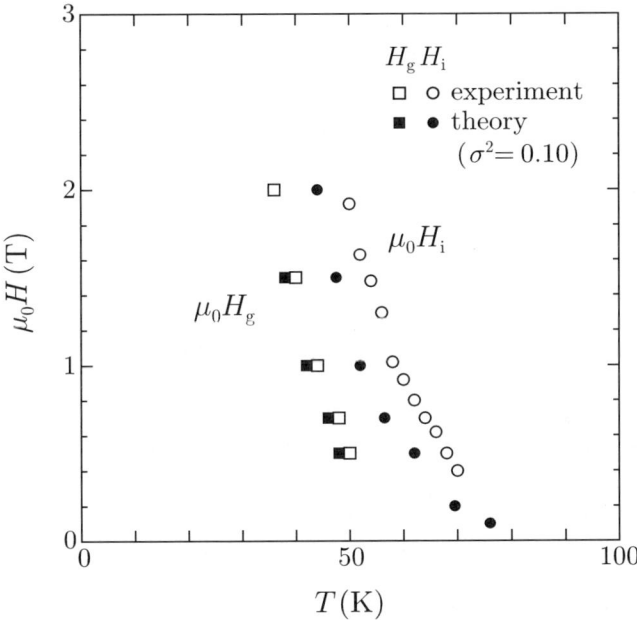

Fig. 8.44. Irreversibility line and G-L transition line of a Bi-2223 tape in a magnetic field along the c-axis [53]. *Open* and *solid* symbols are experimental results and theoretical results from the flux creep-flow model

It has been shown that the electromagnetic phenomena associated with the flux pinning, e.g., the critical indices, ν and z, in the scaling of E-J curves, are strongly influenced by the distribution of flux pinning strength and the electric field. This fact casts a doubt on the theoretical assumption of Fisher et al. [54] that the index μ, representing the state of flux lines in the glass state, is an inherent parameter dependent only on the dimensionality of the superconductor. The fact that the transition point, as well as the two critical indices, changes dramatically with the electric field strength also shows that such a simple assumption is not correct. Although the framework of the theory itself using the index μ might be valid, μ must be a parameter which is influenced by the distribution of pinning strength and the electric field.

8.6 Flux Pinning Properties

One of the important parameters for application of high-temperature superconductors is the critical current density, and this is determined by the flux pinning property. In the determination various aspects are involved such as kind of defects to pin flux lines, their pinning mechanism, the summation property described in Chaps. 6 and 7, and the flux creep described in Chap. 3. In this section various features inherent to high-temperature superconductors will be described.

High-temperature superconductors consist of superconducting CuO_2 planes and almost insulating block layers with the superconducting order parameter varying along the c-axis as shown in Fig. 8.27. Hence, the structure of the superconductor itself is proposed to interact with the flux lines, and this pinning interaction, first pointed out by Tachiki and Takahashi [55], is called the intrinsic pinning. This interaction is a kind of condensation energy interaction with an attractive force. However, for a realization of this pinning interaction, each flux line must fall in the potential of a block layer for a very long length, and hence, it would be observed only when a magnetic field is applied exactly parallel to the a-b plane of a highly perfect superconductor. When the magnetic field is even at a slight angle from the a-b plane, each flux line will have a step-wise structure, and positive and negative pinning forces will appear alternately, resulting in a cancellation of the pinning force except at the two edges. This pinning property will be mentioned later in connection with Y-123 coated conductors. In the following the pinning properties of the main high-temperature superconductors will be described.

8.6.1 Y-123

Y-123 superconductors have the most three-dimensional properties, and the flux pinning is strong. These superconductors are transformed from a tetragonal crystalline structure to an orthorhombic one when cooled from the reaction temperature. In this transformation twin boundaries are formed so as to reduce the distortion by equating the fractions of the a- and b-axes in each direction. It has been found that the flux pinning strength of a twin boundary is strongly influenced by the oxygen deficiency at the boundary, and the pinning is weak when the oxygen deficiency is low [56]. Hence, the pinning mechanism of the twin boundary is the condensation energy interaction of a local region with a lower T_c.

Figure 8.45 shows the variation in the torque density on a single crystal Y-123 slab specimen with twin boundaries aligned in two directions normal to each other when the magnetic field is rotated in a plane perpendicular to the

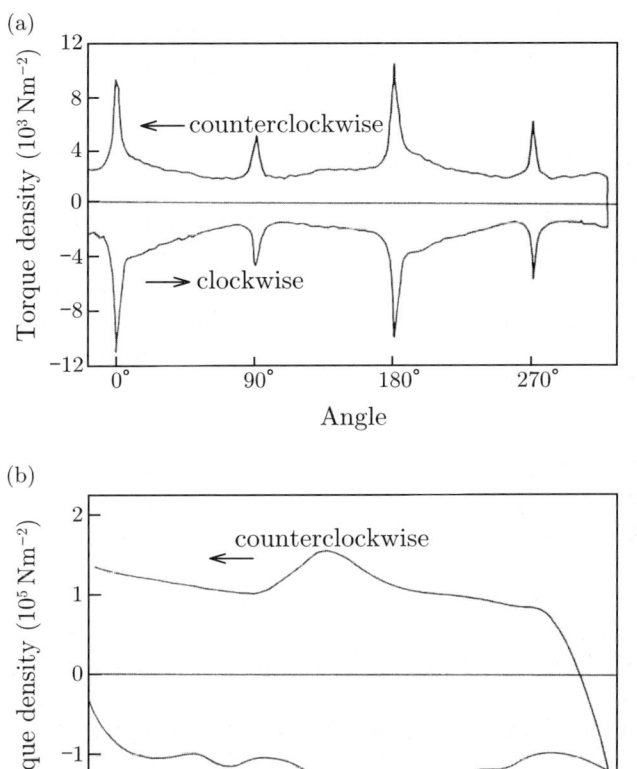

Fig. 8.45. Variation in the torque density on a single crystal Y-123 slab specimen with twin boundaries aligned in two directions normal to each other in a magnetic field rotated in a plane perpendicular to the common c-axis [57]. While a sharp peak appears when the magnetic field is parallel to twin boundaries at 76 K in (**a**), these peaks disappear at 27 K in (**b**)

common c-axis [57]. The torque takes a sharp peak when the magnetic field is parallel to the twin boundaries at 76 K as shown in (a). This strongly implies that the twin boundaries work as pinning centers. If the mean value of the peak of the torque density and the magnitude of the magnetization due to the pinning interaction of twin boundaries are denoted by $\Delta\tau$ and M, respectively, we have $\Delta\tau \simeq BM/2$. Here it is taken into account that approximately half of the twin boundaries in the specimen contribute to each peak. If the mean size of the specimen in the rotating plane of the magnetic field is w, we have $M \simeq J_c w/2$. From these relationships the critical current density due to the pinning by twin boundaries is estimated as $J_c \simeq 4 \times 10^7$ Am^{-2} at 76 K and 0.67 T. This result is approximately the same as the result, 4.5×10^7Am^{-2} at 77.3 K and 1.0 T, obtained from the field angle dependence of the magnetization critical current density of a melt-processed Y-123 bulk specimen [58]. It can be concluded from these results that the pinning by twin boundaries is not very strong. The elementary pinning force per unit length of flux line is estimated as $f'_p \simeq 1.6 \times 10^{-6}$ Nm^{-1}, even if the mean spacing of twin boundaries is overestimated to be approximately 1.0 μm. This value is only about 1/20 of the pinning strength of grain boundaries of Nb$_3$Sn and only about 1/80 of the pinning strength of grain boundaries of V$_3$Ga shown in Fig. 7.18.

On the other hand, the peak of the torque due to twin boundaries cannot be observed at 27 K because it is embedded in the torque of the background pinning as shown in Fig. 8.45(b). This is caused by the different temperature dependence of the pinning force for twin boundaries from that of background pinning centers. Namely we expect $m' = 3/2$ for planar defects with the condensation energy interaction, while $m' \simeq 4$ is obtained for background pinning from Figs. 8.45(a) and (b). Therefore, the background pinning by point defects, etc. is superior at low temperatures.

It is known that a high density of nonsuperconducting Y$_2$BaCuO$_5$(211) phase particles smaller than several micrometers in size is distributed in bulk Y-123 superconductors. The pinning mechanism of these particles is the condensation energy interaction, similarly to the normal precipitates mentioned in Subsect. 6.3.1. From the size of the particles it is likely that the surface works effectively, and a characteristic similar to that shown in Fig. 7.11 is expected. The relationship between the critical current density of Y-123 bulk superconductors and the surface area of 211 phase particles at 77.3 K and at 1.0 T parallel to the c-axis is shown in Fig. 8.46 [59]. The strong correlation shows that the surface of 211 phase particles works effectively.

A broad peak effect of the critical current density is observed in Y-123 single crystals and bulks in a medium field range at around 77.3 K as mentioned in Sect. 7.6. The pinning centers responsible for the peak effect are oxygen deficient regions including twin boundaries. A similar peak effect is also observed for RE-123 superconductors containing rare earth (RE) elements with a large ionic radius, and the pinning centers are substituted regions where Ba sites are occupied by RE elements. Since the substituted regions have lower T_c than the surrounding matrix phase, it is considered that the same mechanism brings about the peak effect as that of oxygen deficient regions. The mechanism which has been considered to cause the peak effect is a field induced one [60]: these regions become normal state at a magnetic field higher than their H_{c2} at the ambient temperature and may work as strong pinning centers.

However, the assumption that the peak effect directly arises from the elementary field-induced pinning mechanism has some problems: for example, the fact that the peak effect disappears when the particle size of Sm-123 superconducting powders is reduced below the pinning correlation length cannot be explained by this mechanism (see Sect. 7.6). Nd-123 provides another example.

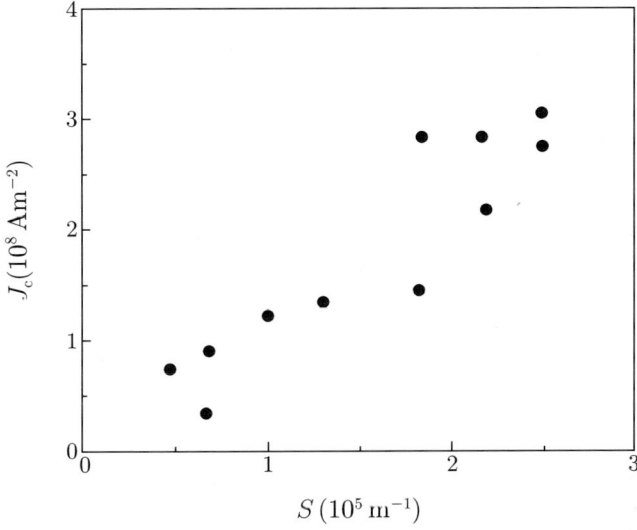

Fig. 8.46. Critical current density of Y-123 bulk superconductors vs. surface area of 211 phase particles in a unit volume at 77.3 K and at 1.0 T parallel to the c-axis [59]

The variation in the critical current density of Nd-123 superconductor in a magnetic field along the c-axis due to an addition of 422 phase is shown in Fig. 8.47 [61]. This result shows that J_c in the medium field region decreases and the peak effect also decreases or even disappears after the addition of 422 phase, while J_c is increased at low and high magnetic fields. The same behavior of J_c has been observed also for the addition of 211 phase to Y-123 bulks, while a strong correlation exists between J_c and the surface area of 211 phase particles at low and high magnetic fields [46].

It is clear that this variation cannot be explained by the field-induced pinning mechanism. If the lower T_c regions contribute to the attractive condensation energy interaction as well as 211(422) phases, J_c should be increased by the addition as observed in the low and high fields. To explain the opposite result a mechanism involving the kinetic energy interaction with the lower T_c regions has been proposed [62]. T_c of oxygen deficient regions is approximately 60 K, and hence, the lower T_c regions are considered to be in the normal state at 77.3 K. However, it seems to be reasonable to assume that the lower T_c regions are in a weakly superconducting state due to the proximity effect, since their size is expected to be sufficiently smaller than ξ_{ab}, etc. Thus, the order parameter around the lower T_c phase may be like the one shown by the solid line in Fig. 8.48. If a magnetic flux penetrates this phase, the order parameter will change as shown by the broken line. Since the coherence length of this phase is longer than that of the surrounding region, the kinetic energy would appreciably increase, and these phases would act as repulsive pinning centers

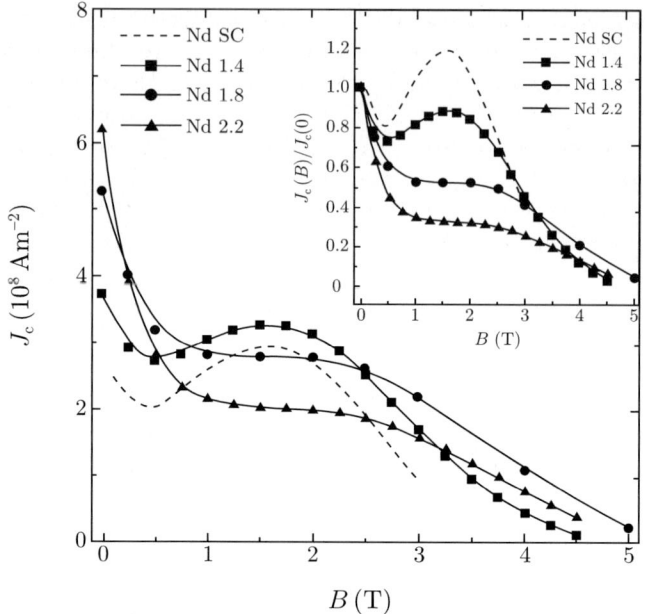

Fig. 8.47. Variation in magnetic field dependence of critical current density of Nd-123 superconductors at 77.3 K due to addition of Nd-422 phase [61]

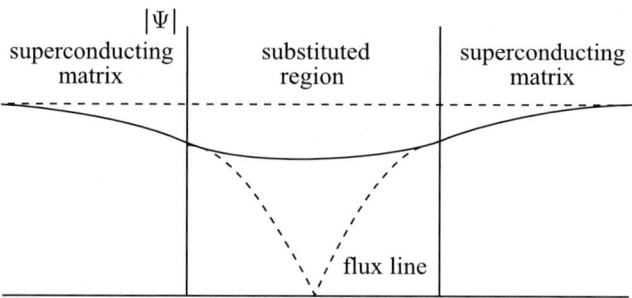

Fig. 8.48. Spatial variation in the order parameter around the lower-T_c region [62]. The *broken line* represents the order parameter when a flux line penetrates the lower-T_c region

against the flux lines. This is the same mechanism (kinetic energy interaction) as that of artificial Nb pinning centers in superconducting Nb-Ti described in Sect. 6.6. Thus, the disappearance of the peak effect in medium fields can be explained by the interference between the negative pinning energy of 211(422) phases and the positive pinning energy of the lower T_c regions.

The important thing to be explained in this case is the reason for the increase of J_c without the above interference in the low and high field regions. In the low field region, since the spacing between the individual flux lines is sufficiently long, it is proposed that direct interference is not significant. That is, the individual pinning forces on flux lines are almost independent of each other, and the resultant global pinning force tends to increase due to the contribution of the two kinds of interaction. On the other hand, every flux line cannot occupy its preferred position because of the very short spacing at high fields. Hence, flux lines must penetrate the lower $T_{\rm c}$ regions, and the free energy increases so long as the induced superconductivity is present. Then, the superconductivity in the lower $T_{\rm c}$ phase and its surrounding region is proposed to diminish to reduce the total free energy. The kinetic energy interactions cease, and $J_{\rm c}$ is determined only by the condensation energy interactions from 211(422) phases. This explains the above experimental results. This argument can also explain a commonly known experimental result that the occurrence of the peak effect and the irreversibility field are independent. A superconductor with a significant peak effect contains a large volume fraction of lower $T_{\rm c}$ regions, and the superconductivity is considered to be appreciably degraded at high fields, resulting in the deterioration of the irreversibility field.

It should be noted that the artificial Nb pins in Nb-Ti do not bring about the peak effect by themselves as shown in Sect. 6.6. The peak effect in high-temperature superconductors such as Y-123 originates from the order-disorder transition of flux lines as mentioned in Subsect. 8.2.3. The reason why the peak effect does not arise from artificial Nb pins in Nb-Ti is that there is no room for the transition, since the pinning efficiency is already high enough. On the other hand, 211 phase does not cause the peak effect, although any kind of pinning center can do so. This is because the usual size of 211 phase particles is so big that the pinning force does not change appreciably as a result of a slight displacement of flux lines. Hence, this small change in the pinning force does not bring about a further displacement of flux lines, and hence, this kind of feedback does not develop to the point of a transition. In summary, for the occurrence of the peak effect, the superconductor should be of a sufficient size, the pinning centers should not be too strong, and their size should be comparable to or smaller than the flux line spacing so that their force could appreciably change after a small displacement of the flux lines.

From the above argument, the following can be said about applications of Y-123 superconductors: those which contain lower $T_{\rm c}$ regions but not 211 phases are desirable for medium-field applications, and those which contain a large amount of 211 phases but not lower $T_{\rm c}$ regions are desirable for high-field applications.

Recently, RE-123 coated conductors deposited on various substrates have been developed, and the critical current densities have been improved. There are mainly two fabrication methods for substrates: the IBAD (Ion Beam Assisted Deposition) method in which a crystal-axis aligned intermediate layer is deposited on a nonaligned metallic substrate and the RABiTS (Rolling Assisted Biaxially Textured Substrate) method in which an aligned substrate is fabricated by rolling and heat treatment and an intermediate layer is deposited on it. There are many deposition methods for the RE-123 layer such as PLD

(Pulsed Laser Deposition) and TFA-MOD (TriFluoroAcetate-MetalOrganic Deposition) in which a coating solution with trifluoroacetate is deposited on a substrate and then heat treated, etc.

The magnetic field angle dependence of the critical current density [63] is similar to that shown in Fig. 8.20 for a single crystal thin film. The critical current density is large for a magnetic field in the a-b plane and small for a magnetic field parallel to the c-axis. Over a wide range of magnetic field angle, the critical current density obeys the theoretical prediction based on the random pinning and the anisotropy of the superconducting electron mass. The observed critical current density takes larger values, deviating from the expectation for the field in the a-b plane or along the c-axis. The deviation within a narrow range of field angle parallel to the a-b plane is attributed to the pinning centers parallel to the a-b plane, and candidates are intrinsic pinning centers, i.e., the block layers themselves, stacking faults and the surface of specimens. The deviation within a relatively wide region of field angle along the c-axis is considered to be caused by defects directed along the c-axis, and candidates are twin planes and screw dislocations nucleated during the deposition of thin films. In this case the cores of dislocations are considered to be in the normal state due to the large strain. Hence, the pinning mechanism of dislocations is not the elastic interaction described in Sect. 6.4 but the condensation energy interaction with a strong force.

Figure 8.49 shows the thickness dependence of the critical current density in self field at 77.3 K for optimized Y-123 thin films fabricated by various methods [64]. The dotted line represents the dependence predicted by the

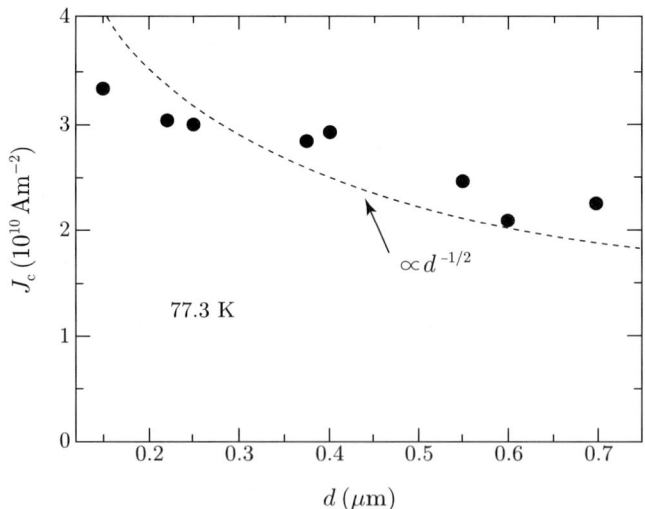

Fig. 8.49. Thickness dependence of critical current density in the self field at 77.3 K of optimized Y-123 thin films [64]. The *line* represents the prediction of the two-dimensional collective pinning mechanism where $J_c \propto d^{-1/2}$

two-dimensional collective pinning mechanism for random point pins: $J_c \propto d^{-1/2}$. The experimental results are close to this prediction, and the agreement seems to support the assumptions about the pinning mechanism. Similar thickness dependence has been observed for Y-123 coated conductors fabricated by the PLD process [65].

The two-dimensional collective pinning occurs when the layer thickness is smaller than the pinning correlation length. Figure 8.50 shows the thickness dependent critical current density at 0.1 T parallel to the c-axis for Y-123 coated tapes prepared by the PLD method on IBAD substrates [66]. It can be seen that the critical current density varies with the thickness as predicted by the collective pinning theory for the thickness below 1 μm. This dependence does not change over a wide temperature range of 5 to 60 K. The change at higher temperatures is a degradation due to the flux creep. The pinning correlation length estimated from Eq. (8.15) with the observed critical current density of the thickest specimen which seems to be in the three-dimensional pinning regime is 0.12 μm at 5 K and increases to 0.37 μm at 60 K. This claims that the flux pinning mechanism is not two-dimensional but three-dimensional at 5 K. Hence, the degradation of the critical current density with increasing thickness in PLD processed tapes is not caused by the collective pinning mechanism. This seems to be consistent with the result that the critical current density is almost independent of the thickness over a wide range of thickness up to 1.4 μm for TFA-MOD processed tapes [67]. In fact, it was shown that

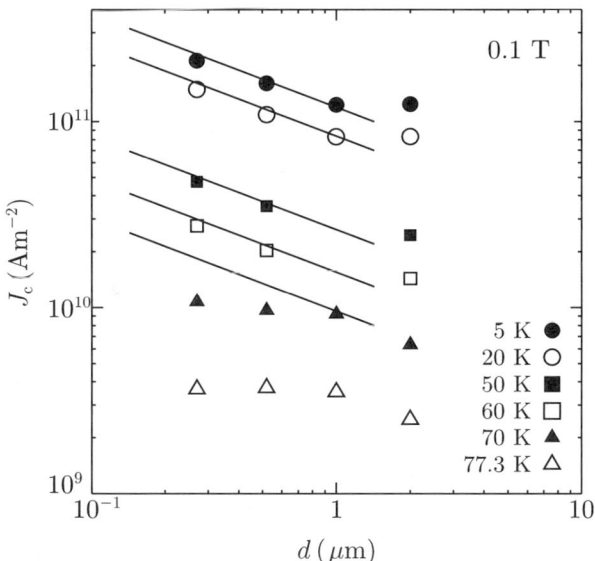

Fig. 8.50. Dependence of magnetization critical current density on the thickness of superconducting layer at 0.1 T parallel to the c-axis for PLD processed Y-123 coated tapes [66]. The *straight lines* show the relationship $J_c \propto d^{-1/2}$

the degradation in thick PLD processed tapes is caused by a change in the structure of superconducting layers such as nucleation of voids and growth of a-axis oriented grains [68].

However, the critical current density at high fields is higher for a thicker superconductor because of the higher irreversibility field [69, 70]. In addition, although the effect of thickness seems not to be large at low temperatures, the n value in the very low electric field region is found to be larger for a thicker specimen [70]. Hence, thicker superconductors are advantageous for applications such as magnets used in a persistent current mode. On the other hand, thinner superconductors are desirable for applications at low fields because of the higher critical current density as well as the lower AC loss.

In these days the introduction of artificial pinning centers into Y-123 thin films or Y-123 coated tapes is being attempted to improve the critical current density at high fields. There are two approaches: one is an introduction of pinning centers by making defects on a substrate, and the other is a precipitation of different phase particles by changing the chemical composition. Matsumoto et al. [71] deposited Y_2O_3 on single crystal $SrTiO_3$ (100) substrates by the PLD process such that small islands of about 25 nm in diameter are formed, and then Y-123 layers were deposited. These islands are useful to nucleate one-dimensional defects such as screw dislocations normal to the substrate. In this case the critical current density in a magnetic field parallel to the c-axis was dramatically enhanced as shown in Fig. 8.51.

Haugan et al. [72] alternately deposited Y-123 and Y-211 layers on single crystal substrates using the PLD process and formed 211 particles of about 10 nm in size after heat treatment. These particles were precipitated in planes parallel to the a-b plane in which 211 layers had existed originally, and the critical current density in a magnetic field parallel to the a-b plane was significantly improved. MacManus-Driscoll et al. [73] deposited Y-123 layers using a target of Y-123 + 5 mol.% $BaZrO_3$ on single crystal substrates or IBAD-MgO substrates. In the films $Ba_2(Zr, Y)_2O_6$ nanoparticles or nanorods of about 10 nm in diameter are found to be dispersed after heat treatment. These were approximately parallel to the c-axis and the resultant critical current density along this field direction was improved.

8.6.2 Bi-2223

Practical Bi-2223 superconductor is now mostly in the form of silver-sheathed tapes fabricated by the PIT (Powder-In-Tube) method. The critical current density of these tapes is much inferior to those of Y-123 thin films or coated conductors. However, the engineering critical current density including substrates or normal conducting matrix is not much different. In addition, taking account of the poor crystal axis alignment in these tapes, the weak link property is not so serious as in Y-123. On the contrary, because of imperfect alignment of the c-axis of individual grains, the current can flow along the direction of the thickness of the tape as assumed in the railway-switch model,

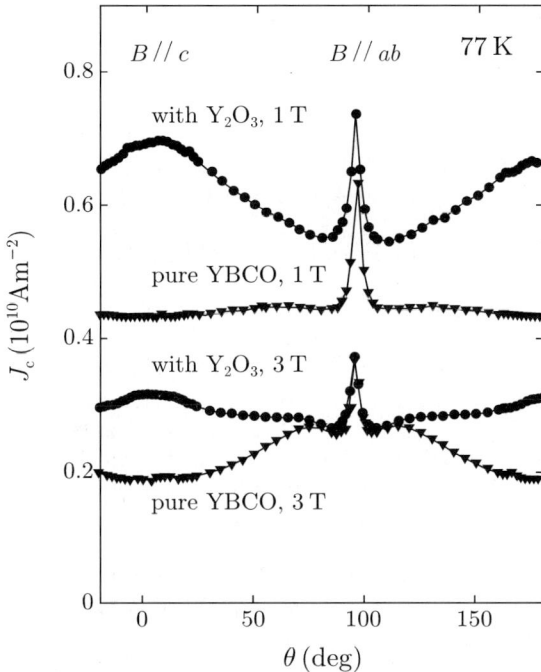

Fig. 8.51. Field angle dependence of critical current density of Y-123 thin films deposited on a single crystal SrTiO$_3$ substrate and a substrate with Y$_2$O$_3$ nanoislands [71]

[74] resulting in a solution of the percolation problem of the transport critical current density to some extent.

The weak pinning property in comparison with Y-123 is the main factor which restricts applications at high temperatures to the low field region. This property is due to the low condensation energy density resulting from the two-dimensional nature of the superconductor. Hence, even for the same defects the pinning energy and the resultant J_{c0} are smaller than in Y-123. In addition, the more serious effects of flux creep make the practical J_c even smaller. Thus, improvement in the dimensionality of the superconductor is more effective for increasing J_c than the introduction of defects. However, improvement of the pinning characteristics by oxygen doping, which is effective in Bi-2212, has not yet been observed in Bi-2223 silver-sheathed tapes.

The temperature dependence of the condensation energy density is shown in Fig. 8.52 for a Bi-2223 single crystal specimen #2 heat-treated at 350°C for 24 hours in an oxygen atmosphere of 1 atm. The condensation energy density was estimated from the analysis of the pinning force of columnar defects nucleated by heavy ion irradiation. Specimen #2 has a much higher condensation energy density than specimen #1 without oxygen treatment. It is found that, although the condensation energy density of specimen #2 is

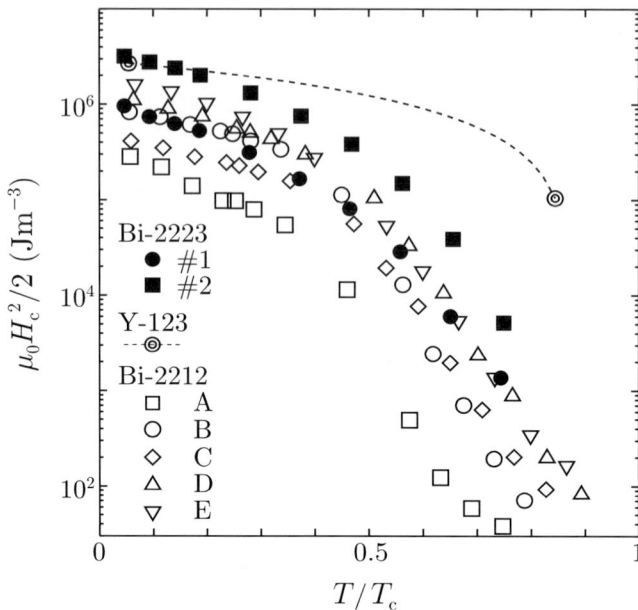

Fig. 8.52. Reduced temperature dependence of condensation energy density of various high-temperature superconductors [75]

much inferior to that of most three-dimensional Y-123 at high temperatures, it increases abruptly with decreasing temperature and exceeds that of Y-123 at 5 K. This shows a high potential of Bi-2223 superconductor for improvement of the critical current density at low temperatures by introduction of strong pinning centers. In fact, a surprisingly high critical current density of 2.8×10^{11} Am^{-2} was attained at 5 K and 0.1 T by introduction of columnar defects of about 10 nm in diameter with the matching field of 1 T to the same single crystal specimen [75]. For comparison the condensation energy density of Bi-2212 is also shown in Fig. 8.52. This with the difference in the critical temperature shows that Bi-2223 has the advantage of Bi-2212 for application at middle and high temperature regions.

The n value of Bi-2223 tapes is low due to the weak links and the strong effect of flux creep, as described in Subsect. 8.4.4. For this reason the pinning force density has a long tail in the vicinity of the irreversibility field. Figure 8.53(a) shows the scaling law of the pinning force density in a magnetic field along the c-axis of a single-core Bi-2223 tape determined by the offset method for the compensation for the current sharing with the silver sheath [43]. Figure 8.53(b) shows the theoretical results of the flux creep-flow model.

The anisotropy of the transport J_c of a Bi-2223 tape with respect to the field angle is smaller than the anisotropy of the magnetization J_c [76]. These anisotropies are very much smaller than what is expected from the anisotropy of the effective mass of the electrons. The smaller anisotropy in

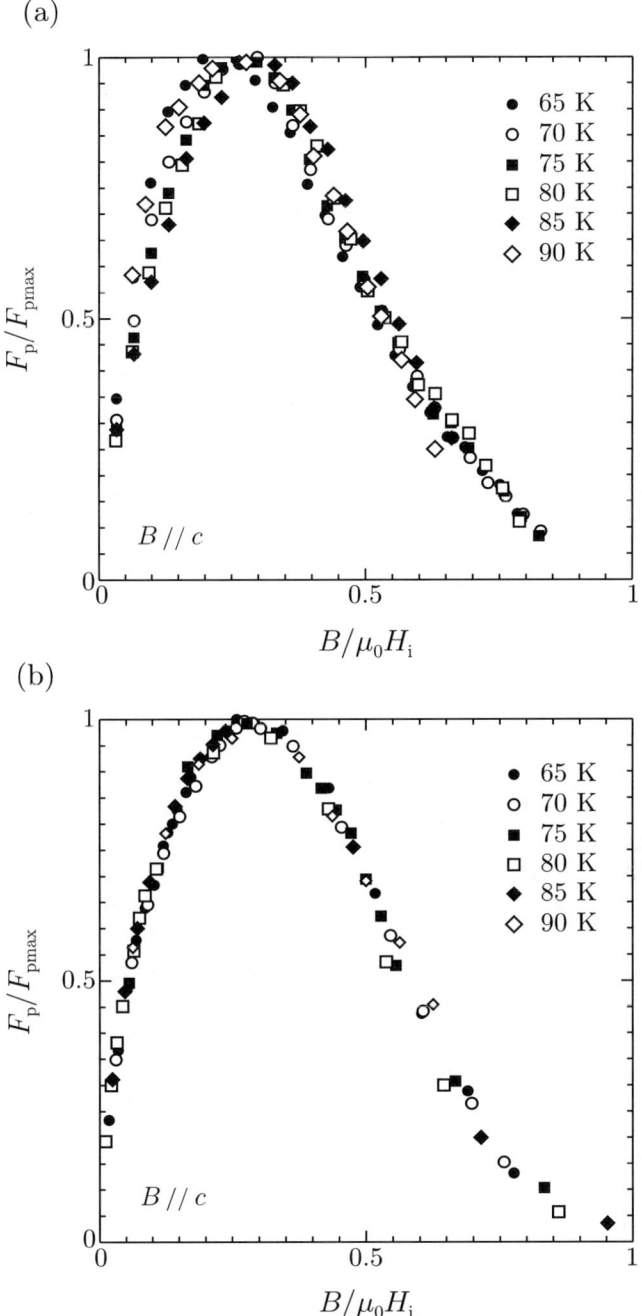

Fig. 8.53. Scaling law of the pinning force density of a single-core Bi-2223 tape in a magnetic field parallel to the c-axis [43]: (**a**) experimental result and (**b**) theoretical result of the flux creep-flow model

the magnetization J_c compared to the intrinsic anisotropy is caused by the imperfect alignment of the a-b plane of each grain and the small fraction of grains with their a-b planes exactly parallel to the tape surface, which contribute to a large intragrain critical current in a magnetic field parallel to the tape, as discussed in Subsect. 8.5.3. The smaller anisotropy of the transport J_c than that of the magnetization J_c is caused by the fact that such high intragrain critical currents of isolated grains cannot contribute to the transport J_c due to the percolation property of the superconducting current in the same field direction.

Application of overpressure more than 20 MPa to tapes at the final heat treatment is effective to densify the superconducting region by elimination of voids. This leads to a significant enhancement of the critical current density and even to an enhancement of the critical current in spite of the reduction in the cross-sectional area of superconducting region. This process also brought about enhancements of the irreversibility field and the n value [77]. These improvements seem to be attributed to a sharpened distribution of J_{c0} with improvement of small J_{c0} values. In practice, a better c-axis alignment is obtained by this treatment and this seems to contribute to an elimination of weak links of grains. The optimization of this overpressure process is desired for further improvement of the critical current properties.

8.6.3 Bi-2212

Practical Bi-2212 superconductor is also in the form of silver-sheathed tapes or round wires fabricated by the PIT method. The critical current density is very low at high temperatures. This is caused by the weak pinning and the strong effect of flux creep, both of which originate from the very two-dimensional properties of Bi-2212. Hence, it seems to be quite difficult to drastically improve the present properties at high temperatures, and a significant improvement of the irreversibility field cannot be expected, as predicted by Eq. (8.40).

For improvement of the pinning properties, improvement of the dimensionality of the superconductor is most effective, similarly to the case of Bi-2223. For this purpose oxygen doping and substitution at Bi sites by Pb atoms are effective [78]. In fact, the critical current density is significantly improved by these treatments, although the critical temperature is depressed because the superconductor changes to an overdoped state. When a large amount of Pb is introduced, a lamellar structure composed of regions with a high density of Pb and those with a low density of Pb appears [79]. This structure is expected to contribute to the strong flux pinning.

The characteristic feature of this superconductor is that the critical current density and the irreversibility field increase dramatically with decreasing temperature below the critical point. The temperature dependence of the irreversibility field of Bi-2212 single crystals with various anisotropy parameters is shown in Fig. 8.54 [80]. The physical origin of such a temperature dependence

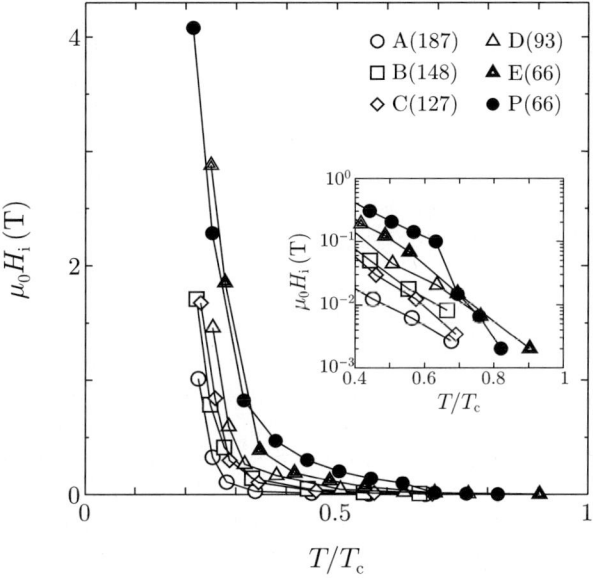

Fig. 8.54. Temperature dependence of the irreversibility field of Bi-2212 single crystals with different anisotropy parameters [80]. Numbers attached to each specimen are anisotropy parameters

of the irreversibility field is not clear. This dependence is directly associated with a significant increase in the condensation energy density with decreasing temperature as shown in Fig. 8.52.

Figure 8.55 shows the temperature dependence of the thermodynamic critical field H_c estimated from the condensation energy density [81]. H_c decreases approximately linearly with increasing temperature and the characteristic temperature T^* lower than T_c is obtained by the linear extrapolation to zero. H_c decreases exponentially above T^*, suggesting that the superconductivity in block layers may disappear there. The normalized temperature T^*/T_c is maximum at the optimally doped state and decreases as the condition deviates from the optimal one [81]. Such a degradation is speculated to be caused by increasing chemical inhomogeneity in the a-b plane.

Figure 8.56 shows the relationship between the condensation energy density and the anisotropy parameter at 5 K [81]. It is found that the superconductivity in block layers is improved and the condensation energy density increases as γ_a decreases. Hence, the introduction of strong pinning forces is possible in the low temperature region for such Bi-2212 superconductor with improved anisotropy. In practice, the high critical current density of 1.7×10^{11} Am^{-2} was attained at 5 K and 0.1 T in an overdoped Bi-2212 single crystal irradiated by a low dose of heavy ions with a matching field of 1.0 T [81]. Thus, application of this superconductor at low temperatures and at high fields is expected to take advantage of this property.

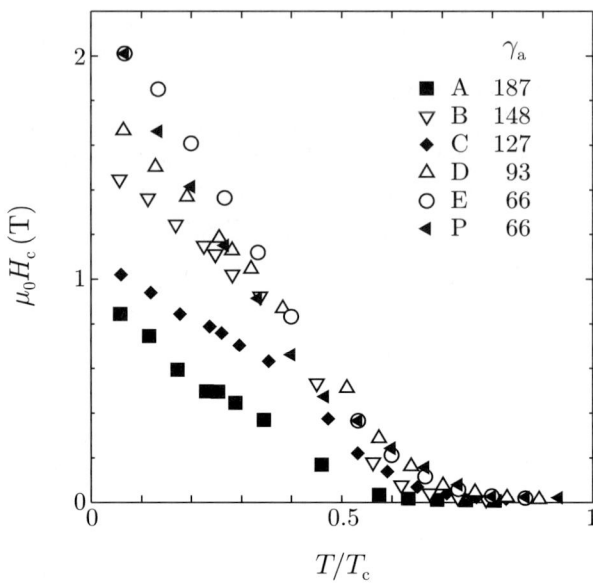

Fig. 8.55. Temperature dependence of the thermodynamic critical field of Bi-2212 single crystals with different anisotropy parameters [81]

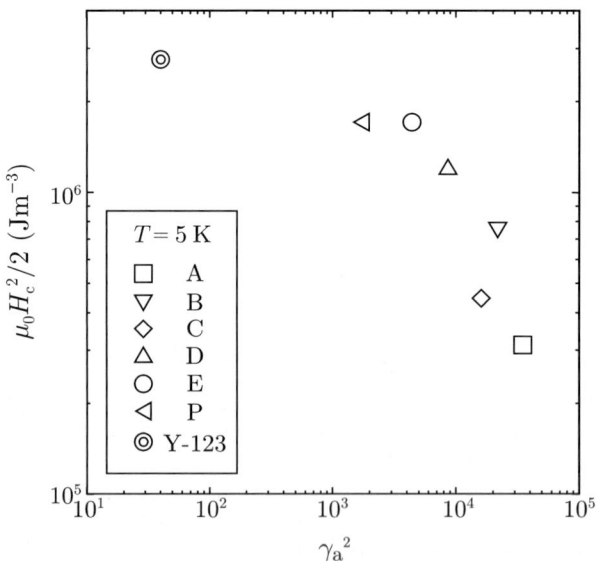

Fig. 8.56. Relationship between condensation energy density at 5 K and anisotropy parameter for Bi-2212 superconductors [81]

Exercises

8.1. Assume a large normal precipitate like a 211 phase particle. Argue the anisotropy of the pinning force of a wide interface between the superconducting and normal regions by considering the cases of magnetic field parallel to the a- and c-axes. It is assumed that the current flows along the b-axis and the geometry of the precipitate is isotropic.

8.2. It is assumed that the scaling law of the pinning force density of Eqs. (7.2) and (7.3) is satisfied at sufficiently low temperatures and that the effective mass model of Eq. (8.12) holds correct for the upper critical field. Show that the critical current density in a magnetic field of arbitrary direction is determined almost entirely by the field component parallel to the c-axis for a superconductor with $H_{c2}^{ab} \gg H_{c2}^{c}$ due to the large anisotropy.

8.3. Derive Eq. (8.18) from Eq. (8.17). (*Hint*: Expand the exponential term in Eq. (8.17) as $e^x \simeq 1 - x$.)

8.4. Derive Eq. (8.41).

8.5. Estimate the irreversibility field along the c-axis at 77.3 K for an Y-123 superconductor which contains a volume fraction of 20% 211 phase particles with a mean diameter of 2.0 μm. For estimation of J_{c0}, use

$$J_{c0}B = \frac{\pi N_p \xi_{ab} D^2 \mu_0 H_c^2}{4a_f}\left(1 - \frac{B}{\mu_0 H_{c2}^c}\right)^2 \quad (8.42)$$

of the saturation-type characteristic instead of Eq. (7.83b) because of the low efficiency of pinning. Use Eqs. (7.12b), (7.93) and (8.42) for estimation of g_e^2. Use the values of T_c and $H_{c2}^c(0)$ in Table 8.3 with the assumptions of $\mu_0 H_c(0) = 1.0$ T and the temperature dependence of $[1 - (T/T_c)^2]$ for these critical fields. The logarithmic term in Eq. (7.96) is assumed to be 14.

References

1. W. E. Lawrence and S. Doniach: *Proc. 12th Int. Conf. on Low Temp. Phys.*, ed. by E. Kanda (Academic Press, 1971) 361.
2. T. Matsushita: Physica C **214** (1993) 100.
3. T. Matsushita and T. Kiss: Physica C **315** (1999) 12.
4. M. P. A. Fisher: Phys. Rev. Lett. **62** (1989) 1416.
5. R. H. Koch, V. Foglietti, W. J. Gallagher, G. Koern, A. Gupta and M. P. A. Fisher: Phys. Rev. Lett. **63** (1989) 1511; R. H. Koch, V. Foglietti and M. P. A. Fisher: Phys. Rev. Lett. **64** (1990) 2586.
6. H. Yamasaki, K. Endo, Y. Mawatari, S. Kosaka, M. Umeda, S. Yoshida and K. Kajimura: IEEE Trans. Appl. Supercond. **5** (1995) 1888.
7. T. Matsushita and T. Kiss: IEEE Trans. Appl. Supercond. **9** (1999) 2629.
8. T. Matsushita, N. Ihara and T. Tohdoh: *Adv. Cryog. Eng. Mater.* Vol. 42 (Plenum, New York, 1996) p. 1011.
9. T. Kiss, T. Matsushita and F. Irie: Supercond. Sci. Technol. **12** (1999) 1079.

10. K. Noguchi, M. Kiuchi, M. Tagomori, T. Matsushita and T. Hasegawa: *Adv. Supercond. IX* (Springer-Verlag, Tokyo, 1997) p. 625.
11. M. Kiuchi, A. Yamasaki, T. Matsushita, J. Fujikami and K. Ohmatsu: Physica C **315** (1999) 241.
12. M. Fukuda, T. Kodama, K. Shiraishi, S. Nishimura, E. S. Otabe, M. Kiuchi, T. Kiss, T. Matsushita and K. Itoh: Physica C **357–360** (2001) 586.
13. T. Nakamura, T. Kiss, Y. Hanayama, T. Matsushita, K. Funaki, M. Takeo and F. Irie: *Adv. Supercond. X* (Springer-Verlag, Tokyo, 1998) p. 581.
14. T. Kodama, M. Fukuda, S. Nishimura, E. S. Otabe, M. Kiuchi, T. Kiss, T. Matsushita and K. Itoh: Physica C **378–381** (2002) 575.
15. T. Matsushita, M. Fukuda, T. Kodama, E. S. Otabe, M. Kiuchi, T. Kiss, T. Akune, N. Sakamoto and K. Itoh: *Adv. Cryog. Eng. Mater.* Vol. 48 (American Inst. Phys., 2002) p. 1193.
16. K. Harada, T. Matsuda, H. Kasai, J. E. Bonevich, T. Yoshida, U. Kawabe and A. Tonomura: Phys. Rev. Lett. **71** (1993) 3371.
17. K. Yamafuji and T. Kiss: Physica C **290** (1997) 9.
18. E. H. Brandt and G. P. Mikitik: Supercond. Sci. Technol. **14** (2001) 651.
19. A. A. Zhukov, S. Kokkaliaris, P. A. J. de Groot, M. J. Higgins, S. Bhattacharya, R. Gagnon and L. Taillefer: Phys. Rev. B **61** (2000) R887.
20. T. Tamegai, Y. Iye, I. Oguro and K. Kishio: Physica C **213** (1993) 33.
21. T. Matsushita, T. Hirano, H. Yamato, M. Kiuchi, Y. Nakayama, J. Shimoyama and K. Kishio: Supercond. Sci. Technol. **11** (1998) 925.
22. R. Cubitt, E. M. Forgan, G. Yang, S. L. Lee, D. McK. Paul, H. A. Mook, M. Yethiraj, P. H. Kes, T. W. Li, A. A. Menovsky, Z. Tarnawski and K. Mortensen: Nature **365** (1993) 407.
23. T. Nishizaki, T. Naito, and N. Kobayashi: Phys. Rev. B **58** (1998) 11169.
24. H. Küpfer, Th. Wolf, R. Meier-Hirmer, M. Kläser and A. A. Zhukov: *Extended Abstract of 2000 Int. Workshop on Supercond.*, June 19–22, 2000, Matsue (Japan) p. 226 (unpublished).
25. K. Kimura, S. Kamisawa and K. Kadowaki: Physica C **357–360** (2001) 442.
26. V. M. Vinokur, P. H. Kes and A. E. Koshelev: Physica C **168** (1990) 29.
27. T. Matsushita, M. Kiuchi, T. Yasuda, H. Wada, T. Uchiyama and I. Iguchi: Supercond. Sci. Technol. **18** (2005) 1348.
28. K. Kishio, J. Shimoyama, S. Watauchi and H. Ikuta: *Proc. of 8th Int. Workshop on Critical Currents in Superconductors* (Singapore, World Scientific, 1996) p. 35.
29. T. Matsushita, T. Hirano, S. Yamaura, Y. Nakayama, J. Shimoyama and K. Kishio: Supercond. Sci. Technol. **12** (1999) 1083.
30. A. Gurevich and E. A. Pashitskii: Phys. Rev. B **57** (1998) 13878.
31. H. Hilgenkamp, C. W. Schneider, B. Goetz, R. R. Schulz, A. Schmehl, H. Bielefeldt and J. Mannhart: Supercond. Sci. Technol. **12** (1999) 1043.
32. D. Dimos, P. Chaudhari, J. Mannhart and F. K. LeGoures: Phys. Rev. Lett. **61** (1988) 219.
33. H. Hilgenkamp, B. Goets, R. R. Schulz, C. W. Schneider, B. Chesca, G. Hammerl, A. Schmehl, H. Bielefeldt and J. Mannhart: *Extended Abstracts of 2000 Int. Workshop on Supercond.*, June 2000, Matsue(Shimane) p. 33 (unpublished).
34. L. Civale, B. Maiorov, A. Serquis, S. R. Foltyn, Q. X. Jia, P. N. Arendt, H. Wang, J. O. Willis, J. Y. Coulter, T. G. Holesinger, J. L. MacManus-Driscoll, M. W. Rupich, W. Zhang and X. Li: Physica C **412–414** (2004) 976.

35. M. Kiuchi, H. Yamato and T. Matsushita: Physica C **269** (1996) 242.
36. T. Matsushita, M. Kiuchi and H. Yamato: *Adv. Cryog. Eng. Mater.* Vol. 44 (Plenum, New York, 1998) p. 647.
37. H. Yamato, M. Kiuchi, T. Matsushita, A. I. Rykov, S. Tajima and N. Koshizuka: *Adv. Supercond. X* (Springer-Verlag, Tolyo, 1998) p. 501.
38. T. Matsushita, H. Yamato, K. Yoshimitsu, M. Kiuchi, A. I. Rykov, S. Tajima and N. Koshizuka: Supercond. Sci. Technol. **11** (1998) 1173.
39. K. Kitazawa, J. Shimoyama, H. Ikuta, T. Sasagawa and K. Kishio: Physica C **282–287** (1997) 335.
40. T. Kodama, M. Fukuda, K. Shiraishi, S. Nishimura, E. S. Otabe, M. Kiuchi, T. Kiss, T. Matsushita and K. Itoh: Physica C **357–360** (2001) 582.
41. M. Tachiki: Physica C **282–287** (1997) 383.
42. Y. Matsuda, M. B. Gaifullin, K. Kumagai, K. Kadowaki and T. Mochiku: Phys. Rev. Lett. **75** (1995) 4512.
43. M. Kiuchi, K. Noguchi, T. Matsushita, T. Kato, T. Hikata and K. Sato: Physica C **278** (1997) 62.
44. T. Matsushita: *Studies of High Temperature Superconductors*, Ed. Anant Narlikar, Vol. 14 (Nova Science, 1995, New York) p. 383.
45. N. Ihara and T. Matsushita: Physica C **257** (1996) 223.
46. T. Matsushita, D. Yoshimi, M. Migita and E. S. Otabe: Supercond. Sci. Technol. **14** (2001) 732.
47. M. Kiuchi, E. S. Otabe, T. Matsushita, T. Kato, T. Hikata and K. Sato: Physica C **260** (1996) 177.
48. T. Matsushita, M. Tagomori, K. Noguchi, M. Kiuchi and T. Hasegawa: *Adv. Cryog. Eng. Mater.* Vol. 44 (Plenum, New York, 1998) p. 609.
49. T. Matsushita, E. S. Otabe, T. Nakane, M. Karppinen and H. Yamauchi: Physica C **322** (1999) 100.
50. T. Matsushita, E. S. Otabe, M. Kiuchi, T. Hikata and K. Sato: *Adv. Cryog. Eng.* Vol. 40 (Plenum, New York, 1994) p. 33.
51. T. Matsushita, E. S. Otabe, H. Wada, Y. Takahama and H. Yamauchi: Physica C **397** (2003) 38.
52. H. Matsuoka, E. S. Otabe, T. Matsushita and T. Hamada: *Adv. Supercond. IX* (Springer-Verlag, Tokyo, 1997) p. 641.
53. T. Matsushita, M. Tagomori, Y. Nakayama, A. Yamasaki, M. Kiuchi and K. Sato: *Proc. 15th Int. Conf. on Magnet Technology*, October 20–24, 1997, p. 966.
54. D. S. Fisher, M. P. A. Fisher and D. A. Huse: Phys. Rev. B **43** (1991) 130.
55. M. Tachiki and S. Takahashi: Solid State Commun. **70** (1989) 291.
56. H. Suematsu, H. Okamura, S. Lee, S. Nagaya and H. Yamauchi: Physica C **338** (2000) 96.
57. E. M. Gyorgy, R. B. van Dover, L. F. Schneemeyer, A. E. White, H. M. O'Bryan, R. J. Felder, J. V. Waszczak, W. W. Rhodes and F. Hellman: Appl. Phys. Lett. **56** (1990) 2465.
58. H. Fujimoto, T. Taguchi, M. Murakami, N. Nakamura and N. Koshizuka: *Adv. Supercond. V* (Springer-Verlag, Tokyo, 1993) p. 411.
59. H. Fujimoto, M. Murakami, N. Nakamura, S. Gotoh, A. Kondoh, N. Koshizuka and S. Tanaka: *Adv. Supercond. IV* (Springer-Verlag, Tokyo, 1992) p. 339.
60. M. Dämling, M. J. Seuntjens and C. D. Larbalestier: Nature **346** (1990) 332.
61. T. Mochida, N. Chikumoto, T. Higuchi, M. Murakami: *Adv. Supercond. X* (Springer-Verlag, Tokyo 1998) p. 489.

62. T. Matsushita: Supercond. Sci. Technol. **13** (2000) 730.
63. L. Civale, B. Maiorov, A. Serquis, J. O. Willis, J. Y. Coulter, H. Wang, Q. X. Jia, P. N. Arendt, J. L. MacManus-Driscoll, M. P. Maley and S. R. Foltyn: Appl. Phys. Lett. **84** (2004) 2121.
64. A. G. Zaitsev, G. Ockenfuss and R. Wördenweber: *Appl. Supercond.* (Inst. Phys. Conf. Ser. 158) (Institute of Physics, Bristol, 1997) p. 25.
65. P. N. Arendt, S. R. Foltyn, L. Civale, R. F. DePaula, P. C. Dowden, J. R. Groves, T. G. Holesinger, Q. X. Jia, S. Kreiskott, L. Stan, I. Usov, H. Wang and J. Y. Coulter: Physica C **412–414** (2004) 795.
66. T. Matsushita, M. Kiuchi, K. Kimura, S. Miyata, A. Ibi, T. Muroga, Y. Yamada and Y. Shiohara: Supercond. Sci. Technol. **18** (2005) S227.
67. T. Izumi, Y. Tokunaga, H. Fuji, R. Teranishi, J. Matsuda, S. Asada, T. Honjo, Y. Shiohara, T. Muroga, S. Miyata, T. Watanabe, Y. Yamada, Y. Iijima, T. Saitoh, T. Goto, A. Yoshinaka and A. Yajima: Physica C **412–414** (2004) 885.
68. Q. Jia, H. Wang, Y. Lin, B. Maiorov, Y. Li, S. Foltyn, L. Civale, P. N. Arendt: *6th Pacific Rim Conf. on Ceramic and Glass Technology*, September 2005, Kapalua, Hawaii.
69. T. Matsushita, H. Wada, T. Kiss, M. Inoue, Y. Iijima, K. Kakimoto, T. Saitoh and Y. Shiohara: Physica C **378–381** (2002) 1102.
70. T. Matsushita, T. Watanabe, Y. Fukumoto, K. Yamauchi, M. Kiuchi, E. S. Otabe, T. Kiss, T. Watanabe, S. Miyata, A. Ibi, T. Muroga, Y. Yamada and Y. Shiohara: Physica C **426–431** (2005) 1096.
71. K. Matsumoto, T. Horide, A. Ichinose, S. Hori, Y. Yoshida and M. Mukaida: Jpn. J. Appl. Phys. **44** (2005) L246.
72. T. Haugan, P. N. Barnes, R. Wheeler, F. Meisenkothen and M. Sumption: Nature **430** (2004) 867.
73. J. L. MacManus-Driscoll, S. R. Foltyn, Q. X. Jia, H. Wang, A. Serquis, L. Civale, B. Maiorov, M. E. Hawley, M. P. Maley and D. E. Peterson: Nature Mater. **3** (2004) 439.
74. B. Hensel, G. Grasso and R. Flükiger: Phy. Rev. B **51** (1995) 15456.
75. E. S. Otabe, I. Kohno, M. Kiuchi, T. Matsushita, T. Nomura, T. Motohashi, M. Karppinen, H. Yamauchi and S. Okayasu: Adv. Cryog. Eng. Vol. 52 (American Inst. Phys., 2006) p.805.
76. M. Kiuchi, Y. Himeda, Y. Fukumoto, E. S. Otabe and T. Matsushita: Supercond. Sci. Technol. **17** (2004) S10.
77. T. Matsushita, Y. Himeda, M. Kiuchi, J. Fujikami, K. Hayashi and K. Sato: Supercond. Sci. Technol. **19** (2006) 1110.
78. J. Shimoyama, Y. Nakayama, K. Kitazawa, K. Kishio, Z. Hiroi, I. Choug and M. Takano: Physica C **281** (1997) 69.
79. Z. Hiroi, I. Choug, M. Izumi and M. Takano: Adv. Supercond. X (Spring, Tokyo, 1998) p. 285.
80. T. Haraguchi, T. Imada, M. Kiuchi, E. S. Otabe, T. Matsushita, T. Yasuda, S. Okayasu, S. Uchida, J. Shimoyama and K. Kishio: Physica C, **426–431** (2005) 304.
81. T. Matsushita, M. Kiuchi, T. Haraguchi, T. Imada, K. Okamura, S. Okayasu, S. Uchida, J. Shimoyama and K. Kishio: Supercond. Sci. Technol. **19** (2006) 200.

9
MgB$_2$

9.1 Superconducting Properties

MgB$_2$ is a new metallic superconductor with a critical temperature of about 39 K that was discovered in 2001. This superconductor has superior properties, such as a critical temperature significantly higher than usual for metallic superconductors and less serious issues of anisotropy and weak links at grain boundaries, which are inevitable in high-temperature superconductors. For these reasons, not only fundamental research in physical properties, but also applied research, has been enthusiastically conducted on this superconductor, and important applications are expected, especially around liquid hydrogen temperature. Thus, wires, including multifilamentary ones, and thin films have been fabricated, and investigations have been intensively carried out on their pinning properties, including the irreversibility field.

This superconductor has a layered crystal structure composed of parallel alternate hexagonal close-packed Mg layer and honeycomb B layer. Hence, the electromagnetic properties show an anisotropy due to the crystal structure, and the coherence length in the a-b plane is longer, resulting in a lower value of the upper critical field for the magnetic field direction along the c-axis normal to the plane, as shown in Fig. 9.1 [1]. Its temperature dependence shows a pronounced upward curvature, which can be seen in the figure. This is caused by the two-gap superconductivity in the π and the σ bands of boron with different critical temperatures. The anisotropy of the upper critical field is shown in the inset. It can be seen that the anisotropy factor is about 2 at the critical temperature and increases with decreasing temperature. The coherence lengths estimated from the upper critical fields are approximately $\xi_c(0) = 9.6$ nm and $\xi_{ab}(0) = 2.0$ nm. These are considerably longer than those in high-temperature superconductors and hence, flux pinning properties similar to those of the usual metallic superconductors are expected.

Figure 9.2 shows the temperature dependence of various characteristic magnetic fields for a polycrystalline specimen [2]. The values of the thermodynamic critical field H_c below 25 K are estimated from the values of the

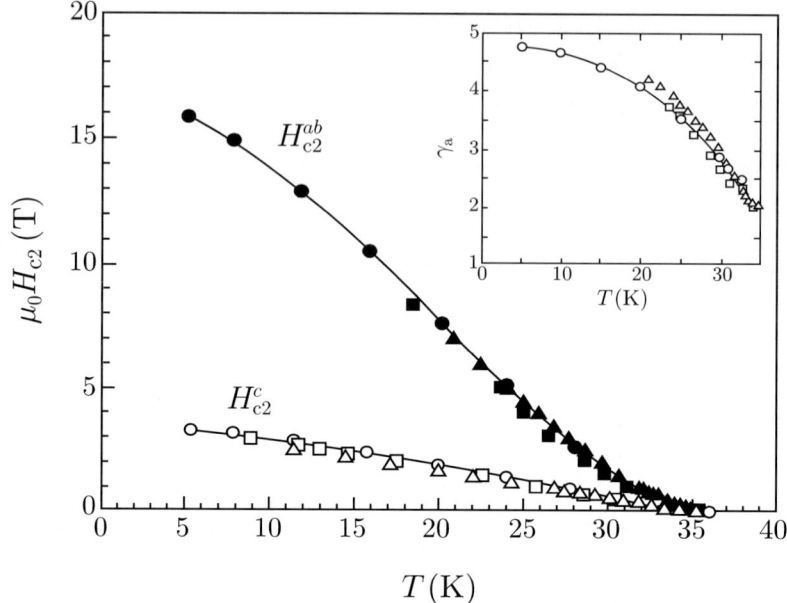

Fig. 9.1. Temperature dependence of upper critical field of MgB$_2$ single crystal [1]. Inset shows the anisotropy of the upper critical field, $\gamma_a = H_{c2}^{ab}/H_{c2}^{c}$

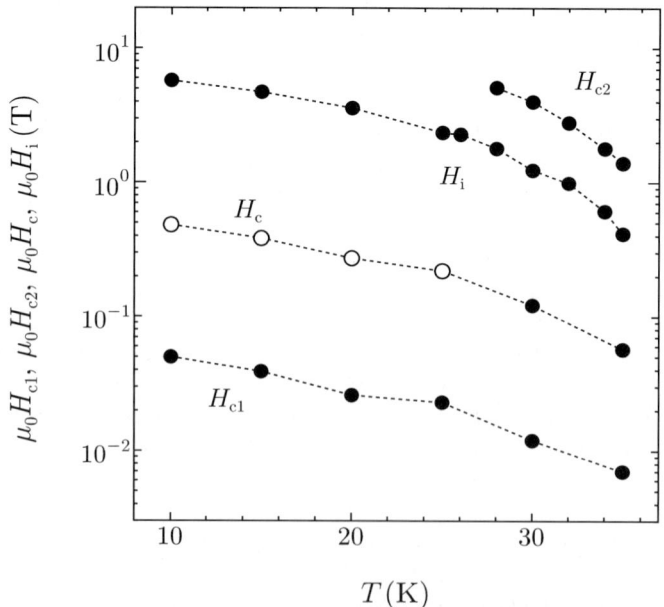

Fig. 9.2. Various critical fields and irreversibility field in MgB$_2$ polycrystalline specimen [2]. The values of the thermodynamic critical field below 25 K (*open circles*) are estimated using extrapolated values of the upper critical field to the low temperature region

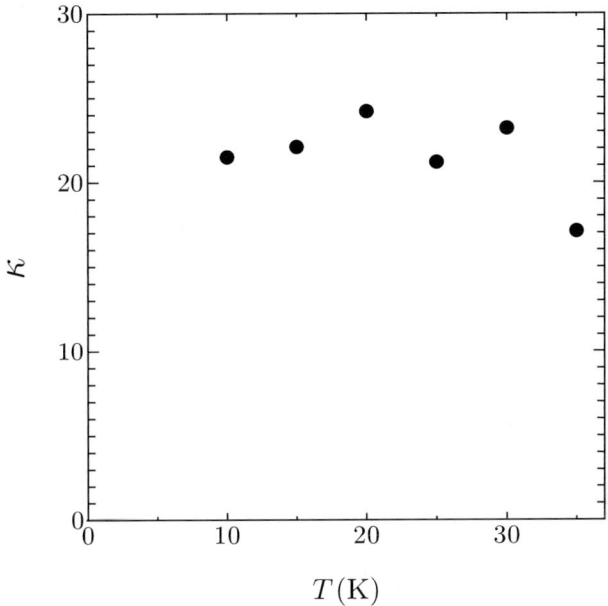

Fig. 9.3. Ginzburg-Landau parameter in MgB_2 polycrystalline specimen [2]

upper critical field H_{c2} extrapolated from the high temperature region and the observed lower critical field H_{c1}. Thus, the values of H_{c1} and H_{c2} are values averaged with respect to the anisotropy. H_c at 10 K is about 0.5 T and is similar to that of Nb_3Sn at 4.2 K. The Ginzburg-Landau parameter, κ, shown in Fig. 9.3 is about 20, independently of the temperature [2]. The coherence length in polycrystalline bulk specimens, wires or thin films can be shortened, resulting in a significant enhancement of the upper critical field from values in single crystals, by choosing a suitable annealing temperature or by addition of carbon atoms. As can be seen from Fig. 9.2, the irreversibility field is remarkably lower than the upper critical field in MgB_2 superconductor. However, the difference is much smaller than in high-temperatures superconductors. Hence, MgB_2 has an irreversible property intermediate between metallic and high-temperature superconductors. The main differences from metallic superconductors are higher temperatures for applications and weaker flux pinning strengths at the present stage.

9.2 Flux Pinning Properties

9.2.1 Wires and Bulk Materials

With respect to the MgB_2 wires being fabricated now, the powder in tube (PIT) process is common, in which a metallic sheath such as iron is packed with powders and then is drawn. There are two methods: one is an *in situ* method in which mixed powders of magnesium and boron or their compounds are used with heat-treatment for reaction after the drawing, and the other is an *ex situ* method in which reacted MgB$_2$ powders are used. The *in situ* wires usually have superior critical current density in high magnetic fields.

MgB$_2$ superconductor in wires is usually polycrystalline, and it is considered that there are no issues of weak links in grain boundaries to influence the transport properties as with metallic superconductors. Thus, the dominant pinning centers are considered to be grain boundaries. In fact, there exists a strong correlation between the critical current density and the grain size over a wide range of grain sizes from wires to thin film, as shown in Fig. 9.4 [3]. This shows that the above hypothesis is correct. Very high critical current densities in thin films in a self field come from the very high density of pinning centers due to small grain sizes. However, the critical current density at 4.2 K and 5 T is only of the order of 1×10^8 Am^{-2}, even for the small grain size of $d_g = 0.2$ μm, which is 2–3% of the critical current density of Nb$_3$Sn under the same conditions. This fact may suggest that the flux pinning strength of grain boundaries is very weak in MgB$_2$.

The reason for the low critical current density in present MgB$_2$ superconductors can be found in the low density of MgB$_2$ phase, which is caused by

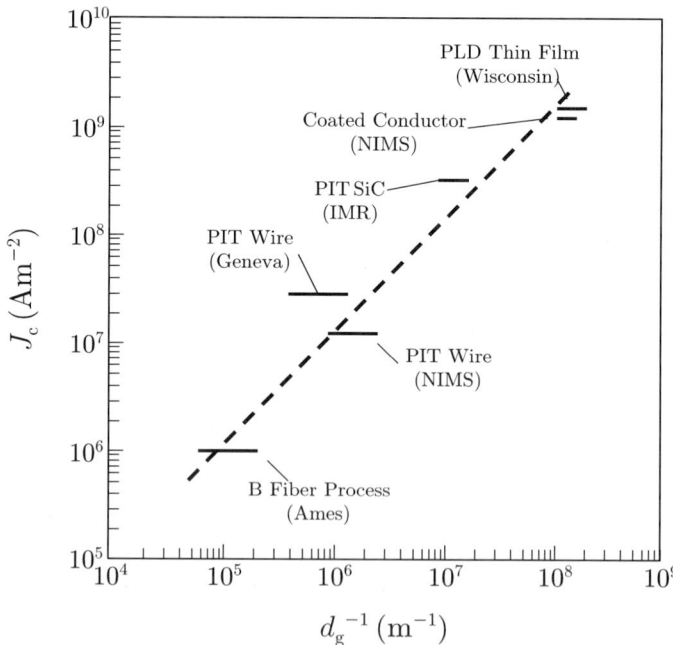

Fig. 9.4. Relationship between the critical current density at 4.2 K and 5 T and the grain size for MgB$_2$ [3]

the low packing density of powders and also by the reduction in volume by the chemical reaction in the *in situ* fabrication process. Nucleation of MgO grains or thin MgO layers at grain boundaries is also a dominant factor in low critical current densities. The grain connectivity is evaluated from the normal state resistivity based on the Rowell model [4]. Yamamoto [5] estimated the connectivity K and the proportion of grain boundaries covered with insulating oxide layers α using the cubic bond-percolation model for various specimens prepared under the same conditions, but with different packing factor P. It was found from the analysis that α is about 0.14.

Then, the grain connectivity, which represents the rate of reduction of the critical current density from a perfect condition, can be estimated as a function of P in terms of the bond-percolation model. The material is approximated by a cubic bond system of grains, each of which has bonds to six adjacent grains. If the number of bonds from one grain is represented by z, the threshold packing density P_c is given by $2/z$, hence we have $P_c = 1/3$. Since the effective packing factor for the transport current is given by $(1-\alpha)P$, the grain connectivity is given by

$$K = \frac{(1-\alpha)P - P_c}{1 - P_c} \ . \tag{9.1}$$

This result reveals that in the case of $P = 0.5$ as in the usual *in situ* processed wires, the grain connectivity is estimated as 0.15. This means that the critical current density is reduced below $1/6$ of the intrinsic value due to the porosity and oxide layers. Thus, the bad influence of oxide layers is remarkable, especially for materials with poor packing density such as *in situ* wires. For further improvement of the critical current density, not only the improvement of the packing density, but also the development of a new process which suppresses the formation of thin oxide layers is necessary. For this purpose it has been reported that removal of B_2O_3 by purification of B powders is effective [6]. In particular, the removal of adhesive B_2O_3 from the surface of B powders is important.

For the improvement of the critical current density it is also important to increase the packing density of MgB_2 as shown above. The *ex situ* process seems to be suitable for this purpose. In fact, it was proved that a high critical current density could be achieved at high fields using this method [7]. In this case special care was taken with the MgB_2 powders sealed in a stainless steel sheath: a tape was fabricated from MgH_2 and amorphous boron in the usual *in situ* PIT process, and the powdered MgB_2 in the core was used. Easy production of clean MgB_2 powders will be a key issue for this process in the future. It was also reported [8] that application of hot pressing to an *in situ* wire was effective for enhancing the critical current density at low fields by reduction of voids in the superconducting region. On the other hand, the PICT (Powder In Closed Tube) process was proposed to produce bulk superconductors with a high volume fraction with the aid of a high vapor pressure of Mg, and a high critical current density was reported [9]. In addition, it was also reported that very high density MgB$_2$ could be synthesized by a diffusion process between an Fe-Mg alloy substrate and a B sheet [10]. Furthermore, a packing factor close to 100% was achieved by a diffusion process for a mixture of commercial MgB$_2$ powder and B powder using the PICT method [11]. Addition of In or Sn was also proposed to fill voids in order to improve the connectivity among grains [12].

Another method for improvement is the strengthening of flux pinning: it is known that addition of SiC is effective. Figure 9.5 shows an example [13]. Added C atoms occupy B sites, and the honeycomb structure of the B atoms is distorted. As a result the unit length of the a-axis is reduced by the addition, while that of the c-axis is not changed [14]. As can be seen from Fig. 9.5, C-addition is effective for the improvement of the critical current

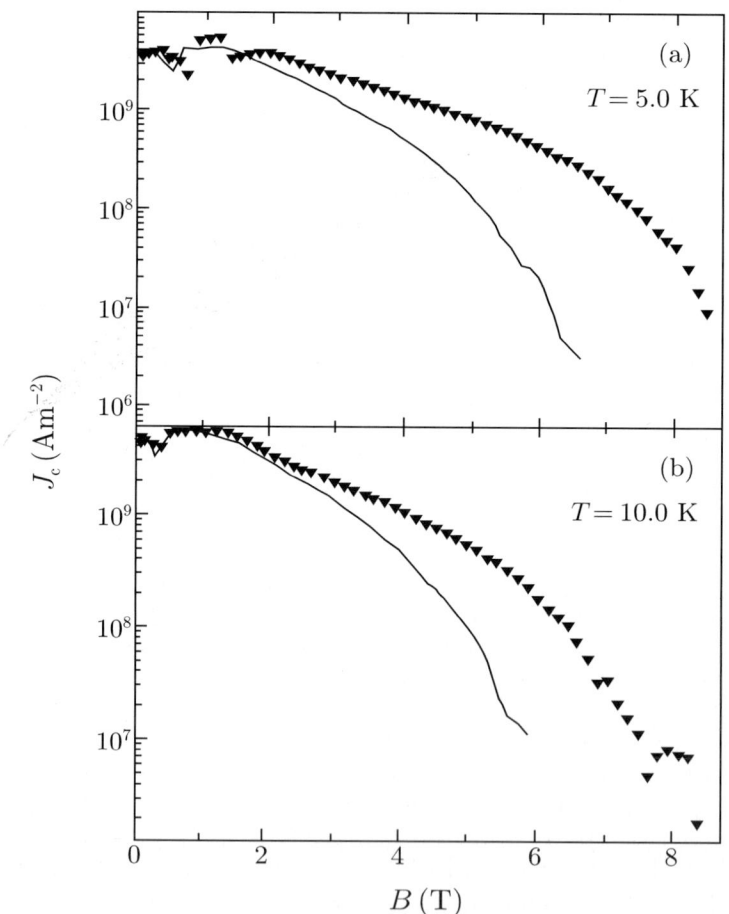

Fig. 9.5. Improvement of critical current density by addition of 10wt% fine SiC powders to MgB$_2$ [13]. *Solid* symbols show the results for the SiC added specimen

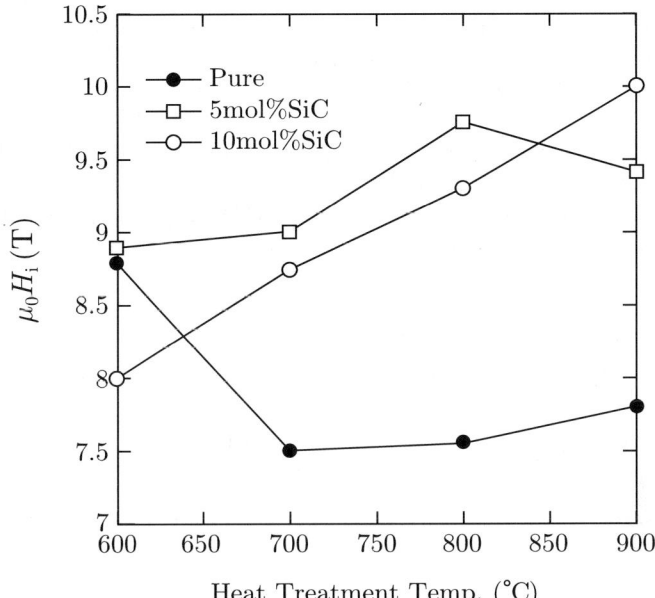

Fig. 9.6. Relationship between the irreversibility field at 20 K in MgB$_2$ specimens with different amounts of SiC addition and different temperatures for the 1 h sintering [15]

density, especially at high fields. It can be concluded that this results from the improvement of the irreversibility field due to the enhanced upper critical field, which is brought about by reduction in the coherence length due to the distortion of the crystal structure by C substitution. Figure 9.6 shows the influence on the irreversibility field at 20 K from the amount of SiC addition and the temperature of a 1 hr heat treatment [15]. For pure specimens the irreversibility field is decreased significantly by a the heat treatment at high temperature. This can be attributed to the reduction in the upper critical field. On the other hand, the irreversibility field is likely to increase with the heat treatment temperature in the SiC doped material. This is because the critical temperature is improved without reduction in the upper critical field by the sintering at high temperatures.

Figure 9.7 represents the relationship between the full-width half-maximum (FWHM) of the (110) X ray diffraction peaks and the irreversibility field at 20 K for pure specimens and specimens doped with B$_4$C or SiC [16]. This shows that the irreversibility field increases with FWHM. The crystal structure is deformed by the C addition, and the deformation causes the change in the pinning property. This change is ascribed partly to the enhancement of the upper critical field and may also be partly due to the enhancement of the flux pinning strength of grain boundaries by the electron scattering mechanism

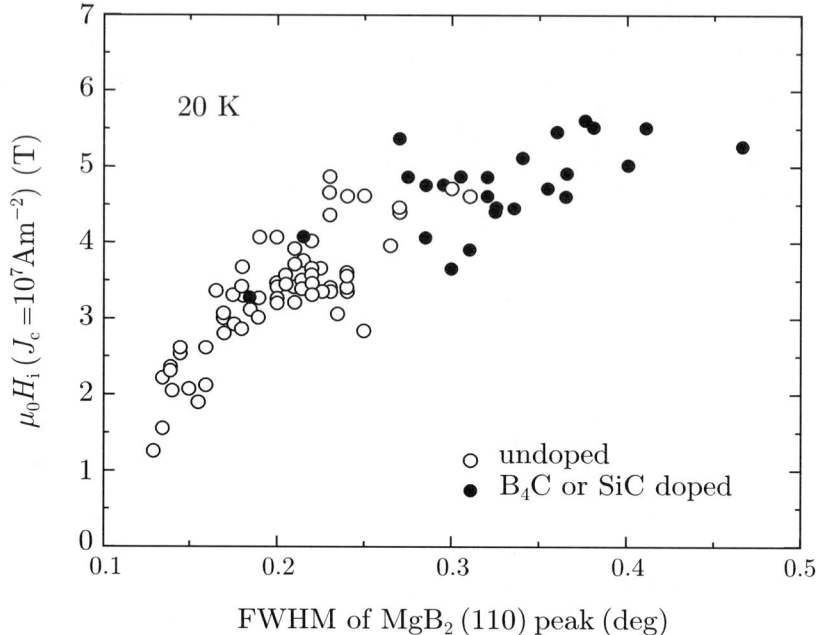

Fig. 9.7. Relationship between FWHM of (110) X ray diffraction peaks and the irreversibility field at 20 K for pure specimens and specimens doped with B_4C or SiC [16]

(see Subsect. 6.3.2). The latter possibility will be clarified by an analysis of experimental results later.

Not only SiC addition but also addition of carbon nanotubes [17], B_4C [18] or aromatic hydrocarbons [19] is also effective for the improvement of the critical current properties at high magnetic fields. Among them SiC is usually best, since the substitution rate at the B sites is highest for SiC. In the future it will be necessary to seek the optimum amount of C substitution so that the impurity parameter takes a value near 1 in order to enhance the flux pinning strength of the grain boundaries.

Recently it has been found that the irreversibility field is also increased by synthesis at low temperatures such as 600°C [20]. In fact, a relationship between the irreversibility field and FWHM of XRD peaks, similar to that shown in Fig. 9.7, is obtained also for pure specimens synthesized at different temperatures [21]. However, the distortion which causes the increase in FWHM is not that of the honeycomb structure of B due to the C substitution but the distortion due to fine grain structures. This structure is also expected to directly contribute to the improvement of the irreversibility field through the increase in the number density of pinning centers. The merit of the low temperature synthesis is that it does not cause deterioration of the critical

9.2 Flux Pinning Properties

Table 9.1. Specifications of specimens, and pinning and superconducting parameters of specimens which fit the experiments [22]

Specimen	1	2	3	4
	MgB$_2$	MgB$_2$	MgB$_{1.5}$(B$_4$C)$_{0.1}$	MgB$_{1.8}$(SiC)$_{0.2}$
Heat treat.	950°C×12h	600°C×24h	850°C× 3h	850°C× 3h
T_c (K)	38.6	38.2	35.4	35.5
$A_m(10^9)$	1.9	3.5	3.1	2.7
m'	1.4	1.7	1.6	1.5
m_1/m_2	2.4/1.6	2.0/1.3	1.4/1.1	1.3/1.1
γ	0.3	0.3	0.4	0.4
δ	2	2	2	2
σ^2	0.0002	0.0008	0.0001	0.0009
g^2	1.0–2.8	1.0–2.8	1.0 1.2	1.0–1.7
$\mu_0 H_{c2}(0)$ (T)	12	15	20	25

temperature. Hence, this procedure is more advantageous than C-addition for applications at temperatures higher than 20 K.

Here the result of quantitative investigation is shown for the effect of C-addition and low temperature synthesis on the flux pinning properties. Specimen 1 was C-free and was synthesized at 950°C for 12 h. Specimen 2 was also C-free but was synthesized at 600°C for 24 h. Specimens 3 and 4 were doped with SiC and B$_4$C, respectively, and were heat treated at 850°C for 3 h. These specimens were prepared by the PICT method. The specifications of the specimens are listed in Table 9.1. It can be seen that the critical temperature is above 38 K for pure specimens 1 and 2, and is degraded by about 3 K due to the C-addition. DC magnetization measurements were carried out on these specimens, and the critical current density was estimated. The obtained results are shown in Fig. 9.8 [22]. It was found that specimen 2 synthesized at low temperature has a considerably higher critical current density at low fields than specimen 1. On the other hand, the critical current density at high fields is significantly improved for specimens 3 and 4 with C-addition. Specimen 4 doped with B$_4$C shows a very high critical current density at low temperatures. The critical current densities of all the specimens at $T/T_c = 0.2$ and 0.6 are compared in Fig. 9.9 to clarify the characteristics of each specimen.

The scaling behavior of the pinning force density of each specimen is shown in Fig. 9.10, where the solid line represents the relationship:

$$\frac{F_p}{F_{pmax}} \propto b_i^{1/2}(1-b_i)^2, \tag{9.2}$$

where $b_i = B/\mu_0 H_i$ is the magnetic field normalized by the irreversibility field and F_{pmax} is the maximum pinning force density. The values of the irreversibility field at low temperatures were estimated from the extrapolation of the relationship of $(J_c B^{1-x})^{1/2}$ vs. B, which was linearized by adjusting the value of x. It is found that this relationship fits well with the result for

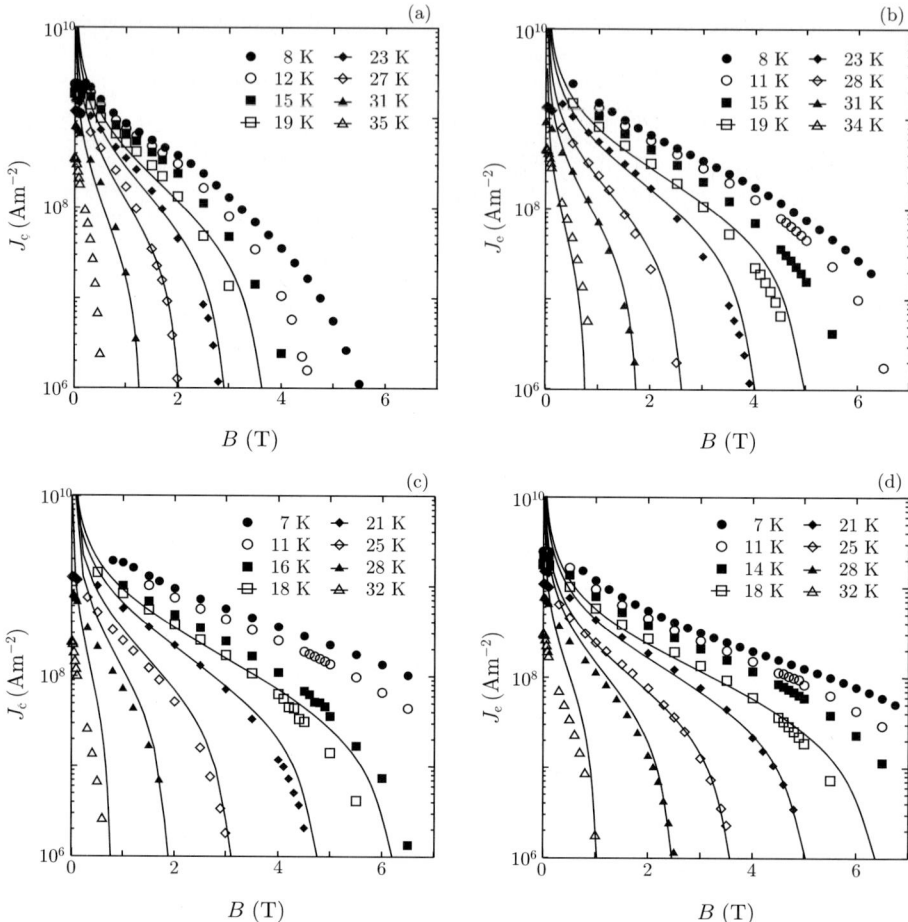

Fig. 9.8. Dependencies of each MgB$_2$ specimen on temperature and magnetic field: (**a**) specimen 1, (**b**) specimen 2 synthesized at low temperature, (**c**) specimen 3 with B$_4$C addition and (**d**) specimen 4 with SiC addition [22]. *Solid lines* show the theoretical predictions of the flux creep-flow model in the high temperature region

specimen 1, which was synthesized at high temperature, over the entire range of measurement temperatures and with the results for other specimens in the high temperature region.

However, the normalized magnetic field at which the pinning force density is at a maximum is appreciably smaller than 0.2, which is predicted by Eq. (9.2), suggesting a different pinning mechanism in the low temperature region for the three specimens other than the specimen synthesized at high temperature.

Figure 9.11 shows the temperature dependence of the irreversibility field for the four specimens. The irreversibility field is significantly enhanced in

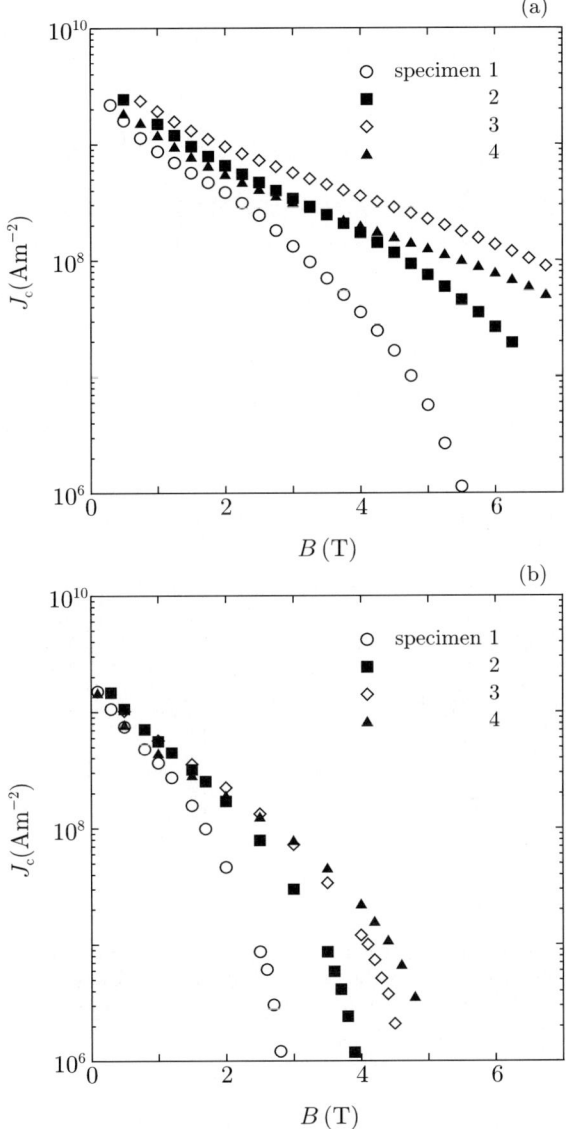

Fig. 9.9. Critical current density of each specimen at (a) $T/T_c = 0.2$ and (b) $T/T_c = 0.6$ [22]

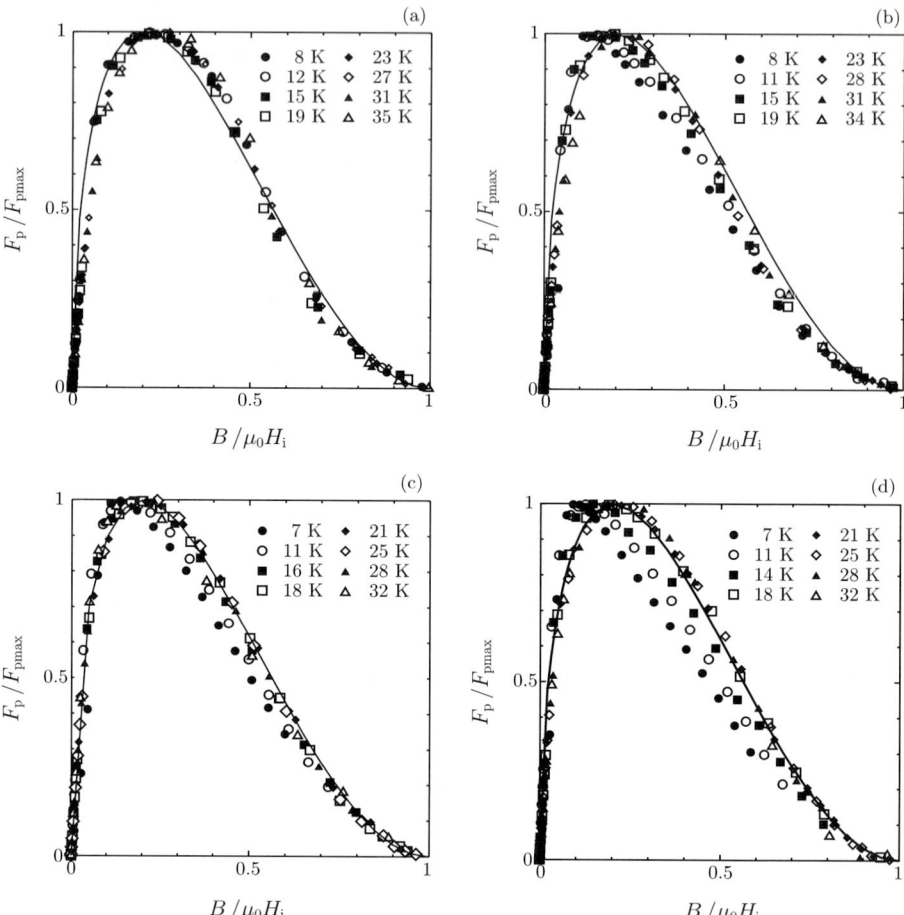

Fig. 9.10. Scaling law of the pinning force density of each MgB$_2$ specimen: (**a**) specimen 1, (**b**) specimen 2 synthesized at low temperature, (**c**) specimen 3 with B$_4$C addition and (**d**) specimen 4 with SiC addition [22]. *Solid line* shows the scaling law of Eq. (9.2)

specimens 3 and 4 with C-addition and is considerably increased also in specimen 2, which was synthesized at low temperature. The critical current density in the high field region is accompanied by such an improvement in the irreversibility field. The temperature dependence of the irreversibility field is almost linear for all the specimens at high temperatures. At low temperatures, although the irreversibility field is somewhat saturated with decreasing temperature in the specimen synthesized at high temperature, it increases remarkably in the C-added specimens. The property of the specimen synthesized at low temperature is intermediate between them.

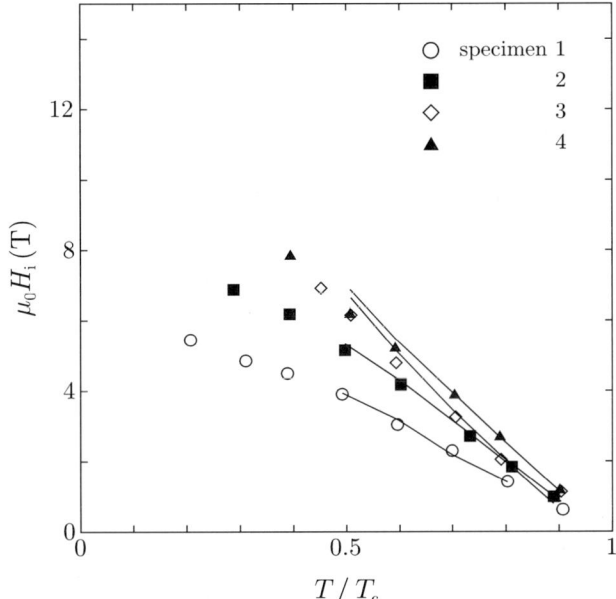

Fig. 9.11. Irreversibility field of each MgB$_2$ specimen [22]. *Solid lines* show the theoretical predictions of the flux creep-flow model

The relationship between F_{pmax} and H_i, i.e., the temperature dependence of the pinning force density, is shown in Fig. 9.12. In the high temperature region F_{pmax} is proportional to the second power of H_i for all the specimens. On the other hand, the dependence of F_{pmax} on H_i is weaker at low temperatures, and this tendency is significant in the specimens with C-addition.

The shift of the maximum of the pinning force density to a low field may be explained by a new contribution from point defects, etc. at low fields. If so, however, F_{pmax} must increase more in the low field region, resulting in a contradiction with the experiments shown in Fig. 9.11. If the slower rate of increase of F_{pmax} in the low temperature region is caused by weak links at grain boundaries, as in Y-123 superconductor, only the high critical current density at low fields is limited, which would result in the shift of F_{pmax} to a higher field.

Another reason for the shift of F_{pmax} to a lower field may be an extraordinary increase in the irreversibility field at low fields, as indicated by the result in Fig. 9.11. When the superconductor becomes extremely dirty, the upper critical field is significantly enhanced at low temperatures because of the two gap superconductivity, resulting in the enhancement of the irreversibility field at low temperatures. On the other hand, the temperature dependence of the thermodynamic critical field, on which the flux pinning property directly depends, would not be so much different from that in fairly clean superconductors. Such a difference in the temperature dependence between the upper

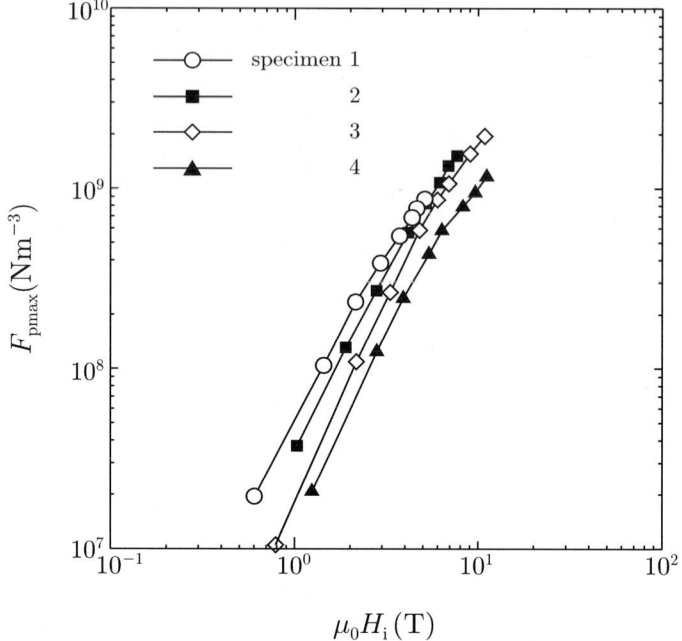

Fig. 9.12. Relationship between the maximum pinning force density and the irreversibility field [22]. Lines are guides for the eye

critical field and the thermodynamic critical field may be the reason for the change in the scaling behavior of the pinning force density at low temperatures.

For a quantitative discussion on the above experimental results, the pinning parameters which match the results in the temperature region above 18 K in Figs. 9.8 and 9.11 were estimated using the flux creep-flow model. As with the scaling law of J_{c0} which represents the flux pinning strength, the following formula was used:

$$J_{c0}(B,T) = Af(t)^{m'} B^{\gamma-1} \left(1 - \frac{B}{\mu_0 H_{c2}}\right)^{\delta}, \qquad (9.3)$$

where

$$H_{c2}(T) = H_{c2}(0)f(t) = H_{c2}(0)(1 - t^{m_1})^{m_2} \qquad (9.4)$$

with $t = T/T_c$ instead of Eqs. (8.27) and (1.2), respectively. The temperature dependence of Eq. (9.4) is adopted here, since this explains the experimental results in a wide temperature regime up to T_c, as in Fig. 9.13. Equation (8.31) is assumed for the statistical distribution of A. The E-J curve is calculated using Eq. (8.36), and the critical current density is estimated using the electric field criterion, $E = 1 \times 10^{-8}$ Vm^{-1}, corresponding to the DC magnetization

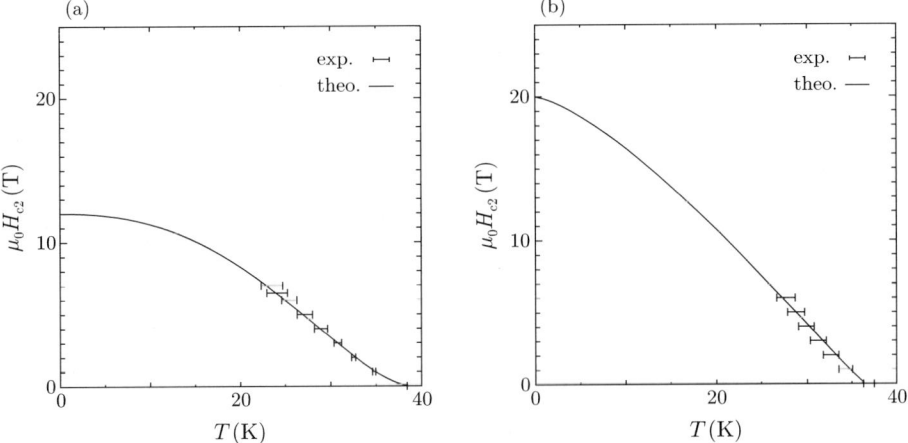

Fig. 9.13. Temperature dependence of upper critical field for (**a**) specimen 1 and (**b**) specimen 3. The *solid lines* show results calculated from Eq. (9.4)

experiments. The irreversibility field was determined with the same criterion as the experiments. The pinning parameters are determined so as to get good agreement with experiment. The parameters used are listed in Table 9.1.

The obtained theoretical critical current densities in the high temperature region are shown by solid lines in Fig. 9.8 [22]. Although the theoretical results show very high critical current density deviating from the experiments at low fields due to small values of γ in the modified Irie-Yamafuji model, the agreement with the experiments is good. Figure 9.11 shows that the agreement is also good for the irreversibility field. Thus, it can be concluded that the parameters listed in Table 9.1 exactly represent the practical situation of each specimen.

The following conclusions can be obtained from the parameters in Table 9.1.

- In specimen 2 synthesized at low temperature A_m expressing the flux pinning strength is increased most, and $H_{\mathrm{c}2}$ is also considerably increased.
- In both the specimens with C-addition A_m is somewhat increased, and $H_{\mathrm{c}2}$ is significantly increased.

The very large value of A_m is the reason for the high J_c at low fields in specimen 2. An increase in the effective number density of pins caused by the fine crystalline structure seems to result in the large A_m. In addition, the enhancement of the flux pinning strength at grain boundaries due to the shortened coherence length, which can be expected from the enhanced upper critical field due to the electron scattering, is also considered to contribute to the increase in A_m. The improvement of the high field properties is attributed to the increases in A_m and $H_{\mathrm{c}2}$.

The significant increase in H_{c2} in specimens 3 and 4 is attributed to the reduction in the coherence length due to the electron scattering by boron sites where there has been carbon substitution. This leads to a significant improvement of the high field properties. The enhancement of A_m is also obtained in these specimens, and this explains the improvement of J_c at low fields. The increase in A_m seems to be attributed to the enhancement of the flux pinning strength at grain boundaries. The improvement of the irreversibility field, which influences the pinning properties at high fields, can be attributed to the enhancement of the flux pinning strength A_m and of the upper critical field H_{c2}.

According to the flux creep theory the irreversibility field H_i is proportional to $A_m^{1/3}$ (see Eq. (8.40)), when H_i is sufficiently lower than H_{c2}. If specimens 1 and 2 are compared, the increase in A_m by a factor of 1.8 is expected to contribute to the increase in H_i by a factor of 1.2. The observed enhancement factor of H_i at $T/T_c = 0.5$ is about 1.73. Hence, it can be said that the increase in H_{c2} has a larger effect than the increase in A_m. The increase in H_i in the other two specimens results mostly from the increase in H_{c2}.

On the other hand, the pinning properties at low fields are mostly determined by A_m. From the comparison between the low temperature synthesis and carbon addition it can be said that the reduction in the grain size is more effective than the reduction in the coherence length for increasing the flux pinning strength.

There are some differences in the pinning properties between specimens 3 and 4. For example, the properties at high fields are better in specimen 4 with SiC, while specimen 3 with B_4C has better properties at low fields and at low temperatures. However, the reason is not clear.

9.2.2 Thin Films

There are mainly two methods for fabrication of MgB_2 thin films: an indirect method in which magnesium is diffused into a deposited thin film by a heat treatment and a direct method to fabricate the compound without heat treatment. Using the former method, the critical temperature can be increased to a value comparable to bulk specimens or single crystals by a heat treatment at high temperatures. However, the upper critical field is not high. On the other hand, using the latter method, whereas the critical temperature is fairly low, the upper critical field can be significantly increased. Hence, the latter method is advantageous for applications at high fields. This is attributed to a short coherence length due to the electron scattering by boundaries of fine grains in a thin film. The reduction in the critical temperature is also caused by the strain in the crystalline structure of fine grains.

The upper critical field and irreversibility field parallel to the c-axis are shown in Fig. 9.14 for thin films fabricated by the two methods [23]. One of them is an *ex situ* thin film, which was fabricated by sealing a boron precursor film wrapped in Ta foil with Mg pellets and Ar gas in a stainless steel tube

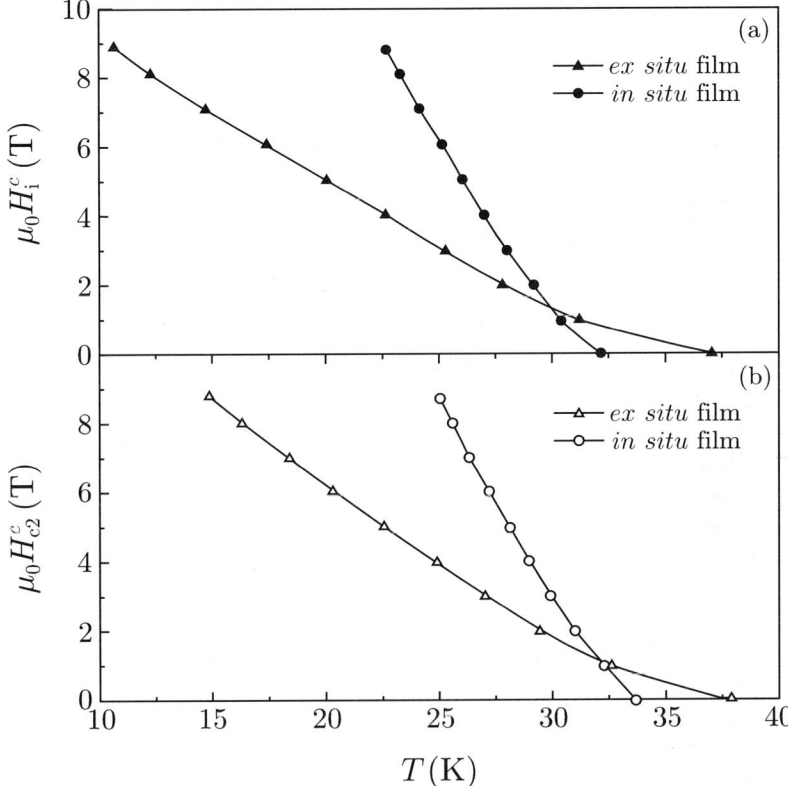

Fig. 9.14. Upper critical field and irreversibility field parallel to the c-axis of two kinds of MgB$_2$ thin film [23]

and then heat treating it at 900°C for 30 min, and the other is an *in situ* thin film, which was directly deposited from a MgB$_2$ target using a pulsed laser technique and then heat treated at 685°C for 12 min. Although the critical temperature is higher, the enhancement of the upper critical field and the irreversibility field with decreasing temperature is not significant for the former thin film. The enhancement of these fields in the latter thin film is significant.

The critical current density in thin films is generally higher than those in bulk specimens or wires. This is owing to the stronger flux pinning strength of grain boundaries at higher density due to the small size of grains. Such a high density of grain boundaries also contributes to the enhancement of the upper critical field, and this results in the better performance of J_c at high fields through the enhancement of the irreversibility field.

Figure 9.15 shows the critical current density for two thin films prepared by the different methods [23], the upper critical fields of which have been shown in Fig. 9.14. The critical current densities of the two films at low fields are approximately the same and are very high. Hence, the flux pinning strength

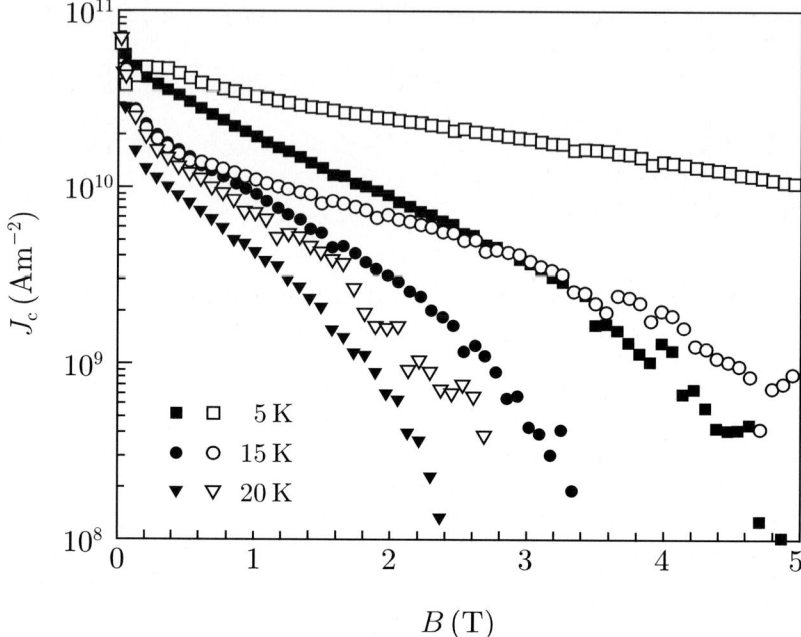

Fig. 9.15. Critical current density in the normal field for *ex situ* thin film heat treated at 900°C (*solid symbols*) and *in situ* thin film heat treated at 650°C (*open symbols*) [23], which are shown in Fig. 9.14

seems to be comparable in both. However, the value of the *ex situ* thin film is lower at high fields due to the lower irreversibility field. This can be attributed to the lower value of the upper critical field. The high critical current density of 2.5×10^{11} Am^{-2} was attained for a similar *ex situ* thin film in self field at 5 K by Kwang et al. [24]. However, J_c quickly decreased with increasing field and was about 1×10^{10} Am^{-2} at 4.5 T.

Such a difference in the magnetic field dependence of the critical current density between *ex situ* thin films made at high temperatures and *in situ* thin films made at low temperatures is fundamentally the same as the difference between bulk specimens made at high temperatures and those made at low temperatures, which have been discussed in Subsect. 9.2.1.

9.3 Possibility of Improvements in the Future

Here we shall discuss the possibility of improvement of MgB$_2$ superconductor in the future. At first it is necessary to remove oxide layers that block the transport current. At the same time, densification of the superconducting phase is also important. For this purpose the PICT method and the diffusion method seem to be useful. For practical applications the long wire fabrication

Table 9.2. Comparison of the value of $(\mu_0 H_c)^2 \xi$ among bulk MgB$_2$, Nb-Ti and Nb$_3$Sn (the data on MgB$_2$ are from [2]).

	MgB$_2$		Nb-Ti	Nb$_3$Sn
	10 K	20 K	4.2 K	4.2 K
$\mu_0 H_c$ (T)	0.48	0.27	0.20	0.50
ξ (nm)	6.04	6.81	5.70	3.90
$(\mu_0 H_c)^2 \xi$ (T^{2}nm)	1.39	0.50	0.23	0.98

with *ex situ* method seems to be promising, although it needs further improvement. These problems will hopefully be resolved by progress in metallurgical technology in the future.

Then, the question is whether or not sufficiently strong flux pinning strength will be achieved in MgB$_2$. This entirely depends on the intrinsic superconducting properties of MgB$_2$. It can be seen from Eqs. (6.7) and (6.25) that the pinning force is proportional to $(\mu_0 H_c)^2 \xi$ for the cases of pinning by normal precipitates and grain boundaries. Hence, from a comparison of this quantity the potential of the superconductor for applications can be estimated. The value of this quantity is compared between MgB$_2$ at 10 and 20 K, in which case the mean value of the bulk material is given, and Nb-Ti and Nb$_3$Sn at 4.2 K in Table 9.2. This shows that MgB$_2$ has an excellent property; its value at 10 K exceeds considerably the value of Nb$_3$Sn at 4.2 K and its value even at 20 K exceeds the value of Nb-Ti at 4.2 K. Hence, it can be concluded that MgB$_2$ is a superconducting material which has a sufficient potential for application and that the critical current density can be significantly improved in the future.

For the improvement of the critical current density it is necessary to enhance the flux pinning strength as discussed in Subsect. 9.2.1, and an increase in the area of grain boundaries achieved by reducing the grain size is effective. Low temperature synthesis will be useful for this purpose. Another important point for the improvement is enhancement of the upper critical field. This leads to the improvement of high field performance through the enhancement of the irreversibility field. This also contributes to strengthening the pinning force at grain boundaries by reducing the coherence length, as discussed in Subsect. 9.2.1. Addition of carbon is effective. It has been made clear recently that suitable addition of nonmagnetic impurities can significantly enhance the upper critical field in bulk and thin film superconductors [25, 26]. Figure 9.16 represents the upper critical field along the c-axis for c-axis-oriented thin films prepared by various methods [25]. It is found that the upper critical field is high for a thin film with a high normal resistivity, and its value reaches about 22 T even at 10 K. The upper critical field normal to the c-axis is about 1.8 times as high as the value along the c-axis in the medium temperature region.

From the above argument it is concluded that the addition of impurities and low temperature synthesis are useful for the improvement of the properties. Hence, it is necessary from now on to find out the best conditions for the

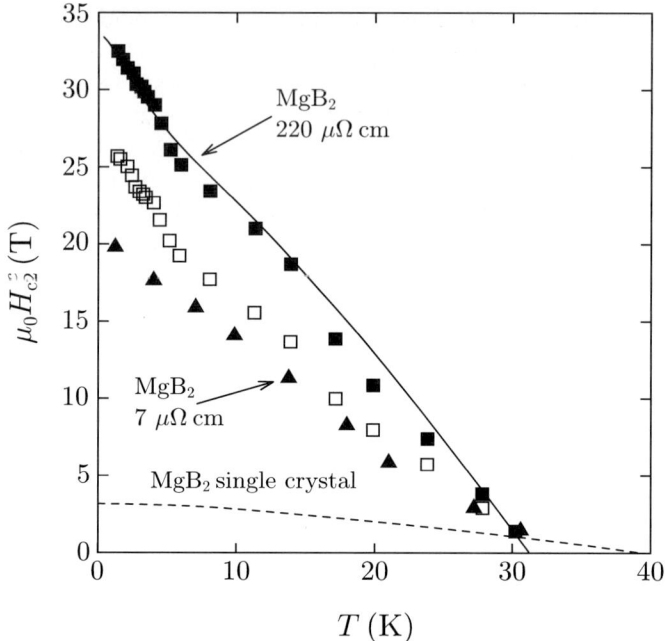

Fig. 9.16. Upper critical field (*solid symbols*) and irreversibility field (*open symbols*) along the c-axis for c-axis-oriented thin films [25]. The *solid line* shows the theoretical prediction in the dirty limit for a superconductor with two energy gaps

kind and amount of impurities to add and the synthesis temperature under the combined situation of these techniques. Mechanical plastic deformation may also be useful as a practical technique to introduce pinning centers into long wire.

Furthermore, if fine normal precipitates like α-Ti in Nb-Ti can be introduced as pinning centers, a high pinning efficiency as in Nb-Ti will be attained. In this case flux pinning properties which surpass Nb$_3$Sn and Nb-Ti will be obtained.

Exercises

9.1. Calculate the virtual critical current density J_{c0} at 10 K and at 5 T in bulk MgB$_2$ superconductor of grain size $d_g = 0.2$ µm. For simplicity grain boundaries are assumed to be parallel to flux lines and perpendicular to the Lorentz force. The impurity parameter is assumed as $\alpha_i = 0.5$ with the elementary pinning force of a grain boundary: $f'_p \simeq 0.14 \mu_0 H_c^2 \xi_0$ from Fig. 6.12. The pinning force density is given by

$$F_{p0} = J_{c0} B = \frac{f'_p}{a_f d_g}\left(1 - \frac{B}{\mu_0 H_{c2}}\right)^2$$

taking into account the correction factor at high fields. Use the value in Table 9.2 for the thermodynamic critical field and $\mu_0 H_{c2}(0) = 25.0$ T for a SiC-doped specimen in Table 9.1. Use Goodman's formula (6.23) for the estimation of the BCS coherence length. Assume Eq. (9.4) with $m_1 = m_2 = 1$ for the temperature dependence of each critical field with $T_c = 35.5$ K.

9.2. Estimate the irreversibility field at 10 K based on the conditions assumed in Exercise 9.1. It is not necessary to assume a statistical distribution of the flux pinning strength. Assume $g^2 = 1.0$ and 14 for the logarithmic term in Eq. (8.29).

References

1. L. Lyard, P. Samuely, P. Szabo, T. Klein, C. Marcenat, L. Paulius, K. H. P. Kim, C. U. Jung, H.-S. Lee, B. Kang, S. Choi, S.-I. Lee, J. Marcus, S. Blanchard, A. G. M. Jansen, U. Welp, G. Karapetrov and W. K. Kwok: Phys. Rev. B **66** 180502.
2. M. Fukuda, E. S. Otabe and T. Matsushita: Physica C **378–381** (2002) 239.
3. K. Togano: private communication.
4. J. M. Rowell: Supercond. Sci. Technol. **16** (2003) R17.
5. A. Yamamoto: private communication.
6. J. Jiang, B. J. Senkowicz, D. C. Larbalestier and E. E. Hellstrom: Supercond. Sci. Technol. **19** (2006) L33.
7. T. Nakane, H. Kitaguchi and H. Kumakura: Appl. Phys. Lett. **88** (2005) 22513.
8. Y. Yamada, M. Nakatsuka, K. Tachikawa and H. Kumakura: J. Cryo. Soc. Jpn. **40** (2005) 493 [in Japanese].
9. A. Yamamoto, J. Shimoyama, S. Ueda, Y. Katsura, S. Horii and K. Kishio: Supercond. Sci. Technol. **17** (2004) 921.
10. K. Togano, T. Nakane, H. Fujii, H. Takeya and H. Kumakura: Supercond. Sci. Technol. **19** (2006) L17.
11. I. Iwayama, S. Ueda, A. Yamamoto, Y. Katsura, J. Shimoyama, S. Horii and K. Kishio: submitted to Physica C.
12. K. Tachikawa, Y. Yamada, M. Enomoto, M. Aodai and H. Kumakura: Physica C **392-396** (2003) 1030.
13. S. X. Dou, A. V. Pan, S. Zhou, M. Ionescu, H. K. Liu and P. R. Munroe: Supercond. Sci. Technol. **15** (2002) 1587.
14. H. Yamada, M. Hirakawa, H. Kumakura and H. Kitaguchi: Supercond. Sci. Technol. **19** (2006) 175.
15. H. Kumakura, H. Kitaguchi, A. Matsumoto and H. Yamada: Supercond. Sci. Technol. **18** (2005) 1042.
16. A. Yamamoto, J. Shimoyama, S. Ueda, Y. Katsura, I. Iwayama, S. Horii and K. Kishio: Appl. Phys. Lett. **86** 212502 (2005).
17. W. K. Yeoh, J. H. Kim, J. Horvat, S. X. Dou and P. Munroe: Supercond. Sci. Technol. **19** (2006) L5.
18. A. Yamamoto J. Shimoyama, S. Ueda, I. Iwayama, S. Horii and K. Kishio: Supercond. Sci. Technol. **18** (2005) 1323.

19. H. Yamada, M. Hirakawa, H. Kumakura and H. Kitaguchi: Supercond. Sci. Technol. **19** (2006) 175.
20. A. Yamamoto, J. Shimoyama, S. Ueda, Y. Katsura, S. Horii and K. Kishio: Supercond. Sci. Technol. **18** (2005) 116.
21. A. Yamamoto, J. Shimoyama, S. Ueda, I. Iwakuma, Y. Katsura, S. Horii and K. Kishio: J. Cryo. Soc. Jpn. **40** (2005) 466 [in Japanese].
22. M. Kiuchi, K. Kimura, T. Matsushita, A. Yamamoto, J. Shimoyama and K. Kishio: to be published.
23. Y. Zhao, M. Ionescu, J. Horvat and S. X. Dou: Supercond. Sci. Technol. **17** (2004) S482.
24. W. N. Kang, E. M. Choi, H. J. Kim, H. J. Kim and S. I. Lee: Physica C **385** (2003) 24.
25. A. Gurevich, S. Patnaik, V. Braccini, K. H. Kim, C. Mielke, X. Song, L. D. Cooley, S. D. Bu, D. M. Kim, J. H. Choi, L. J. Belenky, J. Giencke, M. K. Lee, W. Tian, X. Q. Pan, A. Siri, E. E. Hellstrom, C. B. Eom and D. C. Larbalestier: Supercond. Sci. Technol. **17** (2004) 278.
26. V. Braccini, A. Gurevich, J. E. Giencke, M. C. Jewell, C. B. Eom, D. C. Larbalestier, A. Pogrebnyakov, Y. Cui, B. T. Liu, Y. F. Hu, J. M. Redwing, Qi Li, X. X. Xi, R. K. Singh, R. Gandikota, J. Kim, B. Wilkens, N. Newman, J. Rowell, B. Moeckly, V. Ferrando, C. Tarantini, D. Marre, M. Putti, C. Ferdeghini, R. Vaglio and E. Haanappel: Phys. Rev. B **71** (2005) 012504.

A

Appendix

A.1 Description of Equilibrium State

Here we shall derive the force balance equation on flux lines in a usual transverse magnetic field after the method of Josephson [1] and also discuss the problem in a longitudinal magnetic field.

It is assumed that the magnetic flux density is slightly varied by $\delta\boldsymbol{B}$ in a superconductor, and the corresponding variation in the vector potential is represented by $\delta\boldsymbol{A}$. Thus, we have the relationship $\nabla \times \delta\boldsymbol{A} = \delta\boldsymbol{B}$. The corresponding displacement of flux lines is denoted by $\delta\boldsymbol{u}$. Josephson postulated that $\delta\boldsymbol{A}$ could be written as

$$\delta\boldsymbol{A} = \delta\boldsymbol{u} \times \boldsymbol{B} \tag{A.1}$$

by choosing a suitable gauge. It should be noted that this equation reduces to Eq. (2.17) by differentiating with respect to time. In fact, Eq. (2.17) is satisfied in the transverse field geometry where the magnetic field and the current are perpendicular to each other and there is no problem in the above postulation. However, the postulation of Eq. (A.1) is questionable in the longitudinal field geometry where the magnetic field is parallel to the current, since Eq. (2.17) is not satisfied. This point will be discussed later.

Firstly, Josephson treated an ideal pin-free superconductor. The equilibrium state in this case can be expressed as that in which the work done by an external source is equal to the variation in the free energy inside the superconductor. This is identical to the description that there is no energy loss. If the current density of the external source is represented by $\boldsymbol{J}_\mathrm{e}$, the work done by this current during the variation is given by

$$\delta W = \int \boldsymbol{J}_\mathrm{e} \cdot \delta\boldsymbol{A}\mathrm{d}V\;, \tag{A.2}$$

where the integral is taken over the entire space. On the other hand, the variation in the free energy is written as

436 A Appendix

$$\delta F = \int \boldsymbol{H} \cdot \delta \boldsymbol{B} \mathrm{d}V = \int_{\mathrm{out}} \nabla \times \boldsymbol{H} \cdot \delta \boldsymbol{A} \mathrm{d}V + \int_{\mathrm{in}} \nabla \times \boldsymbol{H} \cdot \delta \boldsymbol{A} \mathrm{d}V . \tag{A.3}$$

In the above a partial integration is done and the surface integral is disregarded. The first and second integrals are taken outside and inside the superconductor, respectively. The condition of $\delta W = \delta F$ is satisfied in the equilibrium state. If we note here that $\nabla \times \boldsymbol{H} = \boldsymbol{J}_{\mathrm{e}}$ outside the superconductor and $\boldsymbol{J}_{\mathrm{e}} = 0$ and $\nabla \times \boldsymbol{H} = \boldsymbol{J}$ inside the superconductor, the above condition reduces to

$$-\int_{\mathrm{in}} \boldsymbol{J} \cdot \delta \boldsymbol{A} \mathrm{d}V = 0 . \tag{A.4}$$

The left hand side is the energy loss in the superconductor. If Eq. (A.1) is used for $\delta \boldsymbol{A}$, Eq. (A.4) is written as

$$-\int_{\mathrm{in}} \boldsymbol{J} \cdot (\delta \boldsymbol{u} \times \boldsymbol{B}) \mathrm{d}V = \int_{\mathrm{in}} \delta \boldsymbol{u} \cdot (\boldsymbol{J} \times \boldsymbol{B}) \mathrm{d}V = 0 . \tag{A.5}$$

Since this is fulfilled for arbitrary $\delta \boldsymbol{u}$, the equation describing the equilibrium state in the pin-free superconductor is obtained as

$$\boldsymbol{J} \times \boldsymbol{B} = 0 . \tag{A.6}$$

In the case where the superconductor contains pins, the equilibrium condition is given by

$$\delta W = \delta F + \delta W_{\mathrm{p}} , \tag{A.7}$$

where δW_{p} is the pinning energy loss. If the pinning force density is denoted by $\boldsymbol{F}_{\mathrm{p}}$, the pinning energy loss density is given by $-\boldsymbol{F}_{\mathrm{p}} \cdot \delta \boldsymbol{u}$. Thus, the equilibrium condition reduces to

$$\int_{\mathrm{in}} \delta \boldsymbol{u} \cdot (\boldsymbol{J} \times \boldsymbol{B} - \boldsymbol{F}_{\mathrm{p}}) \mathrm{d}V = 0 . \tag{A.8}$$

Since $\delta \boldsymbol{u}$ is arbitrary, the force balance equation is obtained:

$$\boldsymbol{J} \times \boldsymbol{B} - \boldsymbol{F}_{\mathrm{p}} = 0 . \tag{A.9}$$

When the magnetic field and the current are parallel to each other, Eq. (A.6) suggests that a force-free current parallel to the flux lines can flow stably even in a pin-free superconductor. However, it should be noted that Eqs. (2.17) and (A.1) are not satisfied in the above geometry. Thus, we have

$$\delta \boldsymbol{A} = \delta \boldsymbol{u} \times \boldsymbol{B} + \nabla \Xi \tag{A.10}$$

from Eq. (4.48). In the above $\dot{\Xi} = \Psi$. Hence, the equilibrium condition for the pin-free superconductor is given by

$$-\int_{\mathrm{in}} \boldsymbol{J} \cdot (\delta \boldsymbol{u} \times \boldsymbol{B} + \nabla \Xi) \mathrm{d}V = 0 . \tag{A.11}$$

Ξ is generally a function of $\delta\boldsymbol{u}$, and hence, we have [2]

$$\boldsymbol{J} = 0 \tag{A.12}$$

so that Eq. (A.11) is satisfied for arbitrary $\delta\boldsymbol{u}$.

The case of simple rotation of flux lines treated in Exercise 4.3 is considered here for an example. When $\boldsymbol{J} \neq 0$, the kernel in the integral in Eq. (A.11) leads to $(B^2\delta\theta/\mu_0)\sin\theta$ with $\delta\theta$ denoting the variation in the surface angle of magnetic field and it can be seen that Eq. (A.11) is not satisfied. This equation is satisfied only when $\alpha_f = 0$, i.e., when Eq. (A.12) is satisfied. Hence, it can be concluded that the force-free state in a longitudinal magnetic field cannot stably exist without the flux pinning effect.

In the cases of the usual transverse magnetic field or a tilted magnetic field where $\boldsymbol{J}$ and $\boldsymbol{B}$ are not parallel to each other, Eq. (A.6) is identical with Eq. (A.12). Therefore, the equation which generally describes the equilibrium state in a pin-free superconductor is only Eq. (A.12).

A.2 Magnetic Properties of a Small Superconductor

The magnetic properties are briefly discussed for a small superconductor, the size of which is comparable to or only slightly larger than the penetration depth λ. For simplicity the case is treated where a magnetic field H_e is applied parallel to a sufficiently wide superconducting slab of thickness $2d$ ($-d \leq x \leq d$). In the beginning a type I superconductor is assumed. The magnetic flux density in the superconductor is

$$B(x) = \mu_0 H_e \frac{\cosh(x/\lambda)}{\cosh(d/\lambda)} . \tag{A.13}$$

Thus, the average magnetic flux density is

$$\langle B \rangle = \frac{\mu_0 H_e \lambda}{d} \tanh\left(\frac{d}{\lambda}\right) \equiv a\mu_0 H_e \tag{A.14}$$

and the average magnetic energy density is

$$\frac{1}{2\mu_0}\langle B^2 \rangle = \frac{1}{4}\mu_0 H_e^2 \left[\frac{1}{\cosh^2(d/\lambda)} + \frac{\lambda}{d}\tanh\left(\frac{d}{\lambda}\right)\right] . \tag{A.15}$$

Thus, the Gibbs free energy density in the superconducting state is

$$G_s(H_e) = F_n(0) - \frac{1}{2}\mu_0 H_c^2 + \frac{1}{4}\mu_0 H_e^2 \left[\frac{1}{\cosh^2(d/\lambda)} - \frac{3\lambda}{d}\tanh\left(\frac{d}{\lambda}\right)\right] . \tag{A.16}$$

On the other hand, the free energy density in the normal state is given by Eq. (1.26). Hence, the critical field of the small superconductor H_c^* is obtained from the condition $G_s(H_c^*) = G_n(H_c^*)$ as

$$H_c^* = \left[1 + \frac{1}{2\cosh^2(d/\lambda)} - \frac{3\lambda}{2d}\tanh\left(\frac{d}{\lambda}\right)\right]^{-1/2} H_c . \qquad (A.17)$$

It can be seen that the critical field is enhanced from the bulk value. Especially in the case of thin limit Eq. (A.17) leads to

$$H_c^* = \left(\frac{15}{2}\right)^{1/2} \left(\frac{\lambda}{d}\right)^2 H_c . \qquad (A.18)$$

Figure A.1 shows the thickness dependence of the critical field in the direction parallel to the surface of thin films of Sn [3]. It can be seen that the dependence is well explained by Eq. (A.17).

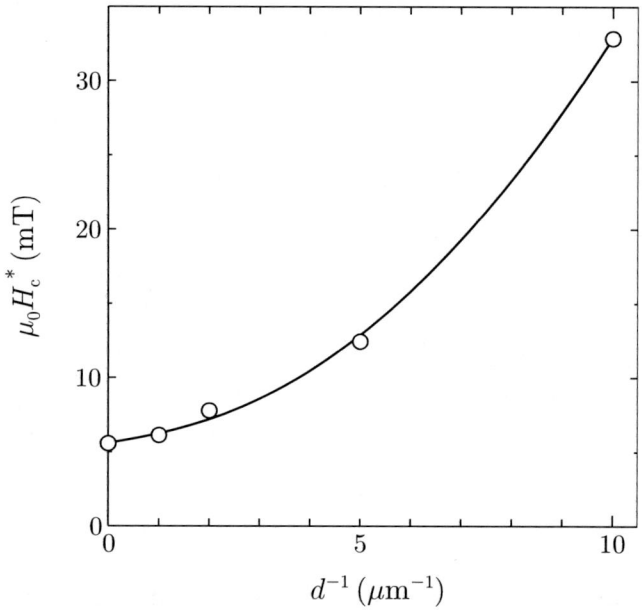

Fig. A.1. Thickness dependence of the critical field in the direction parallel to the surface of thin films of Sn [3] at $(T/T_c)^2 = 0.8$. The *solid line* represents the theoretical prediction of Eq. (A.17) with the assumption of $\lambda = 132$ nm

Secondly the lower critical field is investigated for a type II superconductor. The magnetic flux has already penetrated the superconductor before the quantized flux lines penetrate. Hence, the variation in the average magnetic flux density due to the penetration of flux lines is smaller than in the case of a bulk superconductor. Presuppose a certain region of the cross-sectional area S in which only one flux line exists on average. In this case the magnetic flux which has already penetrated this region is approximately $aS\mu_0 H_e$ in terms of a defined in Eq. (A.14). Since the magnetic flux after the penetration of the flux line is comparable to ϕ_0 from the definition of S, $S\mu_0 H_e \simeq \phi_0$ is obtained. Thus, the increment of the magnetic flux due to the penetration of the flux line is approximately equal to $(1-a)\phi_0$. If the thickness of the superconducting slab is not smaller than λ, the energy of the flux line per unit length ϵ may not be appreciably different from the bulk value. Repeating a similar discussion to that in Subsect. 1.5.2, an approximate expression is obtained for the lower critical field of the small superconductor:

$$H_{c1}^* \simeq (1-a)^{-1} H_{c1} = \left[1 - \frac{\lambda}{d}\tanh\left(\frac{d}{\lambda}\right)\right]^{-1} H_{c1}, \quad (A.19)$$

where H_{c1} is the lower critical field of the corresponding bulk superconductor. This result indicates that H_{c1}^* is proportional to d^{-2} when d is small. Numerical analysis [4] supports this expectation and shows that this proportional relationship holds correct even where the dimensions of d are much smaller than λ.

A.3 Minimization of Energy Dissipation

For simplicity a bulk superconducting slab ($0 \leq x \leq 2d$) is assumed. It is also assumed that the maximum critical current density determined by the flux pinning mechanism is a constant value denoted by J_c and that the practical critical current density is given by νJ_c ($0 < \nu \leq 1$). In fact, it can occur that the current density takes a value smaller than J_c depending on the microscopic arrangement of flux lines, as argued in Sect. 3.7. When the magnetic field applied parallel to the superconducting slab is increased from 0 to H_e in the initial state, the energy loss density inside the superconducting slab is easily calculated as

$$W = \frac{\mu_0}{2\nu J_c d} \int_0^{H_e} H^2 dH = \frac{\mu_0 H_e^3}{6\nu J_c d}, \quad (A.20)$$

where H_e is assumed to be sufficiently low that the penetration depth of the external magnetic field $H_e/\nu J_c$ is shorter than d. Thus, $\nu = 1$ is obtained from the condition of minimum energy loss density. This result agrees with the hypothesis in the critical state model that the flux pinning interaction prevents variation in the flux distribution as strongly as possible. In the above $\nu = 0$ is a singular point and corresponds to an unrealistic situation where there is no energy dissipation. Hence, this case is avoided.

After the magnetic distribution given by $\nu = 1$ is established, when the external magnetic field is further increased resulting in flux motion in the same direction as before, the condition $\nu = 1$ is clearly maintained. This is because, if ν becomes smaller during a slight variation in the external magnetic field, the flux penetration occurs as shown in Fig. A.2, which causes a large energy loss. When the external magnetic field is changed in the opposite direction

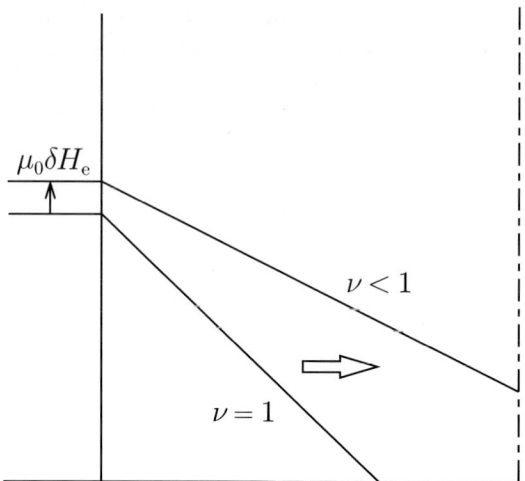

Fig. A.2. Penetration of magnetic flux into superconducting slab

from before, e.g., a variation from increasing field to decreasing field, a similar argument can be repeated to that in the initial state.

As shown above the hypothesis in the critical state model is the same as the condition of minimum energy dissipation. This is similar to the principle of minimum energy dissipation in linear dissipative processes.

A.4 Partition of Pinning Energy

Consider the case where the flux lines move translationally along the y-axis and simultaneously rotate in the plane normal to the y-axis to the direction represented by the angle θ under the influence of the flux pinning interaction. It is assumed that the flux lines are in the equilibrium state at (y_e, θ_e) in the coordinate system (y, θ). When the pinning potential is expanded around this equilibrium state, only terms of even order remain from the condition of symmetry. If only the terms of the lowest order are taken into account, the pinning potential energy density may be written as

$$U = \frac{a}{2}(y - y_e)^2 + \frac{b}{2}(\theta - \theta_e)^2 , \qquad (A.21)$$

where a and b are coefficients. It is assumed that the critical state is attained when U reaches a threshold value U_p. If the coordinates of the flux lines in the critical state are represented by (y_c, θ_c), the threshold value is given by

$$U_p = \frac{a}{2}(y_c - y_e)^2 + \frac{b}{2}(\theta_c - \theta_e)^2 . \qquad (A.22)$$

The position where only the critical balance between the Lorentz force and the pinning force occurs as in a transverse magnetic field is represented by $(y_{\mathrm{cm}}, \theta_{\mathrm{e}})$, and the position in the critical force-free state where only the balance between the force-free torque and the moment of pinning forces occurs is represented by $(y_{\mathrm{e}}, \theta_{\mathrm{cm}})$. Then, we have

$$|y_{\mathrm{cm}} - y_{\mathrm{e}}| = \left(\frac{2U_{\mathrm{p}}}{a}\right)^{1/2} = \frac{F_{\mathrm{p}}}{a}, \qquad (\text{A.23a})$$

$$|\theta_{\mathrm{cm}} - \theta_{\mathrm{e}}| = \left(\frac{2U_{\mathrm{p}}}{b}\right)^{1/2} = \frac{\Omega_{\mathrm{p}}}{b}. \qquad (\text{A.23b})$$

It is possible to define

$$|y_{\mathrm{c}} - y_{\mathrm{e}}| = |y_{\mathrm{cm}} - y_{\mathrm{e}}|\sin\psi, \qquad (\text{A.24a})$$

$$|\theta_{\mathrm{c}} - \theta_{\mathrm{e}}| = |\theta_{\mathrm{cm}} - \theta_{\mathrm{e}}|\cos\psi \qquad (\text{A.24b})$$

in terms of a new variable ψ. Using these expressions, the pinning force density and the density of moment of the pinning forces are respectively described as

$$a|y_{\mathrm{c}} - y_{\mathrm{e}}| = F_{\mathrm{p}}\sin\psi, \qquad (\text{A.25a})$$

$$b|\theta_{\mathrm{c}} - \theta_{\mathrm{e}}| = \Omega_{\mathrm{p}}\cos\psi. \qquad (\text{A.25b})$$

Thus, if the conditions of the balance are written in the forms of Eqs. (4.62a) and (4.62b), we have

$$f = \sin\psi, \qquad g = \cos\psi. \qquad (\text{A.26})$$

A.5 Comments on the Nonlocal Theory of the Elasticity of the Flux Line Lattice

It is considered in the nonlocal theory for the elastic moduli of the flux line lattice that the normal core of a flux line surrounded by the "magnetic flux" can be fairly freely displaced. This leads to small C_{11} and C_{44}, which are originally associated with the magnetic energy. However, it was mentioned at the end of Sect. 7.2 that such a lack of influence between the internal normal core and the surrounding magnetic flux seems to be incompatible with the requirement of gauge-invariance between the order parameter and the magnetic field. This problem is discussed here.

First a deformation of the flux line lattice associated with C_{11} is treated. An arrangement of flux lines parallel to the z-axis and forming a triangular lattice is assumed as shown in Fig. A.3. The flux line spacing is denoted by a_{f}

and the interval between the rows of flux lines along the y-axis is denoted by b_f. This leads to $b_f = (\sqrt{3}/2)a_f$. One of the results that follow from the gauge-invariance is the quantization of magnetic flux. If the magnetic flux density when the lattice is not deformed and its mean value in space are denoted by B_0 and $\langle B_0 \rangle$, respectively, the quantization condition is expressed as

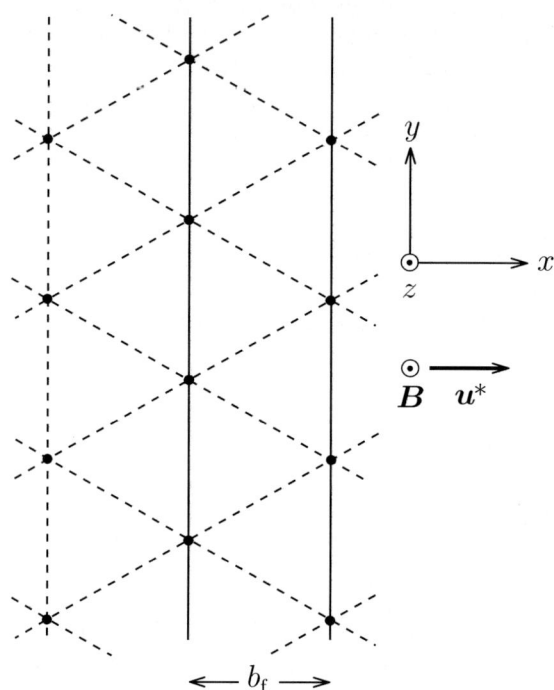

Fig. A.3. Triangular lattice of flux lines

$$\langle B_0 \rangle = \frac{\phi_0}{a_f b_f} . \tag{A.27}$$

When the flux line lattice is deformed and the zero points of the order parameter are slightly displaced by $u^*(x)$ along the x-axis, the new interval of the rows of flux lines is given by

$$b'_f = b_f + \frac{\partial u^*}{\partial x} b_f . \tag{A.28}$$

The magnetic flux density also changes due to the deformation, and the quantization condition is now written as

$$\langle B_0 \rangle + \delta B = \frac{\phi_0}{a_f b'_f} . \tag{A.29}$$

A.5 Comments on the Nonlocal Theory of the Elasticity

Hence, the variation in the magnetic flux density should obey the relationship:

$$\delta B = \frac{\phi_0}{a_f b'_f} - \frac{\phi_0}{a_f b_f} \simeq -\langle B_0 \rangle \frac{\partial u^*}{\partial x} . \tag{A.30}$$

Here it is necessary to define the region in which the magnetic flux is quantized. A candidate may be the region surrounded by two straight lines along the y-axis connecting maximum points of the magnetic flux density (the region between two solid lines in Fig. A.3)[5]. Although the current density is not zero on these straight lines, the direction of the current is perpendicular to the lines and the quantization inside this region can be derived. (This comes from the contribution from the zero points of the order parameter. See Exercise 1.8.)

On the other hand, the variation in the magnetic flux density is deeply connected to the displacement of the "magnetic flux," which is the same as the magnetic field line in electromagnetism. This displacement $u(x)$ can be obtained from the continuity equation for flux lines, Eq. (2.15). In the case of the small variation treated here, this equation leads to

$$\delta B = -\langle B_0 \rangle \frac{\partial u}{\partial x} . \tag{A.31}$$

Comparing Eqs. (A.30) and (A.31), we have

$$u = u^* . \tag{A.32}$$

Namely, the displacement of the "magnetic flux" is identical with that of the structure of the order parameter. This can be easily anticipated from Eq. (1.101), suggesting that the maximum point of the magnetic flux density coincides with the zero point of the order parameter. This means that when the normal core is deformed, the same deformation of the "magnetic flux" occurs too, and hence, the result of the local theory is derived from the corresponding magnetic energy. Therefore, softening of the flux line lattice does not occur for C_{11}.

The above result holds correct for arbitrary $u(x)$. However, the theoretical treatment using Eqs. (A.28) and (A.30) is allowable for a gentle variation, and hence, there is a limit on the wave number k. For a superconductor with a high κ value, however, it is possible to show that the above result is applicable up to a wave number sufficiently larger than the characteristic one k_h, above which the nonlocal property is predicted to be significant. That is, if the wave length of the periodic deformation at $k = k_h$ is denoted by Λ, we have $\Lambda/b_f = 2\pi/k_h b_f = (4\pi/\sqrt{3})^{1/2}[b/(1-b)]^{1/2}\kappa$, where $b = B/\mu_0 H_{c2}$ is the reduced field. It can be seen that the above value is very large. Hence, the above perturbational method can be used even for $k = k_h$. Thus, it is concluded that the elastic modulus C_{11} dose not show the nonlocal property.

In the above, it was shown from the condition of quantization of magnetic flux that C_{11} takes the local value. On the other hand, it is argued in Exercise

7.2 that a nonlocal C_{11} is derived from an assumption similar to that of Brandt [6] and that the quantization of magnetic flux is not fulfilled under this assumption.

Secondly the bending deformation of the flux line lattice associated with C_{44} is considered. It is assumed initially that the flux lines are uniformly distributed and lie in the direction of the z-axis with a magnetic flux density $B_0(\boldsymbol{r})$, where $\boldsymbol{r}$ is a vector in the x-y plane. We assume that the zero points of the order parameter in the flux line lattice are slightly displaced by $u^*(z)$ along the x-axis. Then, the displacement of the "magnetic flux" outside the normal core occurs along the x-axis and an x-component of the magnetic flux density appears consequently. These are denoted by $u(z)$ and $\delta B(z)$, respectively. These quantities are correlated to each other through the continuity equation for flux lines as

$$\delta B = \langle B_0 \rangle \frac{\partial u}{\partial z} \,. \tag{A.33}$$

On the other hand, since the maximum points of the magnetic flux density and the zero points of the order parameter are identical with each other, the condition

$$\frac{B_x}{B_z} \simeq \frac{\delta B}{\langle B_0 \rangle} = \frac{\partial u^*}{\partial z} \tag{A.34}$$

should be satisfied along a line passing through the maximum points of the magnetic flux density. Comparison of Eqs. (A.33) and (A.34) leads to

$$u = u^* \,. \tag{A.35}$$

Thus, the result of the local theory is also derived for C_{44}. In fact, the Lorentz force can be written as

$$\boldsymbol{F}_{\mathrm{L}} = \frac{1}{\mu_0}(\boldsymbol{B}\cdot\nabla)\boldsymbol{B} \simeq \frac{\langle B_0 \rangle}{\mu_0} \cdot \frac{\partial}{\partial z}\delta B \boldsymbol{i}_x = \frac{\langle B_0 \rangle^2}{\mu_0} \cdot \frac{\partial^2 u^*}{\partial z^2} \boldsymbol{i}_x \,, \tag{A.36}$$

and this is required to be identical with the elastic force, $C_{44}\partial^2 u^*/\partial z^2$. Hence, we have again

$$C_{44} = \frac{\langle B_0 \rangle^2}{\mu_0} \,. \tag{A.37}$$

The above deformation is written as $\boldsymbol{u}^*(\boldsymbol{r}) = u^*(z)\boldsymbol{i}_x$. Hence, it satisfies $\nabla \cdot \boldsymbol{u}^* = 0$, and Eq. (22) in the paper by Brandt [7] is written as

$$\boldsymbol{B} = \langle B_0 \rangle \left\{ 1 + \frac{1}{2b\kappa^2}[\langle \omega \rangle - \omega(\boldsymbol{r} - \boldsymbol{u}^*)] \right\} \left[\boldsymbol{i}_z + a(k)\frac{\partial \boldsymbol{u}^*}{\partial z} \right] \,, \tag{A.38}$$

where $b = \langle B_0 \rangle / \mu_0 H_{c2}$, $\omega = |\Psi|^2/|\Psi_\infty|^2$ and $a(k) = k_{\mathrm{h}}^2/(k^2 + k_{\mathrm{h}}^2)$. A small correction was carried out taking account of the deformation of the flux line lattice with a nonuniform magnetic flux density due to the spatial variation in $|\Psi|^2$. Equation (A.38) leads to

A.5 Comments on the Nonlocal Theory of the Elasticity

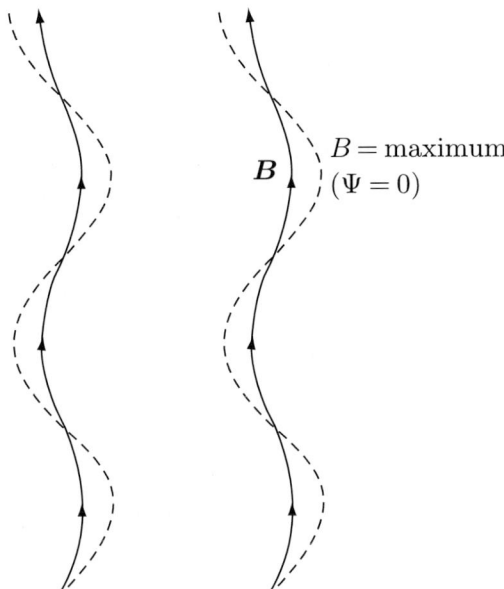

Fig. A.4. Structure of flux lines assumed for derivation of C_{44} in the nonlocal theory. *Solid lines* show "magnetic flux lines" and *broken lines* pass through the maximum points of B. These lines do not coincide with each other

$$\frac{B_x}{B_z} = a(k)\frac{\partial u^*}{\partial z} \neq \frac{\partial u^*}{\partial z} \tag{A.39}$$

on the line through the zero points of $|\Psi|^2$, i.e., the maximum points of the magnetic flux density. This means that the streamline of $\boldsymbol{B}$, i.e. the "magnetic flux line" in electromagnetism, and the line connecting the maximum points of B are not the same (see Fig. A.4). This means that the requirement of $\nabla \cdot \boldsymbol{B} = 0$ is not satisfied [8] according to the theoretical result of Brandt. For the above deformation simply we have $\partial \omega/\partial z = -(\partial u^*/\partial z)\cdot(\partial \omega/\partial x)$. This leads to

$$\nabla \cdot \boldsymbol{B} = -\frac{\langle B_0 \rangle}{2b\kappa^2}\left[\frac{\partial \omega}{\partial z} + a(k)\frac{\partial u^*}{\partial z}\cdot\frac{\partial \omega}{\partial x}\right]$$
$$= \frac{\langle B_0 \rangle}{2b\kappa^2}[1 - a(k)]\frac{\partial u^*}{\partial z}\cdot\frac{\partial \omega}{\partial x}. \tag{A.40}$$

Hence, $a(k)$ should be equal to 1, i.e., the local theory should hold correct so that $\nabla \cdot \boldsymbol{B} = 0$ is satisfied. It is true that any solutions of the London equation satisfy $\nabla \cdot \boldsymbol{B} = 0$. However, Eq. (A.38) is not a good approximate solution.

Brandt [9] derived the nonlocal elastic moduli which depends on the wave number of the distortion using the energy given by Eq. (1.112). It is a general method to directly substitute $\boldsymbol{A}$ and Ψ for a distorted flux line lattice into Eq. (1.21) in order to obtain the energy associated with the distortion of the flux line lattice. Brandt did not do so but started from the simpler Eq. (1.112). To clarify the problems in such a theoretical treatment we shall here show the result from the correct treatment of the same matter.

The expression of the energy which Brandt used is Eq. (1.21) with substitution of Eq. (1.30). It should be noted that $\boldsymbol{A}$ and Ψ corresponding to the distorted flux line lattice do not satisfy Eq. (1.30). This equation is derived so as to minimize the free energy, and hence, so that the distortion is zero. Therefore, it is necessary to use an equation which distorted $\boldsymbol{A}$ and Ψ satisfy to rewrite Eq. (1.21). It is considered that these variables are determined so that the total energy is minimized which includes an additional energy term, i.e., the pinning energy required to distort the flux line lattice. We assume that the pinning energy is of the form $U_\mathrm{p} = \widetilde{U}_\mathrm{p}(\boldsymbol{r})|\Psi|^2$ after Miyahara et al. [10]. Namely, the effect of the pinning centers is likely to appear in the coefficient α in Eq. (1.21). Even in this case Eq. (1.31) is not influenced by the pinning centers. Thus, the equation which Ψ and $\boldsymbol{A}$ satisfy is

$$\frac{1}{2m^*}(-i\hbar\nabla + 2e\boldsymbol{A})^2\Psi + (\alpha + \widetilde{U}_\mathrm{p})\Psi + \beta|\Psi|^2\Psi = 0 \,. \tag{A.41}$$

Substituting this into Eq. (1.21) and taking an average in space, we have

$$\langle F_\mathrm{s} \rangle = \left\langle \frac{B^2}{2\mu_0} - \frac{\mu_0 H_\mathrm{c}^2|\Psi|^4}{2|\Psi_\infty|^4} - U_\mathrm{p} \right\rangle \tag{A.42}$$

after a simple calculation. The third term, the pinning energy is originally introduced to distort the flux line lattice. In order to estimate the energy of distortion the energy necessary to introduce the distortion itself should be omitted. Hence, the energy of the distorted flux line lattice is given by

$$\langle F_\mathrm{s} \rangle = \left\langle \frac{B^2}{2\mu_0} - \frac{\mu_0 H_\mathrm{c}^2|\Psi|^4}{2|\Psi_\infty|^4} \right\rangle \,. \tag{A.43}$$

This gives the same result as Eq. (1.111). It is necessary to use Eq. (1.101) to derive Eq. (1.112) from this equation as Brandt did. However, it is shown in Exercise 7.2 that Eq. (1.101) does not satisfy the condition of flux quantization for a deformed flux line lattice.

The uniaxial compression modulus is derived as

$$C_{11} = \frac{\partial \langle F_\mathrm{s}\rangle}{\partial \beta_\mathrm{A}} \cdot \frac{\partial^2 \beta_\mathrm{A}}{\partial \epsilon_k^2} + \frac{\partial \langle F_\mathrm{s}\rangle}{\partial \beta_\mathrm{m}} \cdot \frac{\partial^2 \beta_\mathrm{m}}{\partial \epsilon_k^2} \tag{A.44}$$

from Eq. (A.43), where ϵ_k is the root mean square value of the distortion of the wave number k and

$$\beta_\mathrm{m} = \frac{\langle B^2\rangle}{\langle B\rangle^2} \simeq \frac{\langle B^2\rangle}{\langle B_0\rangle^2} \,. \tag{A.45}$$

Writing $B = B_0 + \delta B$, we have $\langle B^2\rangle \simeq \langle B_0^2\rangle + \langle \delta B^2\rangle$. If a distortion of

A.5 Comments on the Nonlocal Theory of the Elasticity

$$u(x) = u_{km}\sin(kx) \tag{A.46}$$

with the wave number k is assumed for $u(x)$ in the continuity equation for flux lines, Eq. (A.31), we have

$$\langle \delta B^2 \rangle = \langle B_0 \rangle^2 \frac{k^2 u_{km}^2}{2} = \langle B_0 \rangle^2 \epsilon_k^2 , \tag{A.47}$$

where $\epsilon_k = (\partial u/\partial x)_{\max}/\sqrt{2} = k u_{km}/\sqrt{2}$. Thus, Eq. (A.44) leads to [11]

$$C_{11} = \frac{\partial F_s}{\partial \beta_A} \cdot \frac{\partial^2 \beta_A}{\partial \epsilon_k^2} + \frac{\langle B_0 \rangle^2}{\mu_0} . \tag{A.48}$$

This agrees with the expectation of Eq. (7.22). In the above the first term is a correction of the order of C_{66} to the magnetic interaction. Thus, the local result is obtained again. The local result is also obtained for C_{44} from a similar argument. In this case the variation in the magnetic flux density is very small and the variation in the order parameter is also very small. Hence, the contribution from the first term in Eq. (A.44) can be neglected.

Larkin and Ovchinnikov [12] obtained the same result as Brandt. This is ascribed to the assumption that the displacement of the normal core of a flux line and the variation in the vector potential are independent of each other. Correctly speaking, the term $(\mathrm{rot}\boldsymbol{A}_1)^2$ in Eq. (24) in their original paper corresponds to $\langle \delta B^2 \rangle$ in this book, and the correct local result is obtained from it. However, this term was neglected in their original paper, resulting in the incorrect moduli. The neglect of $(\mathrm{rot}\boldsymbol{A}_1)^2$ is equivalent to the use of Eq. (1.112) instead of Eq. (A.43). $(\mathrm{rot}\boldsymbol{A}_1)^2$ or $\langle \delta B^2 \rangle$ may seem to be safely neglected because it is a small term of the second order. However, this small term is important, since the increase in the elastic energy is proportional to the square of the small distortion.

The effect of nonlocality on the elastic moduli of a flux line lattice is here investigated from the experimental results of flux pinning. Figure A.5 shows the relationship between the contribution to the pinning force density from one pinning center (ordinate) vs. its elementary pinning force (abscissa) for various superconductors. In the figure, vanadium, niobium and Nb-Hf have the lowest, medium and highest κ values, respectively. The nonlocal theory was originally proposed to explain why the normal core of a flux line is easily deformed and pinned. It is derived from this theory that this softening is more effective for a superconductor with a shorter coherence length ξ, and hence a higher κ. That is, it may be expected that the pinning is more effective for a higher κ superconductor. However, Fig. A.5 shows that the pinning efficiency is worse and flux lines are not more easily pinned in a higher κ superconductor. This result does not support the nonlocal theory.

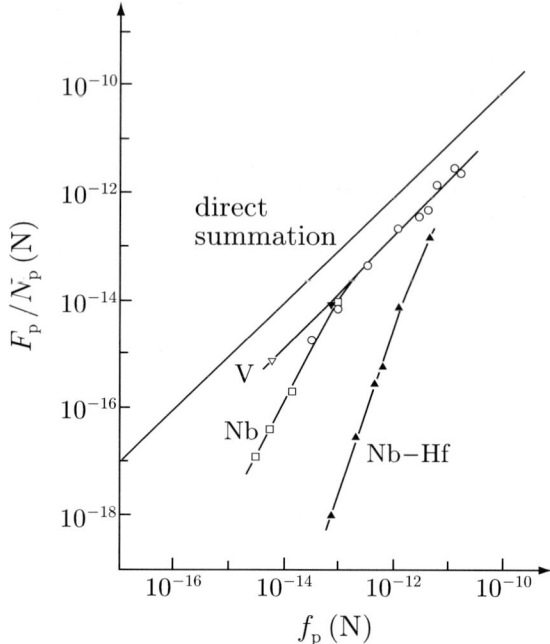

Fig. A.5. Contribution to the pinning force density from one pin F_p/N_p vs. the elementary pinning force f_p for various superconductors. The pinning efficiency is lower for a superconductor with a higher Ginzburg-Landau parameter κ

A.6 Avalanching Flux Flow Model

When the saturation phenomenon of the pinning force density occurs, the interaction distance d_i corresponding to the yielding strain decreases abruptly as the magnetic field or the elementary pinning force increases, as is shown in Fig. 7.32(b). This suggests that the flux line lattice has become brittle. Since d_i is not proportional to the flux line spacing a_f, the beginning of flux flow exceeding the critical state is not simply determined by the limit of the elastic deformation as with the shearing flow assumed in the Kramer model. It is rather considered that the flux line lattice yields before reaching the ideal elastic limit due to some cause resulting in the flux flow state. The cause which seems to be responsible for the yielding is the presence of defects in the flux line lattice. Namely, when the local distortion around a defect reaches a certain limit, a local plastic deformation will occur. Then, a new distortion will be created at the position where a vacancy is brought about by the motion of the flux line. Since the stress will be concentrated on this region, local plastic deformation will be induced again. Thus, the plastic deformation of the flux line lattice will be propagated throughout the specimen and the flux flow state will be the result (see Fig. 7.34).

A.6 Avalanching Flux Flow Model

In such a case the probability of occurrence of the initial local plastic deformation will be proportional to the number density n_d of defects in the flux line lattice, since it is considered that the local plastic deformation takes place around the defect. The nucleation of defects in the flux line lattice originates from flux pinning interactions. Hence, n_d will be expressed as an increasing function G of the elementary pinning force f_p and the number density of pins N_p. On the other hand, n_d is expected to be inversely proportional to the maximum shear stress of the flux line lattice $C_{66}a_f/4$, since the nucleation of defects is depressed when the shear stress is strong. Hence, we expect

$$n_d \propto \frac{G(f_p, N_p)}{C_{66}a_f/4} \; . \tag{A.49}$$

The probability of occurrence of the local plastic deformation increases with the displacement of flux lines u. As a result, it is speculated that the probability of occurrence of the avalanching flux flow is proportional to n_d and u. Here it seems to be reasonable to assume that the avalanching flux flow takes place when the product of n_d and u reaches a certain threshold value. If we note that $u = d_i$ in the critical state, we have

$$d_i \propto \frac{C_{66}a_f}{G(f_p, N_p)} \; . \tag{A.50}$$

On the other hand, the Labusch parameter α_L represents the strength of reaction from pins against the driving force. Hence, α_L is also given by a certain increasing function G' of f_p and N_p. It may be reasonable to assume that n_d is proportional to the number density of edge dislocations n_e in the flux line lattice. n_e is generally related to the current density J through

$$J = \frac{2a_f B n_e}{\mu_0} \tag{A.51}$$

and the pinning force density, $F = BJ$, is given by

$$F = \alpha_L u \; . \tag{A.52}$$

Thus, we have

$$\alpha_L \propto n_e \; . \tag{A.53}$$

This leads to the result that α_L is proportional to n_d. That is, G' is proportional to G and we have

$$\alpha_L \propto G(f_p, N_p) \; . \tag{A.54}$$

This allows us to derive the threshold value of the pinning force density:

$$F_p = \alpha_L d_i \propto C_{66}a_f \propto b^{1/2}(1-b)^2 \; . \tag{A.55}$$

The obtained pinning force density is independent of the pinning parameters, f_p and N_p, and its magnetic field dependence coincides with Kramer's formula given by Eq. (7.88). When the flux pinning is strengthened and G increase, α_L increases and d_i decreases obeying Eqs. (A.54) and (A.50), respectively. That is, the flux line lattice becomes harder but brittle, when the flux pinning is strengthened, as mentioned in Subsect. 7.5.4. These two effects cancel each other and result in a constant yielding stress, i.e., the saturation phenomenon. Since the flux pinning is strong enough even at the saturation phenomenon, a linear summation will hold correct for G. Thus, its magnetic field dependence will be given by

$$G(f_p, N_p) \propto f_p \propto 1 - b . \quad (A.56)$$

As a result, we have

$$\alpha_L \propto 1 - b, \qquad d_i \propto b^{-1/2}(1 - b) . \quad (A.57)$$

The magnetic field dependences of these quantities agree approximately with the experimental results in Fig. 7.32 for Nb-Ti.

When the pinning parameters are increased even more, the defects in the flux line lattice will be increased even more, and the flux line lattice is expected to be in an amorphous state. Both the reduction in the shear modulus C_{66} and the enhancement of α_L cause the reduction in the transverse elastic correlation length of the flux line lattice $(C_{66}/\alpha_L)^{1/2}$. Hence, even if a local plastic deformation occurs as shown in Fig. 7.34(a), it is considered that the surrounding strong pinning interactions prevent the deformation from developing to a catastrophic instability. In other words the brittle flux line lattice containing defects is stabilized by strong pins and can endure until the strain reaches the yielding value determined by the flux pinning strength. Thus, some scaling holds on the state of the flux line lattice, and it is expected that the interaction distance behaves normally as

$$d_i \propto a_f \propto b^{-1/2} \quad (A.58)$$

and increases from the saturation case as in Fig. A.6. In this case the pinning force density is

$$F_p \propto G(f_p, N_p) \propto b^{1/2}(1 - b) . \quad (A.59)$$

This shows that the pinning force density depends on the pinning parameters and is proportional to $(1-b)$ at high fields. That is, the nonsaturation phenomenon is explained. In fact, the interaction distance expressed by Eq. (A.58) is observed for Nb-Ti showing a nonsaturation characteristic [13] (see Fig. A.7). In the figure, d_i decreases in the vicinity of the upper critical field. This seems to be caused by flux creep, since the trend becomes clearer at higher temperatures.

The variation from saturation to nonsaturation with strengthening flux pinning and the brittle property of the flux line lattice at saturation, etc. can be qualitatively explained by the avalanching flux flow model. However, quantitative arguments are necessary for the value of the saturated pinning force density and the value of the elementary pinning force required for the transition from saturation to nonsaturation.

A.6 Avalanching Flux Flow Model 451

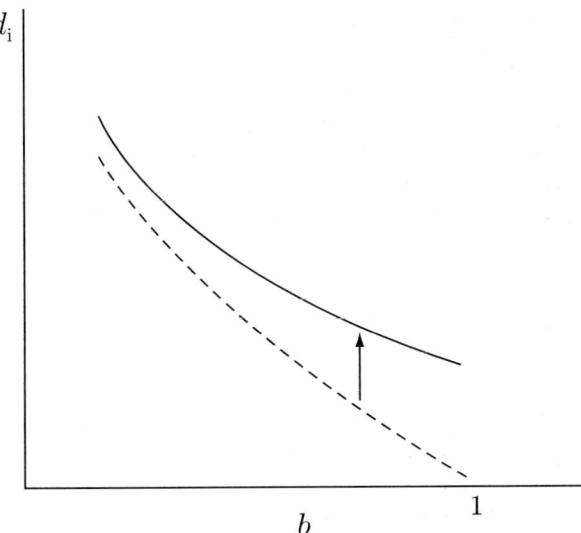

Fig. A.6. Proposed variation in the interaction distance when the flux line lattice is stabilized and nonsaturation is attained by strengthening the flux pinning. The *broken line* shows the interaction distance in the saturation case

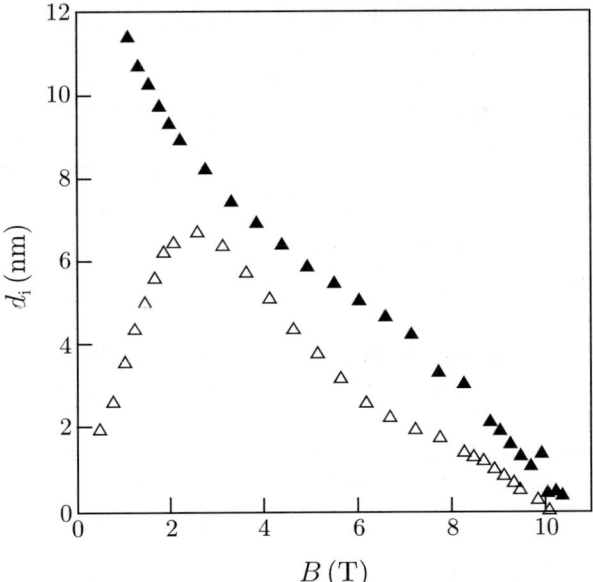

Fig. A.7. Variation in the interaction distance of Nb-Ti during the variation in the pinning force density from saturation ($\triangle$) to nonsaturation ($\blacktriangle$) [13] as shown in Fig. 7.31

A.7 Josephson Penetration Depth

Suppose a long Josephson junction and define the x-axis along the length with $x = 0$ denoting the edge of the junction. The magnetic field inside the junction is not uniform along the x-axis and is represented by h. $\partial h/\partial x$ is equal to the current density flowing across the junction, and Eq. (1.136) expressing the Josephson effect leads to

$$\frac{\partial h}{\partial x} = j_{\rm c} \sin \theta(x) \; . \tag{A.60}$$

In the above $\theta(x)$ is the phase difference given by

$$\theta(x) = \theta(0) + \frac{2\pi}{\phi_0}\Phi(x) \; , \tag{A.61}$$

where $\theta(0)$ is a phase at the edge and $\Phi(x)$ is the magnetic flux in the region of 0 to x given by

$$\Phi(x) = \mu_0 D \int_0^x h(x)\mathrm{d}x \tag{A.62}$$

with D denoting the effective thickness of the junction. Thus, we have

$$\frac{\partial \theta}{\partial x} = \frac{2\pi}{\phi_0}\mu_0 D h \; . \tag{A.63}$$

Hence, the following equation is derived:

$$\frac{\partial^2 \theta}{\partial x^2} = \frac{1}{\lambda_{\rm J}^2}\sin \theta \; , \tag{A.64}$$

where $\lambda_{\rm J}$ is a distance given by

$$\lambda_{\rm J} = \left(\frac{\phi_0}{2\pi\mu_0 j_{\rm c} D}\right)^{1/2} . \tag{A.65}$$

The above equation shows that the magnetic flux penetrates only up to about $\lambda_{\rm J}$ from the edge of the junction when θ is small. This length is called the Josephson penetration depth and is much longer than the London penetration depth.

A.8 On the Transverse Flux Bundle Size

In the collective flux creep theory the transverse flux bundle size is assumed to be equal to the transverse pinning correlation length R_0 given by Eq. (7.93). In the case of strong thermal activation, however, C_{66} is smaller than C_{66}^0 for the ideal perfect flux line lattice given by Eq. (7.12) and becomes almost zero in the vicinity of the melting line. Hence, it varies drastically depending on

A.8 On the Transverse Flux Bundle Size

the state of the flux lines and cannot be foreseen. Thus, it seems to be more effective to determine the transverse flux bundle size by a thermodynamic method.

It is assumed that the flux bundle size is determined so that the critical current density under the influence of the flux creep is maximized. This hypothesis is similar to the principle of minimum energy dissipation in linear dissipative systems.

If the second term in Eq. (3.115) is neglected, the critical current density under the flux creep is given as

$$J_c = J_{c0}\left[1 - \frac{k_B T}{U_0^*}\log\left(\frac{Ba_f\nu_0}{E_c}\right)\right] \quad (A.66)$$

using the expanded approximation of Eq. (3.117). The quantity most difficult to derive in this equation is the apparent pinning potential energy U_0^*. Here, we shall confine ourselves to the flux bundle size in the vicinity of the irreversibility line. In this region the current density is approximately zero, and hence, U_0^* approaches the true pinning potential energy U_0, as is shown in Fig. 3.44. Thus, the theoretical result on U_0 in Sect. 7.7 can be used.

The number of flux lines in the flux bundle is denoted by g^2. It is assumed that g reduces to

$$g = yg_e, \quad (A.67)$$

where g_e is the value of g for $C_{66} = C_{66}^0$ and y is a number smaller than 1. It is assumed here that the flux bundle size along its length dose not change and is given by L_0. This is because the tilt modulus of the flux line lattice C_{44} depends only on the magnetic energy and is not appreciably changed by defects in the flux line lattice. Thus, the correlated volume of the flux line lattice is y^2 times as large as the value when $g = g_e$. From the collective pinning mechanism the virtual critical current density J_{c0} in the creep-free case is y^{-1} times as large as its value J_{ce} when $g = g_e$. Equation (7.97) leads to the result that the pinning potential energy U_0 is $y^{3/2}$ times as large as its value U_e when $y = y_e$. Hence, Eq. (A.66) is written as

$$J_c = \frac{J_{ce}}{y}\left[1 - \frac{k_B T}{U_e y^{3/2}}\log\left(\frac{Ba_f\nu_0}{E_c}\right)\right]. \quad (A.68)$$

By differentiating this with respect to y, the condition of the maximum critical current density is obtained [14]:

$$y = \left[\frac{5k_B T}{2U_e}\log\left(\frac{Ba_f\nu_0}{E_c}\right)\right]^{2/3}. \quad (A.69)$$

Since U_e is proportional to $J_{ce}^{-1/2}$, it is predicted that U_e will become larger, and the decrease of g^2 from g_e^2 becomes more significant as the flux pinning strength is weakened.

Figure A.8 shows the relationship between g^2 and g_e^2 for various Bi-based superconductors in a magnetic field normal to the c-axis [14]. This shows that for a larger g_e^2 due to weaker pinning, the difference from g^2 becomes larger, indicating the same trend as the theoretical prediction.

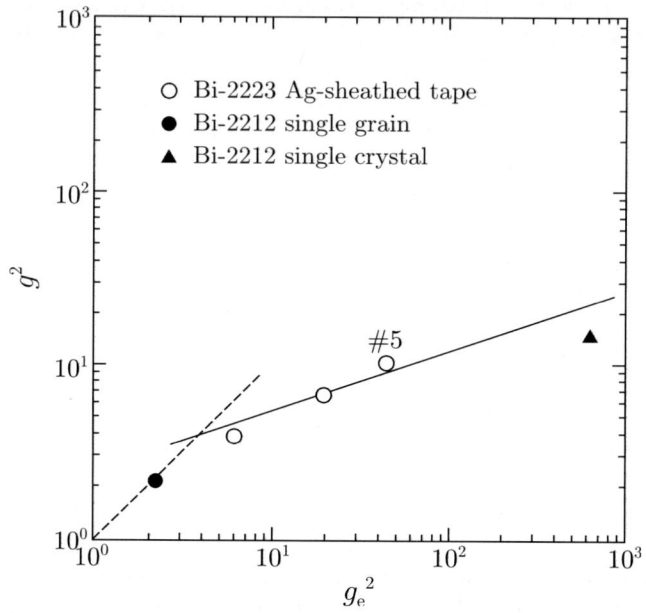

Fig. A.8. Relationship between g^2 and g_e^2 for Bi-based superconductors in a magnetic field normal to the c-axis [14]. The *solid straight line* represents the relationship of Eq. (A.70) for specimen 5 of silver-sheathed tape

The above result is discussed in more detail. g_e^2 is used instead of J_{c0} as a parameter representing the flux pinning strength. Exactly speaking, the value of g_e^2 cannot be obtained unless J_{ce} is known. Here, we approximately use J_{c0} instead of J_{ce} at low temperatures. Then, we have $J_{ce} \propto g_e^{-2}$ and $U_e \propto g_e$. Substitution of these relationships into Eqs. (A.67) and (A.69) leads to

$$g^2 \propto g_e^{2/3} \ . \tag{A.70}$$

The solid line in Fig. A.8 shows this relationship and qualitatively explains the experimental results.

Now the predicted g^2 is discussed quantitatively. The case is considered firstly where the magnetic field is applied normal to the c-axis. We treat for example the results on specimen 5 of Bi-2223 tape wire shown in Fig. A.8. The critical temperature of this specimen is 108.8 K and the parameters in Eq. (8.27) are $A = 4.22 \times 10^8$ Am^{-2}, $m' = 3.6$ and $\gamma = 0.50$. The irreversibility field at 80 K is $\mu_0 H_i = 3.0$ T, and g^2 estimated from this value using

Eqs. (8.27)–(8.29) for the bulk case is 10.3. The flux bundle size is calculated under these conditions. Point-like defects are proposed to be the dominant pinning centers in this temperature region, and hence, we use $\zeta = 2\pi$. If it is assumed again that J_{ce} can be approximately estimated by a value of the critical current density measured at low temperatures as mentioned above, $g_e^2 = 44.5$ is obtained from Eq. (7.93), where the values of $\mu_0 H_{c2}^{ab}(0) = 1000$ T and $\mu_0 H_c(0) = 1.0$ T were assumed. Substituting these values into Eq. (7.97), we have $U_e = 1.26 \times 10^{-19}$ J at 80 K and at 3.0 T. Substitution of typical values into the logarithmic term in Eq. (A.69) leads to $\log(Ba_f\nu_0/E_c) \simeq 14$, and $g^2 = 9.4$ is obtained. This theoretical estimation is close to the value of 10.3 calculated directly from the observed irreversibility field. The straight solid line in Fig. A.8 is an extrapolation of this theoretical value after Eq. (A.70).

Secondly, we shall discuss the case of the same specimen in the magnetic field parallel to the c-axis. The irreversibility field at 70 K is 1.30 T and the obtained pinning parameters are $A = 6.84 \times 10^7$ Am^{-2}, $m' = 3.3$ and $\gamma = 0.66$. Substituting $\log(Ba_f\nu_0/E_c) \simeq 12.5$ and $\zeta = 2\pi$ from the assumption that point-like defects are the dominant pinning centers, we obtain $g^2 = 2.1$. On the other hand, Eq. (7.93) leads to $g_e^2 = 806$, which is tremendously larger than g^2 estimated above. In the above $\mu_0 H_{c2}^c(0) = 50$ T and $\mu_0 H_c(0) = 1.0$ T were assumed. Thus, $U_e = 1.81 \times 10^{-18}$ J and finally $g^2 = 3.4$ are derived at the treated point on the irreversibility line. Although this value is somewhat larger than 2.1 estimated directly from the irreversibility field, it is significantly smaller than g_e^2. Judging from the ambiguity in the superconducting parameters such as $\mu_0 H_{c2}^c(0)$, it may be said that the agreement is rather good.

Therefore, the principle that the flux bundle size is determined so as to maximize the critical current density seems to hold correct for both field directions and to be general. Quantitatively speaking, the reduction in the flux bundle size is more pronounced for the case of magnetic field parallel to the c-axis, suggesting that the effect of flux creep is more severe in this case. From an intuitive argument, the effect of flux creep seems to be smaller in this field direction, since the corresponding coherence length ξ_{ab} and the shear modulus C_{66} are larger, suggesting a larger flux bundle size. However, the result is exactly opposite to this direct speculation. Namely, too large a g_e^2 brings about too large a U_e and results in an enormous factor of reduction in the flux bundle size given by Eq. (A.69). Hence, it is an imaginary mechanism that the flux bundle size is large and the pinning potential energy U_0 takes a large value when the flux pinning is weak. We reach the conclusion that the flux pinning should be made stronger to enhance U_0. The strengthening of flux pinning surely makes g^2 small. However, g^2 never takes a value smaller than 1, and hence, if the flux pinning is made stronger beyond some level, it feeds directly into the enhancement of U_0.

As shown above the significant reduction in C_{66} from C_{66}^0 for the perfect flux line lattice occurs in practice, especially for the case of weak pinning.

Hence, if the expression of C_{66}^0 is substituted, theoretical predictions will significantly differ from experimental results. Very small pinning forces predicted by Larkin-Ovchinnikov theory [12] for the case of weak collective pinning stem from this fact. On the other hand, Eq. (7.81) predicted by the coherent potential approximation theory is not influenced by the reduction in C_{66}. It should be emphasized that the above thermodynamic method only is useful for the derivation of C_{66} because of the positive feedback that the softening of C_{66} caused by the flux line defects and thermal agitation leads to the enhanced pinning efficiency which results in further reduction in C_{66}.

Here we shall also discuss the dependence of the pinning force density on the pinning parameters. It is predicted in the Larkin-Ovchinnikov theory that $J_{ce} \propto N_p^2 f_p^4$ for the case of weak collective pinning. Since J_c is proportional to $y^{-1} J_{ce}$ as above-mentioned and y is proportional to $J_{ce}^{1/3}$ from the relationship $g^2 \propto J_{ce}^{-1}$ and from Eqs. (7.97) and (A.69), we have

$$J_c \propto J_{ce}^{2/3} \propto (N_p f_p^2)^{4/3} . \tag{A.71}$$

This relationship is close to the experimental results of $J_c \propto N_p f_p^2$ obtained for Nb in the region of small f_p shown in Fig. 7.9. Thus, the theory can also be improved from the qualitative viewpoint.

In addition, Eq. (A.69) suggests that the reduction in the flux bundle size is pronounced at low temperatures. This will also be briefly discussed here. If the pinning potential is approximated by a sinusoidally varying one as shown in Fig. 3.43, the activation energy is given by Eq. (3.125). This reduces to $U \propto (1 - J/J_{c0})^{3/2}$ in the vicinity of $J \simeq J_{c0}$. In this case the apparent pinning potential energy derived by Welch [15], Eq. (3.128), leads to

$$U_0^* = 1.65(k_B T U_0^2)^{1/3} . \tag{A.72}$$

Thus, from $U_0 = y^{3/2} U_e$ and Eq. (A.69) we have

$$U_0^* = 3.04 \left[\log\left(\frac{B a_f v_0}{E_c}\right) \right]^{2/3} k_B T \simeq 2.4 \times 10^{-22} T . \tag{A.73}$$

In the above the logarithmic term was estimated as 14. Figure A.9 shows the temperature dependence of the apparent pinning potential energy of a melt-processed Y-123 superconductor [15], and the solid line represents the result of Eq. (A.73). It can be seen that the theoretical prediction agrees approximately with the experimental result. This theoretical result predicts that the apparent pinning potential energy at low temperatures does not appreciably depend on the flux pinning strength except in the case of very strong flux pinning where g_e takes a value smaller than 1. It explains also many experimental results implying that U_0^* takes a value of several tens of meV at 30 K.

In addition, Eq. (A.73) predicts that the logarithmic relaxation rate of the magnetization takes a nonzero value even in the limit $T \to 0$. This temperature dependence is very different from that in Fig. 3.46 for a simple sinusoidal

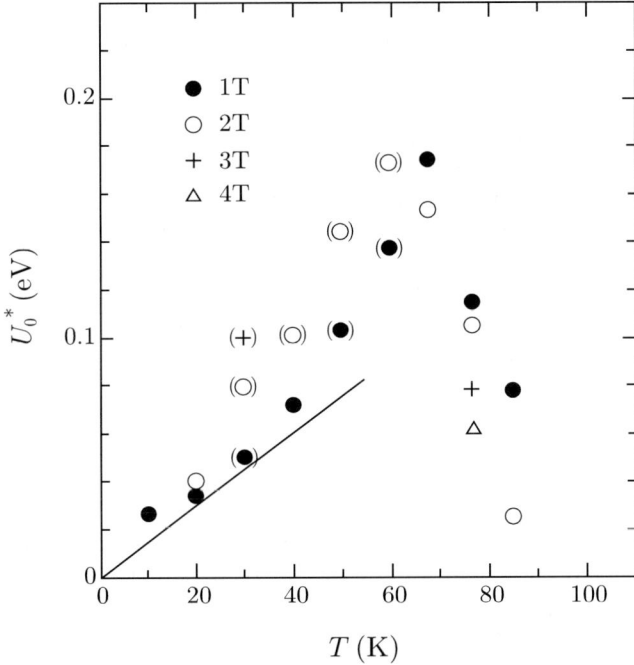

Fig. A.9. Temperature dependence of apparent pinning potential energy estimated from logarithmic relaxation rate of magnetization for a melt-processed Y-123 superconductor [15]. (The data are the same as in Fig. 3.48.) The *solid line* represents Eq. (A.73)

washboard potential. Such an abnormal logarithmic relaxation rate was originally explained only by the mechanism of quantum tunneling [16]. However, the present theoretical prediction shows another possibility to explain this relaxation rate. Hence, a further investigation is necessary on the phenomena at ultra low temperatures to find the true mechanism.

The irreversibility field of Bi-2212 thin films at low temperatures monotonically increases with increasing film thickness as shown in Fig. 8.22. This is caused by the fact that the pinning potential energy increases with the film thickness. However, the thickness dependence of the irreversibility field at $T/T_c = 0.5$ is shown in Fig. A.10, and the irreversibility field is rather found to decrease with increasing thickness contrary to the dependence at low temperatures. It appears to contradict the theoretical prediction of the flux creep model.

Here, we shall try to explain the experimental results on the magnetic field dependence of the critical current density of the 0.5 μm thick specimen at each temperature shown in Fig. A.11. In this case, if the usual assumption of $g^2 = 1$ for a bulk superconductor is used, the magnetic field dependence of the critical current density cannot be explained even by freely adjusting other pinning

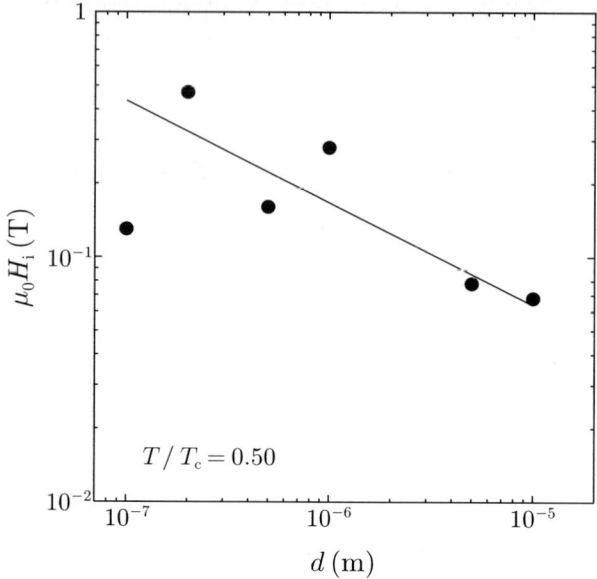

Fig. A.10. Thickness dependence of irreversibility field of Bi-2212 thin films [17] at $T/T_c = 0.5$. The *solid line* represents the prediction of Eq. (A.75) for $\gamma = 0.70$

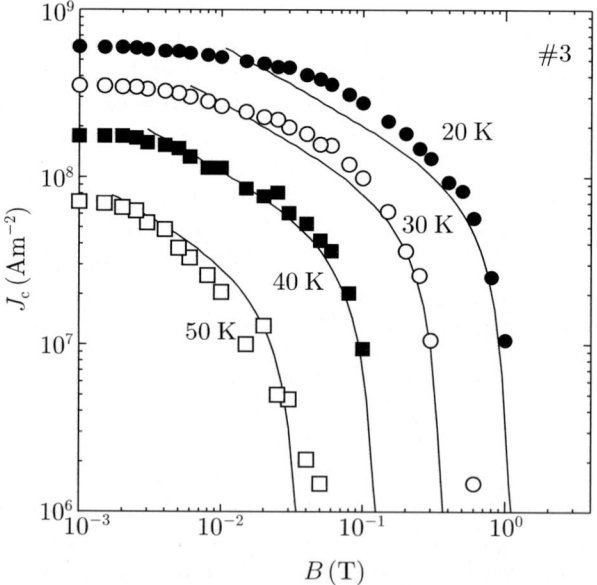

Fig. A.11. Magnetic field dependence of critical current density at each temperature for a Bi-2212 thin film of 0.5 μm thick [17]. The *solid lines* show the theoretical prediction of the flux creep-flow model with g^2 used as a fitting parameter

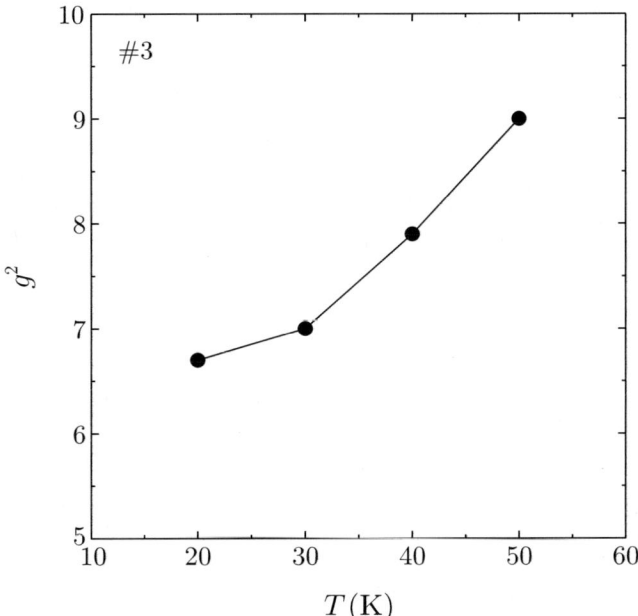

Fig. A.12. Temperature dependence of g^2 for a Bi-2212 thin film 0.5 μm thick

parameters [18]. Hence, g^2 is now also used as a fitting parameter along with other pinning parameters to explain the results in Fig. A.11. The obtained g^2 is shown in Fig. A.12 [17]. The value of g^2 is not necessarily equal to 1 even in two-dimensional Bi-2212 superconductors, but may be larger at high temperatures. The thickness dependence of g^2 at 30, 40 and 50 K obtained for each specimen is shown in Fig. A.13. It is found that g^2 takes on larger values for thinner specimens and at higher temperatures. These results will be explained based on the principle of Eq. (7.96). When the thickness is smaller than the pinning correlation length, the pinning potential is proportional to d. Substituting this dependence of $U_\text{e} \propto d$ into Eq. (7.96) leads to

$$g^2 = C\left(\frac{T}{d}\right)^{4/3}. \tag{A.74}$$

The three straight lines in Fig. A.13 are the theoretical predictions with $C = 2.3 \times 10^{-10} (\text{K}^{-1}\text{m})^{4/3}$ at 30, 40 and 50 K, and it can be seen that these agree with the above theoretical prediction except for the results on specimen 4 where the pinning force is weak. The value of C can also be approximately explained by the theory ($C = 2.0 \times 10^{-10}$ $(\text{K}^{-1}\text{m})^{4/3}$) [17]. Here, compensation for the weak pinning force of specimen 4 is attempted. Abstracting the dependence of C on the pinning strength, we have $C \propto A_\text{m}^{-1}(1 - T/T_\text{c})^{-m}$. The results on specimen 4 compensated by this factor are also shown in Fig. A.13. Thus, the behavior of g^2 obtained from experiments is approximately explained. The above result leads to $U_0 \propto d^{-1/3}$, suggesting that the irreversibility field increases with decreasing thickness. In fact, substitution of Eq. (7.99) with this dependence into Eq. (3.129) leads to

$$H_\mathrm{i}^{(3-2\gamma)/2} \propto d^{-1/3} \ . \tag{A.75}$$

The straight line in Fig. A.10 shows this result for $\gamma = 0.70$ and approximately explains the observed result. Thus, it can be understood that the transverse correlation of flux lines is developed so as to reduce the energy dissipation due to the flux creep at high temperatures for a very thin superconductor.

It should be noted that the same tendency of thickness dependence of g^2 exists also in the low temperature region. In this case, since the thickness region in which g^2 changes becomes narrower as can be expected from Fig. A.13, the influence on the theoretical prediction of Eq. (8.41) is not so significant as in the higher temperature region. The weaker thickness dependence of the irreversibility field in Fig. 8.22 than the simple theoretical prediction is caused by the large g^2 values in thin superconductors.

The discussion has been given on the dependence of g^2 on the dimensionality of the superconductor, flux pinning strength, temperature and the size

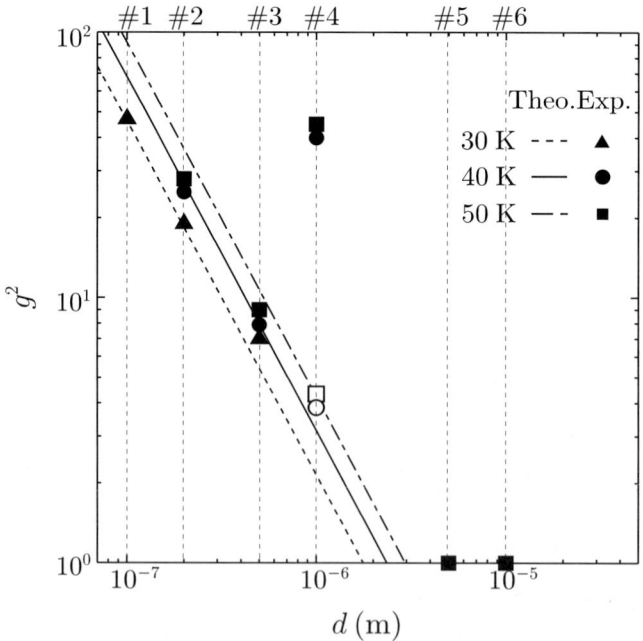

Fig. A.13. Thickness dependence of g^2 at 30, 40 and 50 K [17]. The three *straight lines* show the predictions of Eq. (A.74) with $C = 2.3 \times 10^{-10}$ $(\mathrm{K}^{-1}\mathrm{m})^{4/3}$. The *open symbols* are the results for specimen 4 with $d = 1.0$ μm compensated by a pinning strength factor

of the superconductor. Equation (A.69) predicts that g^2 depends also on the magnetic field through the magnetic field dependence of U_e and the electric field strength. The reason why the thickness dependence on Bi-2212 superconductors shown in Fig. A.13 is emphasized in comparison with the usual bulk case with $g^2 = 1$ is the very weak electric field strength corresponding to the DC magnetization measurement. In the usual transport measurements the value of g^2 is expected to be about a half of the present case. The apparent magnetic field dependence of g^2 was not observed in the above measurements on Bi-2212 superconductors. This can be attributed to the narrow magnetic field region due to the irreversibility field much lower than the upper critical field within the temperature region of measurements: only the J_{c0} characteristics at low magnetic fields influence the g^2 value.

The thickness dependence of the irreversibility field in three-dimensional Y-123 thin films and coated conductors also deviates from the simple theoretical prediction of Eq. (8.41) for the same reason [19]. In this case the thickness dependence is considered to be complicated in comparison with the case of Bi-2212 superconductor. This is attributed to the magnetic field dependence of g^2 due to the high irreversibility field, and also to the thickness dependence of the critical current density in superconductors made by the pulsed laser deposition due to the defect nucleation in thicker superconductors.

References

1. B. D. Josephson: Phys. Rev. **152** (1966) 211.
2. T. Matsushita: Phys. Lett. **86A** (1981) 123.
3. A. C. Rose-Innes and E. H. Rhoderick: *Introduction to Superconductivity*, 2nd Edition (Pergamon Press, 1978) Section 8.4.
4. S. Yuhya, K. Nakao, D. J. Baar, T. Sugimoto and Y. Shiohara: *Adv. Superconductivity IV* (Springer-Verlag, Tokyo, 1992) p. 845.
5. T. Matsushita: J. Phys. Soc. Jpn. **57** (1988) 1043.
6. E. H. Brandt : J. Low Temp. Phys. **28** (1977) 263.
7. E. H. Brandt: J. Low Temp. Phys. **28** (1977) 291.
8. T. Matsushita: Physica C **220** (1994) 172.
9. E. H. Brandt: J. Low Temp. Phys. **26** (1977) 709.
10. K. Miyahara, F. Irie and K. Yamafuji: J. Phys. Soc. Jpn. **27** (1969) 290.
11. T. Matsushita: Physica C **160** (1989) 328.
12. A. I. Larkin and Yu. N. Ovchinnikov: J. Low Temp. Phys. **34** (1979) 409.
13. T. Matsushita and H. Küpfer: J. Appl. Phys. **63** (1988) 5048.
14. T. Matsushita: Physica C **217** (1993) 461.
15. D. O. Welch: IEEE Trans. Magn. **27** (1991) 1133.
16. E. Simánek: Phys. Rev. B **39** (1989) 11384.
17. T. Matsushita, M. Kiuchi, T. Yasuda, H. Wada, T. Uchiyama and I. Iguchi: Supercond. Sci. Technol. **18** (2005) 1348.
18. H. Wada, E. S. Otabe, T. Matsushita, T. Yasuda, T. Uchiyama, I. Iguchi and Z. Wang: Physica C **378–381** (2002) 570.
19. See for example: K. Kimura, M. Kiuchi, E. S. Otabe, T. Matsushita, S. Miyata, A. Ibi, T. Muroga, Y. Yamada and Y. Shiohara: Physica C.

Answers to Exercises

Chapter 1

1.1. If we write $\Psi = |\Psi|e^{i\phi}$, Eq. (1.31) reduces to

$$\boldsymbol{j} = -\frac{2e}{m^*}|\Psi|^2(\hbar\nabla\phi + 2e\boldsymbol{A}) .$$

Hence, the kinetic energy term is written as

$$\frac{1}{2m^*}|(-i\hbar\nabla + 2e\boldsymbol{A})\Psi|^2 = \frac{\hbar^2}{2m^*}(\nabla|\Psi|)^2 + \frac{\mu_0}{2}\lambda^2\left(\frac{|\Psi_\infty|}{|\Psi|}\right)^2 j^2 ,$$

where Ψ_∞ is the equilibrium value of Ψ at zero magnetic field. In the London theory the condensation energy, $\alpha|\Psi|^2 + \beta|\Psi|^4/2$, and the energy associated with the spatial variation in the order parameter, $(\hbar^2/2m^*)(\nabla|\Psi|)^2$, are not considered. In the low field region where Ψ is approximately equal to Ψ_∞, the kinetic energy in the G-L theory agrees with the energy due to the current in the London theory.

The London theory holds correct when the G-L parameter κ is very large as mentioned in Sect. 1.3. At low fields where the flux lines are isolated from each other, the energy originating from the variation in the order parameter inside the core of a flux line, $(\hbar^2/2m^*)(\nabla|\Psi|)^2$, which cannot be derived from the London theory, is only about $3/(8\log\kappa)$ times as large as the energy of the flux line. Hence, this energy is not essential for discussing the superconducting property, although it cannot be neglected for discussing the flux pinning property. Thus, the energy in the London theory agrees with that in the G-L theory at low fields except for the condensation energy which takes a constant value outside the core.

1.2. If we add Eq. (1.30) multiplied by Ψ^* to its complex conjugate, we have

$$i\hbar e \boldsymbol{A} \cdot (\Psi^*\nabla\Psi - \Psi\nabla\Psi^*) - 2e^2 \boldsymbol{A}|\Psi|^2$$
$$m^*(\alpha|\Psi|^2 + \beta|\Psi|^4) - \left(\frac{\hbar}{2}\right)^2 (\Psi^*\nabla^2\Psi + \Psi\nabla^2\Psi^*).$$

Substitute this into Eq. (1.21) and rewrite it using the expression $\Psi = |\Psi|e^{i\phi}$.

1.3. A triangular flux line lattice is considered. If we assume the boundary C around a flux line contains a unit cell as in Fig. 1, we have $\boldsymbol{j} = 0$ on C from symmetry. Thus, Eq. (1.55) holds correct. Hence, it is shown that the magnetic flux is quantized in a unit cell.

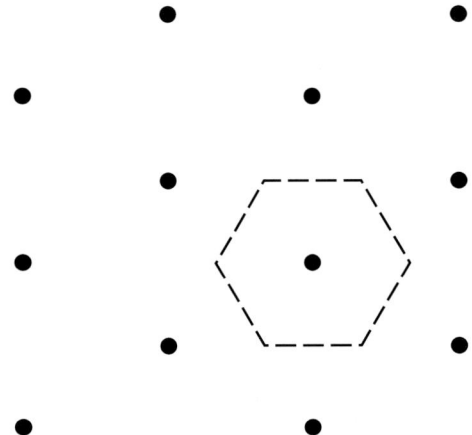

Fig. 1. Unit cell of triangular flux line lattice

1.4. We assume the approximate expression $|\Psi| = |\Psi_\infty|[(3r/2a_0) - (r^3/2a_0^3)]$ for the order parameter inside the core ($r \leq a_0 = (8/3)^{1/3}\xi$). The increment in the energy due to the variation in the order parameter from its equilibrium value is given by

$$\int_0^{a_0} \left[\alpha|\Psi|^2 + \frac{\beta}{2}|\Psi|^4 + |\alpha|\xi^2 \left(\frac{\mathrm{d}}{\mathrm{d}r}|\Psi|\right)^2\right] 2\pi r \mathrm{d}r - \pi a_0^2 \left(\alpha|\Psi_\infty|^2 + \frac{\beta}{2}|\Psi_\infty|^4\right)$$

per unit length of the flux line. After a simple calculation this reduces to $(209/210)\pi\mu_0 H_c^2 \xi^2 \simeq 0.995\pi\mu_0 H_c^2\xi^2$.

From Eq. (1.62a) the magnetic flux density inside the core is obtained as $b \simeq (\phi_0/2\pi\lambda^2)\log\kappa$. Thus, the magnetic energy inside the core is given by

$$\frac{1}{2\mu_0}\left(\frac{\phi_0}{2\pi\lambda^2}\right)^2 (\log\kappa)^2 \pi a_0^2 = \frac{8}{3}\pi\mu_0 H_c^2 \xi^2 \left(\frac{\log\kappa}{\kappa}\right)^2.$$

1.5. If we express $|\Psi|^2$ as

$$|\Psi|^2 = \sum_{m,n} a_{mn} \exp\left[\frac{2\pi i}{a_f}(mX + nY)\right]$$

using the double Fourier series, the coefficient a_{mn} is given by

$$\begin{aligned}a_{mn} &= \frac{1}{a_f^2}\int_0^{a_f}\int_0^{a_f} |\Psi|^2 \exp\left[-\frac{2\pi i}{a_f}(mX+nY)\right]\mathrm{d}X\mathrm{d}Y \\
&= \frac{1}{a_f^2}\sum_{p,q} C_p^* C_q \int_0^{a_f}\mathrm{d}Y\,\exp\left[\frac{2\pi i}{a_f}(p-q-n)Y\right]\\
&\quad\times\int_0^{a_f}\mathrm{d}X\exp\left\{\frac{2\pi i}{a_f}\left(\frac{p-q}{2}-m\right)X\right.\\
&\quad\left.-\frac{\sqrt{3}\pi}{2a_f^2}[(X-pa_f)^2+(X-qa_f)^2]\right\}.\end{aligned}$$

In the above the integration with respect to Y can be written as $a_f \delta_{p,q+n}$ using Kronecker's delta. Since the equation

$$\sum_p \int_0^{a_f} f(X-pa_f)\mathrm{d}X = \int_{-\infty}^{\infty} f(X)\mathrm{d}X$$

holds for an arbitrary function f, the above coefficient reduces to

$$\begin{aligned}a_{mn} &= \frac{|C_0|^2}{a_f}(-1)^{mn}\exp\left[-\frac{\pi}{\sqrt{3}}(m^2-mn+n^2)\right]\\
&\quad\times\int_{-\infty}^{\infty}\mathrm{d}X\exp\left\{-\frac{\sqrt{3}\pi}{a_f^2}\left[X+\frac{na_f}{2}+i\frac{a_f}{\sqrt{3}}\left(m-\frac{n}{2}\right)\right]^2\right\}.\end{aligned}$$

If S is a complex variable, the function $\exp[-(\sqrt{3}\pi/a_f^2)S^2]$ is regular, and there is no pole in the region of the complex plane between the two parallel lines $\mathrm{Im}S = i(a_f/\sqrt{3})(m-n/2)$ and $\mathrm{Im}S = 0$. Thus, using Cauchy's theorem the above integration reduces to

$$\int_{-\infty}^{\infty}\mathrm{d}X\exp\left(-\frac{\sqrt{3}\pi}{a_f^2}X^2\right) = 3^{-1/4}a_f.$$

Thus, a_{mn} is obtained and Eq. (1.97) is derived.

1.6. Using the relationship $C_n = C_0\exp(i\pi n^2/2)$, we have

$$\Psi\left(\frac{\sqrt{3}}{4}a_f,-\frac{a_f}{4}\right) = \sum_n C_0\exp\left[\frac{i\pi n(n+1)}{2}\right]\exp\left[\frac{\sqrt{3}\pi}{4}(2n-1)^2\right].$$

If the summation with respect to n is divided into that from 1 to ∞ and that from 0 to $-\infty$, the latter leads to

$$\sum_{n=1}^{\infty} C_0 \exp\left[\frac{i\pi(n-1)(n-2)}{2}\right] \exp\left[\frac{\sqrt{3}\pi}{4}(2n-1)^2\right]$$

$$= -\sum_{n=1}^{\infty} C_0 \exp\left[\frac{i\pi n(n+1)}{2}\right] \exp\left[\frac{\sqrt{3}\pi}{4}(2n-1)^2\right]$$

by rewriting as $n \to -n+1$. Thus, $\Psi((\sqrt{3}/4)a_\mathrm{f}, -a_\mathrm{f}/4) = 0$ can be proved.

1.7. The kinetic energy density is reduced to

$$\frac{1}{2m^*}|(-i\hbar\nabla + 2e\boldsymbol{A})\Psi|^2$$

$$= \frac{1}{2m^*}\left[\hbar^2 \nabla\Psi \cdot \nabla\Psi^* - 2i\hbar e\boldsymbol{A} \cdot (\Psi^*\nabla\Psi - \Psi\nabla\Psi^*) + 4e^2\boldsymbol{A}^2|\Psi|^2\right].$$

The volume integral of the first term is transformed as

$$\int \nabla\Psi \cdot \nabla\Psi^* \mathrm{d}V = \int \Psi^*\nabla\Psi \cdot \mathrm{d}\boldsymbol{S} - \int \Psi^*\nabla^2\Psi \mathrm{d}V.$$

If we assume a sufficiently large superconductor, the integral on the surface of the superconductor is less important and can be neglected. From a similar treatment we have

$$\int \boldsymbol{A} \cdot \Psi\nabla\Psi^* \mathrm{d}V = -\int \boldsymbol{A} \cdot \Psi^*\nabla\Psi \mathrm{d}V,$$

where $\nabla \cdot \boldsymbol{A} = 0$ was used. Thus, the integral of the kinetic energy is written as

$$\int \frac{1}{2m^*}|(-i\hbar\nabla + 2e\boldsymbol{A})\Psi|^2 \mathrm{d}V = \frac{1}{2m^*}\int \Psi^*(-i\hbar\nabla + 2e\boldsymbol{A})^2\Psi \mathrm{d}V.$$

Using Eq. (1.30), this is found to be equal to $-\int(\alpha|\Psi|^2 + \beta|\Psi|^4)\mathrm{d}V$. Thus, Eq. (1.111) is derived.

1.8. Such a wrong result is brought about by neglect of the fact that $\nabla\phi$ is singular at the center of the quantized flux so that $\nabla \times \nabla\phi$ is given by the two-dimensional delta function as in Eq. (1.67). When $\boldsymbol{A}$ is integrated on the assumed closed loop, there is no contribution from $\boldsymbol{j}$, but this curvilinear integral is identical with the surface integral of $(-\hbar/2e)\nabla \times \nabla\phi$ inside the closed loop. This leads to the correct magnetic flux, $\phi_0/2$.

Here we shall consider a small closed loop composed of a half circle R' and a segment of the straight line L as shown in Fig. 2. The center of the half circle coincides with that of the flux line and its radius is infinitesimal. When

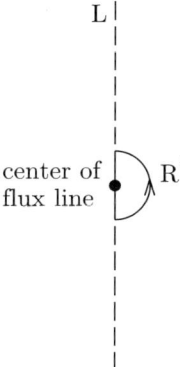

Fig. 2. Integral path composed of a segment of a straight line through the center of a flux line and a half circle R′ with an infinitesimal radius

carrying out a curvilinear integral of $(m^*/4e^2|\Psi|^2)\boldsymbol{j}$ on the closed loop, the integral on the segment of L is zero, and the integral on R′ remains nonzero. On the other hand, this curvilinear integral is divided into the curvilinear integral of $(-\hbar/2e)\nabla\phi$ and that of $-\boldsymbol{A}$ from Eq. (1.54). The former is equal to the surface integral of $(-\hbar/2e)\nabla \times \nabla\phi$ inside this area and gives $\phi_0/2$. The latter gives the magnetic flux inside the small area and can be neglected. Hence, it can be shown from this result and Eq. (1.74) that the current flows as $\boldsymbol{j} \simeq (H_{c2}/\lambda^2)r\boldsymbol{i}_\theta$ in the vicinity of the center of the flux line. It can be derived that the magnetic structure near the center is expressed as $b \simeq$ const. $- (\mu_0 H_{c2}/2\lambda^2)r^2$.

1.9. The first term in the superconducting current density given by Eq. (1.54) is proportional to $\nabla\phi$ and $|\Psi|^2$. The center of the quantized flux is a singular point and $\nabla\phi$ diverges with the order of $1/r$ as given by Eq. (1.66). The current density should approach zero with the order of r as shown in the answer to Exercise 1.8, and this means that the order parameter should approach zero as $|\Psi| \sim r$ in the center of the quantized flux. Thus, the central part of the quantized flux is approximately in the normal state.

Chapter 2

2.1. An elementary vector $d\boldsymbol{s}$ is defined on a closed loop C as shown in Fig. 3, where the direction of $d\boldsymbol{s}$ is chosen so that it and the magnetic flux density $\boldsymbol{B}$ satisfy the right-hand rule. When the velocity of flux lines is $\boldsymbol{v}$, the magnetic flux which comes into C crossing $|d\boldsymbol{s}|$ in a unit time is equal to $(d\boldsymbol{s} \times \boldsymbol{v}) \cdot \boldsymbol{B} = (\boldsymbol{v} \times \boldsymbol{B}) \cdot d\boldsymbol{s}$. Hence, the total magnetic flux coming into C in a unit time is given by

$$\oint_C (\boldsymbol{v} \times \boldsymbol{B}) \cdot d\boldsymbol{s} = \int_S \nabla \times (\boldsymbol{v} \times \boldsymbol{B}) \cdot d\boldsymbol{S},$$

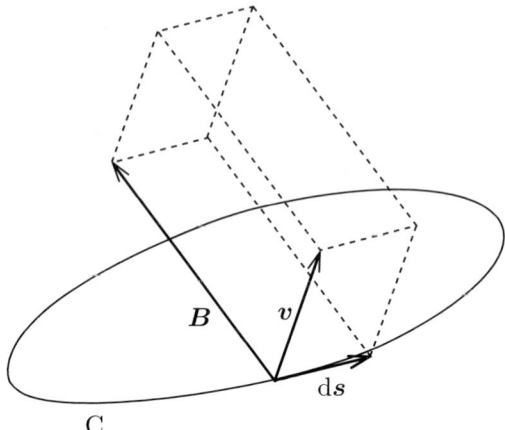

Fig. 3. Magnetic flux density B and velocity of flux lines v on a closed loop

where S is the area of the plane surrounded by C. This should be equal to the variation of the total magnetic flux inside S with time:

$$\frac{\partial}{\partial t}\int_S \boldsymbol{B}\cdot\mathrm{d}\boldsymbol{S} = \int_S \left(\frac{\partial \boldsymbol{B}}{\partial t}\right)\cdot\mathrm{d}\boldsymbol{S}.$$

In the above it is assumed that C and S do not vary with time. Since these two equations are satisfied for arbitrary S, we obtain generally

$$\nabla\times(\boldsymbol{B}\times\boldsymbol{v}) = -\frac{\partial \boldsymbol{B}}{\partial t}.$$

2.2. Using Eq. (2.35), the input power is written as $\langle \eta^* \dot{u}^2\rangle_t + \langle f(u)\dot{u}\rangle_t$. The second term is calculated as

$$\langle f(u)\dot{u}\rangle_t = \frac{1}{T_0}\int_0^{T_0} f(u)\frac{\partial u}{\partial t}\mathrm{d}t = \frac{1}{T_0}\int_0^{d_\mathrm{p}} f(u)\mathrm{d}u = \frac{1}{T_0}[-U(u)]_0^{d_\mathrm{p}},$$

where T_0 is the period in which the flux line meets the pins, d_p is the pin spacing and U is the pinning potential. When the spatial variation of U is repeated with the period d_p, this term is found to be zero from Fig. 4. Hence, it is proved that the input power is equal to the viscous power loss, $\langle \eta^* \dot{u}^2\rangle_t$.

2.3. It is necessary that $\langle (u-u_0)\dot{u}\rangle_t = \langle u-u_0\rangle_t v$ is satisfied so that Eq. (2.38) coincides with $\langle \eta^* v - k_\mathrm{f}(u-u_0)\rangle_t v$. Here we have

$$\langle (u-u_0)(\dot{u}-v)\rangle_t = \frac{1}{T_0}\int_0^{T_0}(u-u_0)\frac{\partial}{\partial t}(u-u_0)\mathrm{d}t = \frac{1}{2T_0}[(u-u_0)^2]_{t=0}^{t=T_0},$$

which reduces to zero under a periodic condition. Hence, the above relationship is satisfied. Thus, some periodic condition is needed, such as the one used

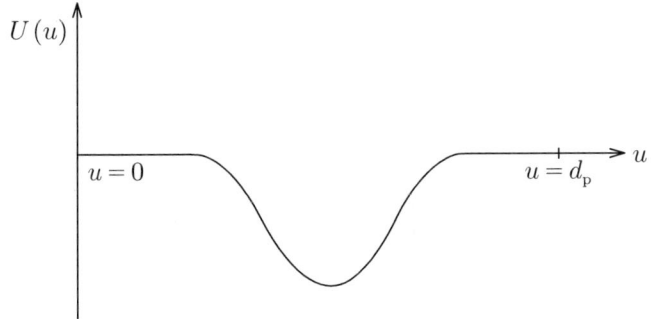

Fig. 4. Variation in the pinning potential along the direction of motion of a flux line

by Yamafuji and Irie, where the pin spacing is assumed to be so long that the strain $(u - u_0)$ relaxes completely until the flux line reaches the next pin.

2.4. The self-field of the current flowing through the superconducting cylinder of radius R has only the azimuthal component in cylindrical coordinates. If the Bean-London model is assumed, the magnetic flux distribution when the current applied to the z-axis is increased is given by

$$rB = R\mu_0 H_\mathrm{I} - \frac{1}{2}\alpha_c\mu_0(R^2 - r^2)$$

(see Subsect. 3.1.1). In the above $H_\mathrm{I}(>0)$ is the strength of the self-field. The critical state is attained, and the flux front reaches the center ($r = 0$) when $H_\mathrm{I} = \alpha_c R/2$. When the current is increased more, the resistive state starts. In this case azimuthal flux rings move to the center of the cylinder ($\boldsymbol{v} = -v\boldsymbol{i}_r$) due to the Lorentz force. The induced electric field is $\boldsymbol{E} = B\boldsymbol{i}_\theta \times (-v)\boldsymbol{i}_r = Bv\boldsymbol{i}_z$ and directed parallel to the current. This electric field is induced by the continuous motion of the azimuthal flux rings (see Fig. 5), and the flux rings are annihilated when they reach the center.

2.5. The macroscopic magnetic flux distribution in the superconductor at zero external magnetic field is like the one shown in Fig. 2.20, but the magnetic flux density outside the superconductor is reduced to zero and the area where $B = 0$ is extended wider. A discussion was given using Fig. 2.22 for the case where the external magnetic field H_e was reduced from $H_{\mathrm{c}1}$. Here we repeat a similar argument for the present case. We assume that, when H_e is reduced slightly below zero, the thermodynamic fields and the magnetic flux densities in the superconducting region just near the surface and in the adjoining superconducting region are like those shown in Fig. 6. The reason why $\mathcal{H}$ is different between the two regions while B is uniformly zero is because of the trapping of flux lines in the pinning region. This is essentially the same argument as in Sect. 2.6. When H_e is further decreased, the distribution on the

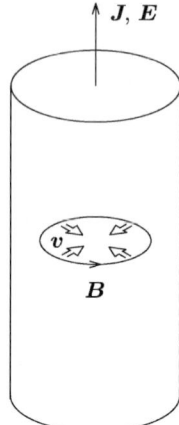

Fig. 5. Motion of flux line rings due to the self field of the current flowing in the superconducting cylinder and the induced electric field

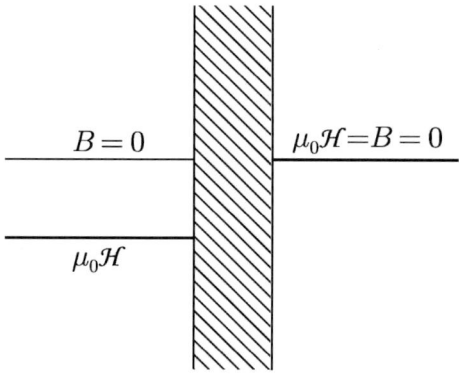

Fig. 6. Thermodynamic magnetic field near the surface of a superconductor when the external magnetic field is reduced slightly below zero

left-hand side is successively shifted to the right. As a result, the macroscopic magnetic flux distribution is considered to vary with decreasing external field as shown in Fig. 7(a) and (b).

2.6. For the case of $H_p < H_m < 2^{1/(2-\gamma)} H_p$ we consider the following four steps in a variation in H_e using h^* as defined below: (1) from H_m to 0, (2) from 0 to $-H_m h^*$, (3) from $-H_m h^*$ to $-H_p$ and (4) from $-H_p$ to $-H_m$. The contribution to the energy loss from each step can be calculated using Eq. (2.79). The energy loss density is given by twice the sum of these contributions and we have

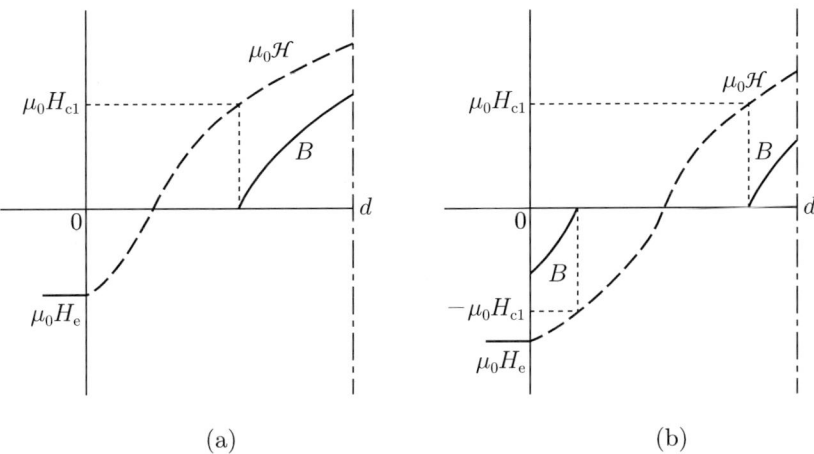

Fig. 7. Macroscopic magnetic flux distribution in the superconductor after the external magnetic field is inverted: (a) for H_e between 0 and $-H_{c1}$ and (b) for H_e below $-H_{c1}$.

$$W = \frac{2(2-\gamma)\mu_0 H_m^{4-\gamma}}{H_p^{2-\gamma}}\left[\frac{2}{4-\gamma} - 2^{-1/(2-\gamma)}\int_0^1 (1+\zeta^{2-\gamma})^{1/(2-\gamma)}\zeta^{2-\gamma}d\zeta\right.$$

$$+ 2^{-1/(2-\gamma)}\int_0^{h^*}(1-\zeta^{2-\gamma})^{1/(2-\gamma)}\zeta^{2-\gamma}d\zeta$$

$$\left.+\int_{h^*}^{h_p}(h_p^{2-\gamma}-\zeta^{2-\gamma})^{1/(2-\gamma)}\zeta^{2-\gamma}d\zeta - \int_{h_p}^1(\zeta^{2-\gamma}-h_p^{2-\gamma})^{1/(2-\gamma)}\zeta^{2-\gamma}d\zeta\right],$$

where $h_p = H_p/H_m$ and $h^* = (2h_p^{2-\gamma}-1)^{1/(2-\gamma)}$.

Four similar steps are considered also for the case of $H_m > 2^{1/(2-\gamma)}H_p$ and the following result is obtained.

$$W = \frac{2(2-\gamma)\mu_0 H_m^{4-\gamma}}{H_p^{2-\gamma}}\left[\frac{2}{4-\gamma} - 2^{-1/(2-\gamma)}\int_{h^\dagger}^1 (1+\zeta^{2-\gamma})^{1/(2-\gamma)}\zeta^{2-\gamma}d\zeta\right.$$

$$-\int_0^{h^\dagger}(\zeta^{2-\gamma}+h_p^{2-\gamma})^{1/(2-\gamma)}\zeta^{2-\gamma}d\zeta$$

$$\left.+\int_0^{h_p}(h_p^{2-\gamma}-\zeta^{2-\gamma})^{1/(2-\gamma)}\zeta^{2-\gamma}d\zeta - \int_{h_p}^1(\zeta^{2-\gamma}-h_p^{2-\gamma})^{1/(2-\gamma)}\zeta^{2-\gamma}d\zeta\right],$$

where $h^\dagger = (1-2h_p^{2-\gamma})^{1/(2-\gamma)}$. These results coincide with Eq. (2.84) for $\gamma = 1$.

2.7. The sign factor of the current is represented by δ_J (when the current flows in the direction of the positive y-axis, $\delta_J = 1$). From the condition that

$B = \mu_0 H_e$ at the surface $x = 0$ the magnetic flux distribution near the surface of half of a superconducting slab ($0 \leq x \leq d$) is expressed as

$$B = -\beta + [(\mu_0 H_e + \beta)^2 - 2\delta_J \mu_0 \alpha_0 x]^{1/2},$$
$$B = \beta - [(\mu_0 H_e - \beta)^2 - 2\delta_J \mu_0 \alpha_0 x]^{1/2}$$

for the regions $B > 0$ and $B < 0$, respectively. The penetration field H_p is given by

$$\mu_0 H_p = -\beta + (\beta^2 + 2\mu_0 \alpha_0 d)^{1/2}.$$

If the maximum field H_m satisfies the condition $\mu_0 H_m > -\beta + (\beta^2 + 4\mu_0 \alpha_0 d)^{1/2}$, the expressions for the magnetization corresponding to the cases from a to e in Eq. (2.55) are:

$$M = \frac{H_e^2}{6\alpha_0 d}(2\mu_0 H_e + 3\beta) - H_e; \qquad 0 < H_e < H_p,$$

$$= -\frac{\beta}{\mu_0} + \frac{1}{3\mu_0^2 \alpha_0 d}\{(\mu_0 H_e + \beta)^3 - [(\mu_0 H_e + \beta)^2 - 2\mu_0 \alpha_0 d]^{3/2}\} - H_e;$$
$$H_p < H_e < H_m,$$

$$= -\frac{\beta}{\mu_0} + \frac{1}{3\mu_0^2 \alpha_0 d}\{2^{-1/2}[(\mu_0 H_m + \beta)^2 + (\mu_0 H_e + \beta)^2]^{3/2} - (\mu_0 H_e + \beta)^3$$
$$- [(\mu_0 H_m + \beta)^2 - 2\mu_0 \alpha_0 d]^{3/2}\} - H_e; \qquad H_m > H_e > H_a,$$

$$= -\frac{\beta}{\mu_0} + \frac{1}{3\mu_0^2 \alpha_0 d}\{[(\mu_0 H_e + \beta)^2 + 2\mu_0 \alpha_0 d]^{3/2} - (\mu_0 H_e + \beta)^3\} - H_e;$$
$$H_a > H_e > 0,$$

$$= -\frac{\beta}{\mu_0} + \frac{1}{3\mu_0^2 \alpha_0 d}\{(\mu_0 H_e - \beta)^3 + 3\beta(\mu_0 H_e - \beta)^2 - 3\beta^3$$
$$+ [2\beta^2 - (\mu_0 H_e - \beta)^2 + 2\mu_0 \alpha_0 d]^{3/2}\} - H_e; \qquad 0 > H_e > -H_p.$$

In the above H_a is given by

$$\mu_0 H_a = -\beta + [(\mu_0 H_m + \beta)^2 - 4\mu_0 \alpha_0 d]^{1/2}.$$

The above results for the case $\beta \to 0$ agree with the values of Eq. (2.55) for the case $\gamma \to 0$.

2.8. While the external magnetic field H_e is reduced from H_m, the magnetization is given by

$$M_- = \frac{H_m^2 + 2H_m H_e - H_e^2}{4H_p}.$$

On the other hand, the magnetization as H_e increases from $-H_m$ is given by

$$M_+ = \frac{-H_m^2 + 2H_m H_e + H_e^2}{4H_p}.$$

Thus, the energy loss density is calculated as

$$W = \int_{-H_m}^{H_m} \mu_0(M_- - M_+)\mathrm{d}H_e = \frac{2\mu_0 H_m^3}{3H_p}.$$

2.9. The force on a single flux line per unit length is $\phi_0 J_c$. Since the variation in the magnetic flux density extends from the surface to $H_m/J_c = x_m$, the number density of moving flux lines is approximately estimated as $(\mu_0 H_e) \cdot (H_m/J_c d)/\phi_0$. If the displacement of flux lines and the variation in the magnetic flux density are represented by u and b, respectively, the continuity equation for flux lines is written as $\partial u/\partial x \simeq b/\mu_0 H_e$. Since b is approximately given by the mean value $\mu_0 H_m$, the mean distance of flux motion over a half period is approximately given by $(H_m/H_e) \cdot x_m/2 = H_m^2/2J_c H_e$. Thus, we have

$$W \simeq \phi_0 J_c \cdot \frac{\mu_0 H_e H_m}{\phi_0 J_c d} \cdot \frac{H_m^2}{2J_c H_e} \cdot 2 = \frac{\mu_0 H_m^3}{H_p}.$$

This value is 3/2 times as large as the theoretical result of the Bean-London model.

2.10. A half process during which the external field varies from H_m to $-H_m$ is considered. (1) When H_e decreases from H_e to H_{c1}, the magnetic flux density inside the superconducting slab is expressed as

$$B(x) = \mu_0(H_e - H_{c1}) + \mu_0 J_c x; \qquad 0 \leq x \leq x_b,$$
$$= \mu_0(H_m - H_{c1}) - \mu_0 J_c x; \qquad x_b < x \leq x_m,$$

where $x_b = (H_m - H_e)/2J_c$ and $x_m = (H_m - H_{c1})/J_c$. Hence, the average magnetic flux density inside the slab is given by

$$\langle B(x) \rangle = \frac{\mu_0}{4J_c d}[2(H_m - H_{c1})^2 - (H_m - H_e)^2].$$

(2) When H_e is varied from H_{c1} to $-H_{c1}$, the average magnetic flux density is constant and given by

$$\langle B(x) \rangle = \frac{\mu_0}{4J_c d}(H_m - H_{c1})^2.$$

(3) When H_e is varied from $-H_{c1}$ to $-H_m$, the average magnetic flux density is calculated as

$$\langle B(x) \rangle = \frac{\mu_0}{4J_c d}[2(H_m - H_{c1})^2 - (H_m - H_e - 2H_{c1})^2].$$

From symmetry between this half period and the other half period the energy loss density per unit cycle is obtained as

$$W = 2\int_{-H_m}^{H_m} \langle B(x) \rangle \mathrm{d}H_e = \frac{2\mu_0}{3J_c d}(H_m - H_{c1})^2 \left(H_m + \frac{H_{c1}}{2}\right).$$

In the limit $H_{c1} \to 0$ this result reduces to the usual formula for the energy loss density for the Bean-London model.

Chapter 3

3.1. The shielding current flows only on the surface of the superconducting cylinder, and hence, the scalar potential ϕ_m can be used to express the magnetic flux density as $\boldsymbol{B} = -\nabla\phi_m$ inside and outside the superconductor except in the surface region. This scalar potential satisfies the Laplace equation $\nabla^2\phi_m = 0$. In the present cylindrical geometry, the situation is uniform along the length and we can separate variables to write ϕ_m as $\phi_m = R(r)\Theta(\theta)$. From the condition of symmetry with respect to θ and the requirements that the potential is finite at $r = 0$ and that $\boldsymbol{B} \to \mu_0\boldsymbol{H}_e$ at $r \to \infty$, we have the solution

$$\phi_m = \sum_{n=1}^{\infty} \alpha_n r^{-n}\cos n\theta - \mu_0 H_e r\cos\theta; \quad r > R,$$

$$= \sum_{n=1}^{\infty} \beta_n r^n \cos n\theta; \quad r < R,$$

except for a constant. It is required that $\boldsymbol{B} = 0$ inside the superconductor. Thus, we have $\beta_n = 0 (n \geq 1)$. The normal component of $\boldsymbol{B}$, i.e. $(\nabla\phi_m)_r$, should be continuous at the boundary $r = R$. This leads to $\alpha_1/R^2 = -\mu_0 H_e$ and $\alpha_n = 0 (n \geq 2)$. Hence, the magnetic flux density in the region $r > R$ is given by

$$B_r = -\frac{\partial\phi_m}{\partial r} = \mu_0 H_e \left(1 - \frac{R^2}{r^2}\right)\cos\theta,$$

$$B_\theta = -\frac{1}{r}\cdot\frac{\partial\phi_m}{\partial\theta} = -\mu_0 H_e \left(1 + \frac{R^2}{r^2}\right)\sin\theta.$$

The shielding current density $\widetilde{J}(\theta)$ flowing on the surface is obtained from the difference in the tangential component of the magnetic field as

$$\widetilde{J}(\theta) = \frac{1}{\mu_0}B_\theta(r = R + 0) = -2H_e \sin\theta.$$

3.2. From symmetry we have only to treat the case of $0 \leq \omega t < \pi (\delta = -1)$ and the contribution to the energy loss from this period is the same as that from the period $\pi \leq \omega t \leq 2\pi$. The average viscous power loss density in the superconductor is

$$\langle p_v\rangle = \frac{\eta\mu_0 H_e}{\phi_0 d}\int_0^{x_{b0}} v^2 dx = \frac{\mu_0 h_0^3 \omega^2}{6J_c d\omega_0}(1 - \cos\omega t)^3 \sin^2\omega t.$$

Thus, the viscous energy loss density is obtained as

$$\frac{2}{\omega}\int_0^\pi \langle p_v\rangle d\omega t = \frac{2\mu_0 h_0^3}{3J_c d}\cdot\frac{7\pi\omega}{16\omega_0}.$$

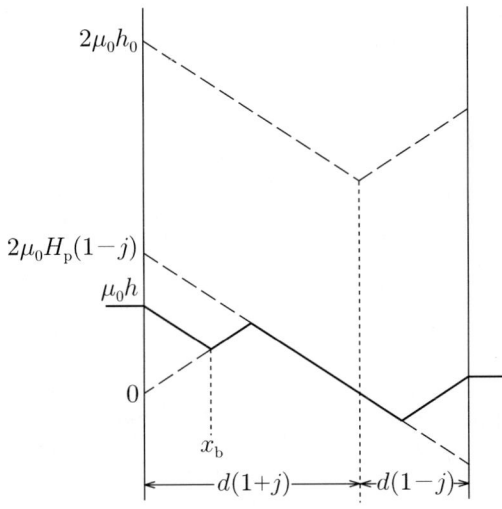

Fig. 8. Magnetic flux distribution inside the superconducting slab when the AC magnetic field increases

3.3. We consider the process where the AC magnetic field increases from the minimum value. The increment of the surface field is represented by h. When $h < 2H_p(1-j)$, the penetration depth of magnetic flux from the surface is $x_b = h/2J_c$ as shown in Fig. 8. From the continuity equation for flux lines we have $Bv = \mu_0(\partial h/\partial t)(x_b - x)$. The energy dissipation occurs at the surface on both sides of the superconducting slab, and the average pinning power loss density is

$$\frac{1}{2d}\int_0^{x_b} J_c Bv dx \times 2 = \frac{\mu_0 J_c}{2d} \cdot \frac{\partial h}{\partial t} x_b^2 .$$

Hence, the contribution to the energy loss density from the period $0 < h < 2H_p(1-j)$ is given by

$$\frac{\mu_0 J_c}{2d}\int \frac{\partial h}{\partial t} x_b^2\, dt = \frac{\mu_0 H_p^2}{3}(1-j)^3 = W_1 .$$

When $2H_p(1-j) < h < 2h_0$, the penetration depth of magnetic flux is $d(1+j)$ and $d(1-j)$ from the left and right surfaces, respectively. Thus, the average power loss density is

$$\frac{\mu_0 J_c}{4d} \cdot \frac{\partial h}{\partial t}[d^2(1+j)^2 + d^2(1-j)^2] = \frac{\mu_0 H_p}{2} \cdot \frac{\partial h}{\partial t}(1+j^2)$$

and the energy loss density is

$$\frac{\mu_0 H_p}{2}(1+j^2)\int \frac{\partial h}{\partial t} dt = \mu_0 H_p(1+j^2)[h_0 - H_p(1-j)] = W_2 .$$

The energy loss in the decreasing field process is the same as that in the increasing field process and finally we obtain Eq. (3.39) from $W = 2(W_1+W_2)$.

3.4. The first term in the first integral in Eq. (3.58) is transformed as

$$\frac{\lambda^2}{2\mu_0} \int_{S_c} [\boldsymbol{b}_f \times (\nabla \times \boldsymbol{b}_f)] \cdot \mathrm{d}\boldsymbol{S} = \frac{\lambda^2}{2\mu_0} \int_{V'} [(\nabla \times \boldsymbol{b}_f)^2 - \boldsymbol{b}_f \cdot (\nabla \times \nabla \times \boldsymbol{b}_f)]\mathrm{d}V$$

from a partial integration of $(\nabla \times \boldsymbol{b}_f)^2$. In the above V' is all space (including the vacuum of $x < 0$) except the region of the normal core. It should be noted that $\boldsymbol{b}_f$ is also defined for $x < 0$. Using the modified London equation, the second term in the integral is written as $\boldsymbol{b}_f^2/2\mu_0$. This term and the first and second terms in the second integral in Eq. (3.58) reduce to

$$\frac{1}{2\mu_0} \int_V [\boldsymbol{b}_f^2 + \lambda^2 (\nabla \times \boldsymbol{b}_f)^2]\mathrm{d}V = \phi_0 H_{c1} ,$$

where V represents all space. Secondly the fourth term in the first integral in Eq. (3.58) is transformed as

$$-\lambda^2 \int_{S_c} [\boldsymbol{H}_e \times (\nabla \times \boldsymbol{b}_f)] \cdot \mathrm{d}\boldsymbol{S}$$

$$= \lambda^2 \int_{V'} [\boldsymbol{H}_e \cdot (\nabla \times \nabla \times \boldsymbol{b}_f) - (\nabla \times \boldsymbol{H}_e) \cdot (\nabla \times \boldsymbol{b}_f)]\mathrm{d}V$$

and the second term on the right-hand side is zero. Using the modified London equation again, the sum of this and the third term in the second integral in Eq. (3.58) reduces to

$$-\int_V \boldsymbol{b}_f \cdot \boldsymbol{H}_e \mathrm{d}V = -\phi_0 H_e .$$

In the fifth term in the first integral in Eq. (3.58), $\boldsymbol{b}_0$ is approximately given by $\mu_0 H_e \exp(-x_0/\lambda)\boldsymbol{i}_z$, where $\boldsymbol{i}_z$ represents the unit vector along the magnetic field. $\nabla \times \boldsymbol{b}_f$ is perpendicular to $\boldsymbol{b}_0$ with its magnitude approximately equal to $\phi_0/2\pi\lambda^2\xi$ and $\boldsymbol{b}_0 \times (\nabla \times \boldsymbol{b}_f)$ is parallel to $\mathrm{d}\boldsymbol{S}$. Hence, this term is given by $\phi_0 H_e \exp(-x_0/\lambda)$. Since $\boldsymbol{b}_i$ is approximately given by $(\phi_0/2\pi\lambda^2)K_0(2x_0/\lambda)\boldsymbol{i}_z$, the second term in the first integral reduces to $-(\phi_0^2/4\pi\mu_0\lambda^2)K_0(2x_0/\lambda)$.

The third and sixth terms in the first integral are almost constant vectors within the region of integration. Thus, these terms go to zero in the limit of zero radius of the normal core. Hence, these terms can be neglected, and Eq. (3.59) is obtained.

3.5. In the temperature range $T_i \geq T > T_0$ the resultant χ is the same as in the case of $m' \neq 2$ and given by Eq. (3.86a). In other ranges of temperature we have

Answers to Exercises   477

$$\chi = -\frac{[\epsilon H_{c2}(0)(1-\delta)]^2}{4dAH_e}\left[\frac{3}{2}+\log\left\{\frac{2dA}{\epsilon H_{c2}(0)(1-\delta)}\left(1-\frac{T}{(1-\delta)T_c}\right)\right\}\right];$$
$$T_0 \geq T > T_{c1},$$

$$= -\frac{[\epsilon H_{c2}(0)(1-\delta)]^2}{4dAH_e}\left[\frac{3}{2}+\log\left\{\frac{2dAH_e}{[\epsilon H_{c2}(0)(1-\delta)]^2}\right\}\right] \equiv \chi_s;\quad T_{c1} \geq T,$$

from a similar calculation [1].

3.6. The magnetic flux distribution in this case is given by

$$B(x) = \mu_0 H_e + \mu_0 M(T) - \mu_0 J_c(T) x$$

instead of Eq. (3.81). If the point at which B is zero is denoted by $x = x'_0$, we have

$$x'_0 = \frac{H_e}{A}f^{-m'}(T) - \frac{\epsilon H_{c2}(0)(1-\delta)}{A}f^{1-m'}(T),$$

where

$$f(T) = 1 - \frac{T}{(1-\delta)T_c}.$$

If the temperature at which x'_0 reaches d is denoted by T'_0, a simple calculation leads to [1]

$$\chi = -1;\qquad\qquad\qquad\qquad\qquad\qquad\qquad T \leq T_{c1},$$

$$= -1 + \frac{H_e}{2dA}f^{-m'}(T) - \frac{\epsilon H_{c2}(0)(1-\delta)}{dA}f^{1-m'}(T)$$
$$+\frac{[\epsilon H_{c2}(0)(1-\delta)]^2}{2dAH_e}f^{2-m'}(T);\qquad\qquad T_{c1} < T \leq T'_0,$$

$$= -\frac{\epsilon H_{c2}(0)(1-\delta)}{H_e}f(T) - \frac{dA}{2H_e}f^{m'}(T);\qquad T'_0 < T.$$

3.7. It is assumed that the magnetic flux distribution is initially in the critical state under the minimum surface field. The variation in the magnetic flux density from this state is denoted by $b(x)$. The displacement of flux lines in a half period from the initial state is given by Eq. (3.108). From the balance between the Lorentz force density and the pinning force density given by Eq. (3.103), we have

$$\frac{db}{dx} = -\frac{b(0)}{\lambda_0'^2}\left(\frac{d_f}{2}-x\right) + \frac{b^2(0)}{4\mu_0 J_c \lambda_0'^4}\left(\frac{d_f}{2}-x\right)^2.$$

This leads to

$$b(x) = b(0) - \frac{b(0)}{\lambda_0'^2}\left[\left(\frac{d_f}{2}\right)^2 - \left(\frac{d_f}{2}-x\right)^2\right]$$
$$+\frac{b^2(0)}{12\mu_0 J_c \lambda_0'^4}\left[\left(\frac{d_f}{2}\right)^3 - \left(\frac{d_f}{2}-x\right)^3\right].$$

Averaging this within the superconducting slab, we have

$$\langle b(x) \rangle = b(0) - \frac{2b(0)}{3\lambda_0'^2}\left(\frac{d_f}{2}\right)^2 + \frac{b^2(0)}{16\mu_0 J_c \lambda_0'^4}\left(\frac{d_f}{2}\right)^3.$$

In the above the first and second terms are linear with respect to $b(0)$, and hence, these terms are reversible responses and do not contribute to the loss. Hence, these are neglected and only the third term is treated. The energy loss density W is approximately given by twice the area surrounded by the above $\langle b(x) \rangle$ vs. $b(0)/\mu_0$ curve and the straight broken line connecting the initial and final points in Fig. 9. Namely, if the value of $\langle b(x) \rangle$ at $b(0) = 2\mu_0 h_0$ is denoted by b_m, we have

$$W = 2h_0 b_m - 2\int_0^{2\mu_0 h_0}\frac{1}{\mu_0}\langle b(x)\rangle \mathrm{d}b(0) = \frac{\mu_0 h_0^3}{3J_c d_f}\left(\frac{d_f}{2\lambda_0'}\right)^4.$$

This coincides with Eq. (3.109).

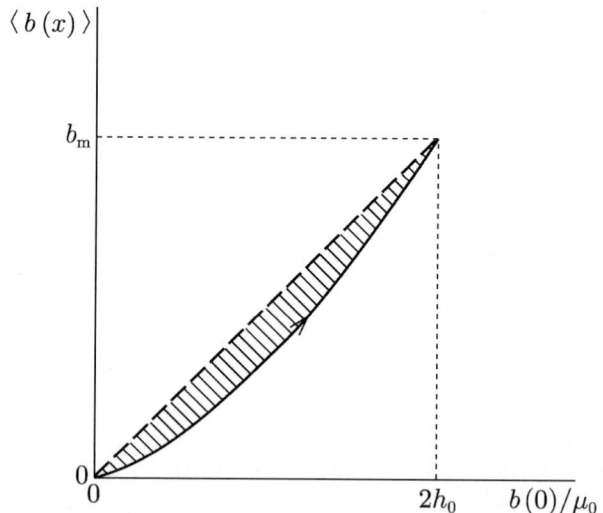

Fig. 9. Variation in the average magnetic flux density during a variation in the surface field of half a period. The area of the *hatched region* gives half of the energy loss density

3.8. Because of $\langle B \rangle \simeq \mu_0 h_0 \cos \omega t$ Eq. (3.99) leads to

$$\mu' \simeq \frac{\mu_0}{\pi}\int_{-\pi}^{\pi}\cos^2 \omega t \, \mathrm{d}\omega t = \mu_0.$$

On the other hand, from Eqs. (3.102) and (3.107) we have

$$\mu'' = \frac{2\mu_0 h_0}{9\pi J_c D}.$$

Substitution of these into Eq. (3.98) leads to

$$\eta_{\rm p} \simeq \left[1 + \left(\frac{9\pi J_c D}{2h_0}\right)^2\right]^{-1/2} \simeq \frac{2h_0}{9\pi J_c D}.$$

Thus, it can be shown that $\eta_{\rm p}$ is proportional to h_0, when h_0 is sufficiently small.

3.9. It is assumed that the external magnetic field is decreased from a certain value after an increasing process on a major magnetization curve. If the decrease in the internal magnetic flux density after starting to decrease the external field is represented by $b(x)$, it is easily calculated as

$$b(x) = b_0 \frac{\cosh(x/\lambda_0')}{\cosh(d_{\rm f}/2\lambda_0')},$$

where b_0 is the decrease in the surface magnetic flux density. Thus, the variation in the magnetization is

$$\delta M = \frac{b_0}{\mu_0}\left[1 - \frac{2\lambda_0'}{d_{\rm f}}\tanh\left(\frac{d_{\rm f}}{2\lambda_0'}\right)\right].$$

From the definition of the characteristic field, $\widehat{H}_{\rm p}$ is given by $(\partial \mu_0 \delta M/\partial b_0)^{-1} H_{\rm p}$ in the limit $b_0 \to 0$. As a result, Eq. (3.111) is obtained. Since $\tanh x \simeq x - x^3/3$ for $x \ll 1$, we have $\widehat{H}_{\rm p} = 3(2\lambda_0'/d_{\rm f})^2 H_{\rm p} = (\sqrt{3}/2)\widetilde{H}_{\rm p}$, where $\widetilde{H}_{\rm p}$ is defined by Eq. (3.110).

3.10. The amplitude of the AC magnetic field applied parallel to the superconducting slab is denoted by h_0. In order to change the current density from J_c to $-J_c$ completely irreversibly in the entire region of the slab, h_0 should be equal to the penetration field $J_c d$ as shown by the critical state model. In this case, the change in magnetic field is maximum at the surface and amounts to $2h_0$. The displacement of flux lines is also maximum at the surface, and this displacement is denoted by $u_{\rm m}$. Then, Eq. (3.88) leads to $2\mu_0 J_c d/B = u_{\rm m}/d$, where the right hand side is obtained from the requirement of symmetry that the displacement is zero at the center. So as to be in the reversible state in the entire slab, $u_{\rm m}$ should be less than $2d_{\rm i}$. This leads to

$$d^2 < \frac{Bd_{\rm i}}{\mu_0 J_c} = \lambda_0'^2.$$

3.11. If the activation energy is expressed as $U(J) = U_0(1 - J/J_{c0})^N$, an expansion of this expression around $J = J_0$ observed at a certain time $(t = t_0)$ in the relaxation process leads to

$$U(J) \simeq U_0\left(1 - \frac{J_0}{J_{c0}}\right)^N - N\frac{U_0}{J_{c0}}\left(1 - \frac{J_0}{J_{c0}}\right)^{N-1}(J - J_0).$$

If we express this as $U(J) = U_0^*(1 - J/J_{c0}^*)$, we have

$$U_0^* = U_0\left(1 - \frac{J_0}{J_{c0}}\right)^{N-1}\left[1 + (N-1)\frac{J_0}{J_{c0}}\right],$$

$$J_{c0}^* = \frac{J_{c0}}{N}\left[1 + (N-1)\frac{J_0}{J_{c0}}\right].$$

Hence, Eq. (3.119) is written as

$$\frac{J_0}{J_{c0}^*} = 1 - \frac{k_B T}{U_0^*}C_0,$$

where $C_0 \simeq \log(2Ba_f\nu_0 U_0^* t_0 / \mu_0 d^2 J_{c0}^* k_B T)$. If we assume $1 - (J_0/J_{c0}) \ll 1$ as observed usually at low temperatures, we have $J_{c0}^* \simeq J_{c0}$ and

$$U_0^* \simeq NU_0\left(\frac{C_0 k_B T}{U_0^*}\right)^{N-1}.$$

This reduces to

$$U_0^* \simeq (NC_0^{N-1})^{1/N}[(k_B T)^{N-1}U_0]^{1/N}.$$

Here we consider, for example, the case of Nb$_3$Sn at $T = 4.2$ K and $B = 1$ T ($a_f = 49$ nm). C_0 is written as $C_0 \simeq \log(a_f \rho_f U_0^* t_0 / \pi \mu_0 d_i d^2 k_B T)$ in terms of the interaction distance d_i. Substitution of typical values of $\rho_f \simeq 1.6 \times 10^{-8}$ Ωm, $d_i \simeq a_f/2\pi$, $d \simeq 1$ mm, $U_0^* \simeq 0.29 \times 10^{-19}$ J (observed value [2]) and $t_0 = 1$ s leads to $C_0 \simeq 16.4$. Thus, in the case of $N = 3/2$ for a sinusoidal washboard potential, we have

$$U_0^* \simeq 3.3(k_B T U_0^2)^{1/3}.$$

Although this result is about twice as large as the theoretical result of Welch [3], these are qualitatively the same. This result predicts that, while U_0 increases slightly with decreasing temperature, U_0^* decreases. This prediction agrees qualitatively with the results in Fig. 3.46(a) and (b). In addition, the above result is written as $U_0^*/U_0 \propto (k_B T/U_0)^{1/3}$. This insists that, even if U_0 is increased by strengthening the flux pinning, the relative difference between U_0^* and U_0 becomes larger, and U_0^* does not increase so much. This also agrees with the results in the figure.

3.12. From Eq. (3.131) the resistivity is written as

$$\rho = \frac{\pi B a_f \nu_0 U_0}{J_{c0} k_B T}\exp\left(-\frac{U_0}{k_B T}\right).$$

Hence, if we use the resistivity criterion of $\rho = \rho_c$, the irreversibility line is given by

$$U_0 = k_B T \log \left(\frac{\pi B a_f \nu_0 U_0}{\rho_c J_{c0} k_B T} \right)$$

instead of Eq. (3.129). Using Eq. (3.114), this is also written as

$$U_0 = k_B T \log \left(\frac{\rho_f a_f U_0}{2 \rho_c d_i k_B T} \right),$$

where d_i is the interaction distance defined in Eq. (3.94).

Chapter 4

4.1. From symmetry we have only to consider half the slab, $0 \leq y \leq d$. If the magnetic flux density in the superconducting slab is represented by $\boldsymbol{B} = (B \sin \theta, 0, B \cos \theta)$, Eq. (4.2) leads to

$$\frac{\partial}{\partial y}(B \cos \theta) = \alpha_f B \sin \theta, \quad \frac{\partial}{\partial y}(B \sin \theta) = -\alpha_f B \cos \theta.$$

From these equations we have $\partial B/\partial y = 0$. Namely, B is uniform in space. Substituting this into the above equations,

$$\theta = \theta_0 - \alpha_f y$$

is obtained, where θ_0 is a constant. The boundary conditions to be satisfied are $B \cos \theta_0 = \mu_0 H_e$ and $B \sin \theta_0 = \mu_0 H_I$, where H_I is the self field in the direction of the x-axis due to the current and $B = \mu_0(H_e^2 + H_I^2)^{1/2}$. Although the current has an x-component, this is canceled by the component in the region of $d \leq y \leq 2d$, resulting in no net current along the x-axis. This force-free state is established from the surface to a depth of $y_0 = \theta_0/\alpha_f$. In the inner region the magnetic flux density has only a z-component and its magnitude is equal to B (see Fig. 10). Thus, the magnetization along the z-axis is given by

$$M_z = \frac{B}{\mu_0 d} \int_0^{y_0} \cos\theta \, dy + \frac{B}{\mu_0}\left(1 - \frac{y_0}{d}\right) - H_e$$

$$= (H_e^2 + H_I^2)^{1/2} - H_e + \frac{H_I}{\alpha_f d}\left[1 - \frac{(H_e^2 + H_I^2)^{1/2}}{H_I} \sin^{-1} \frac{H_I}{(H_e^2 + H_I^2)^{1/2}}\right].$$

When $H_I \ll H_e$, $M_z \simeq H_I^2/2H_e$ and the paramagnetic magnetization is obtained.

4.2. It can be seen from Eqs. (4.8) and (4.9) that the angle θ of helical flux lines measured from the z-axis increases with the distance from the center

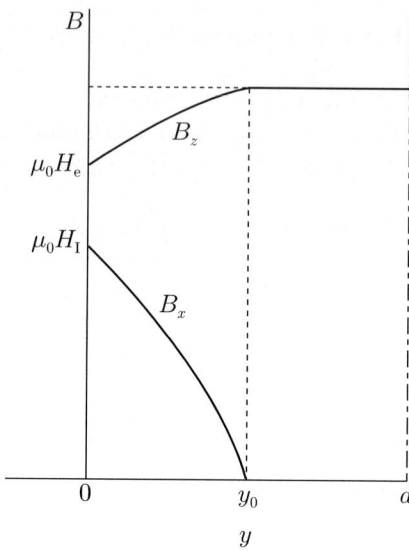

Fig. 10. Distribution of components of the magnetic flux density in the superconducting slab when it is in the force-free state

of the cylinder. Hence, this magnetic structure contains a torsional strain as shown in Fig. 11. When the distance from the center is sufficiently longer than the flux line spacing, this is approximately the same as the strain shown in Fig. 4.14(c).

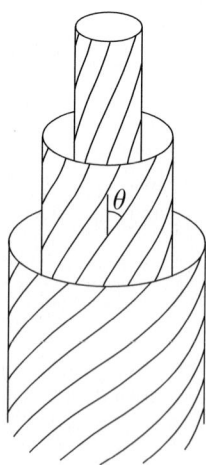

Fig. 11. Torsional structure of flux lines in the superconducting cylinder when it is in the force-free state

From Eqs. (4.3) and (4.4) it is easy to derive $(\partial/\partial r)(B_\phi^2+B_z^2) = -2B_\phi^2/r < 0$. Hence, the magnetic flux density has a higher value in the inner region, and the magnetic pressure works outward in the radial direction. On the other hand, the line tension due to the curvature of the flux lines works inward in the radial direction and is balanced with the magnetic pressure.

4.3. The magnetic flux distribution is expressed as $\boldsymbol{B} = (B\sin\theta, 0, B\cos\theta)$, where B is a constant and $\theta = \theta_0 - \alpha_f y$. If the angle at the surface θ_0 is small enough to satisfy $\theta_0 < \alpha_f d$, the rotation of flux lines penetrates only to the depth of θ_0/α_f from the surface, and the variation in the magnetic flux density in this region is

$$\frac{\partial \boldsymbol{B}}{\partial t} = \boldsymbol{i}_x B \frac{\partial \theta}{\partial t}\cos\theta - \boldsymbol{i}_z B \frac{\partial \theta}{\partial t}\sin\theta .$$

On the other hand, if we assume the velocity of flux lines as $\boldsymbol{v} = (v_x, 0, v_z)$, the continuity equation for flux lines is solved with Eq. (4.36) as

$$v_x = \frac{\partial \theta}{\partial t}\cos\theta(x\sin\theta + z\cos\theta + C) ,$$

$$v_z = -\frac{\partial \theta}{\partial t}\sin\theta(x\sin\theta + z\cos\theta + C) ,$$

where C is a constant and the equation

$$x\sin\theta + z\cos\theta + C = 0$$

represents a straight line connecting the rotation centers of flux lines. If the position of the rotation center of the flux line which we are observing is denoted by $x = x_0$ and $z = z_0$, the velocity of this flux line is written as

$$v_x = r\frac{\partial \theta}{\partial t}\cos\theta, \quad v_z = -r\frac{\partial \theta}{\partial t}\sin\theta ,$$

where $r = (x-x_0)\sin\theta + (z-z_0)\cos\theta$ is the radius of rotation. Hence, it is found that the continuity equation for flux lines describes correctly the rotational motion of flux lines.

The electric field in the region of $0 \leq y \leq \theta_0/\alpha_f$ is given by

$$E_x = \frac{B}{\alpha_f}\cdot\frac{\partial \theta}{\partial t}(1-\cos\theta), \quad E_y = 0, \quad E_z = \frac{B}{\alpha_f}\cdot\frac{\partial \theta}{\partial t}\sin\theta$$

and satisfies

$$\frac{E_x}{E_z} = \tan\frac{\theta}{2} .$$

Since $B_x/B_z = \tan\theta$, it is found that $\boldsymbol{E}$ and $\boldsymbol{B}$ are not perpendicular to each other and do not satisfy $\boldsymbol{E} = \boldsymbol{B} \times \boldsymbol{v}$. In particular, $\boldsymbol{E}$ is parallel to the x-z plane, while $\boldsymbol{B} \times \boldsymbol{v}$ is along the y-axis because of the flux motion in the x-z plane. Thus, if we express the electric field in terms of $\boldsymbol{v}$, Eq. (4.48) is obtained.

4.4. The continuity equation for flux lines is written as

$$v = \frac{1}{\alpha_{\rm f}} \cdot \frac{\partial \theta}{\partial t}, \quad \frac{\partial v}{\partial y} = -\frac{1}{B} \cdot \frac{\partial B}{\partial t}.$$

The first equation reduces to $v = 0$ for $\theta_0/\alpha_{\rm f} \leq y \leq d$ and to

$$v = \frac{\partial H_{\rm I}}{\partial t} \cdot \frac{\mu_0^2 H_{\rm e}}{B^2 \alpha_{\rm f}}$$

for $0 \leq y < \theta_0/\alpha_{\rm f}$. On the other hand,

$$v = \frac{\partial H_{\rm I}}{\partial t} \cdot \frac{\mu_0^2 H_{\rm I}}{B^2}(d-y)$$

is obtained for the entire region of $0 \leq y \leq d$ from the second equation with the condition of $v = 0$ at the center $x = d$. Hence, the two results do not coincide. Thus, the solution of v does not exist under such an incorrect restriction.

4.5. The power density is given by $\boldsymbol{E} \cdot \boldsymbol{J} = (\boldsymbol{B} \times \boldsymbol{v}) \cdot \boldsymbol{J} - \nabla \Psi \cdot \boldsymbol{J}$. In the case of flux motion driven only by the force-free torque, the first term reduces to $(\boldsymbol{J} \times \boldsymbol{B}) \cdot \boldsymbol{v} = 0$ and $\boldsymbol{E} \cdot \boldsymbol{J} = -\nabla \Psi \cdot \boldsymbol{J}$ is obtained. Hence, the important term in Eq. (4.48) is not $\boldsymbol{B} \times \boldsymbol{v}$ but $-\nabla \Psi$.

4.6. For simplicity only the flux lines just inside the surface of the superconducting disk are considered. It is assumed that, when the disk is rotated by an angle Θ with a finite angular velocity, the flux lines near the surface follow by $\delta\theta$ due to the viscous force of the eddy current.

Firstly we argue from the viewpoint of the flux rotation model. According to this model, when the disk is rotated, the internal flux lines are driven by the moment of the viscous force and are rotated until this driving moment is balanced with the force-free torque which prevents flux lines from rotating. The resultant angle is $\delta\theta$. On the other hand, when the external magnetic field is rotated in the opposite direction, the force-free torque acts in proportion to the angle between the external field and the internal flux lines, and the internal flux lines are rotated. But the rotation is retarded by the viscous effect of the eddy current. As a result, the situation is determined by the balance between the force-free torque and the moment of the viscous force in both cases. That is, the two cases are equivalent. In addition, when the rotation is stopped, $\delta\theta$ reduces to zero due to the force-free torque in both cases.

Secondly we argue from the viewpoint of the flux cutting model. When the disk is rotated, the angle between the flux lines following the disk and the external field is represented by θ. It is assumed that this angle exceeds the cutting threshold $\delta\theta_c$. Then, flux cutting takes place, and the angle is assumed to decrease to $\delta\theta$. Hence, it can be seen that this situation is equivalent to the flux rotation model. On the other hand, in the case where the external field is rotated in the opposite direction, the flux cutting is expected to occur when the angle between the external field and the internal flux lines reaches $\delta\theta_c$. Thus, the angle by which the internal flux lines "rotate" from the flux cutting should be $\Theta - \delta\theta_c$ so that the same result is reached. However, this angle is different from $\theta - \delta\theta_c$ in the case of rotation of the disk. That is, the two processes are not equivalent to each other, in so far as the flux cutting has a finite threshold value and is accompanied by a finite energy dissipation. The two processes can be equivalent only when $\theta = \Theta$, i.e., when the flux lines follow the disk completely. However, this is not fulfilled when the angular velocity of rotation is made sufficiently low. In addition, the number of cutting events is independent of the angular velocity of rotation but depends only on the angle of rotation. Such a hysteretic nature is contradictory to the assumption that there is no other mechanism of energy dissipation besides the viscosity. Thus, the rotation of the superconducting disk and the rotation of the external field are not equivalent in the flux cutting model. In this model, when the rotation is stopped, $\delta\theta$ approaches $\delta\theta_c$ and this result is different from the result of the flux rotation model. However, this difference is attributed to the threshold problem and such an argument exceeds the area of the present argument.

Chapter 5

5.1. Using the Bean-London model, we have $\Phi = \mu_0 h_0^2 w / J_c$ for $h_0 < H_p$. Hence, $\lambda' = h_0/J_c$ is obtained from Eq. (5.7), and we have $J = J_c$ from Eq. (5.10). For $h_0 > H_p$, $\Phi = \mu_0(2h_0 - H_p)w$ leads to $\lambda' = d$.

5.2. From Eqs. (5.45), (5.47a) and (5.48a) we have

$$\mu_1 = \mu_0[(\chi_1' + 1)^2 + \chi_1''^2]^{1/2} = \frac{\mu_0 h_0}{2H_p}\left[1 + \left(\frac{4}{3\pi}\right)^2\right]^{1/2}$$

for $h_0 \leq H_p$. While $\Phi = w\mu_0 h_0^2/J_c$, the amplitude of the fundamental component of the AC magnetic flux is $\Phi_1 = 2\mu_1 h_0 wd$. Hence, the penetration depth of AC magnetic flux when Φ_1 is used instead of Φ is

$$\lambda_1' = \frac{1}{2w\mu_0} \cdot \frac{\partial \Phi_1}{\partial h_0} = \lambda'\left[1 + \left(\frac{4}{3\pi}\right)^2\right]^{1/2} \simeq 1.086\lambda'.$$

Thus, the penetration depth of AC magnetic flux is overestimated by about 8.6% and the critical current density is underestimated by about 8.6%.

5.3. Firstly we consider the case where the motion of flux lines is completely reversible. Since the magnetic flux density inside the superconducting slab is given by $b(x) = \mu_0 h(t) \cosh(x/\lambda_0')/\cosh(d/\lambda_0')$ with $h(t) = h_0 \cos\omega t$ denoting the magnetic field at the surface, the amplitude of the AC magnetic flux going in and out of the superconducting slab is given by $\Phi = 2w\mu_0 h_0 \lambda_0' \tanh(d/\lambda_0')$. Hence, the apparent penetration depth of AC magnetic flux is obtained as $\lambda' = \lambda_0' \tanh(d/\lambda_0') \simeq d[1-(d/\lambda_0')^2/3]$ from Eq. (5.7). Thus, the upper limit of the penetration depth is d.

This is qualitatively the same even when the motion of flux lines becomes irreversible. Namely, the penetration depth of the AC magnetic flux at the penetration field is smaller than d. Hence, J_c cannot be correctly estimated but never fails to be overestimated. This is similar to the overestimation of J_c from the peak field amplitude of χ_1''. Figure 12 shows the factor of overestimation, i.e., the ratio of the estimation J_c' from the usual analysis of Campbell's method to the given value J_c, where the Campbell model is used in the calculation [4].

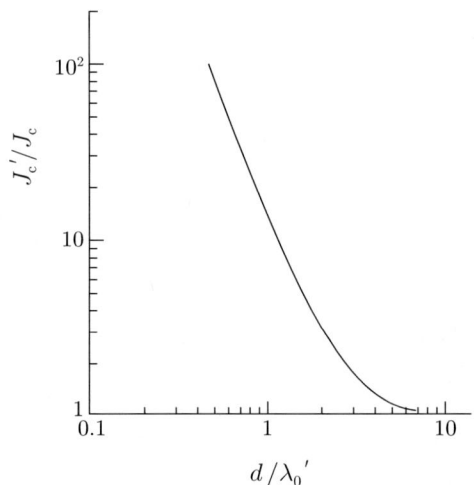

Fig. 12. Factor of overestimation of the critical current density J_c' obtained from Campbell's method

5.4. The average magnetic flux density inside the superconductor in an AC magnetic field of $h_0\cos\omega t$ is

$$\langle B \rangle = \mu_0\left[-h_0 + \frac{H_\mathrm{p}}{2} + \frac{h_0^2}{4H_\mathrm{p}}(1+\cos\omega t)^2\right]$$

for the phase $-\pi < \omega t < -\theta_0$ and

$$\langle B \rangle = \mu_0 \left(h_0 \cos\omega t - \frac{H_\mathrm{p}}{2} \right)$$

for the phase $-\theta_0 < \omega t < 0$. In the above $\theta_0 = \cos^{-1}[(2H_\mathrm{p}/h_0) - 1]$ and a constant contribution from the DC magnetic field was neglected for simplicity. Each integral in Eqs. (5.22) and (5.23) is given by twice the integral in the range from $-\pi$ to 0. After a simple but long calculation we obtain

$$\mu_3 = \frac{2\mu_0 H_\mathrm{p}}{15\pi h_0} \left[20 \left(\frac{H_\mathrm{p}}{h_0} \right)^2 - 44 \left(\frac{H_\mathrm{p}}{h_0} \right) + 25 \right]^{1/2}.$$

5.5. If the DC component of the average magnetic flux density $\langle B \rangle$ is disregarded, each integral in Eqs. (5.43) and (5.44) is given by twice the integral in the range from $-\pi$ to 0 for both cases of $h_0 < H_\mathrm{p}$ and of $h_0 > H_\mathrm{p}$. For this purpose we can assume const. $= \mu_0 h_0^2 / 2H_\mathrm{p}$ in Eq. (5.25) for $h_0 < H_\mathrm{p}$. In the case of $h_0 > H_\mathrm{p}$, $\langle B \rangle$ is given in the answer to Exercise 5.4. The rest is omitted.

Chapter 6

6.1. Since the kinetic energy due to the spatial variation in the order parameter can be disregarded, the flux line energy per unit length is $f_1 = 0$ when the flux line is in the superconducting region. When the flux line moves to the normal region, the energy in the region where the flux line existed before the movement becomes $f_2 = -(1/2)\mu_0 H_\mathrm{c}^2 \pi \xi^2$ per unit length. This energy difference occurs during the movement of the flux line by 2ξ. Hence, the elementary pinning force of the superconducting-normal interface is approximately given by

$$f'_\mathrm{p} \simeq \frac{f_1 - f_2}{2\xi} = \frac{\pi}{4} \xi \mu_0 H_\mathrm{c}^2$$

per unit length of the flux line.

6.2. The order parameter takes the same value in the superconducting and normal regions, and the average free energy density is given by

$$\begin{aligned} F' &= \frac{d_\mathrm{s}}{d_\mathrm{s} + d_\mathrm{n}} \left(\alpha |\Psi|^2 + \frac{\beta}{2} |\Psi|^4 \right) + \frac{d_\mathrm{n}}{d_\mathrm{s} + d_\mathrm{n}} \alpha_\mathrm{n} |\Psi|^2 \\ &= \frac{\mu_0 H_\mathrm{c}^2}{d_\mathrm{s} + d_\mathrm{n}} \left[d_\mathrm{s} \left(-R^2 + \frac{R^4}{2} \right) + d_\mathrm{n} \theta R^2 \right]. \end{aligned}$$

R^2 is determined so that F' is minimized and Eq. (6.14) is derived.

6.3. If θ or α_n becomes too large, the degradation of the order parameter in the superconducting region due to the proximity effect becomes remarkable, resulting in a reduction in the condensation energy. This explains the reduction in the elementary pinning force when θ becomes too large. In this limit ξ_n is very short and the boundary conditions at the interface (continuity of Ψ and its derivative along the normal direction to the interface) used in the analysis in Sect. 6.3 are no longer correct. In this case the pinning interaction is similar to that by an insulating layer. Thus, it is considered that f_p approaches f_{p0}.

6.4. The coherence length in the region with the higher upper critical field is shorter by $\delta\xi = (\xi/2H_{c2})\delta H_{c2}$. Hence, according to the local model, the energy of the flux line per unit length is lower by $(\mu_0 H_c^2/2)2\pi\xi\delta\xi = (\pi\xi^2\mu_0 H_c^2/2)(\delta H_{c2}/H_{c2})$, when the flux line exists in the region with the higher upper critical field. As a result the elementary pinning force of the grain boundary is estimated as

$$f'_p = \frac{\pi}{4}\mu_0 H_c^2 \xi \left(\frac{\delta H_{c2}}{H_{c2}}\right)$$

per unit length of the flux line.

6.5. The shear stress at a position at distance r from the screw dislocation is $\tau = b_0/2\pi r S_{44}$. The interaction energy density is given by $(1/2)\delta S_{44}\tau^2$, where δS_{44} is the variation in the shearing compliance due to the presence of the flux line. Hence, if the distance between the screw dislocation and the flux line is r_0, the interaction energy is

$$\Delta U \simeq \frac{1}{2}\delta S_{44}\left(\frac{b_0}{2\pi r_0 S_{44}}\right)^2 \pi\xi^2$$

per unit length of the flux line, and the corresponding pinning force is given by $f' = -\partial\Delta U/\partial r_0 = \delta S_{44}(b_0/2\pi S_{44})^2 \pi\xi^2/r_0^3$. This increases with decreasing r_0. The elementary pinning force is given in the vicinity of the lower limit of r_0, i.e. ξ, as [5]

$$f'_p = \frac{\pi}{\xi}\delta S_{44}\left(\frac{b_0}{2\pi S_{44}}\right)^2 = \frac{1}{4\pi\xi}\delta S_{44}\left(\frac{b_0}{S_{44}}\right)^2$$

per unit length of the flux line.

Chapter 7

7.1. If the pinning force density under no strain is given by Eq. (7.2), the pinning force density under the strain ϵ changes as

$$F_{\rm p} = AH_{\rm c2}^m(\epsilon)(1+c\epsilon^2)f(b) \simeq AH_{\rm c2m}^m[1-(am-c)\epsilon^2]f(b)$$
$$\simeq \widehat{A}H_{\rm c2}^{\widehat{m}}(\epsilon)f(b)$$

in the range of small a and c. In the above $\widehat{A} = AH_{\rm c2m}^{c/a}$ and

$$\widehat{m} = m - \frac{c}{a}.$$

7.2. Under the condition of Eq. (1.98), the variation in the local magnetic flux density due to the given displacement u^* is

$$\delta B = g\langle B\rangle \left\{ -\cos\left(\frac{2\pi}{b_{\rm f}}(x-u^*)\right) + \cos\left(\frac{2\pi}{b_{\rm f}}x\right) \right.$$
$$\left. -2\sin\left(\frac{2\pi}{a_{\rm f}}y\right)\left[\sin\left(\frac{2\pi}{b_{\rm f}}(x-u^*)\right) - \sin\left(\frac{2\pi}{b_{\rm f}}x\right)\right] \right\},$$

where $b_{\rm f} = (\sqrt{3}/2)a_{\rm f}$ and

$$g = \frac{\mu_0 H_{\rm c2}}{6\kappa^2 \langle B\rangle} \cdot \frac{\langle |\Psi|^2\rangle}{|\Psi_\infty|^2} \simeq \frac{\mu_0 H_{\rm c2}}{6\kappa^2 \beta_{\rm A}\langle B\rangle}(1-b)$$

with $b = \langle B\rangle/\mu_0 H_{\rm c2}$ denoting the reduced field. It should be noted that this assumption is the same as the standpoint of Brandt's theory (Eq. (15) in [6]) from which the nonlocal result is derived. Averaging δB with respect to y leads to

$$\frac{\langle \delta B\rangle_y}{\langle B\rangle} = g\left[\cos\left(\frac{2\pi}{b_{\rm f}}x\right) - \cos\left(\frac{2\pi}{b_{\rm f}}(x-u^*)\right)\right].$$

Hence, the corresponding displacement of flux lines u is obtained from the continuity equation for flux lines (Eq. (A.31) in Appendix A.5) as

$$\frac{\partial u}{\partial x} = -g\left[\cos\left(\frac{2\pi}{b_{\rm f}}x\right) - \cos\left(\frac{2\pi}{b_{\rm f}}(x-u^*)\right)\right].$$

The zero point of the order parameter of the flux line lattice before the displacement is generally expressed as $x_n = (n+1/2)b_{\rm f}$ with n being an integer. Here we displace the $|\Psi|^2$ structure by $u^* = \epsilon\cos kx$, where ϵ is sufficiently small. The resultant magnetic pressure is calculated as

$$C_{11}(0)\frac{\partial^2 u}{\partial x^2}\bigg|_{x=x_n} \simeq -C_{11}(0)g\left(\frac{2\pi}{b_{\rm f}}\right)^2 \epsilon\,\cos kx_n$$

in terms of the local value. On the other hand, the elastic force due to the variation in $|\Psi|^2$ is

$$C_{11}(k)\frac{\partial^2 u^*}{\partial x^2}\bigg|_{x=x_n} = -C_{11}(k)k^2\epsilon\,\cos kx_n.$$

The requirement that these two forces are the same leads to

$$\frac{C_{11}(k)}{C_{11}(0)} = \frac{2\pi}{3\sqrt{3}\beta_{\mathrm{A}}} \cdot \frac{k_{\mathrm{h}}^2}{k^2} \simeq 1.04 \frac{k_{\mathrm{h}}^2}{k^2} .$$

This result is approximately the same as the result of the nonlocal theory, $C_{11}(k)/C_{11}(0) \simeq k_{\mathrm{h}}^2/(k^2+k_{\mathrm{h}}^2)$, in the present range of wave number sufficiently smaller than ξ^{-1}. In the above we did not eliminate the divergence at $k \to 0$, and hence, the derived result is not completely the same as the result of the nonlocal theory.

If the variation in the magnetic flux density assumed in the above really occurs, the spacing of the flux line lattice b_{f} should vary as $b'_{\mathrm{f}} = (1-k\epsilon \sin kx)b_{\mathrm{f}}$ according to the displacement u^*. Then, the magnetic flux in the unit cell between x_n and x_{n+1} is calculated as

$$a_{\mathrm{f}}\left[b'_{\mathrm{f}}\langle B\rangle + \int_{x_n}^{x_n+b'_{\mathrm{f}}(x_n)} \langle \delta B\rangle_y \mathrm{d}x\right] \simeq \phi_0[1 - (1-g)k\epsilon \sin kx_n] ,$$

where the relationship $a_{\mathrm{f}}b_{\mathrm{f}}\langle B\rangle = \phi_0$ is used. Therefore, it is shown that the flux quantization is not fulfilled.

The variations in the magnetic flux density accompanied by the displacement of flux lines assumed in the local and nonlocal theories are schematically shown in Fig. 13(a) and (b), respectively. In the figure the quantization to a half of ϕ_0 is shown for simplicity. In the nonlocal theory the maximum and minimum values are fixed as shown in (b), and hence, the quantization of magnetic flux is not fulfilled when the spacing of flux lines changes. On the other hand, the average magnetic flux density in the unit cell changes when the spacing of the flux lines changes, resulting in the fulfillment of the quantization of magnetic flux in the local theory.

7.3. For $f_{\mathrm{p}} > f_{\mathrm{pt}}$ the characteristic times are reduced to

$$t_1 \simeq \frac{1}{\gamma}\log\left[\frac{(f_{\mathrm{p}} - f_{\mathrm{pt}})^2}{\eta^* v f_{\mathrm{p}}}\right], \quad t_2 \simeq \frac{d(f_{\mathrm{p}} + 3f_{\mathrm{pt}})}{2v f_{\mathrm{pt}}}, \quad t_3 \simeq t_1 + t_2 .$$

By substituting these into Eq. (7.48), Eq. (7.50) is obtained. For $f_{\mathrm{p}} < f_{\mathrm{pt}}$, t_2 is the same as above but the others are

$$t_1 \simeq \frac{d(f_{\mathrm{pt}} - f_{\mathrm{p}})}{2v f_{\mathrm{pt}}}, \quad t_3 \simeq \frac{2d}{v} .$$

7.4. Since the force balance equation is in a separable form, this can be easily integrated:

$$\left(\frac{f - f_{\mathrm{p}}}{f + f_{\mathrm{p}}}\right)^{1/2} \tan\left(\frac{k_{\mathrm{p}}x}{2}\right) = \tan\left[\frac{(f^2 - f_{\mathrm{p}}^2)^{1/2} k_{\mathrm{p}}}{2\eta}(t + t_0)\right] \equiv \tan[c(t + t_0)]$$

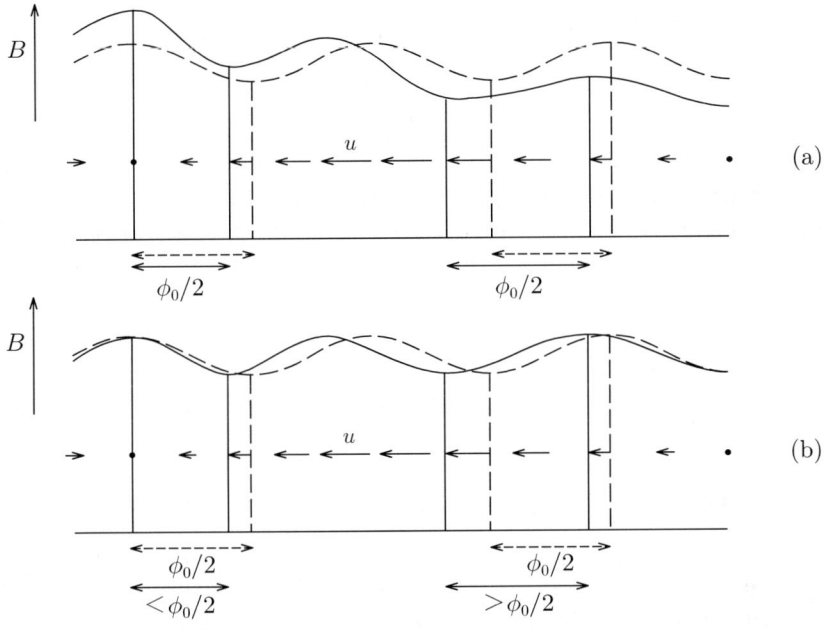

Fig. 13. Expected variations in the magnetic flux density due to the deformation of the flux line lattice from (**a**) the local theory and (**b**) the nonlocal theory. *Broken* and *solid lines* show the magnetic flux densities before and after the deformation, respectively

for $f > f_p$, where t_0 is an integral constant. If we assume that $x = 0$ at $t = 0$, we have $t_0 = 0$. Substituting this into the force balance equation, the mean velocity is obtained as

$$\langle \dot{x} \rangle = \frac{1}{2\pi} \int_0^{2\pi} \dot{x} \mathrm{d}(ct) = \frac{1}{\eta}(f^2 - f_p^2)^{1/2} \ .$$

Using the relationships $\langle \dot{x} \rangle = E/B$, $f = \phi_0 J$, $f_p = \phi_0 J_c$ and $\eta = B\phi_0/\rho_f$, the above equation reduces to [7]

$$E = \rho_f (J^2 - J_c^2)^{1/2} \ .$$

Hence, the E-J characteristic approaches asymptotically $E = \rho_f J$ for $J \gg J_c$ (see Fig. 14) and is different from the usual pinning characteristic such as the ones shown in Fig. 7.10(a).

7.5. It is assumed that the initial statistical distribution of flux lines around pins is in the critical state with the pinning force density $F = -F_p$ shown in Fig. 7.7(b), and then, flux lines are displaced by u in the direction of the negative x-axis (see Fig. 15). The corresponding displacement of flux lines

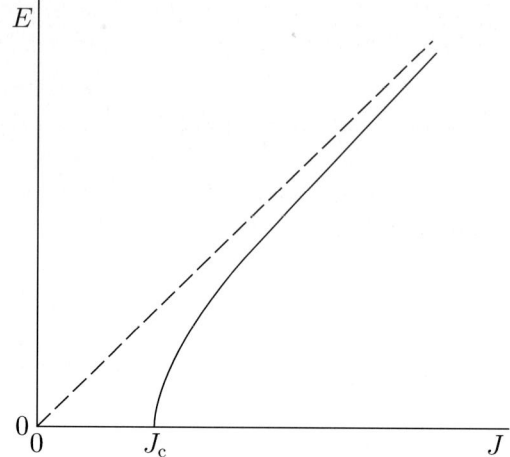

Fig. 14. Current-voltage characteristic for a solid cluster of flux lines or a flux bundle moving in a periodic potential under a constant driving force

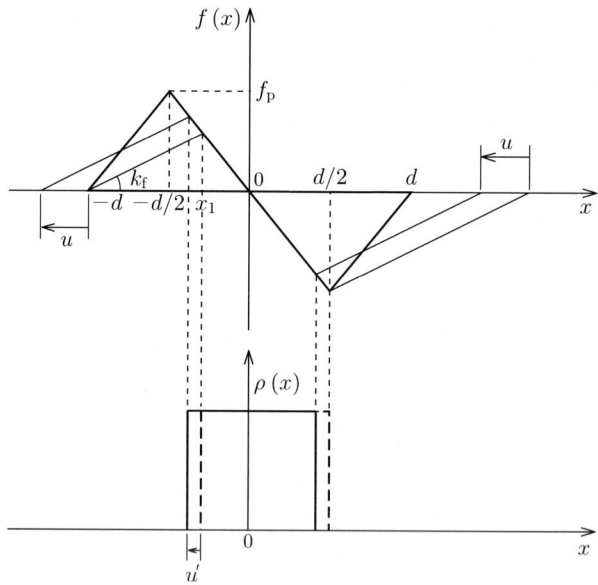

Fig. 15. Variation in the statistical distribution of flux lines on pins (*lower figure*) when flux lines are displaced by u in the opposite direction from the initial critical state

inside pins is $u' = f_{\text{pt}}u/(f_{\text{p}} + f_{\text{pt}})$. Hence, the pinning force density, which is defined to be positive when directed along the positive x-axis, is

$$F = \frac{N_{\text{p}}}{a_{\text{f}}} \int_{x_1-u'}^{d/2-u'} f_{\text{p}}(x)\frac{\partial x_0}{\partial x}\mathrm{d}x = \frac{N_{\text{p}}}{a_{\text{f}}} \cdot \frac{f_{\text{p}}+f_{\text{pt}}}{f_{\text{pt}}} \int_{x_1-u'}^{d/2-u'} \left(-\frac{2f_{\text{p}}}{d}x\right)\mathrm{d}x ,$$

when the displacement is sufficiently small. In the above $x_1 = -f_{\text{pt}}d/(f_{\text{p}}+f_{\text{pt}})$. After a simple calculation we have

$$F = -F_{\text{p}}\left(1 - \frac{u}{d_{\text{i}}}\right)$$

using Eq. (7.33). The interaction distance d_{i} defined by Eq. (3.94) is calculated as

$$d_{\text{i}} = \frac{d}{4}\left(\frac{f_{\text{p}}}{f_{\text{pt}}} - 1\right) .$$

For a displacement longer than $2d_{\text{i}}$ the pinning force takes the constant value F_{p}.

The obtained pinning force density vs displacement characteristic is shown in Fig. 16. When the flux lines are displaced in the opposite direction before they reach $2d_{\text{i}}$, the characteristic is reversible. After they reach $2d_{\text{i}}$, the characteristic is hysteretic. Thus, the pinning force density vs displacement characteristic observed by experiments can be qualitatively explained (see Fig. 3.33). This result is deeply associated with the existence of an unstable region for flux lines inside pinning potentials.

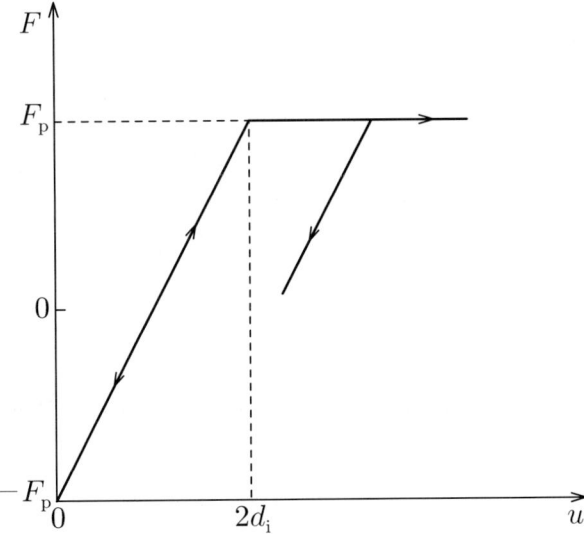

Fig. 16. Pinning force density vs. displacement of flux lines predicted by the statistical theory

7.6. It is assumed that flux lines are strongly pinned by line pins parallel to them which are distributed in the form of a square lattice with spacing $1/\rho_\mathrm{p}^{1/2}$ in the x-y plane as shown in Fig. 7.48. The driving force density F is applied uniformly in the direction of the x-axis on the flux line lattice which is strongly pinned by the line pins distributed on the rows where $y = 0$ and $y = 1/\rho_\mathrm{p}^{1/2}$. The force balance is described as

$$C_{66}\frac{d^2 u}{dy^2} = -F,$$

where u is the displacement of flux lines. From the condition that $u = 0$ at $y = 0$ and $y = 1/\rho_\mathrm{p}^{1/2}$

$$u = \frac{F}{2C_{66}} y \left(\frac{1}{\rho_\mathrm{p}^{1/2}} - y\right)$$

is obtained for $0 \leq y \leq 1/\rho_\mathrm{p}^{1/2}$. Hence, the average displacement with respect to y is $\langle u \rangle = F/12 C_{66}\rho_\mathrm{p}$. On the other hand, the driving force density is written as $F = \alpha_\mathrm{L}\langle u \rangle$. Thus, the Labusch parameter for the shearing deformation is given by

$$\alpha_\mathrm{L} = 12 C_{66}\rho_\mathrm{p}.$$

The obtained Labusch parameter is proportional to $(1-b)^2$ at high fields and takes a maximum value of $3C_{66}/a_\mathrm{f}^2$ at the maximum line pin concentration $\rho_\mathrm{p} = 1/4a_\mathrm{f}^2$. In the critical state F is equal to F_p in Eq. (7.87) and $\langle u \rangle$ is equal to the interaction distance d_i. Hence, d_i amounts to $a_\mathrm{f}/9\pi^2$ at the maximum line pin concentration and is proportional to $b^{-1/2}$.

7.7. Using Eq. (7.59), Eq. (7.56) is written as $R_\mathrm{c} = (8\pi)^{1/4}(C_{66}r_\mathrm{p}/F_\mathrm{p})^{1/2}$. Hence, if we use the relationship $F_\mathrm{p} = \alpha_\mathrm{L} d_\mathrm{i}$ and note that r_p corresponds to d_i, it is found that $R_\mathrm{c} \sim (8\pi)^{1/4} R_0 = 2.24 R_0$. The relationship between L_c and L_0 is also similar.

7.8. From Eqs. (3.94), (5.19), (7.75), (7.95) and (7.98) the pinning potential energy is estimated as

$$U_0 = \left(\frac{2}{\sqrt{3}}\right)^{3/2} \frac{\phi_0^{3/2} dg^2 J_{\mathrm{c}0}}{2\zeta B^{1/2}}.$$

With the numerical equation $(1/2)(2/\sqrt{3})^{3/2}\phi_0^{3/2} \simeq 4.23 k_\mathrm{B}$ Eq. (7.99) is obtained.

Answers to Exercises   495

Chapter 8

8.1. It is assumed that the size of the normal precipitate D is sufficiently larger than the diameter of the normal core of flux lines $2\xi_{ab}$ or $2\xi_c$. At first, the case of low magnetic field is treated. When the flux line is directed parallel to the a-axis, the cross-sectional area of the normal core is $\pi\xi_{ab}\xi_c$, and the pinning energy of the precipitate is $u_p^a = (\mu_0 H_c^2/2)\pi\xi_{ab}\xi_c D$ per flux line. The Lorentz force is directed along the c-axis, and the contribution of this pinning interaction to the force is $f_p^a \simeq u_p^a/2\xi_c = \pi\mu_0 H_c^2 \xi_{ab} D/4$. When the flux line is directed parallel to the c-axis, the cross-sectional area of the normal core is $\pi\xi_{ab}^2$, and the pinning energy of the precipitate is $u_p^c = (\mu_0 H_c^2/2)\pi\xi_{ab}^2 D$ per flux line. The Lorentz force is directed along the a-axis, and the contribution of this pinning interaction to the force is $f_p^c \simeq u_p^c/2\xi_{ab} = \pi\mu_0 H_c^2 \xi_{ab} D/4$, which is the same as f_p^a. Thus, the pinning force is isotropic.

On the other hand, since the irreversibility field changes drastically depending on the direction of the magnetic field, the anisotropy of the pinning force appears at high magnetic fields. That is, the anisotropy of the pinning force arises from the anisotropy of the irreversibility field.

8.2. From Eqs. (7.2) and (7.3) the dependence of the critical current density on the field angle is written as

$$J_{c0}(\theta) = \frac{A}{\mu_0} H_{c2}^{m-1}(T) b^{\gamma-1}(\theta)[1-b(\theta)]^\delta ,$$

where θ is the angle between the field and the c-axis and $b(\theta) = B/\mu_0 H_{c2}(\theta)$. In the case of $H_{c2}^{ab}/H_{c2}^c \gg 1$, Eq. (8.12) leads to $H_{c2}(\theta) \simeq H_{c2}^c \sec\theta$ except in the vicinity of $\theta = 90°$. Thus, the above scaling law reduces to

$$J_{c0}(\theta) = \frac{A}{\mu_0} H_{c2}^{m-1}(T) \left(\frac{B_\perp}{\mu_0 H_{c2}^c}\right)^{\gamma-1} \left(1 - \frac{B_\perp}{\mu_0 H_{c2}^c}\right)^\delta ,$$

where $B_\perp = B\cos\theta$ is the c-axis component of the external magnetic flux density. The above relationship shows that the critical current density is determined only by this component.

8.3. In the case $B \leq \mu_0 H_g$, $J_{cm} \geq 0$ and the electric field is given by

$$E(J) = \int_0^J \rho_f(J - J_c) P(J_c) \mathrm{d}J_c .$$

The exponential term is expanded as

$$\exp\left[-\left(\frac{J_c - J_{cm}}{J_0}\right)^{m_0}\right] \simeq 1 - \left(\frac{J_c - J_{cm}}{J_0}\right)^{m_0}$$

and after a simple calculation the electric field is derived:

$$E(J) = \frac{\rho_f}{m_0+1}\left(\frac{1}{J_0}\right)^{m_0}(J-J_{\rm cm})^{m_0+1}.$$

In the case $B > \mu_0 H_g$, $J_{\rm cm}$ is negative and the electric field is given by the above result minus the contribution from the region $J_{\rm cm} \leq J_c \leq 0$, which does not exist. The latter is formally given by the same formula with the replacements $J_{\rm cm} \to 0$ and $J \to |J_{\rm cm}|$. Hence, the electric field is given by

$$E(J) = \frac{\rho_f}{m_0+1}\left(\frac{1}{J_0}\right)^{m_0}\left[(J+|J_{\rm cm}|)^{m_0+1} - |J_{\rm cm}|^{m_0+1}\right].$$

8.4. In the case $d < L_0$, the left hand side of Eq. (8.41) is rewritten as

$$\frac{3\mu_0\zeta^2 d}{2\phi_0^2 g^2} k_{\rm B} T \log\left(\frac{B a_f \nu_0}{E_c}\right) = \frac{3\mu_0\zeta^2 d}{2\phi_0^2 g^2} U_0 \tag{1}$$

after some calculation with Eq. (8.28), where the original quantities are used for numerical factors and Eq. (3.129) is used. In the above, U_0 is the pinning potential energy for a bulk superconductor given by

$$U_0 = \frac{1}{2}\alpha_{\rm L} d_i^2 (a_f g)^2 L_0 = \frac{2\phi_0^2 g^2}{3\mu_0\zeta^2 L_0},$$

where the equation, $\alpha_{\rm L} = C_{44}/L_0^2$, is used. Hence, the value of Eq. (1) reduces to d/L_0 and Eq. (8.41) is proved to be valid.

8.5. The virtual critical current density at 0 K and 1 T is estimated as $J_{\rm c0}(0\text{ K}, 1\text{ T}) = A = 2.58 \times 10^9$ Am^{-2} with $\xi_{ab}(0) = 2.02$ nm. At 77.3 K we have $\xi_{ab} = 3.85$ nm. We expect $\mu_0 H_i \sim 8$ T at this temperature and $g_e^2 = 2DB\xi_{ab}/\zeta f\phi_0$ is approximately estimated as 74.4 with $\zeta = 4$, where f is the volume fraction of 211 phase and the above expression of g_e^2 is obtained from Eqs. (7.12b) and (7.83a). Then, $U_e = 4.75 \times 10^{-19}$ J is obtained from Eq. (7.97) with $J_{\rm c0}(77.3\text{ K}, 8\text{ T}) = 5.56 \times 10^7$ Am^{-2}. Thus, g^2 is estimated as 2.51 from Eq. (7.96). Hence, $K = 9.39 \times 10^2$ is obtained from Eq. (8.29). Equation (8.28) with $\gamma = 1/2$ and $m' = 3/2$ leads to $\mu_0 H_{i\infty} = 21.7$ T. However, this result is incorrect, since it is quite close to the upper critical field, $\mu_0 H_{c2}^c(77.3\text{ K}) = 22.4$ T. Hence, the irreversibility field should be exactly calculated from

$$(\mu_0 H_i)^{(3-2\gamma)/2} = (\mu_0 H_{i\infty})^{(3-2\gamma)/2}\left(1 - \frac{H_i}{H_{c2}^c}\right)^2$$

in which the effect of the upper critical field is not disregarded. This leads to $\mu_0 H_i = 8.4$ T, which is close to the initial speculation. Thus, the above calculation is consistent.

Chapter 9

9.1. The coherence length at 0 K is estimated as $\xi(0) = 3.63$ nm from the value of $H_{c2}(0)$. Equation (6.23) and this value give us $\xi_0 = 4.89$ nm. Hence, $f_p' = 1.26 \times 10^{-4}$ Nm^{-1} is obtained. The flux line spacing at 5 T is $a_f = 2.19 \times 10^{-8}$ m and the upper critical field is 18.0 T at 10 K. Thus, $J_{c0} = 3.00 \times 10^9$ Am^{-2} is obtained.

9.2. Using the value of J_{c0} at $B = 5$ T obtained in Exercise 9.1, the magnetic field dependence of J_{c0} is given by

$$J_{c0} = 1.29 B^{-1/2} \left(1 - \frac{B}{18.0}\right)^2 \times 10^{10} \text{ Am}^{-2} .$$

From Eqs. (3.129) and (7.97) the equation to estimate the irreversibility field H_i is:

$$\mu_0 H_i = 1.85 \left(1 - \frac{\mu_0 H_i}{18.0}\right)^2 \times 10^3 ,$$

where $g^2 = 1.0$, $\zeta = 2\pi$ and $\log B a_f \nu_0 / E_c = 14$ were substituted. $\mu_0 H_i = 16.3$ T is obtained. This value amounts to about 91% of the upper critical field.

References

1. T. Matsushita, E. S. Otabe, T. Matsuno, M. Murakami and K. Kitazawa: Physica C **170** (1990) 375.
2. D. O. Welch, M. Suenaga, Y. Xu and A. R. Ghosh: *Adv. Superconductivity II* (Springer-Verlag, Tokyo, 1990) p. 655.
3. D. O. Welch: IEEE Trans. Magn. **MAG-27** (1991) 1133.
4. N. Ohtani, E. S. Otabe, T. Matsushita and B. Ni: Jpn. J. Appl. Phys. **31** (1992) L169.
5. A. M. Campbell and J. E. Evetts: Adv. Phys. **21** (1972) 345.
6. E. H. Brandt: J. Low Temp. Phys. **28** (1977) 263.
7. J. E. Evetts and J. R. Appleyard: *Proc. Int. Disc. Meeting on Flux Pinning in Superconductors*, Göttingen, 1974, p. 69.

Index

A

abnormal transverse magnetic field effect 102
Abrikosov vortex state 358
AC current loss
  in ellipsoidal wire 89
  in round wire 86
  in thin strip 89
AC Josephson effect 2, 34
AC loss 76
  Bean-London model 80
  Irie-Yamafuji model 78
  under reversible flux motion 135
AC loss of round wire
  in rotating magnetic field 92
  in transverse magnetic field 90
AC susceptibility 225
Anderson-Kim model 138
anisotropic Ginzburg-Landau equation 342
anisotropy
  in high-temperature superconductor 342, 343
  in MgB_2 413
anisotropy of critical current density 365
anisotropy parameter 342
apparent penetration field 136
apparent pinning potential energy 141, 456
Arrhenius expression 139
artificial pinning center
  in Nb-Ti 259
  in Y-123 402
avalanching flow model 320, 448

B

Bardeen-Stephen model 47
BCS theory 2
Bean-London model 56
Bragg-glass state 352
breaking point (of magnetic flux distribution) 59

C

Campbell model
  reversible flux motion 128
  summation theory 281
Campbell's AC penetration depth 126, 215
Campbell's method 213
clean superconductor 14
coated conductor 399
coherence length 8
  of BCS theory 14
  of G-L theory 14
coherent potential approximation theory 292
  comparison with experiments 300–302, 306–307
condensation energy density 12
  in Bi-2212 407
  in Bi-2223 404
condensation energy interaction 237
continuity equation for flux lines 45, 126, 177, 217, 274
Cooper pair 2, 9

500   Index

Coulomb gauge   13
critical current density   34, 37, 64, 209
   in longitudinal magnetic field   156, 187, 220
   in virtual creep-free case   141, 335
   overestimation due to reversible flux motion   221, 224, 227
critical field   3
   in small superconductor   437
   of surface superconductivity   31
critical point   357, 358, 359
critical size of dimensional crossover of flux lines   368, 370
critical state   44
critical state model   44, 54
C-substitution   419
current distribution   57
current-voltage characteristics   37, 50, 283
   at history effect   327
   in high-temperature superconductor   346, 374

D

DC Josephson effect   32
DC magnetization method   212
DC susceptibility   119
de Gennes model (superheated state)   113
depairing current density   35
Dew-Hughes model   316
diamagnetism   43, 66
dimensional crossover of flux lines   355
direct summation   233, 297
dirty superconductor   14
dynamic critical index   346, 376
dynamic phenomenon   95
Dynamic theory   283
   Matsushita et al.   284
   Yamafuji-Irie   283

E

edge dislocation   254
elastic interaction   253, 256
   ΔC effect (ΔE effect)   253
   ΔV effect   253
   of grain boundary   257
elastic modulus of flux line lattice   271
   bending modulus (local)   271, 444

bending modulus (nonlocal)   273
shear modulus   272
uniaxial compression modulus (local)   271, 443, 447
uniaxial compression modulus (nonlocal)   273
electric field criterion   209
electromotive force   36
electron scattering mechanism   245, 419
elementary pinning force   233, 234
   of edge dislocation   254
   of grain boundary   248, 249, 257
   of normal precipitate   238, 239, 243, 259
   of screw dislocation   255
   of superconducting precipitate   260
energy barrier   139
energy gap   1
energy of flux line   22
Evetts-Plummer model   316
ex situ thin film (MgB_2)   428
ex situ wire (MgB_2)   416

F

field-induced pinning mechanism   324
flow resistivity   50
   in longitudinal magnetic field   206
flux bundle   138
flux bundle size
   longitudinal   333
   transverse   333, 373, 452
flux bundle volume   220, 333
flux creep   138
flux creep-flow model   381
   comparison with experiments   348, 351, 375, 386, 387, 404, 422, 425
flux cutting   182
flux cutting model   161
   cutting threshold   164
   elementary cutting force   164, 189
flux flow   45, 46
flux flow noise   326
flux jump   103
flux line lattice
   triangular lattice   24, 25, 27

flux pinning   36
flux quantum   16, 17
force balance equation   44, 45, 56, 69, 86, 95, 436
force-displacement profile   127, 218
  history effect   330
force-free current   169
force-free model   158, 192
force-free state   158, 168, 436
force-free torque   173, 178, 181, 195
four terminal method   209
free energy density (G-L energy density)   10

G
generalized critical state model   191
Ginzburg-Landau (G-L) equations   13
Ginzburg-Landau theory   2, 9
G-L parameter   15, 30
G-L transition   346, 391
Goodman's interpolation formula   247
grain boundary   245
  pinning in metallic superconductor   252, 301
  pinning in MgB_2 416
grain connectivity   417

H
helical flux flow   198
history effect   327, 352
hysteresis loss   51

I
impurity parameter   248
in situ thin film (MgB_2)   429
in situ wire (MgB_2)   416
interaction distance   129
  at dimensional crossover of flux lines   355
  change during saturation   319
interlayer coupling length   372
intrinsic pinning   394
Irie-Yamafuji model   56
irreversibility field   380
  analytic solution   380
  anisotropy   386
  dependence on
    anisotropy of superconductor   372, 384, 406
  electric field   389
  pinning strength   384
  size   388
  temperature   383
irreversibility line   148, 348

J
Josephson current   17
Josephson junction   32
Josephson penetration depth   344, 452
Josephson plasma frequency   379
Josephson plasma wave
  longitudinal   378
  transverse   379
Josephson's relation   44
  break in longitudinal magnetic field   157, 179, 199
Josephson vortex   343

K
Kim model   56, 82
kinetic energy density   9, 29
kinetic energy interaction   259, 398
Kramer model   313
Kramer's formula   315

L
Labusch parameter   125
  change during saturation   319
Labusch theory   275
Larkin-Ovchinnikov theory   286
  comparison with experiments   298–307
  modification by Wördenweber and Kes   291
  peak effect   289, 325
  two-dimensional pinning   290
Lawrence-Doniach equation   342
line pinning   313
line tension   65
linear summation   267, 297
logarithmic relaxation rate   141
London equations   7, 8
London theory   6, 13
longitudinal magnetic field effect   155
Lorentz force   43
Lowell model   277
lower critical field   3, 22
  in small superconductor   438

M

magnetic flux distribution 57
   Campbell's method 193, 215
magnetic interaction 258
magnetic pressure 65
magnetization 4, 60
   by pins 61, 62
   in the Meissner state 3
   near upper critical field 28
matching mechanism 322
Maxwell equations 41
Meissner current density 35
Meissner effect 1, 7, 13
Meissner state 3
melting transition 345
misorientation angle 362
mixed state 3
modified London equation 18

N

negative electric field 157, 194, 204
nonlocal elastic modulus (see elastic modulus) 272
nonsaturation (of pinning force density) 312
normal core 20, 21
normal precipitate 237
n-value 210

O

off-set method 209
order-disorder transition 326, 330, 352
order parameter 9
order parameter in normal core 20
oscillation frequency 139
oxygen deficiency 394, 396

P

pancake vortex 344
paramagnetic effect 155, 159
partition of pinning energy 440
peak effect 322, 352
penetration depth 7, 13
   in clean superconductor 15
   in dirty superconductor 15
   in Pippard superconductor 15
penetration field 57
percolation model (for E-J curve) 376
perfect conductivity 1

perfect diamagnetism 1
phase diagram 5
phase diagram of flux lines (high-temperature superconductor) 345
   Bi-2212 358
   Y-123 357, 358
pinning correlation length
   longitudinal 289, 333, 348, 368, 401
   transverse(three-dimensional) 289, 333
   transverse(two-dimensional) 290, 292
pinning force density 37, 44, 267
   in reversible state 125
pinning potential 125
pinning potential energy 142, 220, 331
   three-dimensional 335
   two-dimensional 335
pinning power loss density 51, 76
pinning property
   Bi-2212 406
   Bi-2223 402
   MgB_2 415
   Y-123 394
plastic shear (of flux lines) 314
power factor 129
principle of minimum energy dissipation 192, 334, 439, 453, 460
principle of stabilization 107
proximity effect 240, 261

Q

quantization of magnetic flux 16
quantum tunneling 457

R

rectifying effect 97
relaxation of current 141
resistivity criterion 209
reversible flux motion 125
rotational shear distortion (of flux lines) 169
rotation of flux lines (by force-free torque) 178, 181, 182, 195

S

saturation (of pinning force density) 310

Index   503

scaling law of pinning force density
    268
  in Bi-2223 tape   404
  in MgB_2 421
  in Y-123 thin film   376
  strain scaling law   269
  temperature scaling law   268
scaling of E-J curves   346
screw dislocation   255
Shapiro step   34
shear flow   314
Silcox-Rollins model   56
singular point (of phase)   18
size effect
  G-L transition   360
  irreversibility field   368, 388
  peak effect (order-disorder transition)
    326, 359, 367
SQUID   33
static critical index   346
statistical summation   277
statistical theory   275
  Campbell   281
  Labusch   275
  Lowell   277
substituted region by RE element   396
summation problem   233, 267, 275
superconducting volume fraction   123
surface barrier   109
surface irreveresibility   108
  measurement   217
surface pinning   116
surface sheath   109
surface superconductivity   30
synchronization   314, 326

T

thermal energy   139
thermally activated flux flow (TAFF)
    state   149
thermodynamic critical field   4, 16, 30

  in Bi-2212   407
  in MgB_2 413
thermodynamic magnetic field   43, 68
third harmonic voltage method   221
threshold value for elementary pinning
    force   276, 280, 282, 289
  coherent potential approximation
    theory   297
transition (at critical temperature)   11
transition (in magnetic field)   12
transition temperature (G-L transition)
    346
twin boundary   394
two-gap superconductivity   413, 425
211 phase particle   396
type-1 superconductor   3, 4, 5
type-2 superconductor   3, 4, 5, 16

U

upper critical field   3, 15, 16, 30

V

virtual critical current density   141
viscous coefficient   44, 50
viscous force density   44, 95
viscous power loss density   51, 97
vortex glass-liquid transition   346
vortex glass state   352
vortex state   3

W

Walmsley's model   70
wave-form analysis method   216
weak link
  in MgB_2 416
  in high-temperature superconductor
    361

Y

Yamafuji-Irie model   52, 283
Yasukochi model   56